Springer Tracts in Civil Engineering

Springer Tracts in Civil Engineering (STCE) publishes the latest developments in Civil Engineering - quickly, informally and in top quality. The series scope includes monographs, professional books, graduate textbooks and edited volumes, as well as outstanding PhD theses. Its goal is to cover all the main branches of civil engineering, both theoretical and applied, including:

- Construction and Structural Mechanics
- Building Materials
- Concrete, Steel and Timber Structures
- Geotechnical Engineering
- Earthquake Engineering
- Coastal Engineering; Ocean and Offshore Engineering
- Hydraulics, Hydrology and Water Resources Engineering
- Environmental Engineering and Sustainability
- Structural Health and Monitoring
- Surveying and Geographical Information Systems
- Heating, Ventilation and Air Conditioning (HVAC)
- Transportation and Traffic
- Risk Analysis
- Safety and Security

Indexed by Scopus

To submit a proposal or request further information, please contact:

Pierpaolo Riva at Pierpaolo.Riva@springer.com (Europe and Americas)
Wayne Hu at wayne.hu@springer.com (China)

Gang Wu · ZhiQiang Chen · Ji Dang

Intelligent Bridge Maintenance and Management

Emerging Digital Technologies

人民交通出版社股份有限公司
China Communications Press Co.,Ltd.

Springer

Gang Wu
School of Civil Engineering
Southeast University
Nanjing, Jiangsu, China

ZhiQiang Chen
Division of Natural and Built Environment
University of Missouri
Kansas City, MO, USA

Ji Dang
Civil and Environmental Engineering
Program
Saitama University
Saitama, Japan

ISSN 2366-259X ISSN 2366-2603 (electronic)
Springer Tracts in Civil Engineering
ISBN 978-981-97-3826-7 ISBN 978-981-97-3827-4 (eBook)
https://doi.org/10.1007/978-981-97-3827-4

Jointly published with China Communications Press Co., Ltd.
The print edition is not for sale in China (Mainland). Customers from China (Mainland) please order the print book from: China Communications Press Co., Ltd.
ISBN of the Co-Publisher's edition: 978-7-114-17633-3

This Springer imprint is published by the registered company Springer Nature Singapore Pte Ltd.
The registered company address is: 152 Beach Road, #21-01/04 Gateway East, Singapore 189721, Singapore

Preface

Bridges are critical structures as connecting nodes in modern transportation systems. While supporting social and economic activities, bridges are constantly subject to environmental attacks that deteriorate structural materials, hence shortening bridge life; in addition, bridges are threatened by extreme natural and technological events in their life cycles, which, although rare, may lead to catastrophic consequences. Traditionally, civil structural (bridge) engineers and researchers have attempted to develop various sensing, inspection, and monitoring technologies, as reflected in the plethora of literature on structural health monitoring (SHM) and non-destructive evaluation (NDE). Nonetheless, there are many reasons that prevent massive adoption of these (SHM or NDE) technologies for bridge maintenance and management. Not to criticize SHM or NDE technologies, which have greatly contributed to our understanding of the complex behavior of civil and other structures, we assert that they lack three *connectivity* attributes.

First, it is the disconnection between technology components that prevent the realization of *ad hoc* automation.[1] Entering the second decade of the twenty-first century, we are witnessing the movement of industrial automation in many industrial sectors. On the one hand, many 'smart' sensing or monitoring solutions exist in our community; on the other hand, it is evident that the deployment of these technologies is still labor intensive. Furthermore, visual analytics, if rendered, is often remote and latent.

Second, we recognize the disconnection between identifying, quantifying, and integrating local and global structural damage relative to the system-level functionality of bridge structures. To this end, most NDE technologies attempt to detect and localize damage at structural elements. On the other hand, most SHM technologies focus on global structural integrity; if a few SHM technologies are dedicated to local structural integrity, practical implementation and validated solutions are scarce. These technology-based local or global structural damage or integrity information are not systematically represented or synthesized for system prognosis or forecast.

[1] We define ad hoc automation as a specific-purpose automation in an (maintenance or management) process, not a fully or global automation for the integral maintenance and management process.

Last but not least and the most important one, disconnection is more severe between technologies and stakeholders. It is a fact that most existing technologies are still limited either by presenting incomplete visual analytics or overwhelming information to end-users. Added to this is the limitation in physical media for presenting information to end-users. It is noted that the common practice is still based on computer screens, the reporting is still textual or two-dimensional through illustrations or tables, and finally, the decision-making is often centralized, lacking situational awareness.

Toward the next-generation maintenance and management of bridge structures that ensure lifecycle sustainability and resilience, we call for the intelligent integration of traditional and emerging digital technologies. To integrate NDE, SHM, and other inspection and maintenance technologies, we believe that data-driven machine learning, advanced artificial intelligence, next-generation communication (e.g., IoT and 5G/6G), and robotic manipulation and actuation will play critical roles. To augment decision-making efficiency, efficacy, and situational awareness, it is crucial to develop advanced information fusion methods and platforms (e.g., via digital twinning) and novel human-infrastructure interfacing technologies (e.g., augmented reality).

We believe that many researchers and practitioners have recognized the lack of these connectivity attributes as well, or even in a more profound way. However, our humble opinion is that, to this end, there is no systematic exposure of these technologies in one book found in our communities. We admit that it is not necessarily a significant task if the focus is on introducing theories and methods underlying different technologies. We assert that it is the synthesis of theories, implementation of these methods and technologies in practice, and, better yet, the existence of case studies that render preparing a useful book as we will offer a much more challenging endeavor.

In this book, *Intelligent Bridge Maintenance and Management*, we strive to balance between background, theoretical foundation and methods, and practical implementation and case studies. Specifically for the latter, many of the implementation and case studies are from China. This is not incidental but is based on the substantial progress in China's civil infrastructure.

With the comprehensive coverage and the intended balance between theories and practice, this book aims to assist readers in not only understanding emerging digital technologies but also in adopting effective ones for maintaining and managing bridge structures in their life cycle. It is our intention to share with readers the vision that the next-generation bridge maintenance and management should be an intelligently integrated system that embraces physical structures, digital technologies, and stakeholders systematically toward a great degree of automation. The target audience for this book includes bridge engineers, decision-makers in the field, and students and researchers in universities and scientific institutions.

Ten chapters are included. For the interest of readers, they may start with reading Chaps. 1 and 2, then select technologies of their interest that are covered in Chaps. 3–9. Specifically,

- Chapter 1 introduces the background, definition, and main contents of intelligent bridge management and maintenance.
- Chapter 2 analyzes the practical feasibility of establishing an intelligent bridge management and maintenance system and introduces an emerging platform "IntelliBridge."
- Chapter 3 discusses the application of IoT technologies to bridge management and maintenance.
- Chapter 4 focuses on automation technologies and robotic detection for bridge inspection.
- Chapter 5 discusses the importance of cloud computing and other big data technologies applied to bridge management and maintenance, as well as the new opportunities brought by AI technologies to the development of bridge management and maintenance.
- Chapter 6 introduces machine learning for bridge evaluation and early warning.
- Chapter 7 introduces the application of computer vision technology for automating bridge maintenance.
- Chapter 8 presents multi-source data fusion, structural state assessment, virtual reality, and digital twin technologies for intelligent bridge management and maintenance framework.
- Chapter 9 discusses intelligent decision-making for bridge maintenance and management, and the role of AI.
- Chapter 10 summarizes and envisages the future of intelligent bridge management and maintenance.

This book is co-authored by Gang Wu of Southeast University, ZhiQiang Chen of the University of Missouri, and Ji Dang of Saitama University. Professor Gang Wu's team members Shitong Hou, Yujia Zhang, Jiao Dai, Yitian Han, Mida Cui, Tianyu Wang, Haochen Wang, Xiaoxiang Cheng, Fanqi Cong, Hongyu Lu, Jianhua Fan, Zhihong Xiao, Zhao Xu, Xi Chen, Yongcheng Bao, Wei Wang, Lu Zhang, Fengbo Ma, Shizhi Chen, Bin Dong, Huile Li, Shiqing Wang, Jinqiao Chen, Liuzhen Yao, Tianran Han, Xudong Chen, Zhuoran Li, Jianwu Pan, Jingwei Zhao, and Lu Chu contributed much in content writing and editing.

This book is financially supported by the 2021 National Science and Technology Academic Works Publishing Fund, to which we would like to express our most sincere gratitude. Due to the authors' limitations in writing the book, errors and omissions are unavoidable, and we sincerely invite the readers to give us their criticism and correction.

Nanjing, China — Gang Wu
Columbia, USA — ZhiQiang Chen
Saitama, Japan — Ji Dang
November 2022

Executive Summary

This book, *Intelligent Bridge Maintenance and Management*, focuses on the application of digital technologies for enhancing the safety and serviceability of bridge structures. The selected technologies, many of which are emerging for applications in Civil Engineering, include the Internet of things, robotics, machine learning and artificial intelligence, and digital twins, among others. By introducing the theories, methods, case studies, and practical examples, this book aims to assist readers in not only understanding emerging digital technologies but also in adopting effective ones for maintaining and managing bridge structures in their life cycle. In the end, we expect to present the vision that next-generation bridge maintenance and management should be an intelligently integrated system that embraces physical structures, digital technologies, and stakeholders systematically toward a greater degree of automation. The target audience for this book includes bridge engineers, decision-makers, and other professionals in the field, besides students and researchers in universities and scientific institutions.

Contents

About the Authors

Gang Wu winner of the National Science Foundation for Distinguished Young Scholars, Distinguished Changjiang Chair Professor of the Ministry of Education, Lead Talent of Science and Technology Innovation under the National "Ten Thousand People Program," winner of Tencent Science Exploration Award, winner of China Science and Technology for Young Scientists Award, member of Standing Party Committee of Southeast University, Executive Vice President of Southeast University, President of China Young Science and Technology Professionals Association, Director of National and Local Joint Engineering Research Center for Smart Construction, Operation and Maintenance, etc. His research interests include intelligent construction, operation and maintenance of major infrastructure, new materials and new structural systems, and performance improvement of existing engineering structures. He has been ranked a world top 2% scientist and an Elsevier highly cited scholars in China. He has published 193 SCI papers, authored 74 patents (five international patents), published six monographs, and edited eight standards. Additionally, he has led two national key R&D projects, and more than 20 key National Natural Science Foundation projects and major instrument projects (free application). He has also won two second prizes of the National Science and Technology Progress Awards (ranked 1, 2), two first prizes of the Provincial Science and Technology Progress Awards and two second prizes of the National Teaching Achievement Awards.

ZhiQiang Chen is a Professor of Civil Engineering at the Division of Natural and Built Environment, University of Missouri-Kansas City (UMKC). At UMKC, he directs the interdisciplinary research lab of Disaster, Infrastructure, and Geointelligence Technology (DIGIT). Before joining UMKC in 2010, he received his Ph.D. in Structural Engineering from the University of California, San Diego (UCSD). He was a visiting professor at the University of California, Berkeley, in 2020 and a visiting professor at the Saitama University, Japan, in 2022. His research interests, in general, focus on Civil Systems Intelligence and Resilience, including multi-hazard performance and resilience computing for civil structures and the application and development of AI, remote sensing, and human-infrastructure interfacing technologies for disaster response and infrastructure management.

Ji Dang Professor of Department of Environmental and Social Foundation, Graduate School of Science and Technology, Saitama University, since October 2020. Bachelor and Master of Engineering from Southeast University (2002, 2006), Ph.D. from Aichi Institute of Technology (2010), Postdoctoral fellow of Aichi Institute of Technology (2010–2011), researcher of Kyoto University (2011–2013), and teaching assistant of Saitama University (2013–2020). In recent years, his main research directions in intelligent bridge operation and maintenance are: IoT bridge monitoring, development of smartphone applications for vibration monitoring; UAV-based bridge detection, performance-based UAV bridge detection and evaluation, Japan Strategic Innovation Program SIP; and development of damage recognition technology based on AI computer vision. He has published 41 papers and has won awards such as the Outstanding Presentation Award of the Japan Society for Civil Engineering (2016), and the AI and Digital Science Data Award (2021).

Chapter 1
Introduction

Abstract In this chapter, we offer a comprehensive review of the state of the practice with a global perspective and the current technologies for bridge maintenance and management. By discussing that the systematic goal of conducting bridge maintenance and management is to achieve lifecycle system goals, such as sustainability and resilience, we recognize the challenges pertinent to the state of the practice. To resolve these challenges, digital technologies, especially emerging ones, including the Internet of Things, Robotics, Cloud and Edge Computing, and Artificial Intelligence with Big Data, are described. The promise of adopting these technologies to realize the notion of intelligent bridge maintenance and management is then introduced.

1.1 State of the Practice and Definitions

As human beings' growth consists of youth, adolescence, and adulthood, bridges undergo the states of construction, normal service, and degraded services. Similarly analogous is the possibility that a person can be abruptly hurt or become sick, and a bridge can be damaged by extreme events, leading to disrupted services. One important argument is that bridge in-service life is not design life. In Japan, the proposed design life for highway bridges is 100 years; in the UK, it is 120 years; in other European countries, it is 100 years; the United States targets 75–100 years. In fact, the actual life spans of many bridges do not reach their design life due to premature degrading or disaster-induced sudden failure. Statistics from the United States show that the actual service life of bridges, with a design average life of 75 years, is about 40 years. These general statistics manifest that many, if not most, bridges have not been 'cared for' as a human being would be during their active life.

A special example is the I-35W bridge collapse in Minnesota, USA, in 2007. Three months before the collapse of I-35W, the bridge had just obtained its routine evaluation results after a full bridge inspection, leaving the doubt behind over either poor quality of the inspection process or improper use of technologies. This event has facilitated the transformation of bridge maintenance and management throughout the

G. Wu et al., *Intelligent Bridge Maintenance and Management*, Springer Tracts in Civil Engineering, https://doi.org/10.1007/978-981-97-3827-4_1

United States. Other examples of bridge collapse events, or more generally structural failure of different levels, can be found in the literature but are statistically rare due to the physical nature of such failure events. Nonetheless, in this book, we stress the abundance of case-history data related to bridge engineering practice in China, which is not incidental but due to the rapid development of China's civil infrastructure in the last 30 years.

1.1.1 Case-History Data in China

China's bridge industry has developed rapidly in the past 30 years, boasting the largest bridge inventory in the world. Up to the year 2023, China has built more than 800,000 highway bridges, more than 1000 long-span bridges, over 200,000 railway bridges, and over 10,000 km of high-speed railway bridges. However, bridges in China are no exception, as a very large percentage of them have entered their aging stage.

First, up to today, about 40% of China's highway bridges have been in service for more than 20 years. More specifically, there are nearly 50,000 bridges with a safety level of Class III or above on the national, provincial, and county roads, accounting for 16% of the total number of in-service bridges in China. In addition, more than 100,000 bridges are classified as dangerous. In addition, a large number of bridges built in the early 1990s are about to enter their maintenance period. Referring to the bridge maintenance experience of developed countries, numerous county, township, and village bridges will be overhauled or rebuilt in the next 5–10 years. In the next 10–15 years, many national and provincial bridges and long-span bridges will also need significant investment to ensure their safe operation (see Fig. 1.1).

Second, there are general distinctions between bridges of different sizes. It is observed that it is small- and medium-sized bridges that take on most traffic loads in transportation systems. These bridges are constantly faced with progressive degrading from material deterioration, geological foundation settlement, hydraulic scour, and artificial hazards from vehicle and ship collisions. Above all, these bridges are often subjected to traffic loads beyond their designed capacities due to regional economic booming and transportation burdens; structural fatigue or even sudden failure due to overloading is not rare.

As far as long-span (and other large-scale) bridges are concerned, the pursuit of "long, large, high, and exceptional" bridge types brought about much higher construction costs. In contrast, the quality of many long-span bridges was not directly proportional to the increase in project cost. Some bridges were overhauled shortly after they were completed, and some were resurfaced several years after they were opened to traffic. In addition, these structures may encounter extreme events that are beyond their code-based targets of risk levels, for example, high-intensity climatic events (flooding, strong wind, fire, etc.) as witnessed in recent years, which are partially attributed to global climate change. Indeed, risk from extreme climatic events is particularly prominent for long-span cable-stayed or suspension bridges

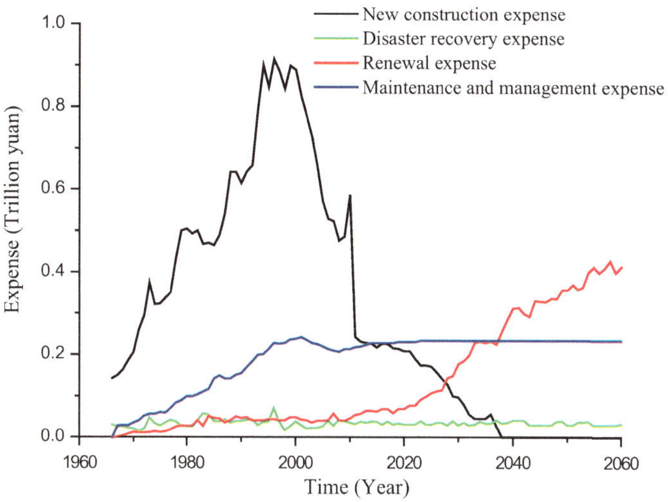

Fig. 1.1 Trend of bridge construction, maintenance, and management expenses in China

with high lateral flexibility, as well as some special-shaped bridges with abnormal static and dynamic responses.

As a result of the bridge condition status in China, it has been recorded that between 2007 and 2012, 37 bridges collapsed in China. On average, more than 6 'fatal' bridge events occurred every year, resulting in more than 180 deaths. It is worth noting that nearly sixty percent of these collapsed bridges were built after 1994, with an average age of less than 20 years.

1.1.2 Technical Standards-Based Practice

In general, technical standards, which are standardized knowledge bodies curated by professional organizations, are found globally related to bridge maintenance and management. These standards provide guidelines and provisions to direct a structuralized bridge maintenance process, including bridge inspection, monitoring, condition evaluation (including load rating), structural repair, and retrofitting. In the following, we comparatively describe the practice of adopting technical standards in China, Japan, and the United States (US).

1.1.2.1 China

Three relevant standards are adopted in China for governing bridge maintenance and management: the Technical Specifications for Urban Bridge Maintenance (CJJ99-2017) [1], the Evaluation Standard for Technical Conditions of Highway Bridges

(JTGT H21-2011) [2], and the Maintenance Specifications for Highway Bridges and Culverts (JTG H11-2004) [3].

With respect to the current bridge maintenance practices, the maintenance and retrofitting of bridges are mainly based on the results of bridge inspection and evaluation, so the rationality and practicality of the inspection and evaluation methods directly decide the maintenance, retrofitting, and management methods adopted by the maintenance personnel. Therefore, bridge inspection and evaluation specifications are the core subjects of bridge maintenance. Among the current specifications for bridge maintenance in China, the Code for Maintenance of Highway Bridges and Culverts [3] has the widest scope of application, while the Technical Code for Maintenance of Urban Bridges [1] has a smaller area of application, which primarily deals with urban bridge maintenance. The Standard for Evaluation of Technical Conditions of Highway Bridges [2] comprises the strengths of the previous two specifications. It classifies different bridge types and describes the distresses of each bridge type [2], and thus is currently the most complete specification for technical evaluations of bridge conditions in China.

As the leading specification, the Code for Maintenance of Highway Bridges and Culverts [3] provides the inspection and evaluation system for highway bridges and culverts, suggestions for maintenance of the upper and lower components, and suggestions for disaster prevention and retrofitting. It involves all stages of bridge maintenance and has the advantages of strong authority, small workload, simple operation, etc. However, there are some shortcomings with this standard, such as the fixed weights for the different components, the failure to divide the components according to the bridge type, and subjectivity and randomness in the technical evaluations of the bridge conditions. The specification is still the version revised in 2004, whose contents are relatively out of date. Subsequently, the Technical Code for Maintenance of Urban Bridges [1] was developed on the basis of the Code for Maintenance of Highway Bridges and Culverts [3]. The layered evaluation method and the weighted average method were adopted for the Technical Code for Maintenance of Urban Bridges [1]. At the same time, the criteria for grading component defects were specified, making the technical evaluations of bridge conditions more objective. However, the arithmetic average method is adopted for the scoring of the upper and the lower structures of the bridge with the potential risk of weakening the structural defects. Moreover, the classification of bridge types and components is insufficient, which is not suitable for the technical evaluation of the conditions of large bridges and complex bridges. In comparison, the Evaluation Standard for Technical Conditions of Highway Bridges [2] adopts the method of combining the layered comprehensive evaluation with five types of single control indicators. First, each component of the bridge is evaluated. Secondly, each part of the bridge is evaluated. Then, the deck system, the superstructure, and the substructure are evaluated sequentially. Finally, the overall technical conditions of the entire bridge are evaluated. The results obtained are highly objective. Besides, this specification proposes different distress types and weights for different bridge types, and for the first time, proposes that the weights of components not mentioned can be proportionally allocated to the existing components.

At present, the main problems with China's bridge inspection and evaluation specifications are (1) the specification uses fixed weights to evaluate bridges, which makes it impossible to correctly reflect the effects of malfunctions of components with small weights on the operational status of the whole bridge under evaluation; (2) the classification of bridge component types in the specification is unreasonably simplified; for some components on complex bridges, their corresponding weights are not clearly specified, and the stipulations are too vague in this respect; and (3) there is no quantitative evaluation method recommended in the specification, and the evaluations of bridge conditions are usually obtained from the subjective experience of the bridge evaluators through a rough qualitative analysis of the bridge. The knowledge and work experiences of the bridge evaluators greatly affect the bridge evaluation results. Besides, in different regions, evaluators have different understanding of distresses and their degrees. In summary, the differences in regions, units, and evaluators directly lead to the differences in computing measures used in evaluation.

1.1.2.2 Japan

Japan built a large number of bridges during its period of rapid economic development after World War II. Similarly, these bridges have entered the stage of maintenance and management, many of which exert an urgent need for these actions. In 2013, Japan passed an amendment to the Road Law, which clearly pointed out that bridge managers have an obligation to carry out preventive maintenance and management for the bridges under their jurisdiction and made clear the necessity of bridge maintenance and management. Subsequently, the Ministry of Land and Communications of Japan issued bridge inspection protocols to standardize the bridge maintenance and management practices under its jurisdiction.

Bridge inspections in Japan include regular inspections and intermediate inspections. The regular inspection is conducted once every five years. Intermediate inspection refers to the inspection conducted before the next regular inspection when damage is found during the previous regular inspection or maintenance or retrofitting, which needs to be confirmed. Intermediate inspection has no regular frequency, and the inspection content is also determined by the inspection parts and the inspection standards. Intermediate inspections are mainly manual on-site visual inspections, supplemented by limited detection tools. According to the regulations, the maintenance rating and the damage appearance rating will be given to each component of the bridge after the inspection.

In order to make the assessment more objective, the Japanese code includes photos of typical distresses of different damage categories and damage levels for reference. After the assessment, the engineers collected the standardized distress detection results and built a proprietary database for subsequent bridge maintenance and management decision-making according to the Japanese standard. Based on the Japanese codes, both the subjective maintenance rating and the objective existing damage rating are independently carried out, which enables the managers to cross-reference the two at the end and improve the credibility of the results.

1.1.2.3 United States

It is noted that the period of massive construction of highway bridges in the United States was much earlier than most other countries. As such, its history of technical standards and standards-based have been scrutinized over a longer period, hence much technically mature to this date. At present, a comparatively complete of private and public industrial sectors have been established that provide bridge inspection, evaluation, maintenance, and retrofitting with proven records of effectiveness.

In the US, related standards are documented in the National Bridge Inspection Standards (NBIS) [4], Guidelines for Bridge Management Systems (GBMS) [5] issued by the Federal Highway Administration (FHWA), and the American Association of State Highway and Transportation Officials (AASHTO). The current bridge technical condition evaluation system in the United States consists of six standards or manuals. According to American law, the National Bridge Inspection Standard (FHWA 2001) [4] is the highest standard, which was formulated by FHWA and passed by Congress and has become a law that must be complied with by all bridge maintenance and management practices. According to the specifications used in the United States, each component of the bridge is first scored on a scale of 0–9, and then the bridge parts are scored. The overall score is directly calculated based on the conditions of each component without weighting. The advantage of this evaluation method is that the decision-makers can easily see the distress of each component and each part of the bridge, which is conducive to bridge maintenance personnel paying more attention to key distresses.

At the same time, FHWA released the Recording and Coding Guide for Structure Inventory and Appraisal of Nation's Bridges (RCGSIANB), which stipulates that all bridge data must comply with this standard for recording bridge maintenance and management information in a standard way. After that, the National Bridge Inventory (NBI) is employed to completely and accurately master the basic composition, structure, function, management, maintenance status and other information related to the highway bridges in real time. FHWA also released a comprehensive manual (Bridge Inspector's Reference Manual, BIRM) on procedures, regulations and technologies for inspection and evaluation of various bridges. AASHTO also released the Manual for Bridge Evaluation (MBE) [6], which can be used to evaluate bridges from both the unit and the structure levels. Finally, in terms of bridge maintenance, repair and retrofitting, the United States adopts the AASHTO Maintenance Manual: The Maintenance and Management of Roads and Bridges [7] which corresponds to China's Code for Maintenance of Highway Bridges and Culverts [3] and Design Code for Strengthening of Highway Bridges [8], etc. As a whole, the US bridge inspection system has clearly classified various typical distresses for different components, so as to reduce the differences in evaluation results from subjective judgments of different inspectors. At the same time, more comprehensive consideration is given to the factors affecting the distress, and more in-depth cause analysis is made by the US system. Finally, more attention is paid to the refinement and standardization of the testing workflow and the corresponding safeguards by related US standards.

1.1.2.4 Remarks

Compared with the standard practice in China, Japan and the United States, it is stated that China is clearly still in the early stage in developing a complete bridge maintenance and management system. In terms of technical completeness, China's specifications and standards are still relatively inadequate. For example, there is no detailed regulation on the specific implementation of manual testing in Chinese standards.

Specific differences are observed. In the process of inspection, practitioners in Japan often carry out the inspection and evaluation of bridges at the component level, but the division of bridge components are not quite the same between the two countries. The main difference is that the United States only classifies the members according to the member type, while Japan makes a more detailed classification considering the location of the bridge span.

Finally, the current bridge maintenance specifications in various countries generally have strict requirements for subjective experiences of the bridge inspectors and a strong overview evaluation process. However, most of them aim to assess bridges qualitatively, and quantitative analysis and reporting of structural integrity or damage is desired but mandated. To overcome the disadvantage of inspectors' subjectivity, Japan and the United States have tried to establish authoritative image libraries of bridge distresses to help on-site inspectors with their identifications, but the objectivity brought in by using this approach is still limited. At the same time, the existing detection methods mainly use visual inspection with some auxiliary instruments. This combination mode will introduce a lot of subjective errors, resulting in limited credibility of the bridge evaluation results.

Last, we observe that in most standards of the three countries, structural inspection is largely limited to visual inspection, and less often, it recommends the use of nondestructive evaluation (NDE). Although researchers have undertaken a lot of investigations over various Structural health monitoring (SHM) technologies, which provide the advantage of producing more quantitative analytics, there is still a lack of relevant specifications to guide and standardize the maintenance works in practice. This void should be filled urgently to improve the efficiency and objectivity of bridge maintenance and management.

1.1.3 Definitions: Bridge Maintenance and Management

Bridges are lifeline facilities that ensure the development of the societal economy and facilitate public transportation. Reviewing the case-history of bridge performance and the current state-of-the-practice, particularly the standards adopted globally, we can state that the existing bridge maintenance and management practices have played an important role in ensuring the safety and uninterrupted services of existing bridges across regions and countries. In this section, formal definitions and descriptions of bridge maintenance and management are given. To characterize the connotation span (or space) of these two notions, we borrow the notion of resilience defined in four dimensions—technological, organizational, social, and economic (i.e., the TOSE dimensions)—originally proposed by Bruneau et al. [9]. Figure 1.2 illustrates how bridge maintenance and management sit in these dimensions.

As illustrated in Fig. 1.2, bridge maintenance is defined as a technology-centered process consisting primarily of four different technology-based means: inspection, monitoring, evaluation, and repair and retrofit (as a whole).

Specifically, bridge maintenance comprises the following phased processes:

- Inspection: A process that uses manual or technological methods to provide data about the condition, including damage or failure modes, of bridge structures. Typical inspection methods include manual (visual) checking and recording, non-destructive testing and evaluation (NDT&E), and imaging or vision-enabled inspection methods.
- Monitoring: A process that uses standalone or networked sensors to collect transient or dynamic responses of bridge structures. Typical methods and technologies are generally explored and developed in the sector of structural health monitoring (SHM).
- Evaluation: A process where engineers conduct empirical rules-based evaluation, structural load rating, or even rigorous structural analysis to identify the safety level of the bridges.
- Repair and retrofit: A process that, upon the decision of structural evaluation, provides solutions to fix, replace, strengthen, or reinforce structural elements in a bridge system.

With the definition and description of bridge maintenance and its four elemental processes above, we assert that maintenance is generally defined in the technological dimension of the multi-dimensional space of system resilience (Sect. 1.3 for

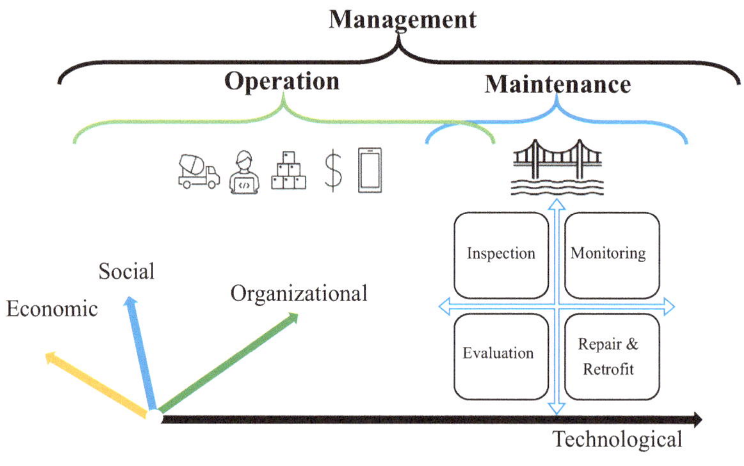

Fig. 1.2 Bridge maintenance and management in the TOSE dimensions

Life-cycle Resilience). Management, however, spans more than the technological dimension.

Broadly speaking, bridge management is a human-centered process that includes personnel training, planning, resource allocation, decision-making, and many other human activities, encompassing means, resources, and other assets, and spanning all TOSE dimensions. If one attempts to define a process describing the activities distinct from technology-centered maintenance activities, the notion of operation is commonly used. Namely, in a bridge management process, operation concerns the planning and distribution of all resources and assets relevant to bridge maintenance.

In this book, we limit the scope of management within the technological realm. Specifically, the notion of management is addressed in two aspects: technologies for various maintenance phases and technologies that contribute to better management for variables in other TOSE dimensions.

1.2 Bridge Maintenance Technologies

1.2.1 Inspection and Monitoring

The purposes of bridge inspection and monitoring are to (1) determine the bearing capacities and service conditions of new bridges; (2) evaluate the service performances and bearing capacities of existing bridges; and (3) study the mechanical behaviors of bridges with new structural forms or new materials and technologies to guide structural design and construction. Bridge inspection and monitoring is a technical discipline that directly serves engineering practices, involving structural design and calculation, testing technology and instrument performance, mathematical statistics and analysis, inspection personnel organization, etc. In terms of the conventional methods, bridge inspection and monitoring can be divided into patrol inspection, static load test, dynamic load test and non-destructive testing.

Patrol inspection does not need to configure any special measurement system on the bridge, but only needs an instrument that can support a variety of measurement operations. On the basis of specifying the inspection route, the data measurement and storage must be undertaken, and then utilized as a basis to subsequently evaluate the health of the bridge. As it is easy to carry out without many equipment, patrol inspection has become the most widely used conventional bridge inspection method. Static load test is a conventional inspection method that examines the static displacement, static strain, crack and other behaviors of the structure after static loads are applied to designated positions on the bridge, which can then infer the serviceability of the bridge under those loads. Dynamic load test is another conventional inspection method that is implemented by exciting vibration of the structure, and then measuring the natural frequency, damping ratio, vibration mode, dynamic impact coefficient, vehicle-induced response and other parameters of the structure to examine the overall structural stiffness and performance of the bridge bearing running vehicles. Non-destructive testing (NDT) is another important bridge inspection technique. It employs visual inspection without damaging the structural materials, or use acoustic, optical, thermal, electrical, magnetic, radiographic or other methods to measure the physical parameters concerning material properties so as to infer the material strength, defects, etc.

Bridge inspection technologies are of great significance for construction quality control, maintenance and management of existing structures, assessment of strength, durability and damage of existing structures, etc. However, the conventional bridge inspection technologies have technical problems themselves that restrict their wide applications. For example, for the static and the dynamic load tests to take place, the traffic need to be closed, and these inspection technologies are also prone to cause structural damages. Without traffic, the testing scenarios are different from the actual bridge operational conditions for these inspection technologies. Besides, the testing results can only reflect the overall mechanical behaviors of the structure, and cannot identify any local damage. NDT is complicated in operation, and its effectiveness largely depends on the damage assumptions of the inspectors.

Bridge structural health monitoring (SHM) practices integrate modern testing and sensing, network communication, signal processing and analysis, structural safety evaluation and decision-making, structural analysis and other technologies from different scientific fields, overcoming many shortcomings of the conventional bridge inspection. A bridge SHM system is usually divided into five parts: online testing module, real-time analysis module, damage diagnosis module, condition assessment module and maintenance oriented decision-making module. Firstly, the inspectors can conduct online testing for the bridge's actual working environments and its responses to various external loads using online testing module that relies on sensing, testing and network communication technologies. Secondly, the inspectors can transfer the measured data to real-time analysis module, and determine the current mechanical state of the whole bridge by finite element model updating and numerical simulation. Thirdly, the inspectors can use damage diagnosis module for damage warning, employing methods based on physical models or machine learning to locate and quantitatively identify the damage. Fourthly, in the condition assessment module, the inspectors can evaluate the safety and durability of various components of the bridge and the whole structure according to the updated indexes. Finally, the maintenance decision-making module will provide suggestions for the current bridge operation, management, maintenance and repair.

Due to the importance of the structure and the economic value, SHM systems are usually installed on selected long-span bridges. The application of bridge SHM systems began in the 1980s, when long-term monitoring instruments and automatic data acquisition systems were installed on Foyle Bridge, UK, to verify the design practice of the time and to study the impact of vehicle, wind and temperature changes on the dynamic behaviors of the bridge. Subsequently, several countries established SHM systems on some newly-built or existing large bridges, incorporating some advanced sensing technologies, computer and communication technologies, and signal analysis and processing technologies.

In recent years, with the enrichment of practical applications and the development of scientific researches, some progresses have been made in bridge SHM technologies which are reflected in: (1) practicing engineers can acquire more comprehensive monitoring information and content; (2) monitoring instruments are more diversified with more complete functions; (3) theory and method of structural damage identification have been improved; and (4) a variety of safety assessment methods has been applied to bridge SHM systems. However, the current bridge SHM technologies are still not perfect, and the following two scientific issues continue to perplex the engineering circle that need to be addressed urgently. Firstly, the structural uncertainties and the bridge's complex working environments have adverse effects on the sensitivity of the structural response, which is difficult to eliminate. Secondly, there is a lack of in-depth research on the changes in the structural characteristics of the bridge during its service life, so it is difficult to establish an objective structural evaluation standard. The conventional monitoring technologies also have the following technical problems: (1) SHM systems based on point contact sensors are of high costs and can only provide scarce data; (2) in terms of communication technologies, there is a lack of networks with large throughput and low latency, and a lack of data transmission protocols with low energy consumption and long life; (3) there is no mature and robust local damage identification method in terms of structural analyses; and (4) SHM systems based on remote sensing technologies lack structural details and elevation perceptions.

1.2.2 Evaluation Methods

Using data obtained from bridge inspection and monitoring, engineers can evaluate the health of the bridge with effective methods. Conventional bridge health evaluation methods can be roughly divided into two categories: deterministic evaluation methods and uncertainty evaluation methods. Deterministic health evaluation methods include those based on appearance survey, methods based on design specifications and methods based on load tests. Uncertainty health evaluation methods include those based on structural reliability theory and methods based on interval analysis.

The method based on appearance survey is the simplest and the most widely used bridge health evaluation method. Based on practical experiences, engineers and technicians have established various types of visual evaluation systems.

The evaluation method based on design specifications has a solid theoretical basis. However, due to differences between the structural design and the health evaluation, it is inappropriate to directly apply bridge design specifications to the evaluation of bridge bearing capacity, and engineers need to correct the structural resistant capability and loading effects according to the information obtained from field investigations. Therefore, the basic steps of bridge health evaluation methods based on design specifications should include: (1) establishing the bridge structure's finite element model according to the measured structural geometric parameters, material properties, support conditions, etc.; (2) applying the design vehicle loads modified according to the field investigation on the finite element model; and (3) comparing the load effects calculated with the updated structural resistance to assess the structure's reliability.

Load test method evaluates the bridge health through field tests. Specifically, this method uses the static and the dynamic responses of the structure under different levels of loads measured on site to identify the geometric parameters, material properties and support conditions of the actual structure. It then uses numerical analyses to estimate the normal service bearing capacity and the ultimate bearing capacity of the actual structure so as to quantify the integrity of the structure. In practice, the reliability of this method largely depends on whether the geometric, material and other parameters of the actual structure are accurately identified by finite element model updating and other technologies.

Health evaluation method based on structural reliability theory uses the failure probability or the reliability index to estimate the safety level of the structure. This method is based on probability and statistics theories, which can effectively take into account the uncertainties of the load and the resistance. Compared with the deterministic evaluation method, this method is proposed within a more reasonable theoretical framework for bridge health evaluation. The integrity evaluation method based on structural reliability theories can be subdivided into the direct evaluation of reliability index method, the sub item safety factor method, the analytic hierarchy process method, etc.

Probability model is not an ideal health evaluation tool for uncertain cases with few statistical data and less accurate computational model. Some scholars proposed a non probabilistic convex set model, which considers uncertain parameters as bounded quantities contained in a convex set, and obtains the range of structural response through set operation. Interval analysis is a representative convex set model-based method. At present, interval analysis method has been introduced into bridge integrity evaluation, but its effectiveness still needs substantial validations by incorporating it into real-world engineering applications.

To sum up, deterministic bridge health evaluation methods are widely used because of their simple processes for easy evaluation of bridge structures under specific conditions. However, these methods all make certainty assumptions for uncertain parameters in the evaluation process, leaving behind potential problems. The uncertain health evaluation methods consider various reasonably uncertain factors in the process of integrity evaluation, which are more mature in theory. However, these methods are not widely used due to their complexity, and their effectiveness in practical use remains to be assessed. In addition, both deterministic and uncertainty evaluation methods have the following common problems: (1) life cycle evaluation has not been realized; (2) visualization and dynamic interaction are not realized for results obtained; and (3) prediction has not been combined with structural health evaluation.

1.2.3 *Repair and Retrofitting*

For the bridges assessed and classified as structures with potential safety hazards, timely repair (e.g., strengthening an existing element) or retrofitting (e.g., adding a new structural element) may be needed to improve the structural capacities and safety levels. Different bridges have different structural forms, serve in different environments and are with different degrees of damage, so targeted maintenances and retrofitting should be provided for each of them. Common bridge repair and retrofitting technologies include steel plate bonding, composite material retrofitting, member section enlargement, prestressed retrofitting, combined steel plate and concrete retrofitting, etc.:

(1) Bonding steel plate retrofitting method is to strengthen the structure by bonding steel plates on the surfaces of the members. This method has many advantages, such as significant improvements of structural load-bearing capacity, low construction cost, and prolonged service life. When using bonded steel plates for retrofitting, the engineers need to select appropriate steel plates through effective assessments of the local working conditions and strengthen the bridge on the principle of maintaining their original structural form.
(2) Composite material retrofitting is an advanced retrofitting method utilizing composite materials such as glass fiber, carbon fiber, aramid fiber, etc. Compared with other fiber materials, carbon fiber has advantages in terms of quality, strength and corrosion resistance. In addition, it is easy to transport and convenient to arrange, which can effectively improve construction efficiency.
(3) A common reason for bridge instability is that its bearing capacity is insufficient, which can be overcome by enlarging the structural members' dimensions. However, due to long construction period and high cost, this method is seldom used.
(4) Prestressed retrofitting is another method for preventing propagation of cracks on bridge members by arranging external prestressed tie rods. The mechanical principle hidden behind this practice is to use high strength components to impose compressive stress on the structure to offset the tensile stress of the original components due to the operational loads, so as to improve the bearing capacity of the structure and prevent its excessive deformation.
(5) Steel plate–concrete composite strengthening is an emerging technology combining the member section enlargement-strengthening method with the bonded steel plate retrofitting method. As an innovative development of the conventional structural retrofitting technologies, the combined retrofitting method not only succeeds in inheriting the advantages from the two original methods, but also effectively avoids their shortcomings. According to the combined retrofitting method, the reinforced part and the original concrete structure are made to work together as a whole by welding studs on the steel plate, planting reinforcement on the surface of the original concrete, and pouring new concrete between the original structure and the reinforced steel plate. In this regard, the new concrete served as the bonding layer between the steel plate and

the original structure. The combined strengthening method makes full use of the properties of both the new and the old materials. As necessary connections between the new and the old concrete and between the steel plate and the new concrete are provided, the overall mechanical performance of the retrofitted structure is reliable. The retrofitted structure has the advantages of high bearing capacity, large stiffness, fast and convenient construction, good durability, etc.

1.3 Lifecyle Engineering Goals

1.3.1 Overview of Lifecycle Engineering

After the bridge is built, it will be in service for decades to tens to hundreds of years. A bridge infrastructure with huge retention is a huge asset for the whole society. How to ensure the good performance of the bridge in its long period of operation while reducing its maintenance cost as much as possible has become an important question for bridge asset managers. Facts have proven that the cost of bridge maintenance after the bridge went into operation is often considerable. Therefore, the maintenance and management of bridges should be carried out under the framework of the whole life cycle, and the needs and costs of each stage in the bridge life cycle should be comprehensively considered in order to effectively manage the huge bridge assets.

After nearly 20 years of research and development, Lifecycle Engineering has been well established as a technical subdiscipline in Civil Engineering for various categories of engineering structure and infrastructure. In general, Lifecycle Engineering concerns the safety and risk of structures and aims to develop effective and optimal methods for maintenance and management. Lifecycle Engineering has become a basic research paradigm to optimize and balance between the lifecycle cost and the structural performance by comprehensively considering the effects of structural performance and degradation on the society, economy and environment during the normal service life of the bridge and in extreme events, and then to find the best bridge maintenance and management strategy.

1.3.2 Lifecycle Performance of Bridges

The allocation and decision-making of bridge maintenance and management funds are directly related to the lifecycle performance of the bridge. The evaluation and prediction of bridge performance are the foundation and key link in the lifecycle maintenance and management framework. Without accurate evaluation and prediction of the bridge performance, unnecessary or untimely maintenance measures may be taken, which will increase the whole life cycle cost of the bridge. In the early lifecycle bridge maintenance and management researches conducted in 1990s, even in the absence of accurate local mathematical models, some global quantitative models were used. Up to now, researches on probabilistic, deterministic or fuzzy models have emerged in large numbers in the field. Different from the original simple subjective scoring, these quantitative studies make the bridge performance evaluation more objective.

The commonly used bridge lifecycle performance indicators include reliability, redundancy, robustness, lifecycle cost, risk, recoverability, carbon emission, pollution, sustainability, etc. These indicators describe the bridge performance from different aspects, but it should be noted that no single indicator has 100% coverage capacity, and each performance indicator has a certain emphasis. In our view, achieving synergistic system sustainability and resilience should be treated as the two highest overarching goals of conducting lifecycle engineering-based bridge maintenance and management. For instance, the general resilience as a capacity indicator involves the conjunct effects of reliability, robustness, and recoverability; and the sustainability needs to be combined with lifecycle cost, carbon emission, and pollution indicators. In bridge maintenance and management, it is necessary to guarantee that the structure meets one or more of the above indicators in operation, and maintain the necessary performance level through detection, maintenance, repair, and other activities.

Uncertainty should be considered when evaluating and predicting the performance of the bridge in its life cycle (including initial performance indicators, damage occurrence, degradation rate, performance improvement by maintenance, etc.). With the passage of time, material properties, structural degradation, load, etc. evolve with uncertainties. In addition, uncertainties also exist in bridge maintenance and cost analysis (such as variables related to discount rate). Uncertainties can be divided into two categories: random uncertainty and cognitive uncertainty. The multiple combined effects of traffic load, environmental effect, material aging, extreme events and so on, which cause the degradation of bridge performance, bring about the abovementioned random uncertainties. This uncertainty is objective and cannot be easily eliminated. Cognitive uncertainty is caused by the lack of knowledge or the inadequate mathematical models employed. It can be reduced by the improvements of scientific researches with an emphasis on the accuracy of bridge performance prediction models. At present, although there are several mathematical methods to deal with uncertainty, the measures based on probability still play major roles in the analysis of life cycle maintenance and management.

Physics based time-varying performance evaluation and prediction of bridge structural systems are usually complicated, and naturally accompanied by various uncertainties. However, bridge SHM and inspection can provide an effective solution to reduce the uncertainty in the bridge performance evaluation with random changes, and to modify and improve the structural evaluation and performance prediction model. By using real-time monitoring of bridge loads, structural static and dynamic responses and geometric positions, SHM provides reliable data for confirming and correcting the load coefficient, the resistance coefficient and the load combination of extreme events.

In recent years, the development of new smart[1] sensing technologies, such as wireless sensing and optical fiber sensing, has further reduced the cost of data acquisition and enhanced the application potential of SHM techniques under the lifecycle framework. On the other hand, the information provided by bridge inspection can also be used to update the degradation model of the bridge structure, thus reducing the uncertainty effects.

1.3.3 Lifecycle Maintenance and Management

The bridge life cycle maintenance and management based decision-making needs to comprehensively consider and balance the various structural performance objectives described above. These structural performance objectives sometimes conflict with each other because there are few maintenance and management plans that can improve all performance levels simultaneously. In view of this, optimization is an essential tool for formulating optimal maintenance and management strategies under the lifecycle framework, and it is also the core link in the bridge maintenance and management process. However, as mentioned above, if the optimization is based on minimizing the maintenance and management cost under constraints of certain performance level, the single optimal scheme obtained may not meet the overall requirements of the maintenance and management for the life cycle performance of the bridge. Therefore, it is necessary to consider the bridge performance evaluation, monitoring, inspection results, etc. as the input, and find the best solution that can simultaneously meet multiple performance objectives and constraints from the possible maintenance and management plans through the optimization process. To find such an optimal solution, we need to use multi-objective optimization techniques to compromise the above-mentioned conflicting performances.

The conventional optimization algorithms have difficulties in dealing with the above mentioned multi-objective optimization problems because most of the conventional methods can only solve single objective optimization problems and a single final optimal solution is provided by the conventional methods accordingly. In order to carry out multi-objective optimization, it is necessary to transform the multi-objective problem into a series of equivalent single objective optimization problems. There are two common transforming methods. One approach is to consider only one target at a time, and other targets are converted into constraints. The difficulty of this method is to determine the upper and lower bounds of the constraint value. The other is the weighted summation method where a single composite objective function is

[1] In Civil Structural and Infrastructure Engineering, 'smart' is used frequently as a personified description of technologies that provide data and analytics. Examples include smart sensors, phones, devices, vehicles, robots, and sometime for structures with such smart devices, smart structures, which embed algorithms for processing data and providing analytics. In this book, the adjective 'intelligent' is used to imply a stronger version of maintenance and management, where connected human-level or strong artificial intelligence is involved, featuring inference and reasoning for decision-making. In the literature, we recognize that the two words are used interchangeably.

obtained by weighted summation of the original multiple objectives using a group of weight coefficients.

Classic artificial intelligence (AI) algorithms, such as evolutionary algorithm, particle swarm algorithm, ant colony algorithm and artificial neural network optimization, can be used to solve multi-objective optimization problems. Multi-objective optimization of bridge life cycle maintenance and management is often solved by genetic algorithm, which is an evolutionary method. The algorithm, first formally presented in the 1960s, is a general numerical tool of simple implementation and good performance. Genetic algorithm can deal with multiple objectives directly and simultaneously. It can also deal with discrete design variable cases because gradient is not required. This is particularly useful for the formulation of bridge maintenance strategies, as different maintenance schemes are usually carried out in discrete years.

The bridge maintenance plan obtained through optimization includes information such as the start time and the type of maintenance, more precisely, whether it is the preventive maintenance or the necessary maintenance. Preventive maintenance practices can help to slightly improve the performance level of the structure and delay the performance degradation process, which include replacing small parts, repairing concrete, repairing cracks, cleaning and painting. Preventive maintenance can be further divided into two categories: state based maintenance and time based maintenance. The former is carried out according to the bridge's operational state, while the latter is carried out based on the predetermined time interval. When the structural performance drops to the specified minimum threshold level for the targeted bridge, it is necessary to perform maintenances to restore the bridge to function normally again, including repair, retrofitting, and component replacement.

When formulating maintenance plans, bridge maintenance and management personnel usually need to consider the maintenance of a group of bridges under their jurisdiction, such as many bridges on a highway or a section of highway. Therefore, it is necessary to undertake lifecycle maintenance and management analysis at the bridge group level, which can be achieved by a series of separate analysis for each bridge, but such a practice leads to huge computation works. Analysis efficiency for the bridge group maintenance and management can be improved by assuming that all structural forms of bridges in a bridge group are similar and the maintenance strategies for these bridges are also analogous.

In fact, all bridges are in the traffic network, which consists of nodes (destinations), connecting lines (roads) and bridges along the connecting lines. The performance of grouped bridges on the network level is related to the performance of each single bridge in the network. Because the performance of a single bridge changes with time and is with uncertainty, it is difficult to accurately evaluate and predict the performance of the bridge network. The key to solving this problem is to establish an appropriate relationship between the reliability of a single bridge and the performance of the bridge network. For example, if a bridge in the network fails to open to traffic, the vehicle path will change accordingly, which becomes one of the possible modes of the bridge network. By comprehensively taking all possible modes into account, the network performance can be evaluated. At the same time, it is also necessary to consider factors such as the lifecycle maintenance costs, the failure costs and the user costs of the bridge network in the process. The user costs refer to the increased traffic costs due to the bridge failure. The above costs are all related to the status of each bridge in the road network.

In general, lifecycle maintenance and management represent the developing tendency of the next-generation bridge maintenance and management. Although many related research achievements have been made, they have not yet been transformed into corresponding clauses in codes and standards for practical use. The increase in the accuracy of structural performance prediction requires a disproportionate increase in the cost of computation. In reality, from the perspective of probability, the use of finite element analysis and other computational techniques to predict the life cycle performance of complex bridge structural systems is still not practical. In addition, the data reflecting the influence of maintenance behaviors on bridge performances is still limited, and the research on integrating SHM into the lifecycle maintenance and management of bridges is still at its initial development stage.

1.4 Recognized Challenges from Practice

1.4.1 Lifecycle Challenges

According to the predictive data given by the American Society of Civil Engineers (ASCE), if all the defective bridges similar to the I-35W Bridge in Minnesota, USA, were repaired after the accidents, the average annual costs for bridge maintenance in the United States will be 9.4 billion dollars in the next 20 years. Thus, such post-event maintenance strategy will inevitably bring either huge safety crises or economic costs when placing the complete inventory of bridges are in mind.

Since the 1980s, the lifecycle cost concept has been applied to the transportation industry. Engineers have begun to pay attention to the lifecycle cost and optimization methods for engineering project management. During this period, the United States drew lessons from the previous serious economic losses caused by insufficient understanding of infrastructure durability and took the lead in proposing the concept of "Lifecycle Cost Analysis" (LCCA) for bridge engineering. Based on this concept, engineers estimated the total cost of the bridge from the beginning to the end of its service life in the bridge's design and construction stages, and selected the optimal scheme based on economic budget comparison. In this way, engineers can be responsible for the whole life performance of the bridge and avoid the problems and economic losses caused by short-term behavior. The Federal Highway Administration of the United States stated in its report that the main purpose of "life cycle cost analysis" is to ensure that the bridge comprehensively costs the least in its whole service life. From the perspective of investment decision-making, design scheme, project implementation, ecological and environmental protection, the engineering design and maintenance practices based on "lifecycle cost analysis" are more scientific and reasonable than the conventional methods.

Similar strategies came later in China. Since 2001, a concept that 'construction is development, maintenance and management is also development' has been promoted in China. However, practitioners are still reluctant to reject the old concept of "focusing on construction, ignoring maintenance and management". At present, maintenance and management of bridges are mostly undertaken in the "post event" stage in China. The direct consequence is that minor bridge distress condition can gradually develop into major or severe ones and eventually lead to bridge failures.

In addition, by adopting the conventional engineering design and maintenance methods, many bridges in China are currently suffering from serious deterioration in performance, low durability, high maintenance and repair costs, and heavy burden on operation and management, which is inconsistent with China's sustainable bridge engineering development strategy. It is an inevitable trend to introduce lifecycle decision-making into the maintenance and management of existing bridges in China, so as to reduce the overall resource consumption and to bring significant social and economic benefits to the country.

1.4.2 Challenges of Current Technologies

In the bridge maintenance and management process, inspection and monitoring technologies are important methods to assess the bridge's health. At present, relatively backward manual inspection and semi-automatic detection predominate bridge inspection in China. Manual inspection has the problems of tedious work, slow speed, low efficiency and high rate of missed detection due to the fatigue of inspectors. Even if simple auxiliary detection equipment are used, there are still issues such as low precision, heavy workload and unreliable information with the inspection. For example, the conventional inspection requires photo-taking, which is affected by lighting, resolution and shooting angle. When taking photos, engineers must work at precarious elevated locations, and the inspection results so obtained might not be able to comprehensively reflect the integrity of the whole bridge. Bridge inspection vehicles are expensive and therefore difficult to popularize, and they pose a threat to the driving safety of passing vehicles and causing traffic congestion. Additionally, the inspectors usually must walk on the gears, which can be dangerous.

SHM technologies have overcome many shortcomings of the conventional inspection technologies, such as strong subjectivity, poor real-time, impact on normal traffic, lack of data accumulation, etc. They have served many long-span bridges and large buildings in China, providing basis for the maintenance and management of these structures. Bridge SHM integrates modern testing and sensing, network communication, signal processing and analysis, safety evaluation and decision-making, structural analysis and other technologies from different scientific domains. Their main difference from the inspection technologies is that SHM technologies are real-time.

Each sensor in the SHM system is installed on the structure for a long duration so that the system can continuously collect and transmit various types of information under operating conditions in real time, and the health of the structure can be evaluated in real time. However, the current bridge monitoring technologies in China are not perfect, and there are problems on both the theoretical and technical levels.

Inspection and monitoring are to obtain information from the bridge, and systematic and effective data analyses and managements are measures to grasp the long-term development of the bridge. Bridge management systems have been applied to bridge maintenance and management in some developed countries for a long time. The earliest bridge management system appeared in the United States in 1968, which was mainly used to store national bridge archival data. Subsequently, engineers developed the bridge management systems PONTIS and BRIDGIT, which are widely used in the United States. These two systems have the functions of bridge data storage, state assessment, degradation prediction, cost–benefit analysis and optimized decision-making. In Europe, the bridge management systems currently widely used are DANBRO in Denmark, Edouard in France, NATS in Britain, etc. In Asia, Japanese engineers developed the public bridge management system, and Korean engineers developed the SHBMS bridge management system.

Development of bridge management system in China began in the early 1980s, for example, CBMS2000 developed by the Ministry of Communications. The system is mainly aimed at the management of medium and small bridges. It includes of six subsystems: data management module, statistical query module, evaluation and decision-making module, cost model module, maintenance plan module and GIS application module. It has multiple functions of information collection, evaluation, cost calculation and decision-making. From 1995 to 1998, Tongji University developed the Shanghai Bridge Management System based on a GIS platform. This system is the first urban bridge management system in China, and some of its functions address some intractable bridge maintenance and management issues. For example, maintenance demand analysis can be made, and bridge maintenance plans can be formulated based on some bridge technical status prediction models. Generally speaking, the development of bridge management systems in China has gone through three stages: (1) the initial bridge management system simply used electronic databases to replace the paper-based bridge management information archives; (2) in addition to the bridge database, the subsequent management system collects bridge inspection, maintenance and repair information, covering the inspection details and detailed classifications and maintenance histories of each bridge component; and (3) in recent years, the management system has added the maintenance decision-making function, which can now formulate and optimize maintenance strategies. Currently, China is gradually moving forward to be a large bridge maintenance country. However, in terms of theoretical development and technology innovation, the bridge maintenance and management practice in China still falls behind in terms of core software development and basic research. The core bridge maintenance technology and the research and development of bridge maintenance and management systems in China are still in the exploratory stage with many key technical problems to be solved.

At present, bridge inspection and evaluation are performed as two independent stages in China, hence it is difficult to ensure the accuracy and consistency of information, resulting in strong subjectivity, scattered data and qualitative evaluation of bridge safety. In addition, at all stages from inspection to maintenance and retrofitting, the computer systems used in the bridge maintenance and management industry in China are isolated from each other. Information often needs to be input repeatedly, resulting in data redundancy, waste of resources, inability to share information, and situations of "information islands". The information exchanges among various subjects in the industry (testing party, evaluation party, maintenance party, management party, etc.) are still based on paper forms resulting in information transfer gaps between different parties, difficulties to reuse information directly, and delayed transmission or even data loss.

1.4.3 Summary of Technical Challenges

As we argue in the Preface, toward realizing the notion of IBM&M, the lack of three connectivity attributes constitutes the root causes to the technical changes toward realizing the notion of IBM&M. Briefly, they include.

- The disconnection between technology components that prevents the realization of ad-hoc automation.
- The disconnection between identifying, quantifying, and integrating local and global structural damage relative to the system-level functionality of bridge structures.
- The disconnection between technologies and stakeholders.

Taking the state of the practice in China to illustrate these points, the bridge maintenance and management is carried out in a form that is post-event, labor intensive, and lacking all connectivity attributed above. For instances, data, information, analytics, and project progress at each stage are independent of each other, and their information is input repeatedly, making it difficult to share information across the different stages. A large amount of information is stored in paper forms, which is difficult to retrieve and reuse directly, and the information exchanges are therefore prone to fault and loss. All kinds of problems have led to the failure of different parties to work together, low efficiency, unnecessary re-work, decreased work quality, increased management costs, and delayed maintenance. As a result, the returns in term of maintenance efficiency and structural safety are not proportional to the investments in human, financial and material resources. On the other hand, China is calling for seamless integration of emerging sensing, communication, AI, and other innovative digital technologies. This will enable the establishment of a set of innovative solutions suites that enrich the Lifecycle Engineering paradigm, achieving the overarching goals of lifecycle sustainability and resilience for bridge structures.

1.5 Emerging Digital Technologies and Promise

At present, the new generation of digital technologies have transformed from conceptual innovation and theoretical research to practical and cutting-edge applications. These emerging technologies, represented by IoT, robotics, cloud and edge computing, and AI and big data, are driving a new round of scientific and technological advancement at an unprecedented rate in various fields.

In this section, we review these technologies and briefly discuss their applications in China's bridge engineering industry. Thorough and systematic introduction to these technologies are found in the latter chapters.

1.5.1 Internet of Things—Connecting Everything

At the 2005 World Summit on "The Information Society", the International Telecommunication Union formally presented the concept of "Internet of Things", and prophesied that in the era of Internet of Things, communications and information exchanges of all kinds of objects in the world would be realized through the Internet.

The integration of bridge maintenance and management with the IoT technology is a strong booster for the entire industry in China. First of all, the new practices replace the conventional manual data collection and the limited data transmission, leading to improvements of the field operation efficiency and management ability. By using long-term low-power high-frequency low-delay sensors to monitor the condition and the environment of infrastructures, engineers can realize the remote collection of large amounts of data. By using a secure online control panel integration and visualization can be attempted. Secondly, while sensing the existing structures' mechanical status, bridge maintenance personnel is enabled by the IoT technology to quickly collect data. Based on the IoT technology, engineers can collect information in a digital automated way, greatly reducing human workload and related costs. With regard to large-scale bridge infrastructure group managements, massive data interfaces are the key to avoid possible "data overload". The various emerging communication networks and IoT platforms met such massive bridge full-dimensional data management requirements. Thirdly, the IoT technology promoted the digitalization of bridge assets, increased operational efficiency, and improved project performance. At the same time, it provides appropriate platforms and processes for managing the data to better serve bridge maintenance.

1.5.2 *Robotics—Taking Over Human Workers with Robots or Humanoids*

Robotics has been constantly discussed for their potentials in structural maintenance and management in Civil Engineering. In 2019, the topic of "operation and inspection robots for long-span bridges" has put forward detailed requirements for robot inspections of the main beams, cables, main cables and underwater structures of bridges, which pointed out a clear direction for the introduction of robots to bridge inspections in the next few years. It can be learned from online media such as "Super Project" that enterprises and bridge management units have begun to introduce robots to carry out daily "health examinations" for bridges. At present, smart equipment such as the bridge inspection robots, unmanned aerial vehicles (UAVs) with high-definition camera, and crawling robot incorporated with infrared NDT technology can overcome the shortcomings of conventional manual inspection via rapid inspection, and directly improve the efficiency of bridge maintenance and management. With the gradual maturity of new robotic technologies in the field of intelligent construction, more autonomous robots will serve various bridge maintenance scenarios, replace manual inspection, and more efficiently ensure the safe operation of bridges.

1.5.3 *Cloud and Edge Computing—Computing on the Go*

The IoT and robotic technologies enhance the data acquisition approaches for China's bridge maintenance, and meanwhile put forward higher requirements for subsequent data management. However, the conventional data storage and management methods obviously cannot fulfill the requirements. As an internet-based computing method, cloud computing has evolved into a technique with a complete technical architecture that can realize sharing of the software and hardware resources and information rationing on demand. In recent years, cloud computing has been put into use in the field of bridge inspection and monitoring. The general monitoring cloud platform can provide data integration and centralized sharing services for the "24-h" monitoring of large bridges, and small and medium-sized bridge groups across China.

While cloud computing provides information sharing services, it brings pressures to the conventional centralized data storage and processing mode. The network bandwidth and the computing throughput have become the bottleneck of the cloud computing technology. With an architecture that provides network edge data processing capabilities and services, edge computing has become the new driving force to promote development of the ICT industry. Edge computing can be regarded as a collaborative and supplementary technique to cloud computing. The two techniques can be compared to two items in a nervous system with cloud computing being the nerve center (coordinator) and edge computing being the nerve tentacle (executor). As a coordinator, cloud computing is responsible for analyses of long-term data, and can serve the periodic maintenances, business decisions, etc.. As an executor, edge computing focuses on the analyses of real-time and short-term data, so as to better support the timely processing and executions of local businesses. In the future, edge computing will significantly enhance the analysis capabilities of various sensors placed on the bridge, relieve the pressures of data managements at the cloud, and provide reliable computing services coordinated with high-quality data acquisitions.

1.5.4 *Artificial Intelligence on Big Data—Making Machines Predict*

As a new resource to tap on, big data with huge potential benefits has penetrated all walks of the national economy. The government, enterprises and research institutions have all paid close attention to it. From the user behavior analysis undertaken by internet enterprises to the emotional analysis of social products, from the tracking and prediction of major events to the precise delivery of medical products, big data technology has already been applied in various industries. It has brought business improvements and asset appreciations to the industries, and gradually changed the conventional working modes of different industries. In the context of development of the intelligent transportation infrastructures, bridge big data has become an important asset. In the field of bridge maintenance and management, big data technology have been used to determine the vehicle loads on bridge decks and evaluate the health of existing bridges. It is a long-term issue to master big data technology and effectively use them to serve bridge maintenance and management.

AI is a branch of computer science which attempts to understand the essence of intelligence and produce a new machine that can respond in a way similar to human intelligence. Since the birth of AI, its theory and technology have gradually matured, and its applications have also been expanding. The recent advances in Generative AI, trained on very large datasets with generative mechanism, further provide zero-shot predictive capabilities or work as non-human knowledge base (i.e., an assistant). With such AI components of different modalities (e.g., language, image, and graph generation) as *agents*, it can greatly enhance the practice of bridge management.

1.6 Intelligent Bridge Maintenance and Management

1.6.1 *Concept*

We present the definition of Intelligent Bridge Maintenance and Management (IBM&M) herein. IBM&M is a process or system that employs conventional and emerging digital technologies to integrate the bridge structure systems and various resources into a comprehensive cyber-physical-human system, controlled by a standardized, intelligent, and decentralized bridge maintenance and management platform, wherein end-users (e.g., bridge engineers) are in the loop for higher-level intervention and decision-making.

This definition is provided with a premise that a fully automated maintenance and management process without or with minimal human intervention is out of the reach given the state-of-the-art technologies as of today and in a foreseeable near-term time frame (e.g., 10–20 years).[2]

If merely in terms of the data and information flow, such systems solution or a platform starts with automatic or even autonomous real-time data collection, goes though the phase of data mining and fusion, integrates with existing information and models (e.g. via a digital twin modeling process), then presents end-users with meaningful analytics. It is up to the level of automation, end-users may choose to conduct optimization of resource allocation or leave it to the system, but make higher-level decision-making, such as the need and method of repair, retrofitting, or demolishing towards new construction.

From an interfacing and connectivity perspective, an IBM&M system needs to equip with:

(1) Comprehensive sensing and monitoring that interfaces with structures to collect maximal coverage of bridge structures.
(2) Deep data processing, information fusion, and modeling methods that connect different system components, including the application of system identification, machine vision and machine learning, other general AI methods, and finally physics-based or information simulation process, if possible, towards a digital-twin modeling solution
(3) Intelligent even immersive interface with the digital system and the end-users to enable scientific decision-making by achieving comprehensive data collection, deep integration, and intelligent connection in (1) and (2).

1.6.2 Connecting Digital Technologies

Intelligent maintenance and management of bridges requires the introduction of advanced technologies from many disciplines, such as computer science, data science and communication science, and their seamless integration with the conventional knowledge in bridge maintenance and management. Among these advanced technologies, IoT, robots, UAVs and other technologies can be used in the monitoring and inspection process to achieve intelligent perception of bridges; machine learning, AI and other technologies can find correlations among the massive collected data and assist the bridge maintenance and management managers to make scientific

[2] Toward bridge maintenance and management automation, we provide some vision in Chap. 10, and we believe that this largely relies on the arguable realization of Artificial General Intelligence (AGI).

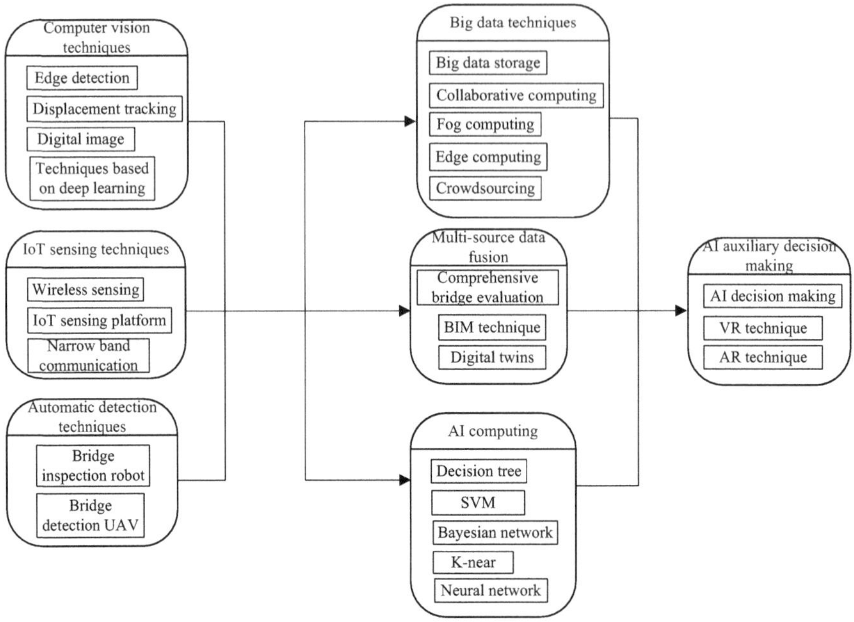

Fig. 1.3 Representative emerging technologies for IBM&M realization

decisions; big data technology can solve the storage and retrieval problems of the massive collected data, build the bridge maintenance and management core database, and provide data support for subsequent evaluation and decision-making. Figure 1.3 shows the interrelationships among various emerging technologies.

1.7 Summary

This chapter first provides a synthesis of bridge maintenance and management practice and technologies that are found several nations. Further, it promotes the notion of intelligent bridge maintenance and management. Data and facts from China are emphasized due to the last 30 years of rapid development and rapid accumulation of bridge failure instances. Characteristics of modern digital technologies and their current application for bridge engineering practice are introduced. This chapter ends with a discussion of the promise of performing intelligent bridge maintenance and management and how they will mitigate the connectivity gaps recognized in the state of the practice.

References

1. Ministry of Housing and Urban-Rural Construction of PRC (2017) Technical code for urban bridge maintenance: CJJ99-2017. China Architecture & Building Press, Beijing
2. Ministry of Transport of PRC (2011) Standard for evaluation of technical conditions of highway bridges: JTGT H21-2011. People's Communications Press, Beijing
3. Ministry of Communications of PRC (2004) Code for maintenance of highway bridges and culverts: JTG H11–2004. People's Communications Press, Beijing
4. FHWA (2001) National bridge inspection standards: FHWA-2001-8954. Washington DC
5. American Association of State Highway and Transportation Officials (AASHTO) (1993) Guidelines for bridge management systems: AASHTO GBMS-1-1993. Washington DC
6. American Association of State Highway and Transportation Officials (AASHTO) (2011) The manual for bridge evaluation: AASHTO MBE-2011. Washington DC
7. American Association of State Highway and Transportation Officials (AASHTO) (1999) AASHTO maintenance manual: the maintenance and management of roadways and bridges: AASHTO MM-3-1999. Washington DC
8. Ministry of Transport of PRC (2008) Code for design of highway bridge reinforcement: JTG/T J22-2008. People's Communications Press, Beijing
9. Bruneau M, Chang SE, Eguchi RT, Lee GC, O'Rourke TD, Reinhorn AM, Von Winterfeldt D et al (2003) A framework to quantitatively assess and enhance the seismic resilience of communities. Earthq Spectra 19(4):733–752

Chapter 2
Intelligent Bridge Maintenance and Management System

Abstract This chapter focuses on the necessity and feasibility of building an intelligent bridge maintenance and management system. It further introduces various existing bridge maintenance and management systems and analyzes their advantages and limitations. Toward developing such a system solution as a digital platform, the general functional requirements and capability requirements are presented. Finally, we introduce IntelliBridge, a software system developed by the authors' team. IntelliBridge provides a complete system architecture for the intelligent bridge operation and maintenance, which can meet the basic needs of current bridge operation and maintenance and can be further enhanced in the future.

2.1 Necessity and Feasibility

2.1.1 Necessity

Intelligent bridge maintenance and management requires the new generation of digital technology coupled with the domain knowledge to achieve comprehensive perception, deep integration, intelligent interaction and scientific decision-making. At present, the amount of data generated in the process of bridge operation and maintenance is ever increasing, and the data types are becoming more abundant. Hence, digitalization level of bridge operation and maintenance is getting higher. With the building of various information systems, a large amount of paper-based and manually filled information is gradually replaced by digital records. In order to meet the needs of increasing data volume, richer data formats, and higher digitalization of bridge operation and maintenance, it is necessary to establish a software system to store and manage the data. Therefore, a software system is needed as the platform for bridge intelligent maintenance and management.

Currently, there are many bridge maintenance and management software systems developed to solve the digital issues and to communicate with the other related systems. However, most of them are single system developed for specific area to solve specific problem, which are unable to accommodate the rapid development of

G. Wu et al., *Intelligent Bridge Maintenance and Management*, Springer Tracts in Civil Engineering, https://doi.org/10.1007/978-981-97-3827-4_2

computer science and communication, and to meet the requirements of new generation of technology in intelligent bridge maintenance and management. In addition, in order to evaluate the health of a bridge and to make decisions in the process of bridge maintenance, it is necessary to introduce professional algorithms to bridge maintenance and to continuously optimize them. Most of the previously developed systems with fixed algorithms failed to provide sufficient cooperative computing power to realize pluggability and replaceability, and cannot meet the requirements of scientific decision-making in bridge intelligent maintenance and management. Although the currently developed information systems can be used to solve the targeted problem, the software systems with the existing system functions must be upgraded and combined into a system platform to enhance their expansivity, connectivity and cooperative computing capabilities.

2.1.2 *Feasibility*

At present, the technology of various management systems based on relational data storage is very mature. Many attempts have been made to solve the digital problems of bridge operation and maintenance, and for the systems to communicate with other related systems. Many software systems have been developed at home and abroad for such uses, and in the process gained much experiences to realize their specific functions such as basic information management, bridge health monitoring, bridge inspection management and bridge maintenance management. Hugh volumes of bridge operation and maintenance related data have also been generated in the process.

In the field of sensing technology, the conventional strain gauges, fiber gratings, accelerometers, etc. are still developing, and non-contact sensing technologies such as camera chains, digital picture correlation, photogrammetry and GPS have also been used. In addition, the ultrasonic, acoustic emission, thermal imaging and other non-destructive sensing technologies are also maturing, which can provide panoramic, high-quality data collected for bridge intelligent operation and maintenance.

In the field of data transmission, 5G transmission and IoT technologies are developing rapidly. 5G transmission can provide higher data transmission rate and lower network latency; and the IoT can provide a sensing cloud by networking multiple sensing devices. 5G transmission and IoT technology can provide various transmission solutions to the problems concerning data transmission in intelligent bridge operation and maintenance, such as "high frequency and low latency" and "low frequency and low power consumption" etc.

Recent advancements in computer science have provided new software implementation strategies for platform development. The theory of distributed systems has taken shape and used to establish the Internet e-commerce platform, etc. In the process of upgrading bridge management systems to intelligent bridge maintenance and management systems, distributed system related technologies will be applied to enhance expansivity; the proposed REST (Representational State Transfer) interface style provides a more consistent and precise definition of data connectivity between systems, hence enhancing connectivity of the intelligent bridge maintenance and management systems.

Big data and artificial intelligence technologies are both fast growing and introducing novel ways to many traditional industries. Big data-related technologies can be used to manage the massive amounts of data generated during bridge operations and maintenance; artificial intelligence technologies can analyze the correlations among big data, and provide new ideas and methods for applications in bridge maintenance and management, such as damage identification and aided decision-making.

To build an intelligent bridge maintenance and management system, it is important to combine the domain knowledge of bridge operation and maintenance with the latest digital technology tools. From the early digital technology exploration, bridge operation and maintenance practitioners have accumulated much experience; and meanwhile a new generation of digital technology tools is rapidly developing to be applied to all aspects of intelligent bridge maintenance and management. In order to accomplish the objective of intelligent bridge maintenance and management, each individual bridge management system should be combined into a unified system with expansivity, connectivity and cooperative computing capabilities based on a good architectural design.

2.2 Overview of Existing Software Systems

2.2.1 Global Trends

In the 1960s, several bridges in the United States entered their aging period, which led to frequent safety accidents, such as the collapse of the Silver Bridge in West Virginia in 1967. To this end, the FHWA established "the National Bridge Inventory (NBI)" in 1968 to store data for all bridges in the United States, including identification information, type specification, bridge health status and geographic location.

In the 1980s, the U.S. developed the PONTIS system and the BRIDGIT system, which upgraded the storage and query system of single bridge information to an integrated management system with the functions of recording, storage, updating and statistics, and unified the different formats of bridge information among states used at the beginning of the application. So far, the PONTIS system is the longest serving and most experienced bridge information management system, providing data supports and recommendations for bridge maintenance in many states. It aimed at predicting the future health states of bridges by using bridge inspection and other data, and making maintenance recommendations in order to achieve better maintenance with limited budgets.

In 2008, the Office of Infrastructure Research and Development (OIRD) of the FHWA and state departments of transportation and other federal agencies launched the Long-Term Bridge Performance Program (LTBP), which planned to unify NBI,

PONTIS and the other databases over a 20-year period, integrating GIS, traffic and climate information to establish a more standardized bridge health database. It is intended for improving the safety, serviceability and service life of U.S. highway bridges, and achieving whole life cycle maintenance and management. Currently, the data collection and standardization has been completed by the FHWA through the LTBP program. The basic information of all bridges in the U.S., including the data on basic bridge units and the partial results of long-term bridge performance are publicly available in the InfoBridge web site (https://infobridge.fhwa.dot.gov, as shown in Fig. 2.1).

In Japan, the bridge inspection data system (MICHI) was developed in the 1990s to collect basic information and inspection data on bridges, nationwide. Based on the data from this system, the J-BMS system was subsequently developed, which can diagnose the damages and predict the health conditions of bridges based on the basic data, and assist the managers to make maintenance decisions so as to reduce overall expenses.

European countries have also developed their own systems to meet their bridge operation and maintenance requirements, such as the Danish DANBRO system, which focuses on the typical problems of low temperature and corrosion, and has

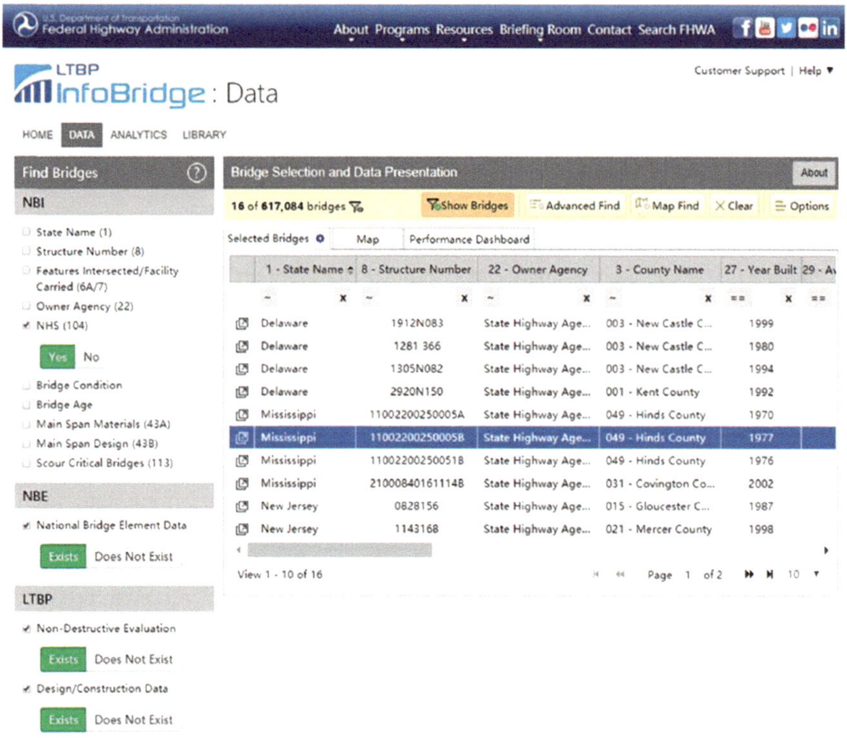

Fig. 2.1 InfoBridge web developed in the US

been used for more than 20 years with significant results. The UK has established its own NATS system; and France has established its own EDOUARD system.

2.2.2 *Trends in China*

Digitization of bridge maintenance and management in China began belatedly. In the 1990s, the Ministry of Transport of the People's Republic of China (MOT) developed the China Bridge Management System (CBMS), which was designed with the B/S architecture of a high level of maturity, and was updated several times. The system was designed according to Specifications for Maintenance of Highway Bridges and Culverts (JTG 51200-2021) and Standards for Technical Condition Evaluation of Highway Bridges (JTG/T H21-2011), having the functions of inspection information management, maintenance decision information management, bridge evaluation, etc. CBMS is the most widely used bridge management system in China, which has stored data from nearly 200,000 bridges up to this point. The main research direction at this stage is how to tap the value of the data obtained through big data analysis technology to guide the bridge maintenance work.

2.2.3 *Analysis of Existing Software Systems*

Currently, a great deal of work has been done at home and abroad on the digitization of bridge O&M, and several excellent bridge management systems such as NBI and CBMS have been developed to fulfill the needs of basic bridge information management. Researches in the fields of bridge analysis, bridge evaluation and bridge management strategy have also been undertaken based on the systems in different countries. In addition, bridge O&M-related consultants, designers and constructors have also developed their own special software systems, such as bridge health monitoring systems and bridge inspection management systems to assist in their bridge O&M.

However, the various implemented Chinese and other countries' software solutions do not quite meet the present increasing requirements of intelligent operation and maintenance of bridges. The existing data formats adopted by various software systems are not uniform, e.g., a region or even a bridge may have its own data structure and coding system, bringing difficulties to any system data exchange. The current software systems are expected to solve this local problem. However, the lack of maintenance means and the expansion and upgrading interface make it unable to embed the advanced software to replace the old. In addition, some systems simply collected data, and the bridge operation and maintenance management were not practically implemented, resulting in the use of such software systems to be more of a burden.

2.3 Basic Functions, Platform Capabilities and Prospects

To enhance information management ability for bridge operation and maintenance, the owners, the consultants and the constructors have successively built some management systems, health monitoring systems, inspection management systems and other software systems, achieving some specific purposes each. However, the systems are isolated from each other, with weak openness and scalability. This problem leads to the severe constraints of subsequent maintenance and upgrades, which are not properly linked with the bridge management work flow, and cannot meet the demand of bridge intelligent operation and maintenance.

In order to solve the problems of data connectivity, openness and expansivity, it is necessary to upgrade the existing systems and establish a bridge intelligent operation and maintenance computing platform to integrate all existing systems and realizes data interoperability between them. Meanwhile, it can provide open interfaces for the expansivity to facilitate easy maintenance and subsequent upgrades, and can continuously accept emerging technologies and new algorithms for the improvements of overall platform capacity.

As a superset of the existing platforms, the bridge intelligent operation and maintenance computing platform must first realize the basic functions of the existing systems and the exchange of data between the subsystems and submodules in the platform. On this basis, it is necessary to develop the overall expansivity, connectivity and cooperative computing capabilities of the platform so that the platform can continuously expand its own functions, connect with external data and integrate emerging technologies and algorithms. Finally, the platform can be interconnected with other smart cities and smart government systems for joint improvements of the infrastructure operation and urban traffic management. Architecture of a bridge intelligent operation and maintenance system platform is shown in Fig. 2.2.

2.3.1 Basic Functions

(1) Uniform Coding Specifications for Bridge-Related Types

Bridge-related type coding that refers to the types of bridges, defects and other information related to bridge operation and maintenance is given to the computer in easily recognized symbols, and forms a code collection. In the early information collection, due to the lack of unified coding specification, different systems used their own different coding forms causing difficulties in data interfacing which cannot satisfy the maintenance needs for the whole life cycle of the bridge group.

In the NBI system developed and maintained by FHWA, the relevant regulations regarding coding are given and updated from time to time to meet the development of the bridge industry [1]. The developers of other related systems refer to

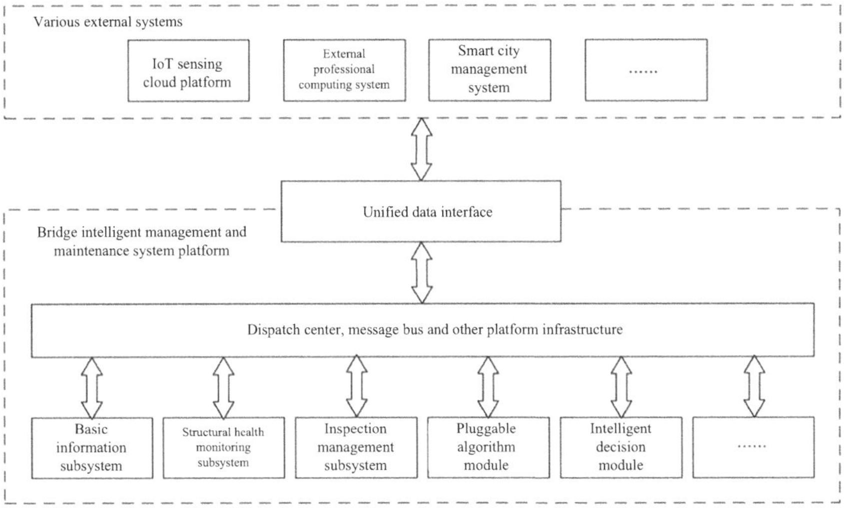

Fig. 2.2 Architecture of bridge intelligent operation and maintenance system platform

this coding specification to complete their own developments, and the cost of data communication between systems is therefore reduced by a great extent.

The general bridge intelligent operation and maintenance computing platform involves a large number of sub-systems and sub-modules, which require a large amount of data exchange between each sub-system or sub-module. The platform also needs to interoperate with external systems such as government systems, so it is necessary to formulate a set of unified coding specifications to reduce data redundancy and inconsistency, and ensure smooth information exchange.

Bridge-related codes should include codes for various bridge types and specific events, involving various aspects of bridge information, inspection information, monitoring information, maintenance information, etc., as well as bridge identification information, geographic location information, functional material information, geometric dimension information, etc. Meanwhile, the types of components and elements in various types of bridges, and the methods to solve different types of defects should be likewise uniformly coded. In addition, the specific examples of individual bridges, components, elements, defects, inspection, maintenance and construction also should be clarified according to the coding format.

(2) Bridge Basic Information Management Function

The basic information of bridges includes their administrative identification information, structural and technical information, hydrological information, archival data and other basic data required for bridge operation and maintenance. The most basic function of bridge intelligent operation and maintenance computing platform is management of the basic bridge information, and a sub-system in the platform is supposed to realize the management of basic bridge information.

In addition to basic information data management function, the basic information management subsystem should also open the data interface to other systems and modules in the platform. The open data interface enables the entire bridge intelligent operation and maintenance system to reuse the bridge's basic information and reduces the workload of repeated development of the platform functions. In addition, management of the basic data is exclusively undertaken by the subsystem, which facilitates the overall authority control during the user interaction and prevents any inconsistent information been recorded due to data redundancy in running the system.

(3) Health Monitoring Data Management and Processing Functions

The data collected in the process of bridge health monitoring possess the following characteristics: the data are written much, read little and rarely changed; the data are in obvious time sequence; the data are generated at a smooth and continuous rate; and the data are read and queried in a time range that requires multi-dimensional statistical analysis. These characteristics are in line with the requirements of "time-series data", and can be used for the dynamic data oriented special processing by referring to the relevant technologies in the Internet and the Internet of Things.

In bridge health monitoring, message queues can be used to store the massive data reported to the platform from sensors and collection devices. Since the monitoring data points are independent of each other, the parallel computing technique can be used to process these data in the queue by multiple consumers to increase the processing capacity of the system. At the same time, a time series database can be used to store the data. Time series database is a type of database optimized for time-series data to improve the insertion performance of time-series data, allows fast filtering of data in time domain and provides a large number of common statistical functions.

Bridge health monitoring is not only intended to collect data, but also aims at dynamic computation based on the collected data to achieve real-time analysis and early warning. The platform can use the technologies related to streaming computing to continuously and dynamically analyze the data according to the presented rules and algorithms.

Finally, the health monitoring sub-module should be able to output results for the references of other modules, such as the evaluation module, the decision making module and the early warning module, which necessitates the exposure of certain interfaces. Particularly, it is vital to establish the event-driven interfaces to provide early warning and alarm information for other sub-modules to subscribe to, so that they can be notified in a timely manner when warnings and alarms occurred and to perform the function of bridge health monitoring.

(4) Inspection Data Management and Inspection-Related Functions

Daily inspection, regular inspection, periodic inspection and special inspection are all types of bridge inspection. The bridge inspection data consists of project management information such as the time and personnel of each inspection, as well as particular inspection results such as the site photographs, defects detected and evaluation results of the bridge. Structure of the inspection data is constant, and the data amount is relatively small in comparison to the monitoring data; therefore, it is suited for relational data storage.

In addition to managing inspection data, the bridge intelligent operation and maintenance system platform must also be able to optimize the inspection process and analyze the inspection results. After collection of the inspection data, the platform can provide algorithms to process and analyze the inspection results, e.g. to automatically compute the inspection evaluation score, generate the inspection reports and other reports, and to track and compare the defects.

(5) Management of Bridge Maintenance Data and Maintenance-related Functions

Bridge maintenance includes the formulation of maintenance and repair plans, and the prompt treatment of distresses detected during inspection. The bridge maintenance data includes the engineering management information, such as the time and personnel of the prior maintenance activities, as well as the particular maintenance information such as photographs of on-site operations, defect treatments and treatment outcomes. In addition, the practice of bridge maintenance is focused on early-stage difficulties, thus it is important to exchange the maintenance data with data from the other systems. The management of bridge maintenance data should also incorporate the formulation of common maintenance procedures, so that the maintenance advice can be provided for the operators' reference when problems are identified.

In addition to managing the data generated from the bridge maintenance practice, the system should be able to optimize the maintenance operation's workflow. The platform should provide a unified secondary development interface and unified coding for various maintenance-related information types, so that each user can develop his/her own intelligent terminal applications based on the actual workflow and report to the platform via the interface. This can ensure applicability and unify data formats, providing good data support for subsequent decision analysis, etc.

2.3.2 *Platform Capabilities*

As a superset of existing software systems, the bridge intelligent operation and maintenance system platform should be empowered and upgraded to achieve the effect of "$1 + 1 > 2$" with regard to the fundamental functions of bridge operation and maintenance information management.

Extensibility, connectivity, cooperative computing capability and dynamic feedback capability are the most crucial aspects of platform empowerment. Extensibility indicates that the system platform can evolve and the service level can improve. Connectivity suggests that the platform can interoperate with other systems and the system boundary can expand. Cooperative computing capability and dynamic feedback capability indicates that the algorithm of each professional computing function is pluggable, thereby facilitating the adoption of more advanced and appropriate algorithms for the targeted problems. For specific problems, more advanced and suitable algorithms can be implemented.

Extensibility and Scalability

An intelligent bridge operation and maintenance computing platform should be extensible and scalable.

Extensibility in the fields of software engineering and system design denotes the infrastructure of a software system does not need to be modified frequently, and its sub-modules are less dependent and coupled, allowing them to respond quickly to changes on demand. In the intelligent bridge operation and maintenance computing platform, minimal number of modifications allow the platform to respond rapidly to changing operation and maintenance requirements, as well as to add new demand functions, access new hardware data and integrate new application algorithms.

Scalability in the field of system design refers to the capacity of a system to handle the development of business volume, primarily through horizontal hardware expansion. The intelligent bridge operation and maintenance computing platform may manage bridge groups consisting of dozens to tens of thousands of bridges, and the amount of data generated varies substantially from bridge to bridge due to different inspection, monitoring and maintenance tactics. Therefore, the platform should be able to develop its functions to accommodate the management requirements of diverse bridge widths and data sizes.

Connectivity

Connectivity of intelligent bridge operation and maintenance computing platform is mainly reflected in its ability to exchange sufficient information with systems outside the platform. Although intelligent bridge operation and maintenance computing platforms are designed to consider the basic information management, inspection, monitoring, evaluation, decision-making and other aspects of data and process management, it cannot solve all the problems confronted. In the process of application, the platform needs to provide sufficient interconnection capabilities to prepare its development according to specific needs or needs to export data to other systems. The connectivity of intelligent bridge operation and maintenance computing platform requires that the system can easily and safely provide data to and receive data from outside.

The REST style interface specifies resources through Uniform Resource Identifier (URI) and provides operations of access, creation, modification and deletion of resources. In the intelligent bridge operation and maintenance computing platform, each bridge, each member, each distress or each monitoring data should be used as a resource, and the platform can describe these resources through the REST style interface, so as to provide a variety of resources with reading and writing capabilities. Security of the connectivity largely depends on the authorization control when reading and writing data in the platform, including both operation permission and data permission. Operation permission refers to whether there is permission to perform certain operations, such as to allow editing of bridge information, etc., which should be determined by the interface address and the operation methods. Data permission refers to whether there is permission to operate certain data, for example, the administrator of a department cannot edit the information of bridges under the jurisdiction of other units. To achieve security of connectivity, the permission allocation should be carefully designed and the operation of user requests should be identified through various identification tokens to realize the authentication of operation and data, so as to guarantee interconnection capability of the intelligent bridge operation and maintenance system platform.

Capabilities for Cooperative Computing and Dynamic Feedback

Cooperative computing capability of the intelligent bridge operation and maintenance computing platform is primarily manifested by the platform's ability to quickly incorporate emerging technologies and algorithms, and to continuously enhance its overall computing capabilities.

The system platform should primarily complete the data connection and integration, provide input and output interfaces, as well as data and runtime infrastructure for the specific algorithm program, without concerning the algorithm's implementation. Diverse algorithms for bridge operation and maintenance problems are dynamic and pluggable, and platform users can therefore add new algorithms at any moment, and flexibly compare and choose between the new and current algorithms. The platform is isolated from the specific algorithm implementations. In addition to the inbuilt algorithms, the platform must also be able to seek other external algorithms for use. The external algorithm typically has fixed input and output, and the platform itself may have heterogeneous data structures and programming languages; therefore, the platform's interconnection capability must be used to provide data access, format conversion and other functions in order to invoke the external algorithms.

The dynamic feedback capabilities of intelligent bridge operation and maintenance system platform enable the algorithm caller to receive positive feedback on the decision-making process and compute results automatically or semi-automatically during algorithm use. The algorithm caller examines the effect of the method, and the platform aids in recording and comparing, and can generate statistics after many computations to aid bridge operators in picking more suitable algorithms for a variety of situations. For artificial intelligence-related algorithms such as bridge-assisted decision making, the algorithm caller's feedback can provide correct discrimination for some machine learning algorithms and promote the enhancement of algorithm model capability via the dynamic feedback capability of the platform.

2.3.3 Prospects

With extensibility, connectivity, cooperative computing and dynamic feedback, the intelligent bridge operation and maintenance computing platform can continuously expand its own functions and improve its performance. Meanwhile, it can interact with external software systems, participate in other life cycles such as bridge design and construction, and collaborate with other intelligent infrastructure systems to jointly contribute to smart city construction.

The problems of bridge operation and repair are extremely complicated, and the requirement of each department is unique. With the open interface offered by the intelligent bridge operation and maintenance computing platform, each department can design its own web terminal application or mobile terminal application based on its own requirement in order to continuously expand the platform's capabilities.

Various application layer data can be unified into the central database by the unified coding specification and data organization form. These data can then be used for training and tuning of various algorithms, which allow bridge operation and maintenance personnel to select more appropriate algorithms to replace the existing algorithms for accuracy improvements and to continuously enhance the intelligence of the platform.

The intelligent bridge operation and maintenance computing platform is developed in the bridge operation and maintenance phase, but can be utilized in other phases of a bridge's life cycle, such as design, construction and demolition. By incorporating the data generated in the different phases, a bridge group or a single bridge's entire life cycle digital file can be created.

Bridges, highways, tunnels, etc. are all interconnected components of a transportation infrastructure. Therefore, the intelligent bridge operation and maintenance system platform should be capable of interacting with various infrastructure management systems, exchanging data such as GIS, weather and traffic events, and realizing the unified intelligent management of transportation infrastructure.

The new concept and manner of city development aims to improve the quality of urban living environments, optimize urban management, production and lifestyle, and increase urbanite happiness. As an essential component of urban transportation infrastructure, bridges must be incorporated into the smart city building process. With expansion and interconnection capabilities, the intelligent bridge operation and maintenance computing platform can continuously expand its own functions and fully exchange data with traffic management, government services and other urban management systems to improve the level of urban management intelligence.

2.4 Introduction of IntelliBridge

To prove the concept of intelligent bridge operation and maintenance, the authors' team developed a workable system platform for intelligent bridge operation and maintenance, the IntelliBridge. In terms of its basic functions, the platform can undertake basic bridge information management, health monitoring data collection and analysis, inspection process management and data analysis, bridge technical condition assessment, and medium and long-term maintenance planning and decision-making. Meanwhile, other advanced applications and programs are also adopted, including management terminal, safety warning and management dashboard, etc. In addition, in order to achieve expansivity, connectivity and cooperative computing capabilities of the platform, its architecture as well as its open data interface are meticulously designed. This allowed the platform to be connected to the government services systems, smart city systems, etc. Furthermore, users can also design appropriate algorithms and application terminals themselves according to their actual needs to further enhance the platform's capabilities. Currently, the primary platform has been established and implemented in the operation and maintenance of a number of Chinese bridge groups.

2.4.1 *IntelliBridge Architecture*

IntelliBridge is a software computing platform that is "internally integrated and externally accessible." The term "internal integration" refers to the integration of basic bridge information management module, health monitoring module, inspection and construction module, maintenance management module, owner management module, safety warning module, medium- and long-term maintenance planning module, and other sub-systems and sub-modules to achieve the basic functions of bridge operation and maintenance management. The term "external integration" refers to the exposure of various APIs of this platform to the public to allow other platforms to realize their functionalities accordingly.

Data access layer, infrastructure layer, data layer, application service support layer and application layer are included in the platform's overall architecture. The data access layer is responsible for accessing data from App, sensing system and third-party system. The infrastructure layer provides IT infrastructure such as servers and operating systems for platform operation. The data layer is used to store and process various types of data on the platform. The application service support layer uses various software system development tools to develop various specific functions required by the platform. The application layer composes of three subsystems for data management, domain specialized analysis, and management and decision making. In addition, users can use the application service support layer's numerous open APIs to interact with the platform according to their own requirements or to implement their own functions of various application layers.

Figure 2.3 illustrates the detailed architecture of the platform.

Data Access Layer

The platform data access layer is mainly responsible for the access of external data. Among the external data, bridge inspection and construction-related data can be automatically reported through various mobile terminal apps during inspection, and can also be entered from electronic inspection reports through natural language processing and other means, or can be transferred from the existing third-party bridge inspection management system through a transfer program. Bridge health monitoring data can be reported directly from various sensors and collection devices or indirectly through the Internet of Things (IoT) sensing cloud, or from the existing health monitoring system through transfer program and can be entered into the existing test and

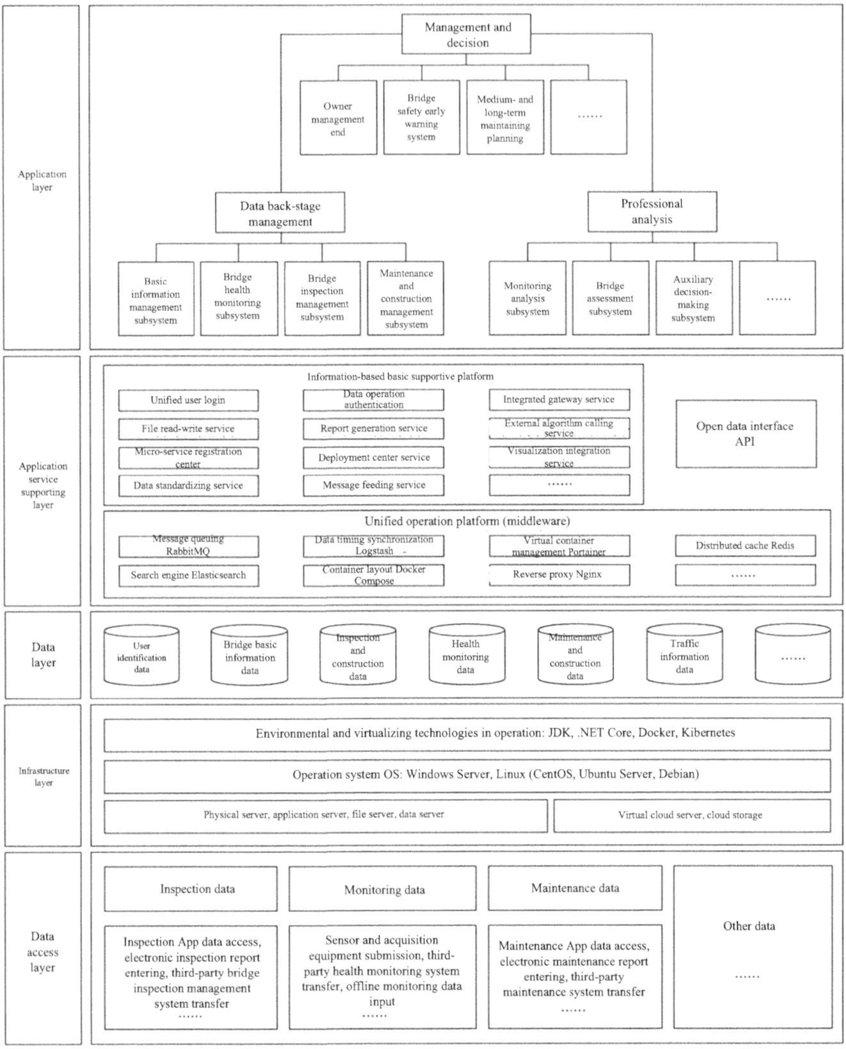

Fig. 2.3 Architecture of IntelliBridge

inspection data. Bridge maintenance and construction data are similar to the bridge inspection data, which can also be reported through the APP, electronic report entry or third-party system transfer.

Infrastructure Layer

The platform infrastructure layer primarily offers the IT resources to support the platform, ensuring its operation as a whole. The whole platform is a distributed system, and physical servers and virtualized cloud servers are used to ensure data security and computing performance. Application servers, file servers and data servers are differentiated to make better use of different hardware configurations. Regarding operating systems, the platform combines various Windows and multiple Linux distributions to meet a variety of requirements. The platform employs the operating environment of JDK and.NET Core, and Docker virtualization technology to eliminate differences in the system layer, improve transferability of the system, increase the degree of system operation automation and provide a runtime environment for the upper layer software.

Data Layer

The data layer of the platform is primarily responsible for data storage and management, as well as data support for higher-level applications. The platform maintains data pertaining to user identity, fundamental bridge information, inspection and construction, health monitoring, maintenance and construction, traffic information, etc. Structured databases (MySQL, SQL Server) are utilized for special optimization of monitoring and other time-series data. Time series databases (TimescaleDB) and document-oriented unstructured databases (MongoDB) are utilized to realize special optimization for monitoring and other time series data. In addition to utilizing mature database products, the platform implements functions for specific requirements such as hot-backups, hot and cold data separation, and read–write separation, which guarantees stable and high-performance data be extracted by the upper layer applications of the platform.

Application Service Support Layer

The application service support layer of the platform communicates with the data layer below. It consists of multiple infrastructures required for system development, and provides specialized functions for each subsystem of the application layer above. The application service support layer consists of the unified operation platform, the digital technology foundation support platform, and the supply of external data interface.

Unifying the computing platform is irrelevant to the domain knowledge, which is primarily for system development of the required infrastructure. Examples include message queue, search engine, web server, reverse proxy server, caching, etc. The unified operation platform makes extensive use of the advanced Internet technologies and architecture, which not only meets the current work requirements but also has robust expandability.

Each function in the digital technology foundation support platform is the lowest logical functional unit in the platform, realizing specified platform subsystem functions. The unified user login function provides the login center of the entire platform. The data operation authorization function can authenticate the user's request based on the their tokens. The platform has multiple functions and exposes the data interface to the outside world at the same time, so that the integrated gateway service can unify the routing control of all types of requests and integrate the authorization-related functions. The file service provides file-related path allocation, backup processing, distributed storage and other functions for other services in a unified manner. The report Generation Service provides encapsulated tools for generating reports in various formats, simplifying the logic of specific report generation. The external algorithm invocation service integrates encapsulated calls to external systems, shielding other services from language and data structure differences in algorithm implementation. The microservice registry effectively manages microservices, decouples microservices from each other and provides horizontal scalability. The configuration center service can centrally manage the common configuration among services, allowing each microservice to obtain configuration information dynamically. The visualization integration service provides the basic toolbox for various customized visualization applications, such as the WebGL-based display engine, the conversion of 3D models, the storage of modeling information, etc. The data standardization service can unify all types of data processing, and prepare for subsequent data analysis and storage. The message push service includes various push mechanisms such as email sending, SMS notification, web message push, etc., making it convenient for higher-level applications to call.

In addition to the unified operation platform and the digital technology foundation support platform, the application service layer offers open data interface to empower the platform with adequate interconnection capacity. The platform provides sufficient API authority control to maintain integrity of the platform data. Users can freely utilize the API to read and write data within the platform or construct new application layer tools by integrating their own workflow with the API.

Application Layer

The application layer of the platform consists of a number of sub-systems and sub-modules to satisfy the requirements of intelligent bridge operation and maintenance, which are targeted for addressing data management, domain-specific analysis and decision-making issues in the process of intelligent bridge operation and maintenance.

The bridge data management is based on low-level data management operations of various metadata of the system, which can be used to manage the fundamental bridge-related information, health monitoring-related metadata, monitoring collection data, bridge inspection-related metadata, inspection result data, maintenance construction-related data, maintenance operation data, etc.

IntelliBridge has been designed and constructed with a monitoring and analysis sub-system, a bridge evaluation sub-system and an auxiliary decision-making sub-system so as to meet the demand of expert analysis for bridge operation and maintenance. The monitoring and analysis subsystem provides data analysis functions related to bridge health monitoring and bridge condition assessment based on monitoring. The bridge assessment subsystem can undertake assessment of the overall bridge condition and specific usage requirements. The auxiliary decision-making subsystem offers the single component maintenance, single bridge maintenance, bridge group maintenance, cycle maintenance and emergency preparedness.

The platform has been designed and created with the owner management subsystem, safety warning subsystem, and medium- and long-term maintenance planning subsystem for management and decision making. Among them, the management sub-system of owners provides statistical and prompt information for the management personnel, including the key information pushing, status dashboard, basic information statistics, maintenance work volume statistics, data comparison analysis, workflow management, etc. The safety warning sub-system provides real-time dashboard display, ship collision warning, vehicle collision warning, emergency response and disposal, and mobile alerts. The medium- and long-term maintenance planning subsystem offers recommendations for medium- and long-term maintenance planning, comparative study of various maintenance plans, etc.

2.4.2 *Basic Features*

IntelliBridge integrates the current states and business requirements of bridge operation and maintenance with the development of various subsystems by offering three functions of data management, specialized analysis and decision making in the application layer of the system architecture.

Classification and codification of bridge operation and maintenance management information.

Existing information classification and coding standards are mostly intended for the construction industry, and the bridge industry has fewer relevant standards and specifications. In developing IntelliBridge, the authors' team referred to the necessary theories, and classified and coded bridges, components, distresses and other aspects in accordance with the principles of scientization, systematization, consistency and scalability.

Typical categorization of entity elements is the sub-division kinds pertaining to certain general categories (Is-a relationship). The bridge-related bridge types, member types and distress types require additional attentions. Certain types of bridges might have specific sorts of members, and specific types of defects can affect the specific types of members. By articulating this Has-a link, the relationship between the several layers of concepts in the bridge O&M management process may be described with greater clarity.

According to the bridge type, member type and defect type, the first level type (e.g., girder bridge, arch bridge, cable-stayed bridge, suspension bridge, rigid bridge, combined structure bridge, etc.) is divided, followed by the second level type (simply supported girder bridge, cantilever girder bridge, continuous girder bridge, etc.). The secondary type is a subdivision of the primary type, hence the relationship is a tight Is-a. Figure 2.4 depicts the classification of bridges.

The member type and bridge type, as well as the distress type and member type, have a Has-a relationship. There may be numerous member types in each bridge type, and several distress types in each member type. Although these member types and distress types share the same name, they should be considered as distinct types. A set of unified coding standards should be utilized to separate the different types in order to address this issue.

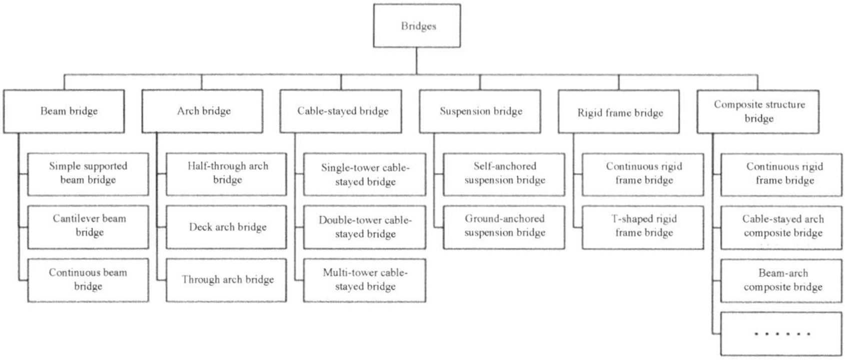

Fig. 2.4 Classification of bridges

The unified coding of bridge type, member type and distress type uses a 12-digit fixed-length numeric code, with "0" to compensate for the lack of 12 digits. The bridge type, member type and distress type are listed in descending order. Figure 2.5 shows that the bridge type, member type and defect type are all expressed by a 4-digit primary classification and a 4-digit secondary classification. Classification of the bridge type, member type and defect type can be indicated in a uniform form and code so that the types with identical names can be recognized precisely. This allows for the storage of high-quality bridge O&M information, which gives useful data for analysis and decision making in later phases of bridge O&M.

Bridge Basic Information Management Subsystem

The primary objective of establishing the bridge basic information management subsystem is to meet the fundamental bridge information management requirements for bridge operation and maintenance, as well as the usual platform usage requirements from the personnel and the authority.

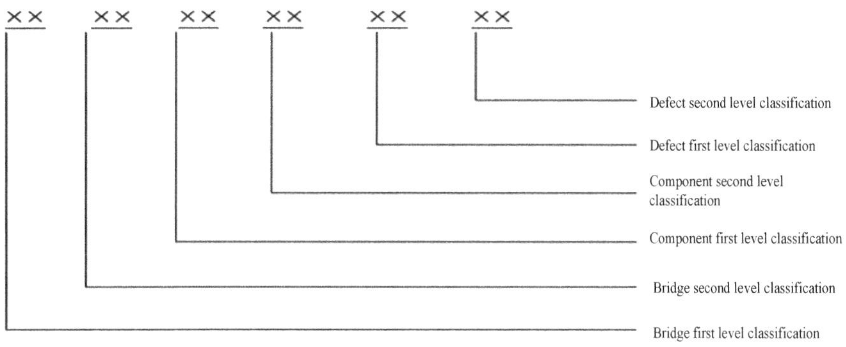

Fig. 2.5 Uniform coding format for classification

The department unit, personnel and authority management function can realize the differentiation of system users into departments, and establish the hierarchy of data management through the department-sub-department relationship. Users belong to a department, which limits the data authority of users. At the same time, users hold a certain position to stipulates the operation authority of users, and security of the system can be ensured by the double limitation of data authority and operation authority. Simultaneously, the system will record the user's login and significant operations, and the administrator will be able to examine the operation time, IP address and other data to assist the study of anomalous operations.

In the basic information management subsystem, the bridge group can be displayed in two different ways. The geospatial distribution of bridges can be displayed dynamically and intuitively via geographic information positioning on a map, or the list of bridges can be displayed precisely and methodically in the form of tables. In both display styles, the detailed information entrance of individual bridges is accessible, and the fundamental information of each bridge can be viewed and edited after entering. Figure 2.6 illustrates the fundamental information management subsystem.

Bridge Condition Monitoring Program

Access, storage, analysis and retrieval of sensor data, etc., are provided by the bridge health monitoring subsystem, and the supporting apps or applets for monitoring overview, dynamic weighing management and other mobile-friendly content have also been developed.

The monitoring overview comprises two levels: overview of monitoring for bridge groups and overview of monitoring for individual bridges. In the bridge monitoring overview of a bridge group, the geographical position, the alert state, and the abnormal ratio of each monitored bridge may be examined and shown logically and vividly on a map using distinct markers. The monitoring overview of individual bridges displays the current alarm and abnormality status, the sensor distribution, and the

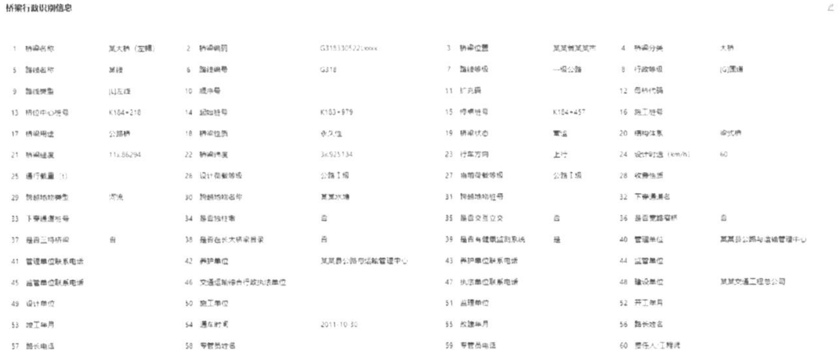

Fig. 2.6 Single bridge detail page

Fig. 2.7 Monolithic bridge monitoring profile

score based on monitoring data, and also permits the export of monitoring-required reports. Figure 2.7 depicts the monitoring overview page for each bridge.

Monitoring management involves the management of monitoring data, such as the data types and the sensors to generate the data, as well as the management of monitoring tasks and alert identification procedures. Data column is a type of data that sensors can report, such as the name, physical unit and individual calculation method, etc. Specific data types should be attached to the management of sensors, which includes the primary kind, acquisition index, deployment position and maintenance records of each sensor installed on the bridge.

In bridge health monitoring, numerous types of sensors and acquisition devices are utilized, and the data structure and transmission techniques of products purchased from various manufacturers are likewise distinct. The data access layer of the platform should be able to pre-process the diverse monitoring data, standardize the data format and then utilize the universal transmission method of the platform to contact the monitoring system in order to process the monitoring data in a unified manner. After data access, the platform may store, cleanse, analyze and display the data by using a uniform processing procedure. To deal with the robust real-time monitoring data and the large volume of data, the system provides real-time display, historical classification query, comparison query and other functions (see Fig. 2.8 for the historical data comparison view function) and enables the visual output of query results. In addition to the conventional sensor data access, the system also incorporates the bridge weigh-in-motion system, which enables data viewing, over-limit determination and other operations. Meanwhile, the system incorporates a portion of the algorithm for the monitoring data analysis, and is able to deliver modal analysis and other professional analysis results.

Monitoring of bridge health is more digital and automated than the inspection and maintenance construction, and the processing of monitoring results in real time is more demanding. A monitoring management applet is designed to examine the real-time monitoring condition of the bridge group at any time on smart phones for users to quickly and easily obtain useful monitoring data. In order to simplify the

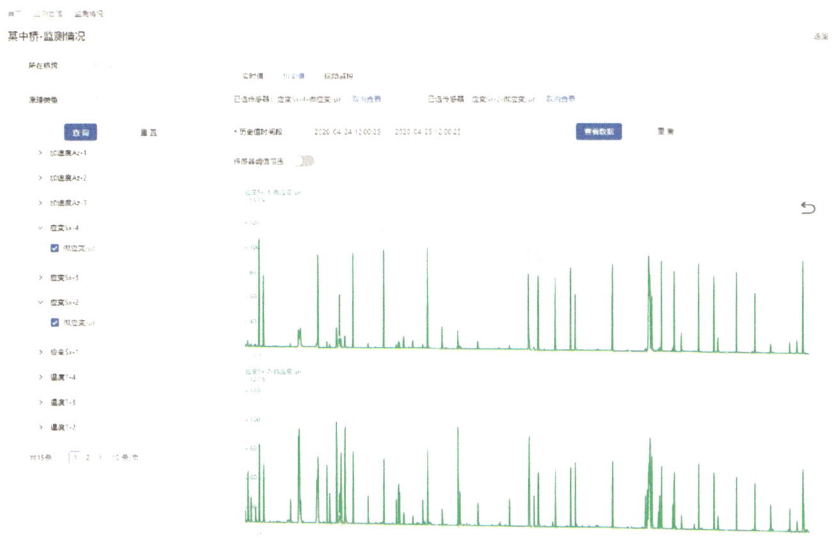

Fig. 2.8 Comparative view of monitoring history data

cooperation with traffic control departments to control high loads, the system has also built a weigh-in-motion applet (see Fig. 2.9 for page impact) that can display the current and the historical records of bridge traffic weighing.

Human-System Interfacing

Bridge inspection administration subsystem provides the functions of daily inspection management, regular inspection management, periodic inspection management, special management of bridge inspection works, management of supporting inspection equipment and the inspection results for viewing and comparison analysis. The relevant APP has been developed to meet the inspection requirements for site operation.

The inspection project contains inspection of numerous bridges, which are often grouped and distributed among the inspection departments. In the management of inspection projects, the system recording information, such as the inspection departments, project start date, current status, etc., is classified on a list. Each project can be referred to in detail in view of the overall project information, the list of inspection task of each bridge within the project and the specific content of current inspection.

For the periodic inspection, the focus is on escalation of existing and uncovered flaws. The system collects the related data, such as the component to which each defect belongs, the type of defect, the location of defect, the severity of defect, the description of defect, etc., and stores the images of the defect scenarios. In addition, each defect can be linked to the defect discovered during prior inspection in order to track the structural health condition. In addition to normal inspection, the system provides inspection records and viewing functions that apply to underwater

a)监测管理小程序 b)车辆称重小程序

Fig. 2.9 Monitoring related WeChat applets

sonar inspection, non-destructive testing of steel pipe arch welds and non-destructive testing of reinforcing steel corrosion.

Traditional inspection of bridge construction needs the inspectors to make paper records on-site, perform internal work after inspection, enter the records manually into the information system and generate inspection reports manually. Professionalism of the inspector largely determines the quality of the data acquired by this traditional operating procedure, and subsequent processing of information is of low efficiency and poor data quality. In order to reduce the inspectors' workload and improve the quality of inspection data, the system developed a tablet computer application for inspection of construction work (see Fig. 2.10), using which the inspectors can take pictures of the defect on-site and select the appropriate defect level from the defect database. In developing the inspection app, the authors' team gathered and adopted the opinions of front-line inspectors, and developed the functions of report description template entry, automatic replication of historical distresses, and automatic generation of inspection reports, which can improve the efficiency of on-site inspections as well as quality of the data.

Fig. 2.10 App for inspecting construction works

Management Subsystem

The management subsystem is a management support system for the owner's middle and senior management employees to operate and maintain the workflow of bridges. The entire management subsystem consists of three modules: work platform module, information statistics module and data analysis module. It also provides a link to other subsystems, such as the safety alert subsystem and management screen subsystem.

The "work platform" module of the management subsystem displays the current works in the Kanban board (see Fig. 2.11), including the daily inspection plan progress, project lag, key information, annual inspection works and maintenance works progress. "Details" for each section of the board can provide the work content and the current process of certain work processes, allowing the managers to control time and work nodes, and increase the efficiency of bridge operation and maintenance management in a targeted manner.

The "Information Statistics" module of the management sub-system gives statistics of the total information of the bridge group. The bridge group can be divided by route, structure system, distress type, maintenance state, bridge label, etc. The system will analyse selected bridges by linkage, and automatically classify them by bridge size, construction shape and technical condition. Subsequently, the management subsystem can designate the bridges as key items based on their tags and favorite marks, and prioritize their presentation in the bridge list for its flexibility and convenience for use.

The management subsystem can be accessed to inquire the fundamental information, technical condition rating, major flaws, project time and monitoring and inspection data for a single bridge. Moreover, utilizing Web BIM technology, the

Fig. 2.11 "Workbench" Kanban board

management subsystem can load the lightweight BIM model in the browser to rightly input the inspection, monitoring, and maintenance related information into the model, in accordance with the location of the components and sensors deployed in the BIM model, so as to realize a centralized and interactive view of all types of bridge information in the 3D model (see Fig. 2.12).

The "Data Analysis" module of the management subsystem provides statistical data comparison, and analysis for bridge operation and maintenance management. Comparison analysis function of the traffic flow and the technical condition is

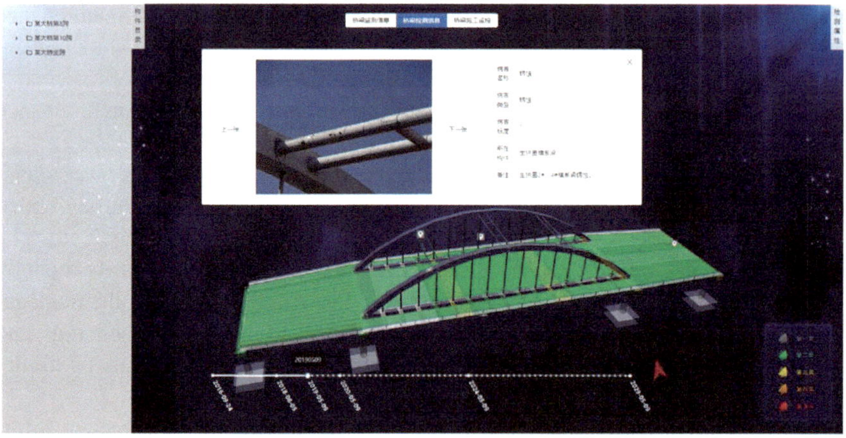

Fig. 2.12 Integrated inspection and evaluation information in the 3D model

intended to compare the traffic flow information and the technical condition rating of a bridge group at a specific time in the same geographical location. Comparison analysis function of the bridge group technical condition is intended to compare the technical condition of the individual bridges in a bridge group at different time points in the same geographical location. Load segmental comparison analysis function is intended to compare the load distribution and the technical condition assessment information of bridges in the same geographical location during a specific time period. Single bridge technical condition comparison analysis function is intended to compare the technical condition assessment level, the key defects and the component information of a specific bridge at different time points. Figure 2.13 depicts the effect of the comparative analysis function for individual bridges at various times.

In order to assist the managers in efficient bridge operation and maintenance management, the platform also developed the WeChat applet, which implements the management of monitoring, maintenance, daily inspection project progress and key information, etc., and offers the interaction function, same as the web version. Remote management is possible for the WeChat applet on a variety of mobile terminals.

Warning Subsystem

During the operation and maintenance of bridge group, the safety warning subsystem monitors the structural health and safety, possible ship and vehicle collision, and other occurrences, and displays the warning information on the large screen.

Safety warning and treatment recommendations (see Fig. 2.14 for display of the large screen) focus primarily on structural safety issues in the process of bridge operation and maintenance, and display the detection, monitoring and warning information in the form of a large visualization screen. The first screen displays the distribution map of the single bridges in the bridge group and uses long link technology to obtain the background warning information in real time. When the warning information is

Fig. 2.13 Comparative analysis of monolithic bridges at different time points

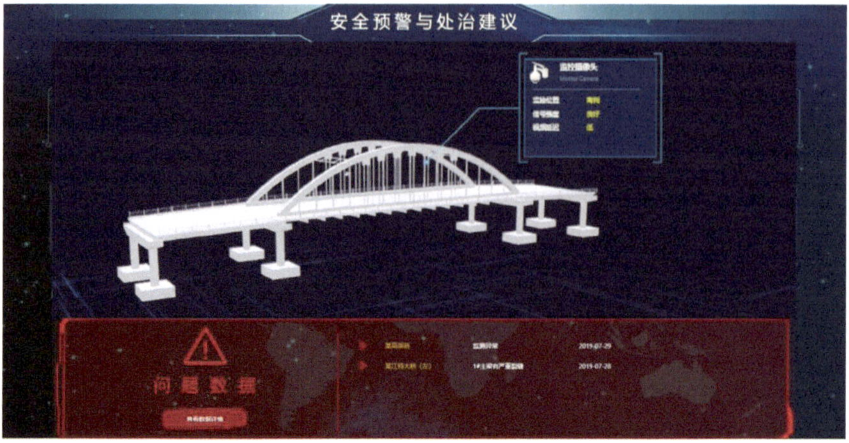

Fig. 2.14 Security warning and treatment advice large screen

sent out, it is displayed in an eye-catching manner to remind the user of the message, making it possible for the command center to monitor and execute other applications.

Medium and Long-term Maintenance Planning Subsystem

To meet the needs of medium- and long-term maintenance of bridge groups, the medium- and long-term maintenance planning subsystem uses the Markov model to predict the bridge condition degradation, establishes a dynamic maintenance measure library, and develops a maintenance decision optimization model using a reinforcement learning algorithm. The defect library and the maintenance measure library can be configured by the user to reflect the economic impact of each maintenance measure. The system may provide natural deterioration forecasts for individual bridges, maintenance suggestions for individual bridges, and maintenance advice for bridge groups.

For single bridges, the system computes the progression of natural deterioration based on the present technical state and current condition of the bridge, provides the performance development curve and generates a performance evaluation report. On the basis of natural deterioration, the system can combine the defect database and the maintenance measures database to generate a single bridge maintenance decision and maintenance measures for the bridge in each coming year. In addition, the system can compare the effect of maintenance according to the recommendations with the effect of natural deterioration (see Fig. 2.15) for the management staff's decision-making reference.

For a bridge group, the system first computes and analyzes the individual bridges in the bridge group, then ranks the maintenance priority of each bridge based on economic and other constraints, provides maintenance plan for the bridge group, and computes the maintenance budget for each bridge in the bridge group. Additionally, the user can read the detailed maintenance suggestions for a particular bridge.

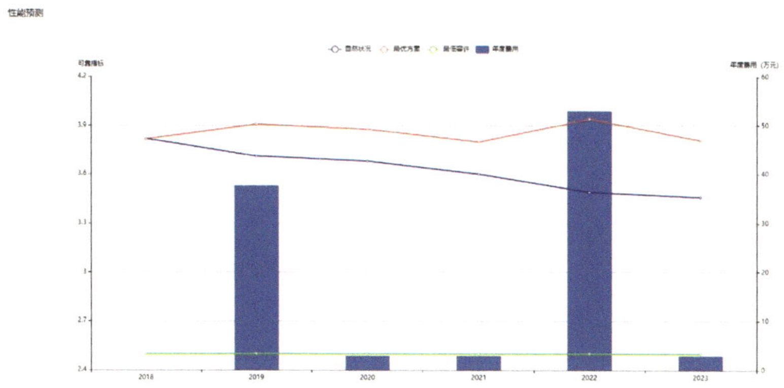

Fig. 2.15 Comparison of natural deterioration and adoption of program maintenance

2.4.3 IntelliBridge Capabilities

In addition to the development of various subsystems for data management, specialized analysis, management and decision making, the IntelliBridge system combines many advanced technologies, such as computer 3D image processing and artificial intelligence, with the specific technical requirements of bridge operation and maintenance, together with devices such as underwater sonar, inspection robots, drones and vehicle weighing scales, in order to develop specialized functions for bridge inspection. The specialized functions for each phase of intelligent operation and maintenance of a bridge are described below.

Access to and processing of operational and maintenance historical data.

Prior to the establishment of the digital system for bridge operation and maintenance, a substantial amount of bridge operation and maintenance data, such as the bridge basic information cards, bridge inspection reports, etc., was stored on paper or in electronic files. Combining Optical Character Recognition (OCR), Natural Language Processing (NLP), etc., the system is designed to provide a comprehensive solution for the entry of data in various formats, such as paper data, Word documents, PDF documents, etc., in order to permit the historical data to more effectively fulfill their role in bridge operation and maintenance. Figure 2.16 depicts the data entry method for historical records.

(1) Integrate Parametric Modeling Device

The majority of bridges on most traffic routes in the real projects are simple girder bridges with only slight varied span lengths; hence, the 3D depiction for bridge O&M does not require an unduly comprehensive 3D model. The present tool firstly automatically creates each span of the bridge and then merges them into a complete bridge model, and correlates the information of each span and each member in the

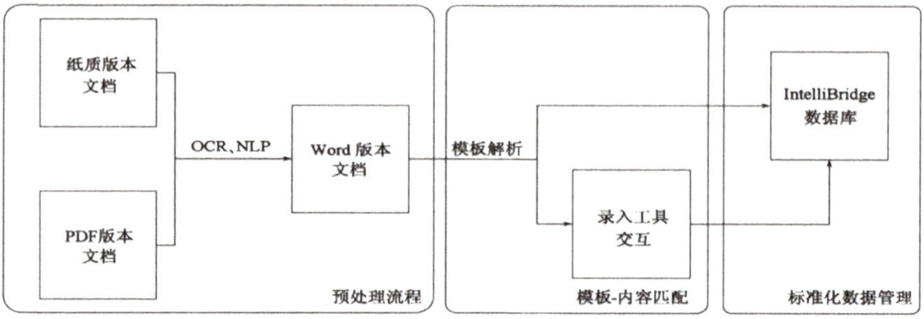

Fig. 2.16 Various versions of historical data entry process

database. Finally, it allows the user to produce a useable 3D model through fine tuning. Figure 2.17 illustrates the result of the modeling tool.

(2) Large Display for Construction Management System Monitoring

During repair and construction, many sensors are generally installed on the bridge to closely monitor the operation, particularly during maintenance and construction without interruption of traffic. Due to the enormous number of monitoring indicators and monitoring data generated during bridge maintenance and construction, the authors' team designed a construction monitoring management system large screen as a part of the IntelliBridge system (see Fig. 2.18). The huge screen employs charts and other forms to display the monitoring indicators and statistical information collected during the construction process, as well as to remind the construction command of critical warning information in real time.

Fig. 2.17 Modeling tool

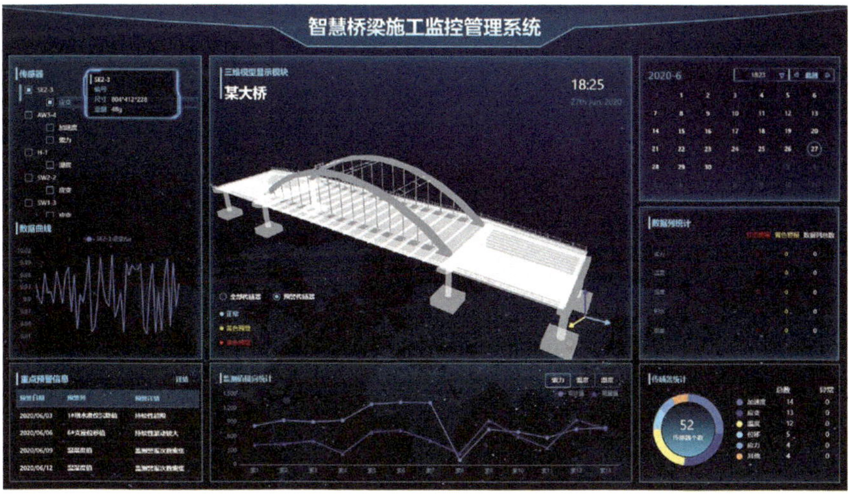

Fig. 2.18 IntelliBridge construction monitoring management system large screen

(3) Single bridge operating screen monitoring

In the operation and maintenance of important bridges, health monitoring systems are typically employed to monitor the environmental conditions, stresses and deformations of the bridges, as well as, when necessary, to acquire real-time images using cameras. To meet the demand of an integrated display of important bridge health monitoring, the authors' team developed a single bridge operation monitoring screen as a component of the IntelliBridge system (see Fig. 2.19). The large screen is designed with visualization ability to display various monitoring indicators and is scientifically combined with a 3D model as its core, which can display environmental monitoring indicators, mechanical and deformation monitoring indicators at key locations, key events of bridge operation, and real-time camera image data.

(4) Video Image Acquisition and Processing

The IntelliBridge system employs live streaming technology and adopts "backward penetration" network architecture, which requires less network bandwidth to support multiple live video data and reduces the live streaming delay effectively. At the same time, a deep learning algorithm related to image processing is introduced to analyze the live video stream in real time. This algorithm can classify the vehicles passing the bridge (cars, buses, trucks, etc.), as well as analyze the speed of each vehicle and the distance between vehicles. The obtained data is sent by the message queue technology, which can be utilized by other system modules such as traffic flow load calculation module and heavy vehicle identification module. The analysis result video can also be sent to the monitoring screen again by the streaming technology for real-time display, which can aid the monitoring management personnel in performing relevant management tasks. Figure 2.20 depicts the outcome of the recognition process.

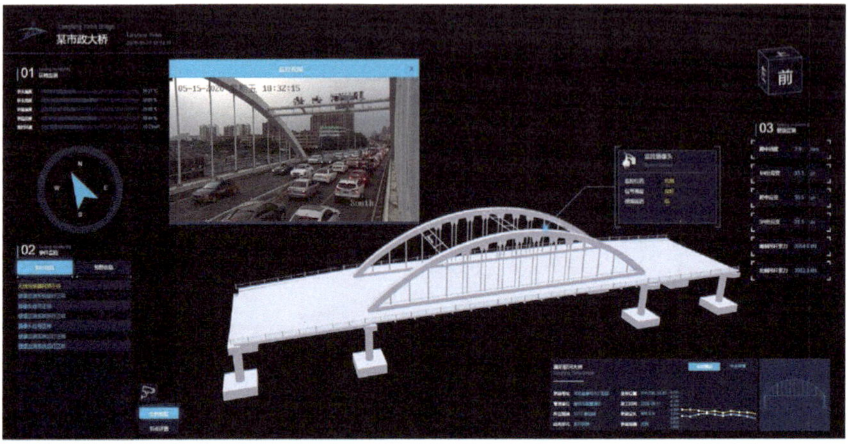

Fig. 2.19 IntelliBridge single bridge operation monitoring large screen

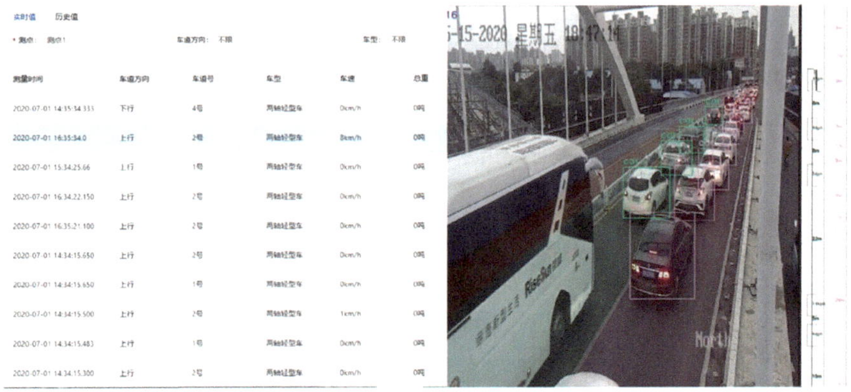

Fig. 2.20 Real-time recognition of vehicle type and traffic flow in a bridge operation

Visualization of Bridge Substructure and Underwater Inspection

In recent years, sonar technology has been increasingly used in the underwater construction of bridges as the technology makes it simple to spot faults and scouring in underwater bridge components. In a 3D model, the technology may concurrently depict the above- and the below-water damages to the foundation of the bridge. To meet the actual needs of sonar scanning operation, an automatic picture recognition function is built for the IntelliBridge system to assist the constructors in identifying the underwater damage. Figure 2.21 depicts the visualization of the substructure and the underwater inspection.

Identification of cable force, detection of apparent damage and visualization of bridge cables.

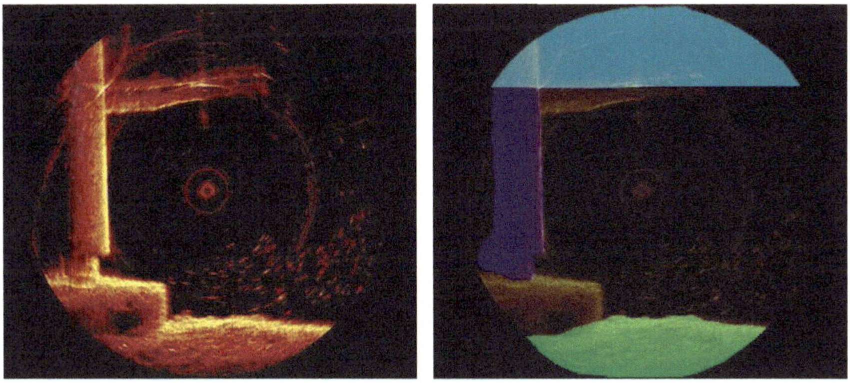

Fig. 2.21 Sonar scan image and deep learning recognition

The IntelliBridge system can use data from a robot capable of climbing cable. During the operation of the robot, acceleration and picture can be captured, and the 3D coordinates of each acquisition point are simultaneously recorded. After the acceleration acquisition, the cable force recognition algorithm can be employed in real time to compute the time-varying cable force. After the image acquisition, the image recognition algorithm can also be utilized in real time to identify the apparent damage on the cable, and the relative position of the damage in the bridge can be determined by using 3D coordinates. During the cable inspection process, the time-varying position of the robot, the acceleration and force detected at each place, and the location of the damage can be vividly and intuitively presented in 3D on the platform.

In addition, the authors' team utilizes the smart lasso product developed by Liuzhou OVM Machinery Co., Ltd. to import the smart lasso's sensing data. The 3D visualization module displays the real-time cable force and anchor head beeps with the help of various cable force identification algorithms. Currently, monitoring of the main arch of the "world's first arch" Guangxi Pingnan Three Bridges utilizes data processing capabilities of the smart lasso connected to the platform. Figure 2.22 depicts the smart lasso.

UAV-Based Bridge Inspection and Visualization

In recent years, the performance of small unmanned aerial vehicles (UAVs) has vastly improved in terms of autonomous hovering and smooth control, and their price has decreased as a result of technological progress. These small UAVs may be controlled by controllers or even cellphones, and when used for bridge inspection, the controllers can operate them from ground stations without having to physically arrive at the inspection site, which help to enhance safety and reduce labor costs. However, there are still issues that must be resolved during bridge inspections. For instance, it is difficult to dynamically arrange the inspection path based on problems

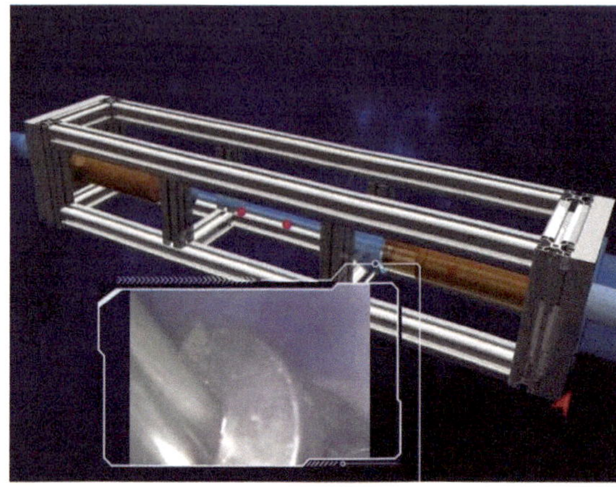

Fig. 2.22 Application of IntelliBridge system with Ovim Smart Lasso

discovered for the UAVs, and it is difficult to identify damage when the UAV is flying under a bridge or in areas with weak GPS signals.

The authors' team utilized the UAV SDK (Software Development Kit) for customized development in conjunction with artificial intelligence technologies to identify the photos acquired during flight in real-time, and direct the design of the next inspection path. Additionally, the use of self-assembled wireless sensors for UAV-assisted positioning can, to a certain extent, overcome the problem of UAV positioning in areas with poor GPS signals. During the inspection process, the flight path of the UAV and the location of the discovered bridge damages can be included in the IntelliBridge system for unified data management and 3D display. Figure 2.23 depicts the results of a damaged bridge inspection and artificial intelligence automatic identification using a UAV.

Weighing system and video monitoring combined for heavy vehicle identification and visualization.

Vehicle load is one of the primary loads of the bridge, and vehicles, particularly huge trucks, have a significant impact on the bridges, causing damages to the road

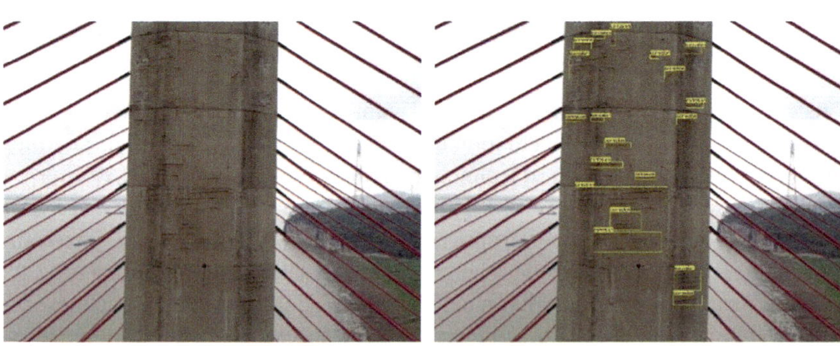

Fig. 2.23 UAV bridge inspection and artificial intelligence automatic identification results

surface and even the structure. Currently, certain bridges are equipped with dynamic weighing devices that are buried beneath the roadway to assess the weight of passing vehicles in real time. However, when the traffic flow is complicated, the use of weighing devices alone can lead to inaccurate estimates, and subterranean scales are susceptible to damage during pavement maintenance, resulting in high overall expenses.

The authors' team improved an existing dynamic weighing system that sends its data by using message queue technology for real-time display and historical query on the IntelliBridge system. In addition, using streaming media technology and deep learning image recognition technology, the bridge monitoring video may be processed frame by frame in order to analyze the bridge's traffic flow data. Ultimately, the IntelliBridge system fuses weighing data and image analysis data in a multi-source heterogeneous data fusion way, enabling more accurate identification and central visualization of heavy vehicles on a 3D platform.

2.4.4 Unique System Features

In order to realize the concept of intelligent bridge operation and maintenance, the authors' team developed an intelligent system platform, IntelliBridge, that realized the basic functions of bridge operation and maintenance management, and empowered the platform to improve its expansivity, connectivity and cooperative computing capabilities, so that IntelliBridge can be expanded in function and performance, and can exchange data with other systems. IntelliBridge can employ interchangeable and pluggable professional algorithms to promote the development of relevant theories and continuously enhance the level of bridge operation and maintenance management.

Expansivity

The platform offers both functional expansivity and performance expansivity by its layered design and microservice cluster architecture, respectively.

To achieve functional expansivity, the architecture design of the platform follows a layered design paradigm, separating the application service support layer from the application layer, as shown in Fig. 2.24. The application service support layer connects to various databases downward and provides fine-grained services to the upper layer applications via open data interface APIs. The application layer combines with the bridge operation and maintenance business logic to freely combine the use of various data interfaces in order to realize a variety of professional applications. Based on this strategy, when there is a new demand of function, it only needs to expand the API to form new apps for the platform to be able to expand its function capabilities.

To achieve performance expansivity, the application service support layer of the platform employs a microservice cluster architecture. Microservice refers to the division of an application into numerous essential functions, each of which is referred

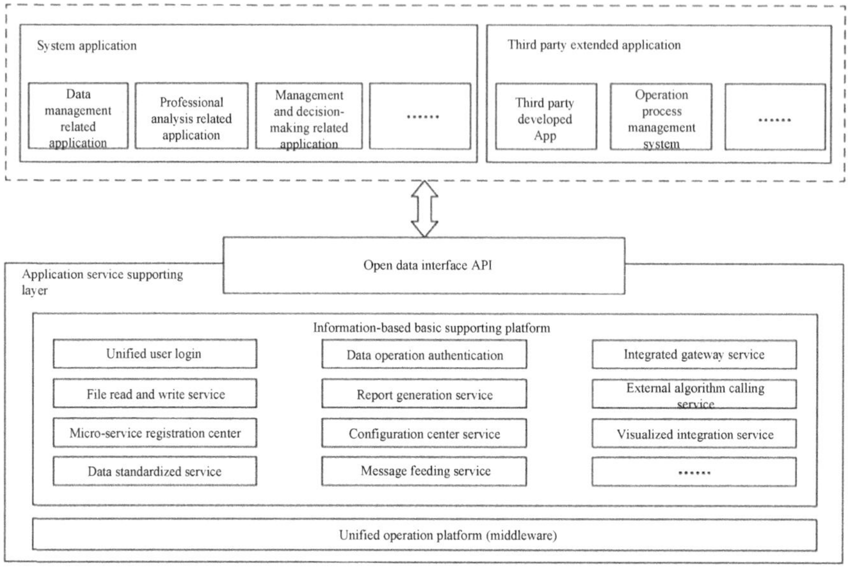

Fig. 2.24 Separation of application service support layer from application layer

to as a service. Microservices can not only be independently developed and deployed, but can also be invoked via an application programming interface (see Fig. 2.25). Each service can run several nodes to provide services to the public in a clustered fashion, and each service node will register with the registry when it is started and be controlled and scheduled by the registry in a uniform manner. The application of service support layer composes of numerous microservices, including authentication service, bridge information service, detection service, monitoring service and file service, among others. When the performance pressure of a service is heavy, it is sufficient to raise the number of its copies, and the registry will automatically balance the nodes of the unified service, enabling the platform to scale its performance.

Connectivity

Connectivity stands for simple and secure communication with external data, and the IntelliBridge system platform ensures both a clear and concise REST-style interface and a dual authentication approach of operation and resource verification. The IntelliBridge intelligent bridge operation and maintenance management system is designed to support connectivity, and the application service support layer exposes various clear and concise REST-style interfaces that can be used for both data support at the application layer and for external data exchanges.

As shown in Fig. 2.26, the intelligent bridge operation and maintenance system employs a dual verification and authentication strategy that combines operation

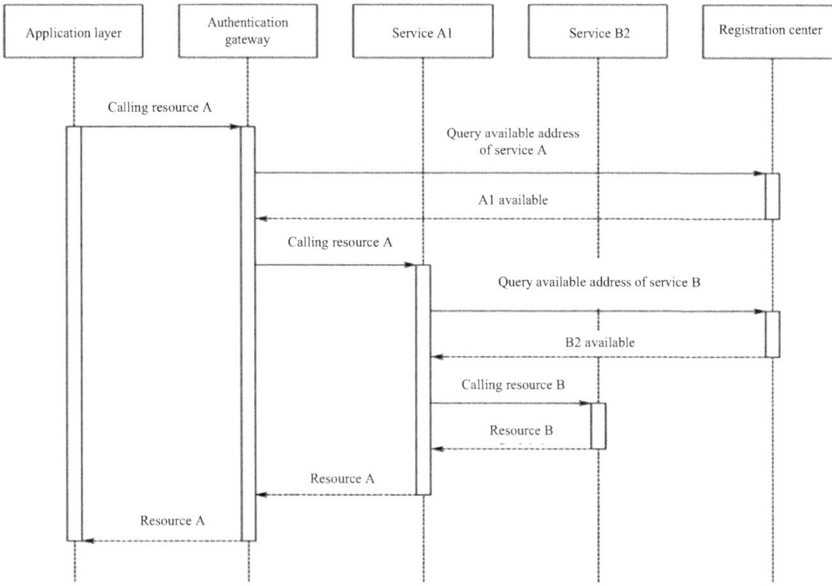

Fig. 2.25 Possible sequencing of a request in a microservice cluster

rights and resource rights to safeguard connectivity. Each user has a certain function and belongs to a unit. The system contains distinct operation authority definitions for different roles, and the authentication gateway can decide if a request is a valid operation based on the URI and access methods. The resources of this system belong to a particular department, which might be subordinated to its parent department. The hierarchical relationship of resource attribution can be formed by this technique, and specific operations may be compared between the user department and the resource department; however, only authorized resources can be operated. Combining operation authority and resource authority, the dual verification and authentication technique provides sufficient security for connectivity.

(3) Capability for distributed computing

By separating interfaces and implementations, and partitioning the professional modules, the IntelliBridge system platform ensures pluggability and replaceability of cooperative computing capacity.

Pluggability of the platform is achieved by encapsulating each professional function in a separate application layer module that is independent of other businesses and has no effect on them. In addition, the functions and algorithms are separated, and the functions are solely specified via language-independent APIs, while the platform incorporates the algorithms via the underlying runtime environment to ensure replaceability.

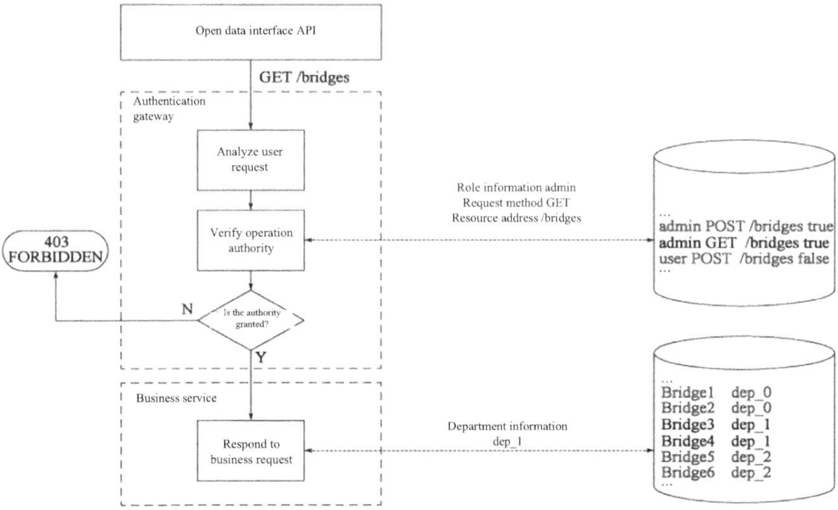

Fig. 2.26 Dual verification authentication combining operation rights and resources

For instance, in response to the inconvenience that bridge drones cannot identify in real time and must upload manually afterwards, the platform developed a drone image recognition App (see Fig. 2.27) on the DJI flight control platform that can take photos and report image data during flight and automatically call the distress identification algorithm. The application is not dependent on other applications and can run independently, and the image recognition interface is simple, requiring only input and output file paths. The authors' team has changed the algorithm and compared it with other algorithms multiple times. Replacement of the algorithm is transparent to the display and background calls, reflecting the algorithm's pluggability and replaceability in the cooperative computing capability of the platform.

2.5 Summary

This chapter focuses on intelligent bridge operation and maintenance system platform. First, the requirements and practicability of constructing a platform for intelligent operation and maintenance of bridges are assessed. The bridge operation and maintenance stage will generate a large amount of data that is difficult to use effectively with the conventional non-digital methods. Besides, the existing management system lacks expandability and upgrading is difficult, plus poor data circulation makes it easy to generate data silos. Consequently, it is important to establish a system platform. With the advancement of computer science, communication technology, artificial intelligence, big data and other technologies, the conditions for

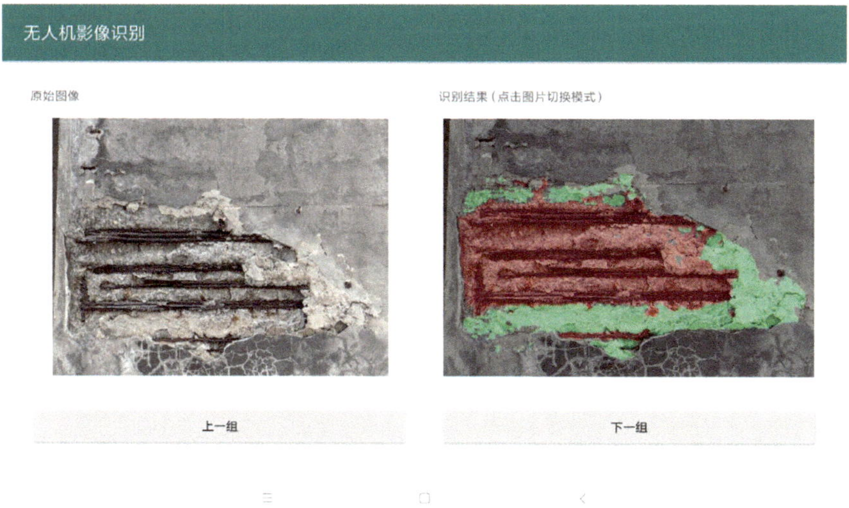

Fig. 2.27 UAV Image Recognition App

establishment of an intelligent operation and maintenance platform for bridges are mature.

The present Chinese and international systems are then introduced and assessed. There are currently a large number of large bridge management-related systems in the United States and other countries, such as PONTIS and CBMS. Existing systems, however, are either not standardized for specific technical points, such as inspection and monitoring, nor are customized for a particular region. In addition, the existing systems lack expandability and are of poor data exchange. They cannot be effectively connected with the workflow, making them cumbersome in use. In the design and development of intelligent operation and maintenance computing platform for bridges, it is vital to learn from the advantages of various existing systems and to avoid the numerous problems that have been encountered thus far.

General requirements for the design of intelligent bridge operation and maintenance system platform are recapitulated. In terms of platform functionality, it should unify the relevant types of codes, and provide basic information management function, monitoring function, detection function, maintenance and other operation and maintenance process management functions, and data processing related function. These functions cover the functions of existing systems, and the realization of these functions can complete the platform. In addition, the platform should have the capacities of expansivity, connectivity and cooperative computing. Through these capabilities and functions, the platform integrated into the system can eventually interface with other systems and platforms to facilitate the development of smart cities.

The final section introduces the IntelliBridge system platform, which is an intelligent bridge operation and maintenance system platform developed by the authors' team. IntelliBridge also possesses the capabilities of expansivity, connectivity and cooperative computing, able to actualize pluggability of modules and algorithms, and expose the external data interface. It has been put into use in numerous Chinese provinces.

Reference

1. FHWA. National Bridge Inventory (NBI)[EB/OL]. https://www.fhwa.dot.gov/bridge/nbi.cfm.

Chapter 3
IoT Sensing Technology

Abstract Using sensing technology to obtain physical and environmental data is the basic function of the bridge management system. Therefore, in the process of bridge operation and maintenance, data acquisition is particularly important. However, the conventional sensing technologies are independent of each other, and cannot work together to obtain the health status of the bridge groups from a comprehensive time and space perspective. In addition, bridge maintenance and management using the conventional technologies is expensive. Aiming for intelligent civil engineering upgrade, Internet of Things (IoT) sensing has become a research focus in the academic circle in recent years as it can surpass the limits of the conventional sensing technology. This chapter introduces the development and application of IoT sensing technology in bridge maintenance and management from four aspects: the conventional structural sensing technology, the intelligent IoT technology, the IoT cloud management platform and the application of IoT technology in bridge maintenance and management.

3.1 Conventional Structural Sensing Technology

Structural sensing refers to information acquisition that perceives the state and features of the structure utilizing physical, chemical and biological effects, and transforms them into usable representative signals according to certain rules. Civil engineering structures feature large scale structural complexity, harsh service environment, and strong randomness of load actions. Whether a set of highly feasible sensing technology can be implemented on site determines the success rate of the structural health diagnosis and the management, maintenance and operation undertaken on the actual bridge. This section classifies the conventional structure sensing technologies into two categories (contact and non-contact), and expounds them separately.

G. Wu et al., *Intelligent Bridge Maintenance and Management*, Springer Tracts in Civil Engineering, https://doi.org/10.1007/978-981-97-3827-4_3

3.1.1 Contact Sensing Technology

Contact sensing refers to the method that obtains information by installing various sensors on different parts of the measured structure. Contact sensing is currently the most widely used and mature sensing method in the field of bridge health monitoring, and it is the leading approach to realize the long-term monitoring of bridges. This kind of sensing technology is mature, but there are disadvantages such as point-to-point signal transmission, wired transmission with electric source, large volume, large mass and so on.

In the bridge health monitoring system, contact sensors can be roughly divided into contact sensors for environmental incentive parameters and contact sensors for structural response parameters, as shown in Table 3.1. By recording the time-space evolutionary histories of all parameters, the sensors can provide information for evaluation of the bridge reliability and durability, and the basis for evaluation of the bridge safety and applicability.

3.1.2 Non-contact Sensing Technology

Non-contact sensing refers to the measurements that obtain information of the target object without directly contacting the object. Having the advantages of being independent of the measuring position and without adding loads to the structure, non-contact sensing technology has been widely used in bridge monitoring. Non-contact sensing can be classified into passive sensing and active sensing. Passive sensing can only collect signals converted from the sunlight reflected by or the self-radiation emitted from the measured object, such as the visual measurement technology, the conventional geodetic technology, etc. Active sensing is a sensing technology that emits a signal to the target object and then collects the signal fed back by the object, such as the synthetic aperture radar interferometry.

(1) Visual Measurement Technology

Visual measurement technology is mainly used for measuring the geometric shape, spatial position and attitude of the target object. Usually, after the characteristic image of the object is obtained, the image can be analyzed according to the prior knowledge in accordance with the technology [10, 11]. Visual measurement technology has outstanding features such as high precision, high degree of automation, low cost, full field measurement, small size, and low power consumption. According to the number of cameras used in the measurement, the visual measurement systems are usually classified into monocular visual measurement and binocular (multi-eye) visual measurement [12, 13] (Fig. 3.1).

As a branch of visual measurement technology, Digital Image Correlation (DIC) can directly process digital images with certain grayscale distributions. The pixels in the image can record the grayscale information. By comparing the grayscale images

Table 3.1 Classification of contact sensors

	Environmental incentive parameters/structural response parameters	Contact sensors	Classification of sensors
Contact Sensors Related to Monitoring of Environmental Incentive Parameters	Wind speed and direction	Anemometer	Ultrasonic and mechanical anemometer, according to the working principle
	Temperature	Temperature Sensor	Fiber grating, digital, thermocouple, thermistor type, etc. according to the working principle
	Vehicle load	Weigh-in-motion system	Bridge Weigh in Motion and Pavement Weigh in Motion [1, 2], according to the measurement principles
	Others (humidity, air pressure, rainfall, sunshine, water level, corrosion, etc.)	Temperature and humidity meter, barometer, rain gauge, sunshine intensity meter, water level meter, corrosion meter, etc.	
Contact Sensors Related to the Monitoring of Structural Response Parameters	Displacement [3–8] (translational displacement, angular displacement, crack)	Displacement sensors, inclinometers, extensometers, etc.	Potentiometer displacement sensor, magnetostrictive displacement sensor, grating displacement sensor, LVDT displacement sensor, laser displacement sensor, eddy current displacement sensor, capacitive displacement sensor, etc. according to the working principle

(continued)

Table 3.1 (continued)

	Environmental incentive parameters/ structural response parameters	Contact sensors	Classification of sensors
	Strain and stress	Strain gauges/strain sensors	Conventional point sensors such as resistance strain gauges and vibrating wire strain gauges, as well as fiber optic sensors such as fiber Bragg grating strain sensors and distributed optical fiber strain sensors [8, 9]
	Acceleration	Accelerometer	Piezoelectric, piezoresistive, capacitive, servo, fiber grating type, etc. according to the principle of different sensing elements

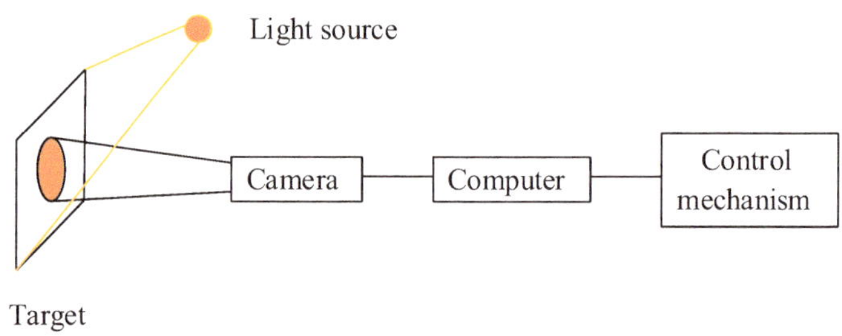

Fig. 3.1 Working principle of visual measurement technology (modified based on [11])

of the structure before and after deformation, and using related algorithms to obtain the full displacement and strain fields, the accuracy of displacement measurement can reach up to 0.01 pixel for DIC [14, 15].

Visual measurement technology has been widely used in the field of civil structural monitoring, mainly in the measurements of structural displacement and strain [16–20], and also dynamic response of the modal parameters [21, 22]. However, vision measurement technology has the disadvantages of high environmental requirements, difficult accuracy assurance, complex image processing, time-consuming calculation, and inability to display in real time.

(2) Geodetic Technology

Conventional geodetic technology mainly uses the levels, theodolites, and total stations to detect the angle and the distance of structural objects to generate structural three-dimensional deformation. Conventional geodetic technology is relatively mature, with measurement accuracy reaching sub-millimeter level. Reliable observation data can thereby be obtained using this technology at low costs. Conventional geodetic technology is mainly used for the deployment of deformation monitoring networks and periodic observations.

However, geodetic technology has high flexibility but low detection efficiency, and it is easily affected by the on-site terrain condition, weather, visibility and other conditions. Besides, it is difficult to realize synchronous monitoring of multiple measuring points for the overall deformation of large structures. In addition, geodetic technology is mainly intended for the measurement of overall structural deformation, and cannot be used for measuring local deformations. Generally speaking, the conventional geodetic technology is easily affected by the environment and climate. With the development of advanced modern surveying and mapping instruments, such as the photogrammetry technology, laser scanners and measuring robots, the new geodetic technology is more widely used in the field of deformation monitoring, which reduces the field workload and also can be automated [23].

(3) Synthetic Aperture Radar Interferometry

Synthetic aperture radar interferometry technology refers to the acquisition of complex image pairs of target objects under the condition of knowing the radar sensor system parameters, attitude parameters and geometric relationship between orbits, and the subsequent accurate measurements of the three-dimensional position of a certain point on the earth's surface and its small changes [23, 24].

Radar interferometry technology can be used for micro-deformation monitoring, employing the radar in the air (airborne radar or spaceborne radar) and the radar on the ground, as shown in Fig. 3.2. Among them, aerial radar is mainly used for monitoring static events, such as surface deformation, earthquake deformation and landslide monitoring [25]; while ground radar can be used for both static monitoring and dynamic monitoring of slopes and bridges, with accuracy of sub-millimeter.

In the field of bridge health monitoring, ground radar is currently used in the dynamic and micro-deformation monitoring of large structures, while air radar is the most effective way to obtain data in areas being cloudy and rainy all year round [26], which is also mainly used in dynamic deformation monitoring and micro-deformation measurement of bridges [24, 27].

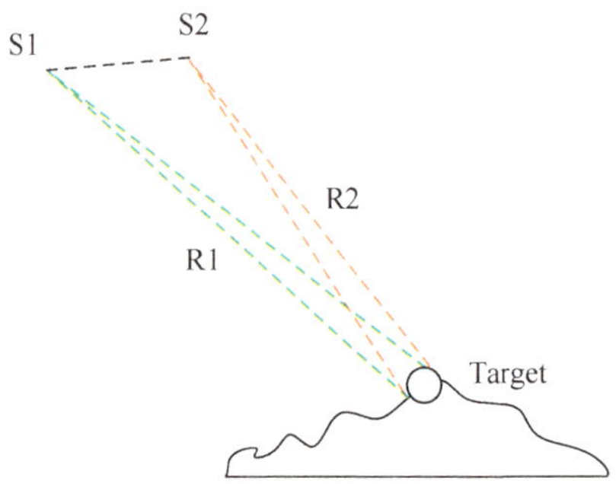

(a) Radar in the air

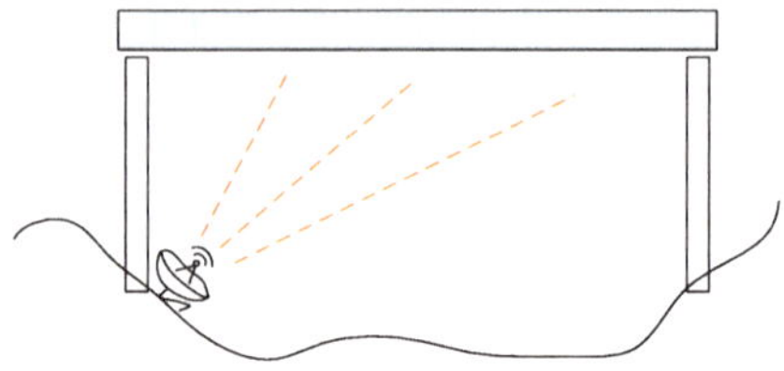

(b) Radar on the ground

Fig. 3.2 Synthetic aperture radar interferometry (modified based on [24, 25])

3.1.3 Limitations of the Conventional Structural Sensing Technologies

The conventional structural sensing technologies have developed vigorously and have been widely used in the health monitoring of bridge structures, but they still have some limitations. From the perspective of installation and operation, bridge maintenance and management systems based on the conventional sensing technologies are mostly arranged on large-scale bridges or bridges on important traffic arteries. Their data acquisition mode is mostly contact sensing, their data transmission mode is wired or wireless, and their data storage and calculation mode is local or centralized. For the large number of small and medium-sized bridges whose health conditions should also be guaranteed, high installation and operation costs will inevitably be incurred by using the conventional structural sensing technologies. In addition, the bridge maintenance and management systems based on the conventional sensing technologies may also include the following technical disadvantages:

(1) The conventional SHM system usually takes the structural vibration data in response to the environmental actions, the loadings or the disasters as the main analytical object, supplemented by the environmental monitoring. The main features of this monitoring mode are the low density of sensing locations from the spatial view, and the low event coverage rate from the time view (for example, monitoring the traffic conditions, disasters, emergencies, and structural/environmental responses at the same time). Using a specific algorithm to analyze this kind of small-scale spatio-temporal coverage data may lead to false positive or false negative identifications.
(2) The forms of data collected by various sensing technologies are not uniform, and the data are therefore not interoperable. Most of the existing algorithms employed have proprietary attributes, that is, they are exclusively used for processing specific data and transforming them into useful information and

empirical knowledge. It is difficult to utilize this conventional data management method to effectively manage the monitoring and detection data, and also difficult to query and analyze the data.

(3) The sensing technologies and the equipment are independent of each other. After all kinds of data are transmitted from the sensing layer to the computing layer and the decision-making platform, there is a lack of effective decision-making assistance technology. At the same time, there is a lack of feedback mechanism, as it is difficult to reflect the adjustment of the sensing scheme at the decision-making operation layer to the sensing layer. Therefore, it is necessary to apply the sensing technology and the sensing management technology that can solve the above issues to the long-term intelligent operation and maintenance of bridge groups.

3.2 Intelligent IoT Technology

At present, as the size of the single bridge and the number of bridges in the bridge group increased, parameters that need to be measured on bridges also increased and higher requirements are posed on the sensing technology in bridge maintenance and management. The conventional bridge maintenance and management systems have the disadvantages of low efficiency, low timeliness, high maintenance cost, strong selectivity, large amount of data produced during the maintenance process, high variety and incompatibility of devices, difficulty in performing efficient analyses, incompatibility with the health monitoring of bridge groups, etc. Hence, IoT technology is introduced into bridge maintenance and management, starting from the technical level of the sensing layer to the network layer of IoT system architecture. This section introduces the modern IoT sensing, IoT communication, and sensor and network integration in turn, which resolved the disadvantages of the conventional sensing technology in multi-parameter sensing and data transmission. It also provides reliable technical supports for the big data management and analysis, and the fusion of multi-source heterogeneous data in the following chapters.

3.2.1 Overview of IoT

IoT is widely acknowledged as another generation of emerging digital technology after the Internet, and it is also regarded as the third wave of the world's information industry after the computers and the Internet and mobile communication networks [28]. Its main objective is to establish a real-time sensing system that connects things, manages things, and allocates resources with the overall planning. IoT has become one of the key development technologies these days.

In 1999, the Auto-ID laboratory of the Massachusetts Institute of Technology proposed that by installing tiny identification devices on objects such as books, shoes, and auto parts, you can know the location, status and other information of the objects at any time, to realize intelligent management. As the prototype of IoT, Auto-ID was conceptually based on the wireless sensor network and the radio frequency identification technology [29]. On November 17, 2005, at the World Summit of the Information Society (WSIS) held in Tunisia, the International Telecommunication Union released the "ITU Internet Report 2005: Internet of Things", formally proposing the concept of "Internet of Things". In March 2006, the European Union held a conference "From RFID to the IoT", in which the IoT was further described, and a roadmap was formulated for IoT research strategies in 2009 by the European Union. Since 2008, the Auto-ID laboratory, which had developed into a number of research institutions around the world, organized the "Internet of Things" international annual conference. In 2009, IBM CEO Samuel J. Palmisano proposed the concept of "Smart Planet", aiming at embedding sensors into power grids, railways, bridges, tunnels, highways, buildings, water supply systems, dams, oil and gas pipelines and other infrastructures, to achieve a smart state through intelligent processing. On June 16 of 2016, the 72nd meeting of "The 3rd Generation Partner Project (3GPP) Radio Access Network plenary session" was held in Busan, South Korea. At this conference, 3GPP announced the freezing of the mobile communication technology Narrowband Internet of Things (NB-IoT) standard for the IoT. This marked the exclusive official communication technology developed for the IoT. Since then, development of the IoT industry has entered an unprecedented prosperous stage.

IoT refers to the basic network connecting all items in the world via the Internet through information sensing equipment (such as radio frequency identification equipment, infrared sensors, global positioning systems, laser scanning equipment, sound and light, and gas sensors, etc.) in accordance with standard communication protocols. Information communication and data exchange are so carried out that intelligent identification, positioning, tracking, monitoring and management of items, and interconnected and interoperable communication between things and things, people and people, and people and things are thereby realized.

To meet the above requirements, IoT should possess the following key elements:

(1) Dynamic self-adaptation. IoT devices and systems should be able to dynamically adapt to the changing processes and take appropriate actions based on their operational needs, user needs or perceived environment.
(2) Interconnected and interoperable communication. IoT devices can support multiple interoperable communication protocols and can communicate with other devices and infrastructures.
(3) Unique identification. Every IoT device is uniquely identified with the unique identifier (such as an IP address or URI).
(4) Being integrated into the information network. IoT devices are usually integrated into the Internet to communicate and exchange data with other devices and systems.

(5) Edge computing capability. The nodes of IoT should have certain computing and decision-making capabilities, in particular the ability to perform preliminary processing and to even make decisions based on the perceived data.
(6) Intelligent processing ability. Using various computing technologies such as data fusion and processing, cloud computing, etc., IoT can analyze, integrate and process massive multi-source heterogeneous data, and provide supportive information to users afterwards.

At present, IoT does not have a unified architecture. Different researchers have proposed different architecture designs. As shown in Fig. 3.3, in most cases, IoT can be divided into three parts: the sensing layer, the network layer and the application layer [30, 31].

Sensing layer, the front end in the IoT architecture, is the eyes of the IoT. It is the channel for information exchange between the IoT and the external world. It is usually composed of different sensors. The sensing layer needs to integrate and transmit the acquired data, therefore, edge computing capability is also essential for the sensing layer.

Network layer, traditionally regarded as the nervous system of the IoT, is mainly responsible for information transmission, which includes the access network and the core network. The access of many devices in the IoT is ubiquitous and heterogeneous. There are various networks to access, including the mobile network, the

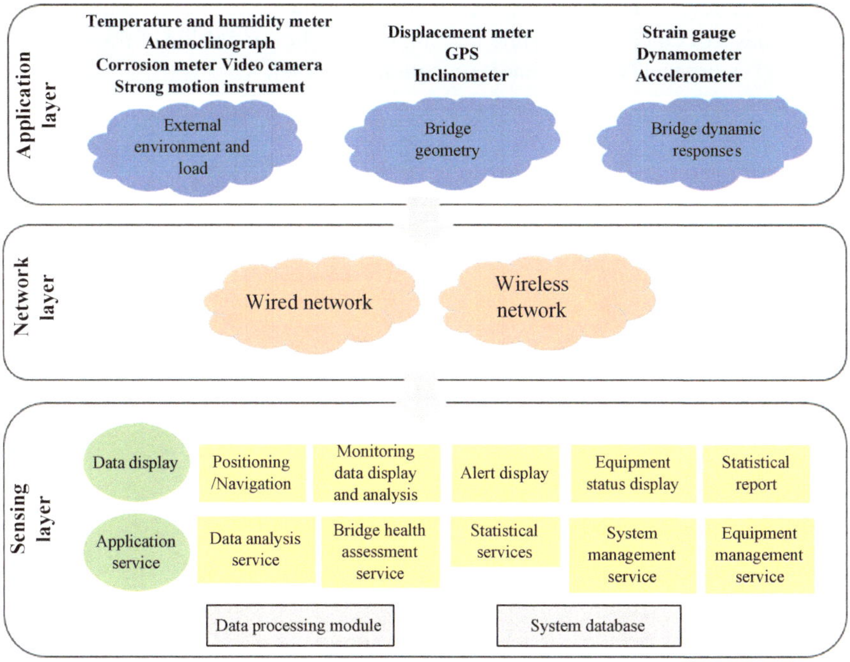

Fig. 3.3 Architecture of bridge health monitoring system based on IoT

wireless access network, the fixed network and the cable TV network. In addition, things are movable, and they should be "online" anytime and anywhere. Therefore, there should be a locally independent network, which connects to a large network, forming a hierarchical network structure. This requires the use of wired and wireless technologies to achieve seamless and transparent access.

Application layer, playing the role of the top layer in the IoT architecture, is responsible for realizing data exchange and processing between the data transmitted from the network layer and the various service platforms or application software. The information interaction between the IoT and humans is implemented by the application layer, so it is the part most closely connected to humans. It defines various application software in which IoT can be deployed, such as the smart home, the smart city, the smart healthcare, etc.

Figure 3.4 shows the difference between the typical structure of a wireless sensor network and that of IoT. A wireless sensor network is a multi-hop self-configured network system composed of a large number of micro-sensor nodes, with computing and communication capabilities. Its purpose is to cooperatively sense, collect, and process the information with the perceived objects in the network coverage and send it to the observer. Because the network structure does not include base stations or switches for information transfer, it is often called an Ad-Hoc network.

The wireless sensor network in the conventional sense is like the sensing layer of the IoT, which does not have the connection capability of T2T (Thing to Thing), and also does not have the ability to control the physical system. Wireless sensor networks merely collect and perceive signals, but do not pay close attention to the identification of objects [32]; while IoT emphasizes the connection of things (wired or wireless), especially interconnection with the Internet.

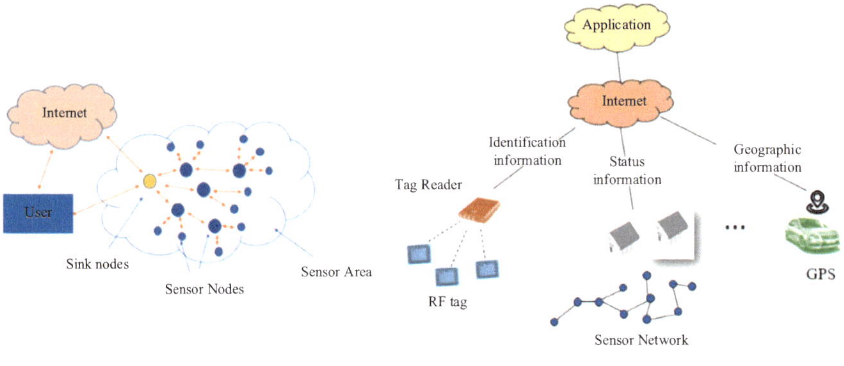

Fig. 3.4 Wireless sensor networks and the IoT. Wang Wei modified based on [1]

3.2.2 IoT Sensing

In the IoT system, the device that acquires and simply processes various information is called an IoT sensor, which can convert various physical, chemical and biological quantities into measurable electrical signals. Generally speaking, the sensor can be treated as the sensory organ of the IoT, which can perceive, detect and collect information about various states of the target object. It is also one of the most essential and most basic components in the entire IoT system.

Through nearly 200 years, sensors have constantly developed with the development of new conductive materials, microelectronics technology, embedded system technology and wireless communication technology. Development of the sensing technology can be divided into three stages. At the beginning, structural sensors rely on signals generated from changes in the structural parameters, such as resistance strain sensors which perceives electrical signals due to changes in the resistance when metal materials undergo elastic deformation. In the 1970s, with the development of integrated technology, microelectronics technology and computer technology, solid integrated sensors have emerged. This kind of sensors is composed of solid components such as semiconductors, dielectrics, and magnetic materials. The thermoelectric effect, the hall effect, and the photosensitive effect of the materials were used to develop the various sensors. In the 1980s, development of the sensing technology entered the period of intelligence. The microprocessors based on the microelectronics technology integrated with memory, communication modules and other aspects, and the concept of embedded system was thus formulated. Then, today's new generation of IoT sensor structure that integrates sensing, processing, communication and energy supply is gradually developed, as shown in Fig. 3.5.

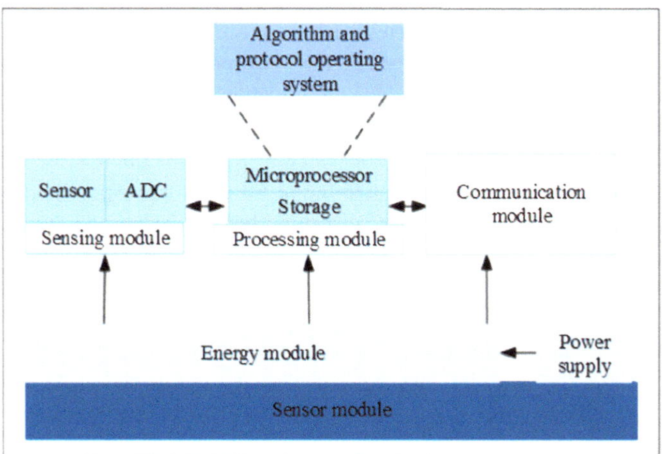

Fig. 3.5 IoT sensor node module

In the developing process towards miniaturization, integration, intelligence, networking and multi-function, IoT sensors mainly take advantage of the development and the application of new materials and technologies. In addition to the conventional circuit integration technology, emerging technologies such as the single-chip microcomputer, the embedded system and the micro-electro-mechanical system (MEMS) have emerged. These emerging technologies have been further applied in the development of IoT sensors.

① Single-chip microcomputer: single-chip microcontroller, which basically is an integrated circuit chip. A single-chip microcomputer mainly comprises the CPU, the read-only memory (ROM) and the random-access memory (RAM). Diversified data acquisition and equipped control system enables the single-chip microcomputer to perform various complex calculations such as controlling the operation symbols and issuing operation instructions.

In the application of IoT sensors, single-chip microcomputer mainly works as the processing and the control modules. The key technical issue with the application of single-chip microcomputer in IoT sensors is the connection between the single-chip microcomputer, the sensing module and the communication module.

② Embedded system: evolved from single-chip microcomputer, it is an application-centric computer system based on the modern computer technology that can flexibly tailor software and hardware modules according to user requirements concerning the function, reliability, cost, volume, power consumption, and the environment, etc. An embedded system is composed of the hardware and the software, and is a device that can operate independently. Its software refers to the software operating environment and its operating system, and its hardware includes the signal processor, the memory, the communication module and so on.

Because of its application-centric and programmable features, embedded systems are mostly used in various industries such as mobile phones, vehicles, industrial automation control, military industry, and the multimedia. At the current stage, in the intelligent IoT application system where people, machines and things are organically connected, physical entities should have good perception, computing and execution capabilities. In order to improve these capabilities, embedded system should be introduced into the IoT system to realize the benefits of these specific functions.

Seismologists at the University of California, Berkeley released a small app called MyShake, as shown in Fig. 3.6. MyShake uses the built-in GPS and acceleration sensor in the mobile phone to measure the earthquake data at different smartphone locations, then transmits these data to the IoT back-end platform, and finally analyzes the possibility of an earthquake in a specific area through intensive data analyses, and finally alerts the users of imminent earthquakes. It is proven that the quality of MyShake data can meet the requirements of many scientific disaster reduction projects, and the test results showed that MyShake can completely record high signal-to-noise ratio waveforms of earthquakes with an epicenter distance of more than 100 km and a magnitude of M5.0 or higher [33].

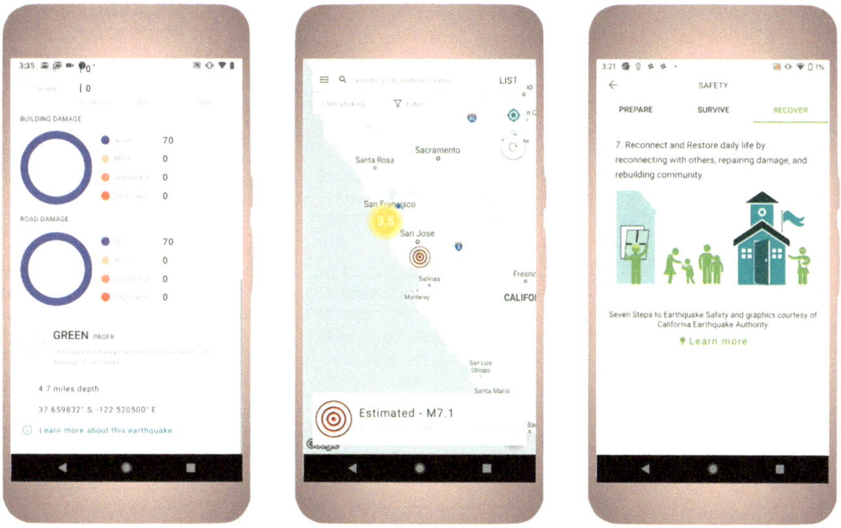

Fig. 3.6 Application interface of MyShake [2] (website: myshake.berkeley.edu)

③ MEMS: High-tech electro-mechanical devices using conventional semiconductor technology and materials, also combined technologies such as lithography, corrosion, thin film, LIGA (Lithographie, Galvanoformung, Abformung), silicon micromachining, non-silicon micromachining, precision machining, etc. It is fundamentally a micro-device or system that integrates micro-sensors, micro-actuators, micro-mechanical mechanisms, signal processing and control circuits, interfaces, communications, and power supplies.

MEMS includes sensors and actuators which are complementary to each other. In a nutshell, MEMS sensor realizes the conversion of mechanical energy to electrical energy, and the actuator is responsible for the conversion of electrical energy to mechanical energy. Working together, they jointly implement the bidirectional operation of MEMS.

As an important branch of MEMS devices, MEMS sensors adopt micromachining and processing technology with typical advantages such as good environment resistance, small size, complete functions, high sensitivity and reliability [34]. There are many types of MEMS sensors according to different classification methods. The MEMS sensors commonly used in civil engineering SHM are shown in Fig. 3.7.

3.2.3 IoT Communication

As mentioned above, the network layer can be regarded as the nervous system of the IoT, mainly intended for information transmission. As shown in Fig. 3.8, the data

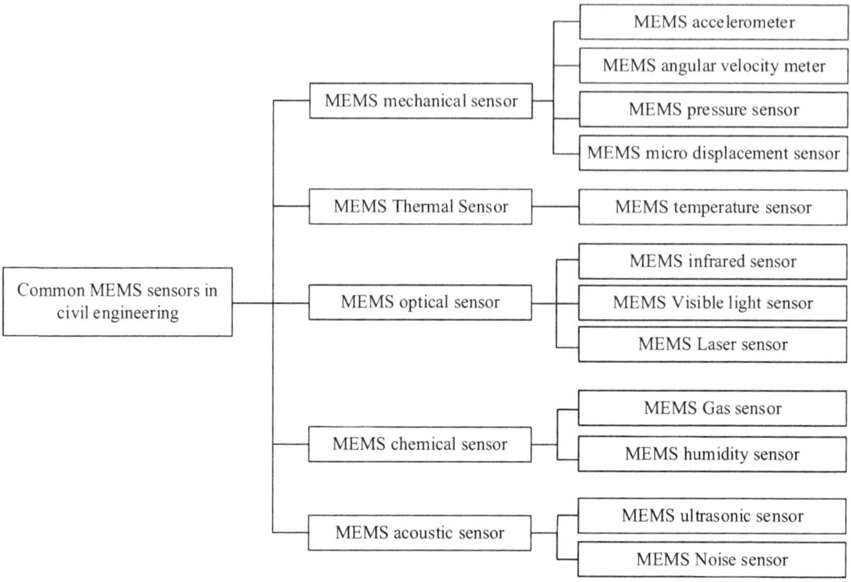

Fig. 3.7 MEMS sensors commonly used in civil engineering SHM

at the sensing end is transmitted to the data processing center through the network layer for data fusion, calculation and processing. In addition, the application layer can control and manage the sensors through the network layer.

IoT communication can be divided into wired communication and wireless communication according to their connection method. Wired communication uses tangible media such as metal wires and optical fibers to transmit information [35]. Wireless communication uses electromagnetic waves to directly propagate signals through space for information transmission, without any tangible media to connect the two ends of communication.

According to the communication distance, IoT communication can be divided into short-distance communication and long-distance communication; where communication within 100 m is short-distance communication and communication exceeding 1000 m is long-distance communication. In reality, there are many communication technologies that can meet various communication needs [35], however, considering factors such as costs, power consumption and efficiency, there is no single communication technology that can meet all communication needs. Transmitting data through longer distances and transmitting more data often bring about higher energy consumption and higher costs.

1. Common Wired Communication Technologies

(1) Serial Port

Serial port is a common interface for communication between devices, and is also widely used for communication between devices and instruments [36]. Common

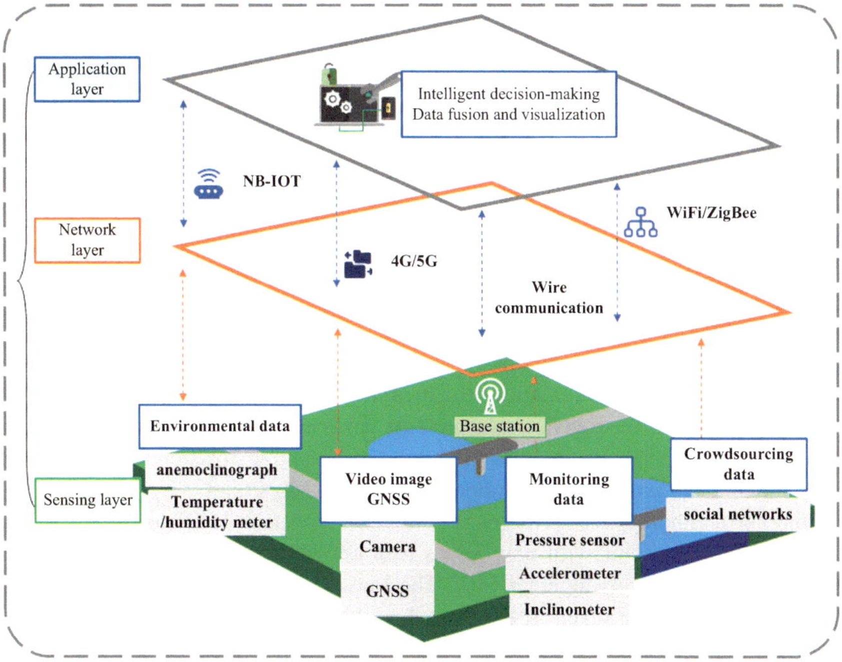

Fig. 3.8 IoT communication

serial ports include half-duplex RS-485 interface, full-duplex RS-232 and RS-422 interface, etc.

(2) Parallel Port

Parallel port is an interface that uses the parallel data transmission standard. There are dozens of parallel ports from the simplest parallel data register or the integrated circuit chip with special interface such as 8255, 6820, etc., to the more complex Small Computer System Interface (SCSI) or Integrated Drive Electronics (IDE) parallel interface. The characteristics of a parallel interface can be described from two aspects: the width of the data channel transmitted in parallel, also known as the number of bits transmitted by the interface; and the characteristics of the additional interface control lines or interactive signals used to coordinate parallel data transmission [37]. However, because interference increases with the transmission distance, the transmission distance is limited for parallel port.

(3) Universal Serial Bus

Universal Serial Bus (USB) is an external bus standard used to standardize the connection and communication between the computers and the external devices. It is also an interface technology applied in the field of PC.

USB has successfully replaced serial ports and parallel ports, and has become one of the standard expansion interfaces and the necessary interfaces for a large number of computers and smart devices in the 21st century, and has now developed to USB 4.0 version. USB has the advantages of fast transmission speed, convenient use, hot-swappable support, flexible connection, independent power supply, etc. It can connect keyboards, mice, large-capacity storage devices and other peripherals to the PC. This interface is also widely used in smart phones [38]. The interaction between smart devices such as computers and external data is also mainly based on network with USB interfaces.

(4) Ethernet

Ethernet is a local area network communication technology. The IEEE 802.3 standard organized by the Institute of Electrical and Electronics Engineers (IEEE) specifies the technical standards of Ethernet concerning the physical layer connections, the electronic signals and the content of the media access layer protocol. Ethernet uses twisted pair wires as the transmission medium, and can cover a range of up to 200 m without any relay. The most popular Ethernet type data transmission rate is 100 Mb/s, and the updated standard even supports 1 and 10 kMb/s rates.

Ethernet technology is currently the most widely used LAN technology. The Internet we are familiar with now refers to a network covering the whole world by connecting all LANs together [36]. The low layer networking technology and protocol, and the upper layer protocol employed by Ethernet are TCP/IP.

2. Wireless Communication Technologies

(1) Short-Distance Wireless Communication

Common short-distance wireless communication technologies include wireless codec chips, wireless data transmission chips, IrDA, RFID, Bluetooth, ZigBee, UWB, Wi-Fi.

Wireless Codec Chip

Wireless codec chip encodes the data for wireless transmission, and the wireless decoder receives the wireless data and performs the fault-tolerant processing. Using wireless codec chip, the physical layer of the wireless network is realized by building a wireless data path and providing several "channels" for the users. It is generally used in wireless remote-control switches for cars, access controls and toys. The common models are PT2262/2272 [39].

Wireless Data Transmission Chip

Wireless data transmission chip allows the users to transmit data through the interface provided by the chip. The common interfaces involved include Serial Peripheral Interface (SPI) and serial ports. The general working frequency bands of wireless data transmission are 315/433/868/915/2.4 GHz. These frequency bands are public frequency bands. Typical wireless data transmission chips include Si4432, nRF24L01, CC1100 [39].

IrDA

IrDA uses infrared to communicate and is a low-cost communication solution. This standard establishes a half-duplex communication system with a communication range of 1–10 m and a transmission angle of 30–60. The disadvantages of IrDA are its relatively high requirements on transmission path (cannot be blocked), limited transmission distance and transmission angle, and poor anti-interference performance. These disadvantages place restrictions on the applications of IrDA. IrDA is commonly used in the conventional home appliance control.

Radio Frequency Identification

Radio Frequency Identification (RFID) technology, also known as electronic tags, is a short-range wireless communication technology using non-contact automatic identification technology based on radio frequency. The working principle is to use radio frequency signal transmitted between the radio frequency tag and the radio frequency reader, and its space coupling and transmission characteristics to realize automatic identification of the target. RFID technology has the advantages of convenient reading, fast identification speed, large data capacity, long service life, wide application range, good security, and dynamic real-time communication [40]. It is widely used in logistics (cargo tracking, warehousing management), transportation (taxi management, bus hub management), identification (various ID cards) and other fields.

Bluetooth

Bluetooth technology is a low-power, short-distance wireless connection technology, which achieves wireless communication between devices at relatively low cost. Bluetooth technology is essentially to establish a general radio air interface and a unified standard for controlling software, so that communication technology and computer technology can be seamlessly combined. Portable devices produced by different manufacturers can communicate with each other over short distances without cable connections. The emergence of Bluetooth technology has expanded the application range of wireless communication [41].

Bluetooth technology has the advantages of fast communication speed, simple connection, universal use, low power consumption and low cost. In addition, with the universal working frequency band 2.4 GHz, it can transmit audio and data at the same time, and has good anti-interference ability. This technology is widely used in mobile phones, computers and entertainment peripherals. Its disadvantages are it can only be used in point-to-point communication and has poor networking ability.

ZigBee

ZigBee technology is an emerging short-distance, low-speed, low-power, and low-cost wireless network technology. It is a technical proposal between the wireless tag technology and the Bluetooth technology. The research and development standard for this technology is IEEE 802.15.4 wireless standards. ZigBee technology is mainly used in short-distance electronic devices that do not require high data transmission rates. Through the deployment of multiple ZigBee nodes, a wireless sensor network is established to achieve data transmission.

The characteristics of ZigBee technology include low data transmission rate (20–250 kb/s), low power consumption, low cost and large network capacity. Each ZigBee device can be connected with up to 254 devices to form a ZigBee network with 255 nodes, covering a range of 10–75 m, which can basically meet the needs of general home or office network coverage. The working frequency band is relatively flexible. Not only the universal 2.4 GHz frequency band, but also the European 868 MHz frequency band and the US 915 MHz frequency band can be applied.

Ultra Wide Band

Ultra Wide Band (UWB) technology is a carrier-free communication technology that does not use a sinusoidal carrier, but directly modulates the communication signal onto a pulse with a pulse width of nanoseconds to form a spread-spectrum UWB signal for information transmission. Communication at a rate of hundreds of Mbit/s to several Gbit/s can therefore be realized within a distance of 10 m. The characteristics of UWB technology include: (a) high rate of up to 0.5–7.5 GHz bandwidth. Even with low transmission power, it can achieve a transmission rate of up to hundreds of megabytes to 1 Gbit/s in a short distance; (b) short communication distance. When the distance between the transceivers is less than 10 m, the channel capacity of the UWB system is higher than that of the conventional narrowband system; (c) good coexistence with the existing narrowband wireless system, high communication security, and strong concealment; (d) high positioning accuracy and strong anti-multipath ability; and (e) small size with low power consumption.

Wi-Fi

As a wireless local area network communication technology, Wireless-Fidelity (Wi-Fi) is widely used in laptops, mobile phones and tablet computers supporting devices to connect to the Internet wirelessly. The communication throughput of Wi-Fi is very high, and it has good compatibility with existing network equipment [39]. Wi-Fi is a wireless local area network communication technology. The IEEE 802.11 standard of the IEEE organization stipulates the technical standard of Ethernet. Wi-Fi terminals refer to the devices that use high-frequency radio signals to send and receive data following Ethernet communication protocols, with communication distance within 50 m.

Usually Wi-Fi technology uses frequency bands of 2.4 and 5 GHz, but the frequency allocation and operating restrictions are not exactly the same around the world.

A comprehensive comparison among the common short-distance wireless communications is shown in Table 3.2.

(2) Long-Distance Wireless Communication

Common long-distance wireless communications include NB-IoT, LoRa, 4G, 5G, etc.

Table 3.2 Comparison of common short-distance wireless communications

Communication technologies	Communication rate	Communication distance	Network topology	Power consumption	Volume	Hardware cost	Application range
Wireless codec chip	Relatively low	10–100 m	Point to point	Low	Relatively small	Relatively low	Data exchange for node devices
Wireless data transmission chip	Relatively high	10–1000 m	Point to point/star	Average	Relatively small	Average	Data exchange for node devices
IrDA	Relatively high	2 m	Point to point	Low	Smallest	Lowest	Data exchange for node devices (one-way)
RFID	Relatively low	1–50 mm	Point to point	Relatively low	Relatively small	Relatively low	Data exchange for node devices
Bluetooth	Average	10 m	Point to point	Relatively low	Relatively large	Relatively high	Data exchange for node devices
Zigbee	Relatively low	10–100 m	Star/net	Relatively low	Relatively large	Relatively high	Node networking
UWB	High	10 m	Star/net	Low	Relatively small	Relatively low	Node networking
Wi-Fi	High	100 m	Star	Highest	Largest	Relatively high	Access layer

NB-IoT

IoT and its communication technology have developed rapidly in recent years. With the development of 2G, 3G, 4G, and the long-term evolution (LTE) CAT-NB1 that evolved from 4G, a new generation of IoT communication technology NB-IoT technology has emerged. NB-IoT is an IoT communication technology that utilizes 3GPP technical standards. It is a major communication technology intended for smart devices with long-distance and low-speed communication, low-power consumption, and multi-terminal services.

LoRa

LoRa comes from the word LongRange, which is a technology for long-distance communication. LoRa technology is based on linear Chirp spread spectrum modulation with low power consumption characteristics of frequency shift keying modulation. However, the communication range has been greatly increased for LoRa. Chirp spread spectrum modulation has the features of long-distance transmission and good anti-interference, and has been used in military and aerospace communications for years. In special cases, a single LoRa gateway or base station can cover an entire city or tens of kilometers.

4G

As the fourth generation of mobile communication technology developed on the basis of 3G, 4G is a high-speed cellular mobile network based on IP protocol. 4G integrated 3G and Wireless Local Area Networks (WLAN), and can quickly transmit high-quality data, audio, video, images and other multimedia information.

5G

Mobile communication has profoundly changed our lives. The fifth-generation mobile communication technology has been developed to cope with the explosive growth of network flow, massive device connections and emerging new business scenarios in the future. 5G communication technology is a new generation of mobile communication technology following the 4th generation mobile communication technology developed to meet the rapid popularization of smart terminals and the rapid development of mobile Internet, and also to meet the requirement of future human information society. The basic characteristics of 5G using millimeter waves are as follows: (a) data traffic increases by 1000 times, and throughput per unit area reaches more than 100 Gbps/km^2; (b) number of networked devices is 100 times that of 4G, and for special cases, the number of devices per unit area can reach 1 million/km^2; (c) minimum peak rate is 10 Gb/s; (d) user available rate can reach 10–100 Mb/s; and (e) its delay is 1/10–1/5 that of 4G [40].

It is predicted that the future IoT communication will be based on 5G, supplemented by other wireless communication technologies. The advantages and disadvantages of common long-distance wireless communications are compared in Table 3.3.

Table 3.3 Advantages and disadvantages of common long-distance wireless communication

Long-distance wireless communication	NB-IoT	LoRa	4G	5G
Networking method	Based on existing cellular networking	Based on lora Gateway	Carrier based	Carrier based
Network disposing method	Node	Node and gateway (node deployment location requirements are higher)	Node	Node
Transmission distance	Long distance (generally more than 10 km)	Long distance (up to more than ten kilometers, 1–2 km in the city, 20 km in the suburb)	Long distance (1–3 km in densely built urban areas, up to more than ten kilometers in the wild)	Medium distance (100–300 m in densely built urban areas, up to 2 km in the wild)
Single network access Node capacity	About 200,000	About 60,000	Less	A lot
Battery life	About 10 years in theory/AA battery	Theoretical about 10 years/AA battery	–	–
Transmission speed	Theoretical 160–250 kbps	0.3–50 kbps	Fast	Very fast
Network delay	6–10 s	Relatively high (depends on frequency, spreading factor, coding rate, etc.)	Low	Very low
Applicable fields	Outdoor scenes, Low-Power Wide-Area Network (LPWAN), large-area sensor applications	Outdoor scenarios, Low Power Wide Area Network (LPWAN), large-area sensor applications, private networks, and places where cellular networks cannot be covered	Widely	Widely

3.2.4 Convergence of Sensors and Networks

Wireless networks have made significant progress with respect to the technological evolution and commercialization. From the perspective of access, low-power wide-area networks, fifth-generation mobile communication technology 5G, and various short-distance communication networks have brought new functions and commercial applications to the industry. The implementation of those emerging technologies not only expanded the number and scope of physical devices that can be accessed, but also made it more convenient, safe and low-cost for physical devices to access the network. As shown in Fig. 3.9, sensors can access the Internet directly through low-power wide-area networks (LoRa, NB-IoT), fifth-generation mobile communication technology (5G), or indirectly through various short-distance networks (ZigBee, Wi-Fi, etc.) for local connection combined with a smart gateway for connection to the Internet [42].

(1) Sensors Connected Directly to the Internet

Long-distance wireless network has the advantages of wide coverage, multiple access, low power consumption and low cost, and can meet the requirements of massive sensor access. Connecting sensors by long-distance wireless networks avoids the perceived data being sent to local or remote servers, and is able to use cloud servers with almost unlimited computing resources. It is also easy to deploy a large number of sensors connected to the cloud nodes, including conventional wireless sensors equipped with IoT interfaces, thus simplifying the network topology.

To deal with the high cost of wired sensor measurements, the transmission distance limitations of conventional wireless measurement and the difficulties in equipment maintenance and data management, Hou [43], a member of the authors' team, designed a low-cost bridge displacement monitoring sensor system which integrated

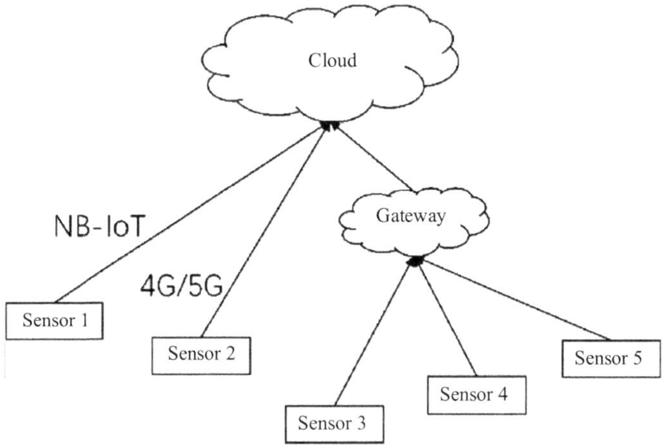

Fig. 3.9 Convergence of sensors and networks

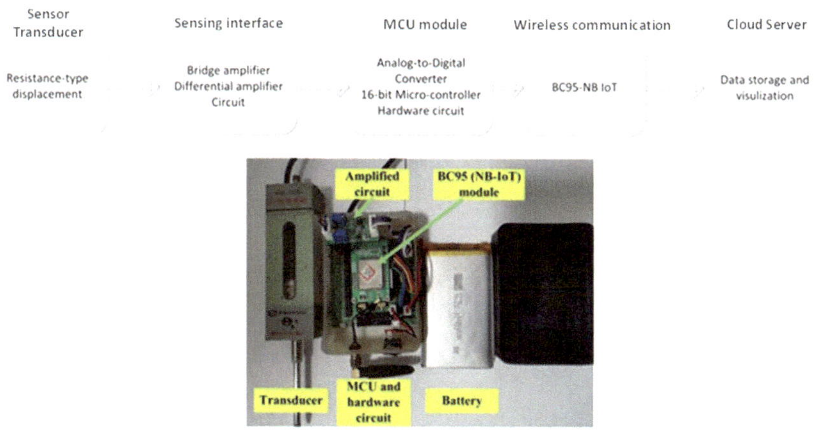

Fig. 3.10 Integration of sensors and NB-IoT [3]

low-power wireless communication technology NB-IoT and conventional sensors to accurately collect displacement data and send the data to the server, as depicted in Fig. 3.10. The system consists of sensors, sensing interface, micro control unit module, wireless communication module and cloud server.

(2) Sensors Connected to Internet Via Gateway

Gateway is also called Internet connector or protocol converter. The gateway implements network interconnection above the network layer, and is a complex network interconnection device which is only intended for interconnection of two networks with different high-level protocols. The bottom networks feed the data collected by the sensors to the gateway through short-distance communication technologies such as RFID, WSN, Wi-Fi, ZigBee and Bluetooth, which connects to the Internet to further transmit the sensory information to the application end.

Tong et al. [44] developed a low-power, low-cost wireless acceleration sensor integrating the self-powered wireless gateways and a cloud platform for data upload, download, storage and visualization based on the IoT protocol. Physical structure of the acceleration sensor is shown in Fig. 3.11.

The wireless acceleration sensor communication module is CC1101 radio frequency (RF) chip of 470 MHz frequency band and Gauss Frequency Shift keying (GFSK) transmission mode. When the sensor is in a dormant state, the gateway can be controlled to send electromagnetic waves from the radio frequency chip to activate the sensor for data collection and communication. The data collected by the wireless acceleration sensor is then transmitted to the cloud platform through the gateway, and uploaded to the cloud platform database.

In addition to applications in SHM, combinations of various IoT communication technologies and sensors have also been applied in various other fields. In different application scenarios, such as mobile phones, smart homes, industries, automobiles,

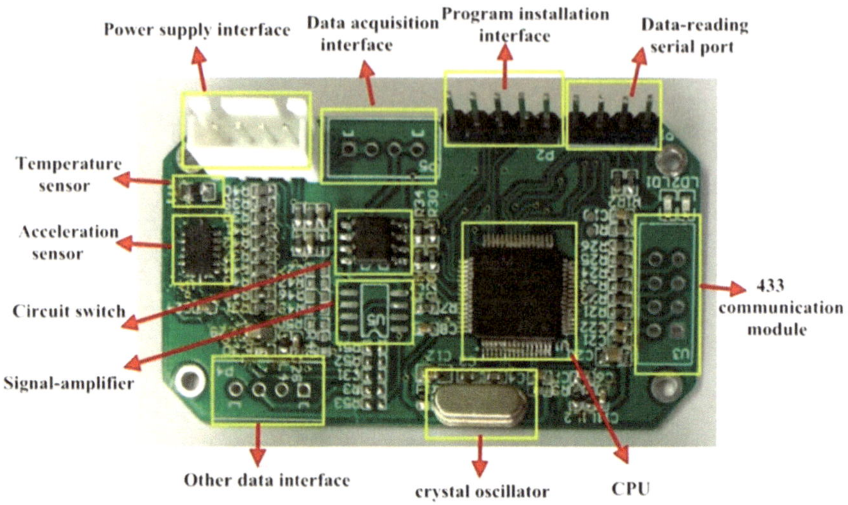

Fig. 3.11 Diagram of wireless acceleration sensor [4]

agriculture, medical care, transportation, finance, drones, etc., different sensing technologies were used according to different requirements. The differences in transmission distance, data type and frequency in the process of information transmission required different combination of sensors and IoT communication technologies.

3.3 IoT Cloud Management Platform

In 2018, Gartner, a world-renowned analysis organization, forecasted that there will be as many as 14.2 billion connected objects in use by 2019, and the total number will reach 25 billion by 2021. The explosive growth of the IoT market will stimulate the related hardware market providing huge market opportunity for the electronic component manufacturers. The main purpose of developing the IoT is not to connect things, but to exchange information so as to provide intelligent services. After the intelligentization of the hardware, it can be loaded through Internet services to form a typical architecture of "cloud + terminal". A large number of IoT device terminals need to be managed centrally. At the same time, a large amount of data generated is stored on the "cloud" platform for centralized computation, analysis and learning, thereby generating cognitive and intelligent analysis results, which can be fed back to the "end" devices, and then generate an endless stream of IoT applications.

With the unprecedented prosperity of IoT, new products and emerging technologies emerge in an endless stream, and combination with the cloud platform is imperative. There are many types of smart hardware, and each smart hardware requires a lot of R&D investment. The development of cloud platform for IoT can resolve this problem. Various things in the IoT are fully connected, and all real-time dynamic information collected can be sent to the computing processing center through wireless and other networks for summary, analysis and processing. The intelligent storage, processing and analysis of the massive data collected in the process is the key to realizing the application of IoT in the true sense. Generally speaking, cloud storage can provide enough storage space for the massive data of IoT; cloud computing can apply different types of device collections through grid technology, distributed technology, etc.; and cloud storage and computing cooperate to provide external data storage, business analysis and so on. Cloud computing will be the core technology to realize the application of IoT. People can use cloud computing model to conduct real-time dynamic intelligent analysis and make management decisions on different businesses in the IoT. Combination of IoT and cloud computing can surely promote the upgrading and leap forward of the entire industrial and value chain through the promotion of various capabilities, resource sharing and in-depth mining of information value.

The cloud management platform for IoT is different from the general cloud platform (see Chap. 5), because of its close relationship with hardware. It is an important part of a large IoT ecosystem that connects and supports all components in the system, helping to facilitate device management, handle hardware and software communication protocols, collect and analyze data, enhance data flow and intelligent application functions. The IoT cloud platform consists of four major components: device access, device management, rule engine, and security authentication and authority management.

(1) Device Access

Device access includes a variety of device access protocols. The mainstream is the Message Queuing Telemetry Transport (MQTT) protocol, in addition to ModBus, Enhanced Dence Protocol (EDP), Hyper Text Transfer Protocol (HTTP), CAN Calibration Protocol (CCP), CNA Correction Protocol), Constrained Application Protocol (COAP) and so on.

(2) Device Management

Device management includes device creation management and device status management. Generally, all devices are managed in a tree, from the product to the device group and the specific device. It mainly includes product registration and management, adding, deleting, modifying and checking the subordinate equipment of the product, equipment news release, Operational Transconductance Amplifier (OTA) equipment upgrade management, etc.

(3) Rule Engine

IoT cloud platforms are built on the basis of existing cloud computing platforms. A mature business of IoT, in addition to the functions provided by the IoT cloud platform, generally also requires the functions provided by the cloud computing platform, such as cloud hosts, cloud databases, etc. The rule engine generally uses Structured Query Language (SQL)-like language to filter and process the data of the

IoT platform, and then forward it to other cloud computing products. For example, the data reported by the device can be forwarded to Alibaba Cloud's TableStore time series database.

(4) Security Authentication and Authority Management

The IoT cloud platform issues a unique certificate for each device, and then allows the device to access. The smallest authorization granularity is generally at the device level. Certificates can generally be divided into product-level certificates and equipment-level certificates. Each device connected to the cloud platform stores a certificate (KEY composed of character strings) locally, and this certificate must be carried every time it communicates with the cloud for verification of cloud security components.

Table 3.4 shows some well-known IoT cloud platforms.

Table 3.4 Some well-known IoT cloud platforms

Platform	Introduction	Features
ThingWorx	The leading IoT platform for industrial companies, providing devices with the functions of easy connection, construction, deployment and expansion, with all the functions and powerful scalability of advanced enterprise solutions, integrated with cutting-edge technologies	Easily connect with sensors, RFID and other devices, the dashboard can pre-build widgets, integrate machine learning, provide simple web design, easy to manage devices, easy to connect; but difficult to manage complex systems
Azure IoT	A fully managed cloud platform can be easily integrated with any existing or future system due to its strong security mechanism and excellent scalability. Allows to connect hundreds of millions of devices, collect, analyze data and perform machine learning	Simple device registry, rich integration with SAP, Salesforce, Oracle, etc., dashboard visualization, real-time streaming media, third-party services, security and scalability, high availability; requires management and the cost is high
AWS IoT	Easily collect data from sensors and connected devices. Interact even when the device is offline. Support and process billions of devices and trillions of messages	Strong device management and security, using device shadow technology, good integration with Infrastructure as a Service (IaaS) products, continuous price decline, flexible and open; steep learning curve
Cisco IoT Cloud Connect	Cloud-based mobile suite that accelerates data transformation and data manipulation. Supports extensive customization for voice and data communications	Data and voice connections, device and IP session reporting, billing customizable, flexible deployment options

(continued)

Table 3.4 (continued)

Platform	Introduction	Features
ThingSpeak	An open platform produced by MATLAB, which provides applications for MATLAB analysis and visualization of data. Arduino development boards and Raspberry Pi single-outlet machines can be used to send sensor data, lightweight development and application	Can be used for private collection of data, application integration, event scheduling, MATLAB analysis and visualization, easy to visualize, provides functions for Ruby, NodeJS, Python; limited uploaded API data
Alibaba Cloud IoT	The device communicates stably with the cloud, global multi-node deployment effectively reduces communication delay, and multiple protection capabilities ensure the security of the device cloud. Provides rich device management functions and stable and reliable data storage capabilities. Using the rule engine, customers can configure simple rules on the Web to forward device data to other products of Alibaba Cloud, obtain full-stack services of data collection, data calculation, and data storage, and realize flexible and rapid construction of IoT applications	Provide secure authentication methods and device-level permission granularity for the device side, provide Topic cross-account authorization, realize interconnection with devices from different manufacturers, and better meet national conditions and usage habits, provide device-side SDK, and quickly connect devices to the cloud. It supports global device access, heterogeneous network device access, multi-environment device access and multi-protocol device access. The long-term connection capability of billion-level devices, the concurrent processing capacity of millions of levels, and the architecture support horizontal expansion service availability are extremely high. Decentralization, no single point of dependence. multi-data center support

3.4 Application of IoT Technology in Bridge Maintenance and Management

Presently, a large number of bridges in service in China and around the world have entered their maintenance stage, including many approaching their end of service lives. For them, the environmental loads and material degradation can result in structural damages, especially in large bridges. Once a bridge collapses, irreparable losses will be caused. So, to reduce or even eliminate the occurrence of such accidents, application of the IoT sensing technology to bridge maintenance and management has been established. This section introduces the development trend of SHM based on IoT technology, the process of connecting self-developed wireless sensors to IoT cloud platform, the construction of cloud platform-based IoT bridge operation and maintenance system by the authors' team, and an application of IoT sensing technology to bridge maintenance and management in turn.

3.4.1 Development of SHM Based on IoT Technology

Rapid developments in both smart sensors and wireless communication technologies have revolutionized SHM in aerospace, automotive and civil engineering. As a key technology in this revolution, wireless sensor network (WSN) technology has greatly improved the way of implementing SHM systems from installation of sensors to transmission and processing of sensor data [41]. With the integration of sensor network and SHM system, the intelligent level of sensors has been improved, and it becomes very important to introduce new effective communication protocols to the sensor networks to control remote sensors and data streams they produced. Consequently, interest in implementing WSN technologies in accordance with Internet architectural standards to SHM has grown significantly over the past few decades.

For effective interaction between smart sensors and remote services, the Internet should be integrated into the SHM system based on the IoT technology. With the integration of intelligent sensing technology, wireless technology, digital technology and the Internet, the SHM system based on IoT brings new opportunities to the development of real-time data acquisition. For example, a data acquisition subsystem can be empowered with the functions of information communication through IP and real-time control of sensors. When an emergency occurs or the amount of data exceeds a certain threshold, the data acquisition subsystem can automatically send a reminder to the monitoring center. Then, engineers or technicians responsible for maintenance and undertaking other monitoring tasks can access the data and information from anywhere using various devices [46].

Currently, WSN technology is considered to be one of the most important technologies in the development of SHM systems. In the WSN, sensors can communicate with the base station through wireless transmission protocols such as Wi-Fi, Bluetooth, Z-Wave, and ZigBee. In recent decades, the trend to connect WSNs with the Internet based on IP protocol suite possibly adopted in WSNs has been developed, with the purpose of interconnecting sensor networks with other domains, as shown in Fig. 3.12 [47]. However, some WSN technologies such as IEEE802.15.4/ZigBee do not currently support IP; therefore, solutions involving the use of IP proxies or gateways are proposed.

WSN technology also plays an important role in the development of the IoT. The concept of IoT is that all sensing devices at the edge of the network can support the IP protocol. In this regard, the new version of ZigBee is specially designed for WSN with IPv6 connectivity to realize the concept of IoT [46].

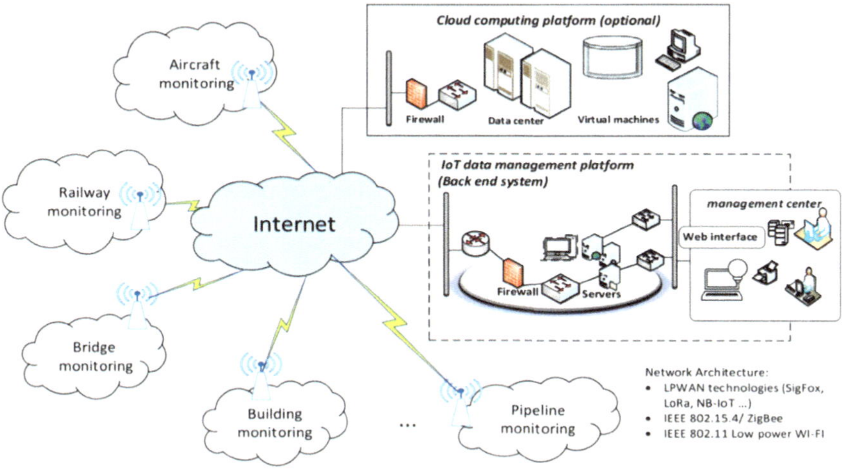

Fig. 3.12 Framework of IoT-based SHM [47]

3.4.2 Construction of IoT Cloud Platform Based on Wireless Sensing

Many enterprises and researchers have focused their attention on wireless sensing technology because of its advantages of low cost, flexibility, easy of use, safety and reliability, and short construction period. The NB-IoT communication method launched in 2016 supports a longer average standby time (up to 3–5 years) and has a coverage rate of over 90% in China. The data carrying capacity has also been greatly improved. The data sending interval officially recommended by NB-IoT is half an hour. The most commonly used communication mode is Power Saving Mode (PSM) for NB-IoT. More precisely, it will start deep sleep state without receiving data for 22 s, and it will restart and be online at the next collection cycle or the arrival of preset heartbeat packet. The sleep state current of NB-IoT communication module is generally at the μA level.

For different collection frequencies and application scenarios, 4G mode, NB-IoT mode and corresponding IoT SIM cards can be used respectively. Among them, the 4G method is mainly used in "high-frequency or low-latency" scenarios, such as sampling the pylon cable at a 50 Hz acceleration and sending the collected data to the server every second; the NB method is suitable for "low-frequency and low-power consumption" scenarios, such as hourly static strain measurements on the girders of bridge structures.

The author's team has independently developed the NB-IoT main control board, 4G main control board, and six types of small acquisition boards, as shown in Fig. 3.13. Each main control board has 4 channels, and each channel is connected to the acquisition board with RS485 cable; the acquisition board is then connected to the conventional sensor or directly integrated with the MEMS module. Using

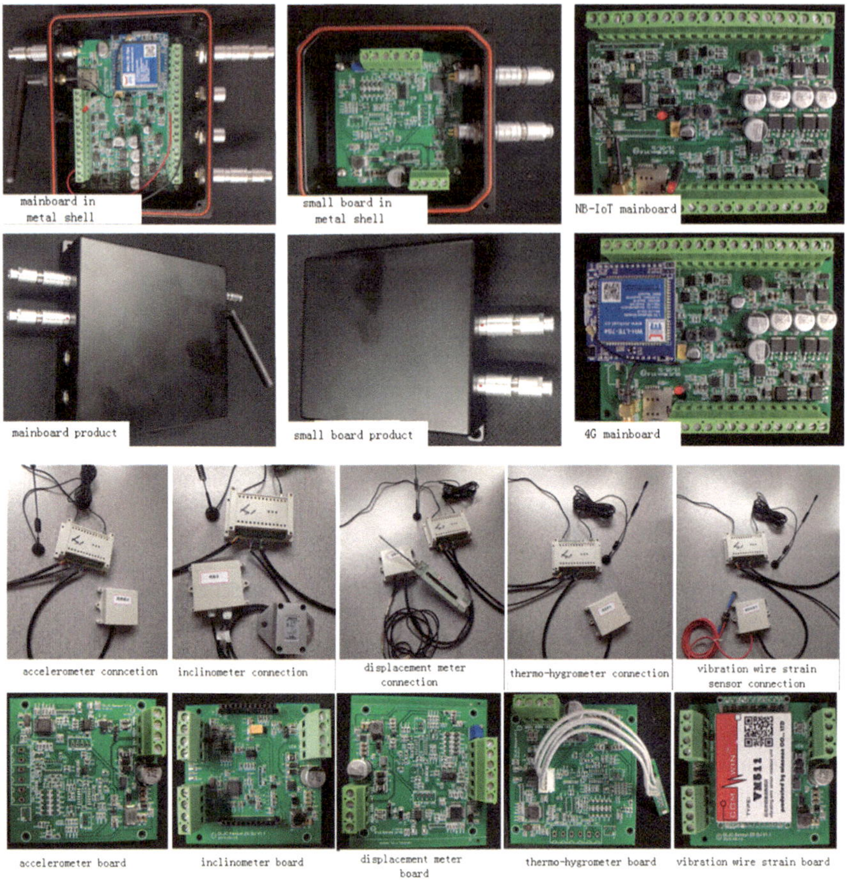

Fig. 3.13 Various printed circuit board assembly developed by author's team

embedded programming, the main control board can aggregate the data uploaded by the small collection boards from each channel, and then send it to the set server through wireless NB or 4G transmission.

For NB-IoT, data can be directly and quickly sent to the server bound to the 4G card using 4G for the server to receive, monitor and process it. NB-IoT follows these five levels in sequence, i.e., "NB equipment—NB base station—NB Controller (core network)—IoT connection management cloud platform—user service application (user cloud)". Accordingly, the data collected by the equipment terminal must first pass to the base station, then to the core network, then to IoT cloud platform, and finally to the top-level user who can forward and use the data; as shown in Fig. 3.14.

As shown in Fig. 3.14, the core is the management of NB-IoT equipment, and the uplink and downlink of information: the southbound (the interface provided downward in the hierarchical model diagram) is connected to the NB-IoT equipment,

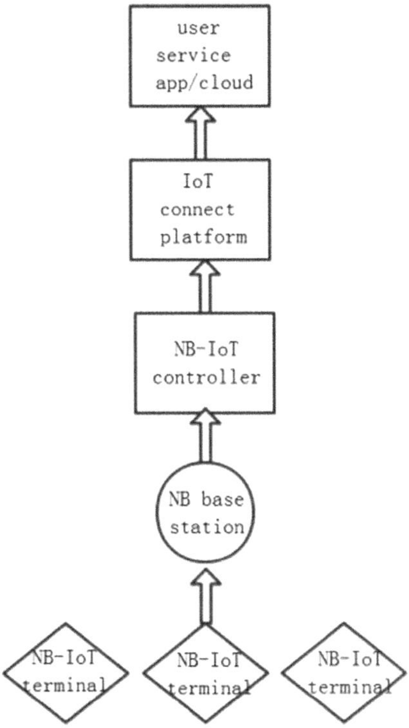

Fig. 3.14 Five-level development model of NB-IoT

and the northbound (the interface provided upward in the hierarchical model diagram) is connected to the user application.

The API documentation provided by the platform can also be used for programming to realize automation of the entire application process. For example, on the basis of the API, the developed WeChat applet (Fig. 3.15) can control the upper limit of the main control board, the start and stop of the sensor channel, the adjustment of sensor collection frequency, the calibration of sensor, the physical quantity associated with the sensor, etc.

3.4.3 Construction of IoT Bridge Operation and Maintenance System Based on Cloud Platform

Using the "Smart Bridge Management Platform" developed by the authors' team as an example (see Sect. 2.4), the overall framework of the IoT bridge operation and maintenance system is shown in Fig. 3.16. The infrastructure service layer includes cloud platform infrastructure and smart bridge infrastructure. This layer provides facility supports for the construction of the smart bridge maintenance and

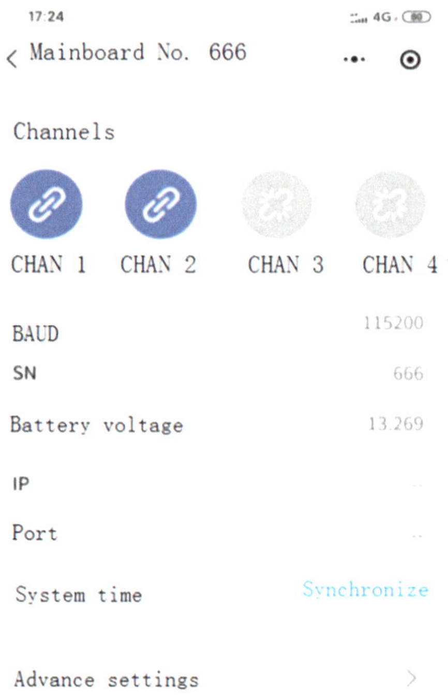

Fig. 3.15 Main control board with WeChat applet

management platform, computing and storage functions, and network resources for upper-layer services.

The platform service layer is divided into two parts: the data management platform and the service platform. The data management platform provides unified management of data. Generally, after completing the planning of database tables and fields, it is built using existing database products for the data management platform. The service platform generally adopts a development framework with corresponding functions for configuration and subsequent development. These database contents were used to program into a variety of system integration application programming interfaces (APIs), for the subsequent SaaS layer to make specific calls.

The application service layer includes many aspects, which is the function expansion and interface packaging on the API of the PaaS layer, and makes it more convenient for the owner to manage and maintain.

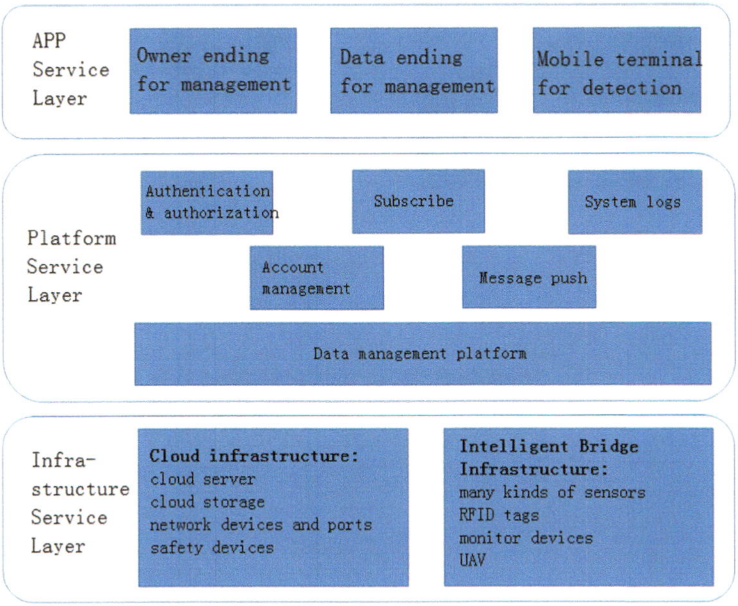

Fig. 3.16 Framework of "smart bridge management platform"

3.4.4 An Engineering Application—Bridge Earthquake Monitoring System Based on Smartphone

In this section, a case study of a smart phone-based bridge earthquake monitoring system developed by the author's team [47] is introduced to demonstarte how a smart phone integrated into a cloud server can be used as an ubiquitous wireless network monitoring platform to collect health monitoring information in a unified database. Smartphones are connected with sensor networks, and cloud servers are used for uploading and accessing data conveniently. A decentralized structural earthquake monitoring network is built using easily installed smartphones as sensing units and smart neurons.

The authors' team developed a measurement software using the Objective-C programming language in the Xcode integrated development environment, which interacts with the smartphone hardware and the operating system. The software enabled the built-in MEMS sensor components to acquire, analyze, store and transmit data to the cloud server. As shown in Fig. 3.17, the measurement system includes an independent smart phone, which can be installed at different locations on the bridge to measure data independently. Connecting these devices to a common network facilitates communication with each other to create a stable measurement system for the entire bridge. Additionally, real-time processing of data using the computing power of a smartphone allows for faster structural diagnostics, damage index reports or emergency alert outputs.

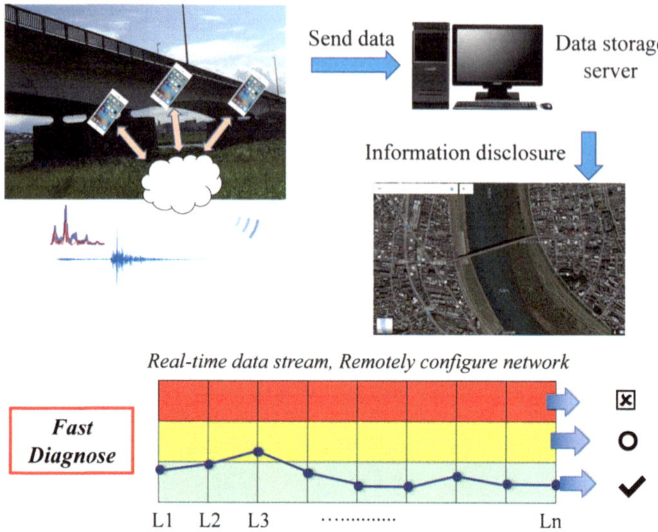

Fig. 3.17 Measurement system using smartphone [47]

The system has undergone a long-term field test on the Takamatsu Bridge in Japan (see Fig. 3.18). The Takamatsu Bridge was built in 1982 and is located in Miyazaki, Japan. It is a 7-span prestressed concrete continuous box girder bridge with a length of 444 m.

Data from the smartphone system is compared with the results of a high-precision accelerometer. To measure the dynamic response of the bridge, two smartphones were

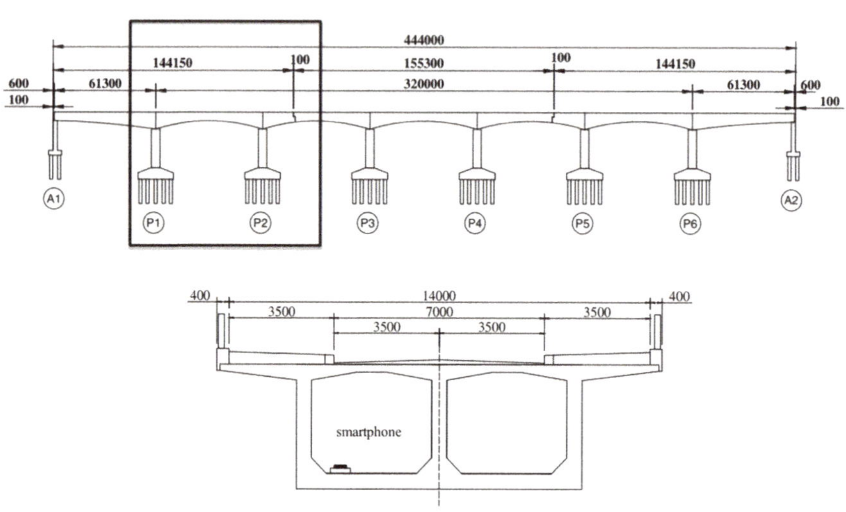

Fig. 3.18 Sectional dimension of Takamatsu Bridge, Japan [47] (all units: mm)

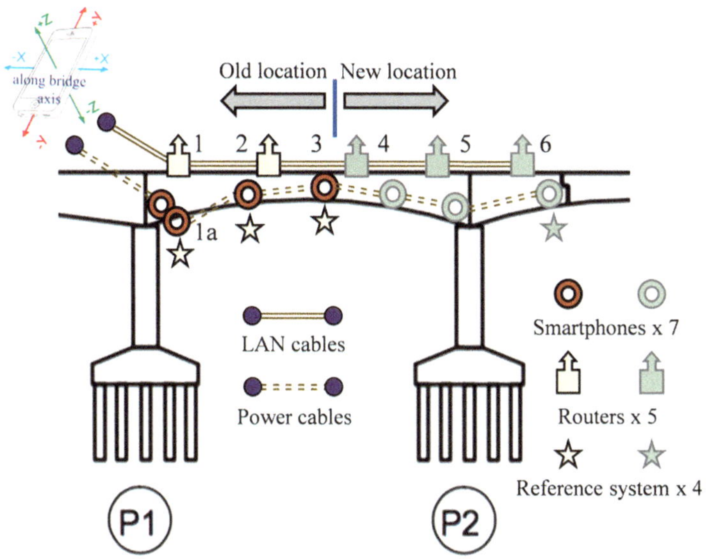

Fig. 3.19 Layout of sensors on Takamatsu Bridge [47]

arranged at the pier, another two at the 1/4 position, and one at the cantilever end of the Gerber beam. The device installation details are shown in Figs. 3.19 and 3.20.

By setting up the Wi-Fi connection of each smartphone terminal, the data was wirelessly transmitted to the cloud server. Power was supplied, and the network connections were established through LAN services installed inside the box girder.

After long-term continuous measurements, the smartphones recorded four seismic events since the day the system was established. It can be seen that the average PGA of the four events near the bridge was about 10 cm/s^2, which means that the amplitudes of the four earthquakes recorded near the bridge were all small. Among the four earthquake events recorded, the acceleration waveforms recorded at three different locations on the bridge on March 2, 2017 were compared with that of the reference system, as depicted in Fig. 3.21.

Acceleration records from a servo accelerometer (Hakusan Industrial SU501) were used to compare with the waveforms and Fourier spectra recorded simultaneously by the smartphones. Since the noise component of the continuous signal obtained by the MEMS sensor was about $\pm$ (2–3) cm/s^2, it can be seen from the records of all four earthquakes that the accurate waveform data of the small-magnitude earthquake was covered by the noise component. Nevertheless, some significant peaks can still be identified to distinguish the signature of the earthquake, and they showed good agreement with those of the reference system.

This case study highlights the potential of smartphones for earthquake monitoring in a rapid, remote, automated and quantitative way. The use of IoT technology based on smartphone monitoring can fundamentally impact the future development of smart and sustainable infrastructure.

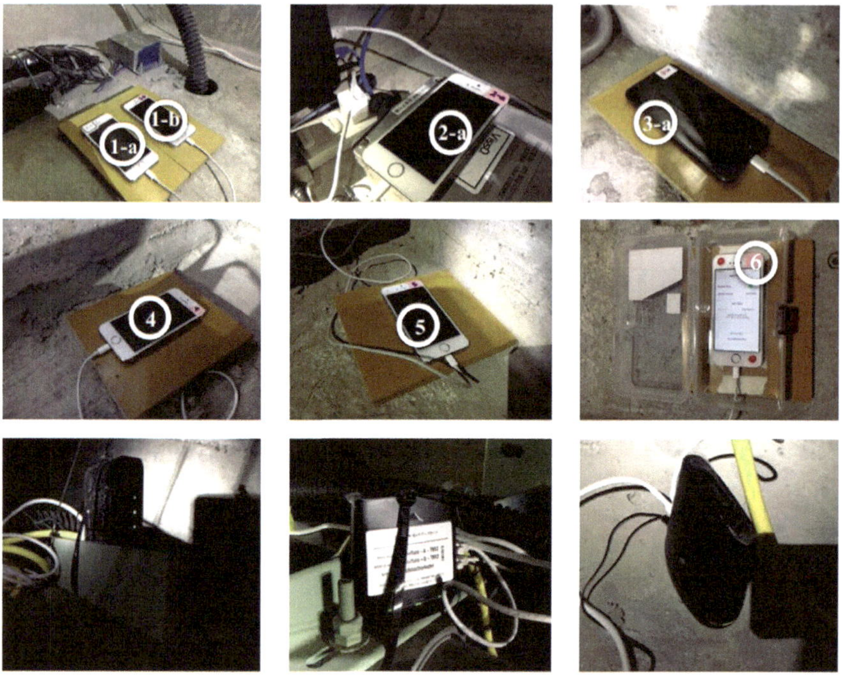

Fig. 3.20 Site installation of smartphones and Wi-Fi router [47]

It can be inferred that IoT technology can play an important role in bridge operation and maintenance, and it is hoped that modern IoT technology can be widely used in the future to make timely assessment and information update of bridge status and operating environment, such as the combination of IoT sensing technology and big data artificial intelligence analysis technology.

3.5 Summary

This chapter first introduces the conventional structure sensing technology from two aspects of contact and non-contact. It then comprehensively evaluates the limitations of the conventional sensing technology, including insufficient interpretation of data, incomplete structural information obtained, incompatibility of data, difficulty in effective analysis, difficulty in data query and analysis, perception technology and equipment are independent of each other, high data transmission costs, no application for small and medium span bridges, etc. It also illustrates that the advance concept of "things connected, interconnected and exchanged" of the IoT technology can be applied in civil engineering to help solve these problems. Then, it introduces the intelligent IoT technology from the aspects of overview of IoT, IoT sensing and IoT

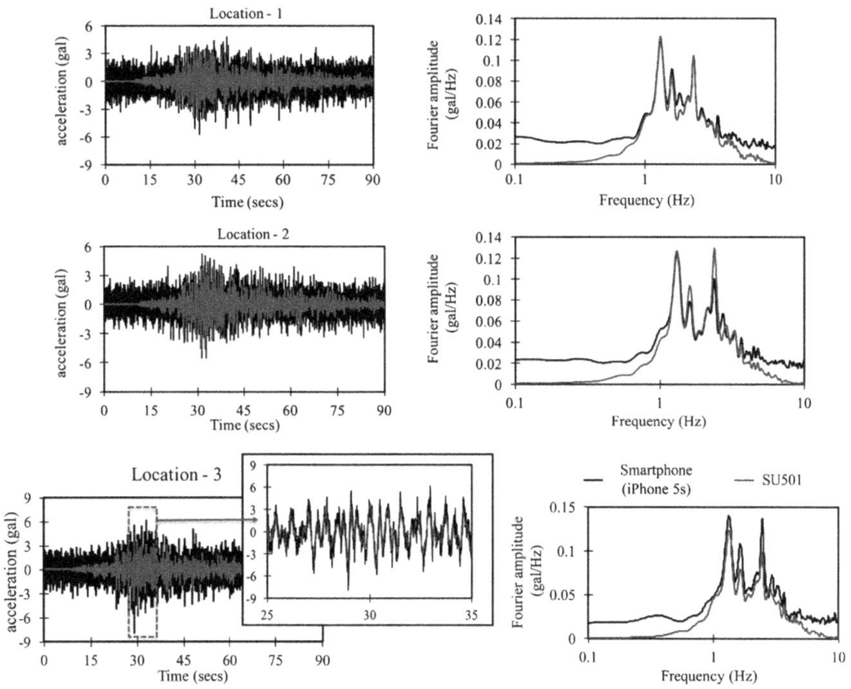

Fig. 3.21 Comparison of waveforms and amplitudes at three different positions [47]

communication technologies, etc., and points out that the IoT technology requires the cloud platform to provide intelligent services like centralized management and interconnection exchange. The difference between the IoT cloud management platform and the general cloud platform are also introduced. Finally, the status and application of IoT sensing technology in bridge intelligent operation and maintenance are introduced through an actual real-life application.

References

1. Lydon M, Su ET, Doherty C et al (2017) Bridge weigh-in-motion using fibre optic sensors. Inst Civil Eng Bridge Eng 170:219–231
2. Cheng L (2008) Research on vehicle weighing in motion technology. Zhejiang University, Hangzhou (in Chinese)
3. Harik I, Peiris A (2013) Case studies of structural health monitoring of bridges. In: Nondestructive testing of materials and structures. Springer, Berlin
4. Deng Y, Li A (2019) Vertical deflection monitoring of main girder. In: Structural health monitoring for suspension bridges. Springer, Berlin
5. Hou XM, Yang XS, Huang Q (2005) Using inclinometers to measure bridge deflection. J Bridg Eng 10(5):564–569

6. Yang XS, Hou XM, Liao ZP et al (2002) A new method of bridge deflection measurement. Chin Civil Eng J 35(2):92–96 (in Chinese)
7. Stefano AD, Ceravolo R (2005) Monitoring and response of CFRP prestressed concrete bridge. In: Sensing issues in civil structural health monitoring. Springer, Berlin
8. Robertson IN, Johnson GP, Wang S (2005) Instrumentation performance during long-term bridge monitoring. In: Sensing issues in civil structural health monitoring. Springer, Berlin
9. Du Y, Sun B, Li J et al (2019) Cable stress monitoring technology based on fiber bragg grating. In: Optical fiber sensing and structural health monitoring technology. Springer, Berlin
10. Grosser H, Beckmann-Dobrev B, Politz F et al (2013) Computer vision analysis of 3D scanned circuit boards for functional testing and redesign. Procedia CIRP 11:229–233
11. Min YZ (2014) Research on vision measurement method of railway subgrade surface settlement with camera chain. Lanzhou Jiaotong University, Lanzhou (in Chinese)
12. Gorpas D, Politopoulos K, Yova D (2007) A binocular machine vision system for three-dimensional surface measurement of small objects. Comput Med Imaging Graph 31(8):625–637
13. Cui EK (2019) Research on key technologies of high precision binocular stereo vision measurement system. University of Chinese Academy of Sciences (Changchun Institute of Optics, Precision Machinery and Physics, Chinese Academy of Sciences), Changchun (in Chinese)
14. Shao XX (2018) High precision, real-time digital image correlation deformation measurement. Southeast University, Nanjing (in Chinese)
15. Lin JY, Xu XJ, Liang P et al (2019) Application of digital image correlation technology in structural damage detection. Constr Qual 37(3):48–52 (in Chinese)
16. Malesa M, Kujawinska M (2013) Deformation measurements by digital image correlation with automatic merging of data distributed in time. Appl Opt 52(19):4681
17. Pan B, Li K, Tong W (2013) Fast, Robust and accurate digital image correlation calculation without redundant computations. Exp Mech 53(7):1277–1289
18. Yoneyama S, Kitagawa A, Iwata S et al (2010) Bridge deflection measurement using digital image correlation. Exp Tech 31(1):34–40
19. Hu ML, Li XX, Ren WX et al (2015) Application research of bridge dynamic displacement measurement based on digital image technology. J Hefei Univ Technol (Nat Sci Edn) 38(12):1661–1667 (in Chinese)
20. Liu C (2011) Research on vibration measurement of continuous beam based on digital image processing technology. Guangzhou University, Guangzhou (in Chinese)
21. Hunaday R, Hagara M et al (2017) A new procedure of modal parameter estimation for high-speed digital image correlation. Mechan Syst Signal Process 93:66–79
22. Trebuna F, Hagara M (2014) Experimental modal analysis performed by high-speed digital image correlation system. Measurement 50:78–85
23. Liu MW (2013) Research on application of GPS based bridge deformation monitoring—taking Donghai bridge as an example. Donghua University of Technology, Nanchang (in Chinese)
24. Liu DY (2009) Comparison and analysis of GPS and microwave interferometry in bridge dynamic deflection measurement. Bridge Constr 3:81–84 (in Chinese)
25. Liu YY (2014) Study on time series InSAR surface deformation monitoring based on multi-source SAR data. Chang'an University, Xi'an (in Chinese)
26. Liu M (2016) Application of ground interferometric radar technology in bridge dynamic detection. Beijing University of Civil Engineering and Architecture (in Chinese)
27. Chen YQ (2013) Research on deformation monitoring technology based on InSAR. University of Electronic Science and Technology of China (in Chinese)
28. Wang BY (2009) Overview of IoT technology research. J Electron Measur Instrum 23(12):1–7 (in Chinese)
29. Liu Q, Cui L, Chen HM et al (2010) Key technologies and applications of the Internet of Things. Comput Sci 37(6):1–4, 10 (in Chinese)
30. Sethi P, Sarangi SR (2017) Internet of things: architectures, protocols, and applications. J Electric Comput Eng 2017:1–25

31. Zhu HB, Yang LX, Zhu Q et al (2011) Progress and application of internet of things technology. J Nanjing Univ Posts Telecommun (Nat Sci Edn) 31(1):1–9 (in Chinese)
32. Ning HS, Xu QY (2010) Some thoughts on the development of the global internet of things and the construction of the internet of things in China. Acta Electron Sin 38(11):2590–2599 (in Chinese)
33. Kong Q, Allen RM, Schreier L (2016) MyShake: initial observations from a global smartphone seismic network. Geophys Res Lett 43(18):9588–9594
34. Wu GW, Wu SL (2019) Development and application of MEMS sensor technology. China Plant Eng 17:200–201 (in Chinese)
35. A figure summarizes the communication methods of the Internet of Things, and an article discusses several common IoT communication methods. (2 Nov 2017). http://www.360doc.com/content/17/1202/16/40903010_709260259.shtml (in Chinese)
36. 6 Technologies of wireless connection of Internet of Things are introduced in detail (22 Nov 2018). https://www.taodocs.com/p-181835668.html (in Chinese)
37. Wang L (2011) Preparation of electrically actuated artificial muscle and its measurement and control technology. Nanchang Hangkong University, Nanchang (in Chinese)
38. Dong P (2019) Design of a multi-channel detection system for automatic hot plug of USB devices. Chem Autom Instrum 46(5):356–360 (in Chinese)
39. Zhang DS (2013) Design of wireless monitoring system for industrial equipment status based on STM32. Dalian University of Technology, Dalian (in Chinese)
40. Dong H, Tang JR, Yu DX et al (2014) Overview of the development status and trend of internet of things communication technology. Commun Technol 47(11):1233–1239 (in Chinese)
41. Xu XM, Cao LY, Zhao YL et al (2015) Several short-range wireless communication technologies and their applications. Internet of Things Technology 5(11):101–102 (in Chinese)
42. Zhi DX (2018) Far more ferocious than 5G! 2018 White Paper of the Internet of Things, the domestic scale has reached 1.2 trillion (13 Dec 2018) (in Chinese). https://baijiahao.baidu.com/s?id=1620646665729335793&wfr=spider&for=pc
43. Hou S, Wu G (2019) A low-cost IoT-based wireless sensor system for bridge displacement monitoring. Smart Mater Struct 28(8):085047
44. Tong XL, Yang HL, Wang LB et al (2019) The development and field evaluation of an IoT system of low-power vibration for bridge health monitoring. Sensors 19(5):1222
45. Lee JL, Tyan YY, Wen MH et al (2018) Applying ZigBee wireless sensor and control network for bridge safety monitoring. Adv Mech Eng 10(7):2072047827
46. Arcadius Tokognon C, Gao B, Tian GY et al (2017) Structural health monitoring framework based on internet of things: a survey. IEEE Internet Things J 4(3):619–635
47. Shrestha A, Dang J, Wang X et al (2020) Smartphone-based bridge seismic monitoring system and long-term field application tests. J Struct Eng 146(2):04019208.1–4019208.14

Chapter 4
Bridge Inspection Automation

Abstract Bridge inspection provides an important basis for bridge maintenance. For a considerable amount of time, the current practice primarily adopted the manual approach of inspecting the bridge appearance, i.e., the apparent bridge defects are measured and documented manually. Such manual inspection largely depends on the environment and on the technical skills of the inspectors, which is inefficient, time-consuming and costly; and the results may not always be accurate. With the development of automation in Civil Engineering, there are now emerging inspection tools that facilitate inspection of bridge decks, high piers, cables and underwater structures to different degree of automation. We introduce two representative technologies in this chapter, underwater inspection robots and unmanned aerial vehicles (UAVs), which have shown their potentials in bridge inspection.

4.1 Overview of Bridge Inspection Automation Technologies

Bridge inspection automation aims to develop a process where the use of sensors, vehicles, and computing devices can be integrated in a way that the process is automatic or automated, or even autonomous. As such, we differentiate bridge inspection automation in terms of three levels:

- Automatic inspection—a process that can be executed in order by the system solution itself with some human intervention. Example includes an unmanned aerial vehicle (UAV) that is piloted remotely by an end-user.
- Automated inspection—a process that can be carried by the system's hardware (machines or computers) without needing human control. Taking the example unmanned vehicle again, it includes that given a pre-defined navigation map, the vehicle can navigate and complete the sensing or scanning for the data collection.
- Autonomous inspection—a highest-level automation wherein the system can execute the inspection task itself with a certain level of 'awareness' about the condition of the bridge. An example includes that a UAV or even a humanoid can

G. Wu et al., *Intelligent Bridge Maintenance and Management*, Springer Tracts in Civil Engineering, https://doi.org/10.1007/978-981-97-3827-4_4

take off, search damaged location, evaluate damage, and go to the next location without a navigation map.

Consequently, bridge inspection automation, even at different levels, aims to minimize the human participation and maximize the use of equipment and machineries for data collection, information processing, analysis and evaluation, operation, and control in accordance with the needs of the bridge inspection. Currently, the technical level of bridge inspection practice cannot reach the level of fully automation (i.e., via an autonomous vehicle or even a humanoid), and the running of machineries must still be controlled by humans. Therefore, the technologies of interest are mostly in the category of automatic inspection.

4.1.1 Conventional Bridge Inspection Methods

During bridge inspection, the objects to inspect consist of all the bridge components, including the bridge deck system, the superstructure and the substructure. So far, a set of more mature implementation procedures has been formulated, such as the Technical Specification for Inspection and Assessment of Urban Bridges (CJJ/T 233-2015) and the Technical Condition Assessment Standards for Highway Bridges (JTG/T H21-2011). Both of these standards are technical specifications for bridge inspection and assessment, and they detailed the inspection contents and the testing indicators. With regard to the conventional methods of bridge inspection, the inspectors manually perform the inspection task in accordance with the applicable codes and standards, which has numerous restrictions.

Figure 4.1 and Table 4.1 illustrate the conventional bridge inspection methods for various inspection objects. The conventional manual inspection methods generally have the following disadvantages:

(1) Low efficiency and high labor intensity: Inspection is tedious and time-consuming, and it is frequently essential to set up and disassemble the scaffolding and other equipment.
(2) Low precision: Inspection relies primarily on human visions, which are heavily influenced by subjective factors.

Fig. 4.1 Typical manual bridge inspection

Table 4.1 Conventional bridge inspection methods

Inspection object	Inspection content	Inspection methods	Description
Bridge deck system	Apparent defects (Smoothness of pavement surface, cracks, deformation of expansion joints)	Inspectors walk on the bridge deck to take photos and record	It is more convenient for inspectors to walk on the bridge deck, making this inspection easier to complete
Beam body	Apparent defects; Material strength; Bearing capacity; Concrete carbonation depth; Location of reinforcement	Inspectors must build brackets to approach and detect inspection objects, or they may require the assistance of bridge inspection vehicles; Load testing is used to determine the bearing capacity	Inspectors cannot easily approach the inspection object, and it is costly to build brackets and rent bridge inspection vehicles
Main cable	External cracks; Cable clamp slipping	Inspectors walk and record on the inspection channel of the main cable	This method is intuitive, but the safety of aerial work is poor
Tension (suspension) cable	External damage; Internal wire breakage	Use hanging baskets to carry inspectors, and some inspectors can be carried by bridge inspection vehicles	Inspectors cannot easily approach the inspection object, and the hanging basket safety may be inadequate
Bridge tower	Apparent defects; Material strength	Utilize telescopes for long-distance observation or bridge inspection vehicles to aid with inspection	Inspectors cannot easily approach the inspection object, resulting in inaccurate measurements
Bearing	Deformation 、rock breaking, etc	Some bearings are easily accessible, while others can only be reached by creating platforms or utilizing bridge inspection vehicles	In the majority of bearing inspections, inspectors cannot easily approach the inspection object, and bridge inspection vehicles are expensive
Pier	Apparent defects; Material strength	Higher piers need to be erected with platforms or assisted by bridge inspection vehicles	Inspectors cannot easily approach the inspection object, and it is costly to build brackets and rent bridge inspection vehicles
Underwater foundation	Apparent defects (Exposed tendons, cracks, damage, etc.); Void	Professionals dive underwater to photograph the damaged components	The professional standards for inspectors are stringent, and underwater operations can be unsafe and inefficient

(3) Low security: During inspection, the inspectors must either go down to the bottom of the bridge or climb up to the main cable and support towers, both of which can be very dangerous.
(4) High costs: Inspection is labor- and equipment-intensive, requiring high capital expenses.
(5) Low level of informationalization: The data collected manually exists in isolation, making it impossible to create an accurate historical database for the bridge and for management to make maintenance decisions.

The issues listed above severely affect the bridge maintenance levels, making automation of bridge inspection technology highly necessary.

4.1.2 Current Status of Automatic Bridge Inspection Technology

In 1946, D.S. Harder, a mechanical engineer at the Ford Motor Company in the United States, coined the term "automation". Today, the development of automatic inspection technology is becoming increasingly systematic, flexible, integrated and intelligent, which enhances the technical level and market competitiveness of industries, including the traditional industry, agriculture and manufacturing industry, such as the robot production line in usual automobile manufacturing plants. However, in bridge inspection, people still rely on manual inspection, which provides much room for the development of automatic inspection technology in this field.

(1) Automatic Inspection of Bridge Decks

With bridge inspection vehicles, the inspectors can use cameras instead of manual observation to photograph and record defects. Such vehicle-mounted inspection equipment can conveniently detect defects at the bottom of the beam. The "Smart Vision" system designed by Beijing Texida Special Equipment Technology Co., Ltd. features a reduced design that allows a single person to carry and operate.

These equipment have been used successfully in actual bridge inspections, but still need sufficient bridge deck or substructure space, and occasional need blocking of traffic lane. Therefore, a more compact and adaptable adsorption inspection robot was developed. This robot was originally created for steel bridges, but it is also capable of performing adsorption on concrete bridges. Not only can it inspect bridge decks, but it can also inspect bridge piers and pylons. This equipment is still in design, and a commercial version has yet to be developed. There are now mature bridge inspection vehicles commercially available for bridge deck system inspection, as well as pavement inspection robots with integrated nondestructive inspection capabilities for bridge internal damages. Figure 4.2 shows the development process of bridge deck inspection equipments.

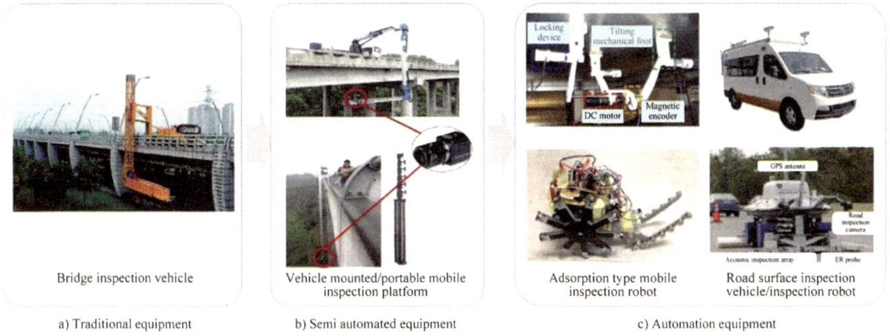

Fig. 4.2 Development process of bridge deck inspection equipment [1–5]

(2) Automatic Inspection of High-pier and Tower

A large bridge inspection vehicle can be used to transport inspectors to solve the problem of accessibility. Nevertheless, its coverage is always limited by the inspectors' arm length. However, the lifting platform is directly mounted on the bridge tower's track and can move along it. This platform can be used permanently on the bridge and requires a specialized traction equipment. As for bridge deck inspection, the inspectors have mobile platforms equipped with cameras. Compared to setting up a bracket platform for the inspectors to walk on, the mobile platform requires less manual work and is of lower costs, but the preparatory work is still cumbersome and inefficient for the mobile platforms. The development process of high-pier and tower inspection equipment is depicted in Fig. 4.3.

To realize efficient inspection of "simultaneous walking and measuring", a series of adsorption-crawling robots have been developed. Bionic adsorption, negative pressure adsorption and electrostatic adsorption are among the many adsorption modes

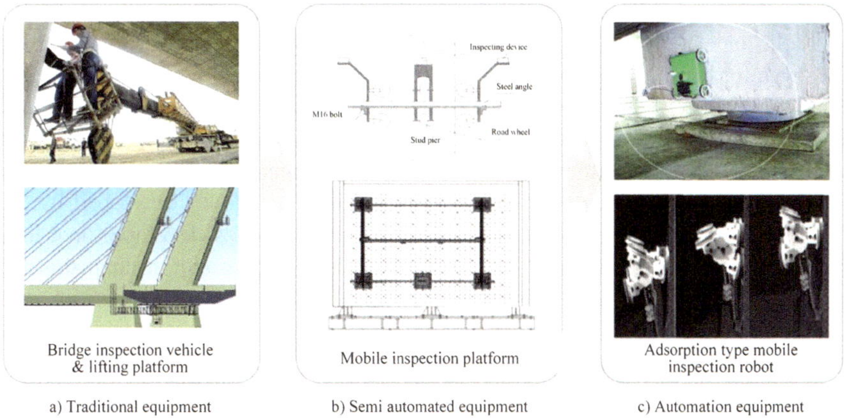

Fig. 4.3 Development process of high-pier and tower inspection equipment [6–8]

Fig. 4.4 Current main cable inspection equipment [9, 10]

for concrete structures that robots possess. There are numerous adsorption robots that performed well in indoor tests, and some have already been used to inspect actual bridge piers.

(3) Automatic Inspection of Main Cables

Conventionally, the inspectors would walk, observe and manually record the defects on the inspection channels of the main cables. Nevertheless, in practice, inspection and repair of the main cables are frequently undertaken simultaneously. To accomplish this, most inspectors would use maintenance vehicles, such as those shown in Fig. 4.4, to transport other operation-relevant workers. However, the mechanical structure is complex for the maintenance vehicle, making it difficult to control and install. Also, the vehicle's weight can be excessive, making it easy to damage the main cable during the inspection process. Currently, there are a few research accomplishments on main cable inspection robot, both in China and other countries. For the autonomous mobile main cable inspection robot, it is a key technical issue to ensure its stability under high-altitude winds, and with traverse cable clamp barriers.

(4) Automatic Inspection of Stay Cables

In addition to the conventional method of observation utilizing telescopes, there are two primary methods to inspect the cables. For small cable-stayed bridges, a hydraulic or electric lifting platform is utilized for inspection of the stay cables. However, for large bridges, a fixed point is set at the top of the tower in advance, and a hanging basket is dragged by steel cables to hoist up the inspectors. In practice, these methods have limited application range so less efficient, and there are possible safety risks associated with working high above the ground. Figure 4.5 illustrates the development process of cable inspection equipment.

Climbing mechanisms of the automatic inspection robots developed in China and other countries are primarily categorized as wheeled climbing mechanisms and multi-legged climbing mechanisms. Multi-legged climbing mechanisms can help to pass the obstacles by making discrete contact with the cable surface, but their speed is severely low, and their safety and load-carrying capacity are inadequate. The stay cable inspection robot developed by Southeast University, China, has been successfully applied on the Hangzhou Bay Bridge, the Tangling Yangtze River Bridge

Fig. 4.5 Development process of cable inspection equipment

and some other large bridges in China, and it currently holds a leading technical position both in China and other countries.

(5) Automatic Inspection of Underwater Structures

Inspectors always have difficulty with inspection of the bridge substructures as the service and environmental conditions of the substructures, particularly the underwater structures and the pillars of the bridge, are usually worse than those of the superstructures. In addition, the piers of some medium- and long-span bridges are primarily located in deep water, making the defects of some underwater structural components more concealed; consequently, often missed.

The conventional underwater inspection methods rely primarily on manual inspection. Up to today, manual inspection results are still subjective and unable to provide targeted solutions to the attachments accumulated over many years on the piers (such as moss, etc.). Additionally, manual inspection possesses significant potential safety hazards and risks, so the cost of inspection rises proportionally. Since the 1960s, underwater inspection equipment have gradually replaced the manual inspection methods because such equipment offers longer operation time, more robust inspection flexibility, deeper operation depth, bigger operation radius and more flexible applicability. Currently, there have been some successful applications of robots in underwater inspection [11]. However, it is still difficult for the conventional general underwater robot to operate steadily and reliably in complex hydrological environment. Its deficiencies are primarily manifested in three aspects: obsolete inspection technology, low performance in operation and poor inspection results.

Existing inspection technology is not designed for use in underwater environments. Current underwater robots primarily consist of the operating equipment and the measuring sensors, with underwater cameras and sonar for underwater inspection. Existing equipment have low stability in terms of working performance. When the underwater robot performs an inspecting task, it must be equipped with positioning and inspection equipment to work with the manipulator extension. The load change sometimes makes evident local change of the underwater robot's mass and moment of inertia, and since the original balance is no longer met, accurate attitude control is difficult to achieve [12]. In terms of inspection results, it is equally challenging to provide precise inspection results. Currently, optical image data plays a significant role in the inspection of defects in underwater structures. Underwater photos are of low resolution, with a large amount of interference information and a small field of vision, which makes subsequent image processing extremely difficult. The development process of underwater structure inspection equipment is depicted in Fig. 4.6.

(6) UAV Automatic Inspection

The semi-automatic and automatic inspection devices mentioned above are mostly targeted for particular application scenarios, such as bridge decks, high piers, high towers, main cables, stay cables, underwater structures, etc. However, the technology of taking images of apparent defects on bridge components by UAVs in flight is less space-restricted and more application-flexible, making it appropriate for comprehensive bridge inspection. UAVs can obtain the apparent information, three-dimensional

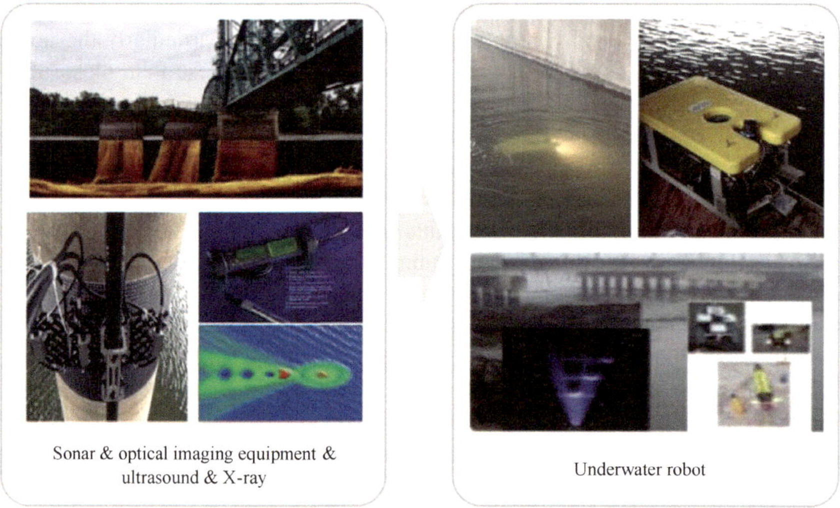

a) Traditional equipment & new technologies b) Automation equipment

Fig. 4.6 Development process of underwater structure inspection equipment information and dynamic response of the bridges, and can also combine with other measuring equipment to fulfill the specific measurement needs.

UAVs can be utilized for bridge structure defects inspection, which normally includes corresponding image collection and subsequent data analysis. There are numerous bridge structural components and some of these components are narrow and concealed. The UAV's obstacle avoidance mechanism prevents it from collection of close-range and high-definition photos of the beam's bottom. Therefore, it is necessary to develop UAVs applicable for specific bridge inspections according to the characteristics of the bridge detection, as depicted in Fig. 4.7. Weak GPS signal under the bridge deck and poor positioning accuracy under the bridge are among the primary issues of utilizing UAVs for bridge inspection. Consequently, it is necessary to develop applicable autonomous navigation technology for UAV bridge inspection.

In general, the development and application of automatic bridge inspection technology and equipment free "people" from the inspection work. Using equipment instead of human eyes and using program computation instead of manual processing significantly increases inspection efficiency, reduces labor costs and improves inspection precision. Simultaneously, computer processing unifies the format and description form of all similar inspection data, which facilitates further computation of big

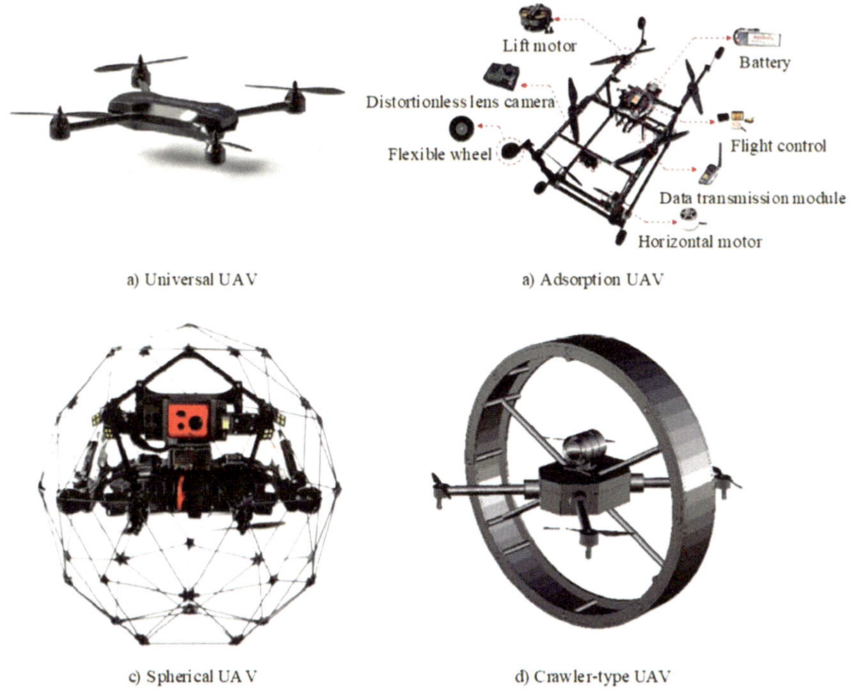

Fig. 4.7 UAVs for bridge inspection [13–15]

data. All data are automatically classified and uniformly archived, and a database of a certain size is subsequently created, which facilitates further data mining.

4.2 Semi-automatic Bridge Contact Inspection Equipment

Research on bridge inspection equipment focuses on designing mechanical devices that can fulfill the requirements of various inspection scenarios and can be equipped with various inspection instruments to achieve automatic defects information gathering. To adapt to complex environments, the mechanical devices must possess specific motion capabilities and coverage, as well as targeted obstacle-crossing and obstacle-avoiding abilities with sufficient stability. It should be noted that automation of bridge inspection refers primarily to automation of defects information gathering, i.e., the use of inspection equipment instead of manual inspection, hence enhancing inspection efficiency and data reliability. However, the selection of inspection range and the control of inspection equipment's movement must still be completed by humans.

4.2.1 Bridge Deck Inspection Equipment

Bridge deck inspection primarily involves the inspection of bridge deck's upper surface (pavement) and lower surface (beam bottom). The primary content of beam bottom inspection includes the inspection of apparent defects such as cracks, damages and corrosion. In contrast, the pavement's inspection content usually includes assessment of the material strength and detection of the internal damage of the beam besides the apparent defects. People have developed bridge deck inspection equipment for various structural forms and of different working mechanisms in accordance with the practical needs. As the level of automation increases for these equipments, their inspection efficiency and precision also continue to improve.

(1) Bridge Inspection Vehicles

The bridge inspection vehicle is a special large-scale equipment for bridge inspection and maintenance. A Class II automobile can be transformed into a bridge inspection vehicle by adding a special working device with a control system to its chassis. The equipment offers numerous advantages, such as convenient inspection, no traffic disruption, flexible work, high operating efficiency, easy operation, safety and reliability. Currently, it is classified mainly according to the folding arm type and the truss type [16].

The folding arm bridge inspection vehicle (Fig. 4.8 left) employs a hanging basket workbench, allowing it to be less affected by the bridge structure and to work with greater flexibility. It can be used to inspect the bridge's lower parts or be elevated to inspect the bridge's upper structures. The truss bridge inspection vehicle (Fig. 4.8

right) employs a channel-type working platform, so as to achieve high stability and loading capacity. Inspectors can easily enter the platform or return to the bridge deck from the platform when it is in use. If elevators are installed, the lower bridge's depth can be increasingly reached. From structural perspective, the folding arm bridge inspection vehicle is simpler than the truss bridge inspection vehicle; however, operation efficiency of the truss bridge inspection vehicle is higher than that of the folding arm bridge inspection vehicle, allowing for better and more stable inspection operation as well as greater working range.

In contrast to the conventional bridge inspection vehicles, the beam bottom inspection vehicle (Fig. 4.9) can also be a movable work platform for defects inspection on the bridge's lower surface. The platform is installed on the track beneath the bridge beam or under the beam body's flange and is equipped with a special track that is used permanently on the bridge. It does not interrupt the traffic during operation and the inspection scope is extensive, making it applicable for inspection of varieties of railway and highway bridges.

Fig. 4.8 Folding arm bridge inspection vehicle (left) and truss bridge inspection vehicle (right)

Fig. 4.9 Beam bottom inspection vehicle

(2) Semi-automatic Bridge Deck Vehicle Inspection System

Based on the bridge inspection vehicle, scholars at home and abroad imitated its working mode and developed a series of semi-automatic intelligent bridge inspection robotic systems. The Hanyang University of Korea first successfully developed an image-based vehicle-mounted bridge crack inspection system [17, 18]. As shown in Fig. 4.10, the system consists of an automobile, a multi-link mechanical arm, a guide rail and the inspection equipment, in which the guide rail is installed at the end of the mechanical arm. The inspection equipment can slide on the guide rail to inspect the defects at the bottom of the beam.

With reference to this system, other institutions developed a series of semi-automatic vehicle inspection robot systems, such as the semi-autonomous mobile bridge inspection robot by Lyon University in France and the universal bridge inspection robot system U-BIROS (Ubiquitous Bridge Inspection Robot System) by South Korea [19], etc. (Fig. 4.11).

Research on intelligent bridge inspection equipment is still in its infancy, despite the existence of a large number of products with practical applications. Future research will focus on modular design, lightweight long-span mechanism design and manipulator stability control.

(3) Portable Bridge Deck Inspection Equipment

Although large-scale vehicle-mounted inspection platforms help to improve the inspection efficiency and accuracy, due to their large size and high cost, it is not economical to apply them for small bridge inspections. To this end, portable inspection equipment is used for small bridge inspections. The "Texida Smart Vision" (see Fig. 4.12) draws experiences from the design of bridge inspection vehicles. It adopted the advanced remote inspection and telemetry technology to formulate a comprehensive inspection means with high efficiency, high quality and low cost

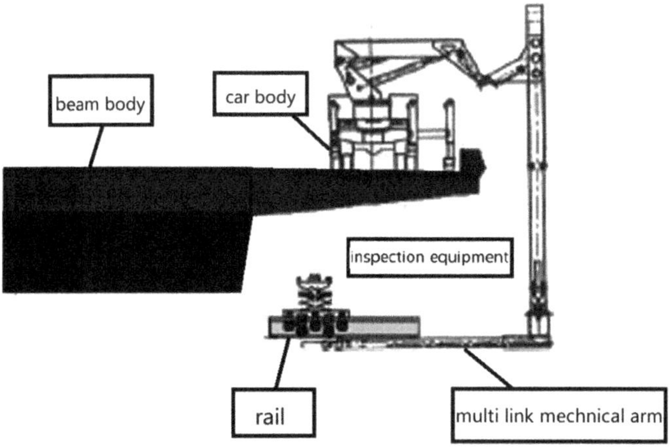

Fig. 4.10 Composition of a bridge crack detection system

Fig. 4.11 The universal bridge inspection robot system U-BIROS [19]

Fig. 4.12 The "Texida Smart Vision"

for bridge appearance inspection. The equipment consists of a stand part, a cloud platform and an operation part.

The hardware-carrying platform adopted two design ideas: the downward probe and the upward support. The downward propping support can be quickly fixed on the bridge deck guardrail, and after extension, the cloud platform can be sent to the suitable inspection position under the bridge. The upward support can be used for dry bridge inspection, which can be erected to adapt to the terrain and effectively observe the T-beam, box girder and ancillary facilities.

The cloud platform (see Fig. 4.13) is equipped with a hardware data acquisition terminal which can realize 360 degrees rotation in horizontal and elevation directions. The platform can realize the functions of defect picture and video acquisition, defect spatial coordinate three-dimensional positioning, defect size remote measurement, automatic defect re-examination, crack width comparison and analysis, etc., by integrating the high-definition image acquisition module, compensated total station module and remote crack width measurement module.

Fig. 4.13 The cloud platform

(4) Road Surface Status Inspection Vehicle

Apart from inspection of the bridge's lower surface defects, inspection of the bridge's upper surface defects, such as pavement defects, is also important. Since the 1920s, research institutions in western countries have developed prototypes of highway inspection vehicles with the camera technology. In the twenty-first century, with the continuous development of computers, sensors, radar technology and digital image processing technology, the related technologies of pavement inspection vehicles have been continuously improved. With the development of high-speed image acquisition and image processing technology, automatic inspection of pavement cracks has become possible. Currently, the mature multi-functional inspection vehicles at home and abroad include PATHRUNNER, CiCS, ARAN, ZOYON-RTM, and so on [1]. Figure 4.14 shows a schematic diagram of the PATHRUNNER multi-functional inspection vehicle.

In addition, FHWA has developed a robotic system called RABIT (Robot Aided Bridge Inspection Tool) for the nondestructive inspection of concrete bridge decks [2] (see Fig. 4.15). Besides identifying pavement surface defects, the robot uses four nondestructive inspection technologies: resistivity method, impact echo method, ground penetrating radar detection and ultrasonic detection to detect three kinds of internal defects of concrete bridges, namely steel bar corrosion, delamination and concrete degradation, and can draw the distribution map of internal distresses in the bridge.

Generally speaking, multi-functional inspection vehicles have been applied in certain scenarios, and various technologies are constantly being updated and improved, but the cost is too high. There is yet to be a mature standard configuration of inspection vehicles globally.

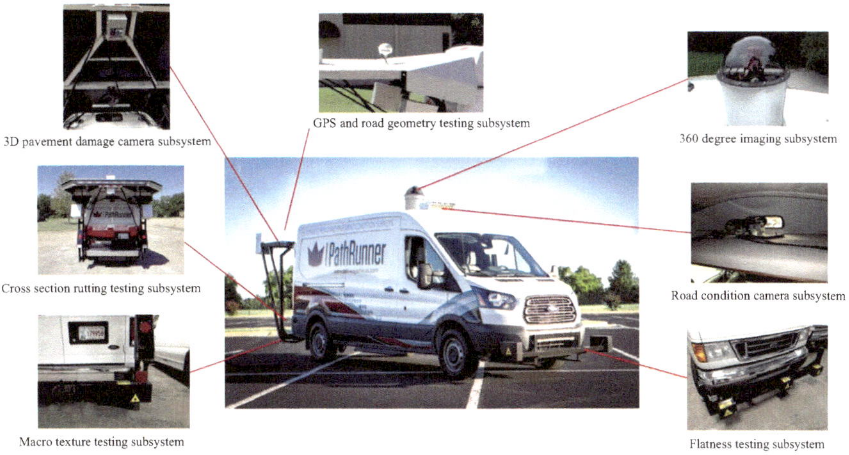

Fig. 4.14 The PATHRUNNER multi-functional inspection vehicle

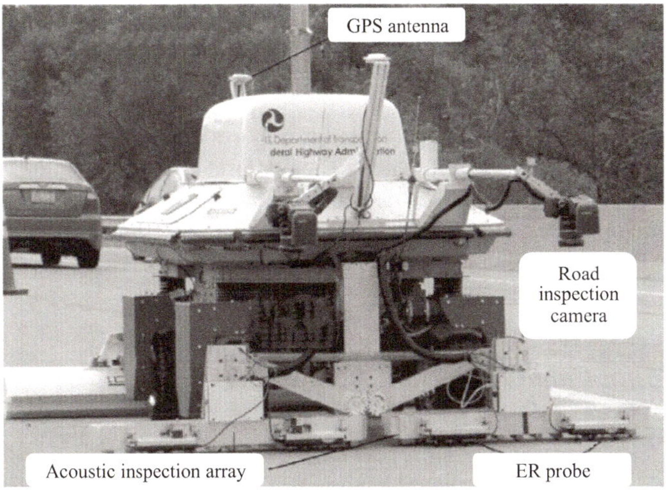

Fig. 4.15 RABIT Bridge nondestructive inspection robot [3]

4.2.2 Inspection Equipment for High Piers and Towers

Because of their heights, high piers and towers are difficult to approach without climbing facilities. Therefore, the conventional inspection method for piers and towers is that the inspectors would observe at a distance with the help of telescopes, which is subjected to large judgment errors and dead observation angles, and is time-consuming and tedious. For close observation, the inspectors would set up working supports for operation; but is also time-consuming, laborious and costly; plus it is not

Fig. 4.16 Lifting bridge tower inspection platform

easy to undertake inspection of objects at high heights. Therefore, it is necessary to realize close inspection of high piers and towers with devices as simple as possible.

Using bridge inspection vehicles to carry inspectors to inspect bridge piers and towers can solve the problem of long observation distances. However, it will occupy much space on the bridge deck and disrupt the normal traffic. Additionally, due to the limited arm span, it is still impossible to realize full-range inspection of high piers and towers.

The lifting bridge tower inspection platform shown in Fig. 4.16 can inspect bridge towers without affecting the traffic flow. The platform is installed on the back of the bridge tower cable track to realize defects inspection, anchorage replacement, anti-corrosion maintenance, etc. The track is arranged permanently on the bridge, which is applicable for inspecting the towers of highway bridges.

4.2.3 Main Cable Inspection Equipment

The main cable is an important load-carrying component of a suspension bridge, which may be eroded due to natural factors such as wind, rain, freezing, temperature and humidity changes over a long period. Therefore, the main cable and the sling must be inspected and maintained regularly. Currently, the main cables of suspension bridges are mainly inspected manually. Because diameters of the main cables are sufficiently large, there are special passages and guardrails for people to walk on them, and the inspectors can walk along the main cables and record the observed defects at any time. However, the main cables of many suspension bridges often have large vertical height differences and steep slopes, and the site environment can be very complex. Therefore, manual inspection can be dangerous and inefficient, and it is also impossible for the inspectors to reach all parts of the main cables. Also, the inspection and maintenance of main cables are often carried out at the same time, which requires special maintenance equipments. According to the climbing mode, the main cable inspection vehicle can be divided into traction and automatic walking types [3, 6]. These two types of equipment are described in detail below.

Traction main cable maintenance equipment is usually driven by wire ropes. This kind of inspection vehicle has a simple structure with convenient operation, but has some shortcomings such as small load-carrying capacity and poor walking stability. Moreover, due to lack of self-walking ability, the applicability of this kind of maintenance vehicle is limited to specific bridges. A typical equipment is the main cable maintenance platform manufactured by Guangxi Jisen Machinery Manufacturing Co., Ltd. (Fig. 4.17), with a walking speed of 8m/min.

In other countries, traction type inspection vehicles are mainly used for main cable maintenance, for example, the main cables of Forth Road Bridge and M48 Severn Bridge in Britain were inspected and repaired in 2004 and 2006, respectively, using towed main cable inspection vehicles [7, 9] (Fig. 4.18).

Unlike traction inspection vehicles, self-propelled main cable inspection vehicles can move autonomously along the main cable without any traction equipment, such as the vehicles developed by Xinzhu Road and Bridge Company. According to the maintenance needs of the different types of suspension bridges, the Company has

Fig. 4.17 The main cable maintenance platform manufactured by Guangxi Jisen Machinery Manufacturing Co., Ltd.

Fig. 4.18 Maintenance of the main cables of Forth Road Bridge

successfully developed several types of main cable maintenance vehicles with self-moving ability, such as crawler-type main cable maintenance vehicles for the main cable maintenance of Jiangyin Suspension Bridge [3].

Generally, according to the shape and structure characteristics of the main cables, the main bearing component of large suspension bridges, at present, most of the main cable inspection vehicles adopt the suspended hanging basket structure, which has slow moving speed, huge shape, complex structural connection and suffers from difficult installation. The main disadvantage lies in the heavy weight of the inspection vehicle. As the main cables are used as the moving track and stress point, it is easy to cause extra damages to the surfaces of the main cables during inspection operation. Currently, there are a few researches on lighter and more automatic main cable inspection robots.

4.2.4 Tension (Hoist) Cable Inspection Equipment

As one of the most important load-carrying components of a cable-stayed bridge, it is important to inspect the cable accurately. Compared with the conventional inefficient and dangerous manual inspection methods, with poor precision, more and more automatic inspection equipment is used in cable inspection. By crawling on the cable surfaces, damages to the PE (Polyethylene) casing on the outer surface of the cable structure and broken wires inside the cable can be inspected. According to the structural form, the cable inspection equipment is mainly divided into wheeled and multi-legged climbing equipment.

(1) Wheeled Climbing Equipment

The Robotic and Artificial Intelligence Laboratory of Tehran University has developed a climbing equipment called UT-PCR [20] (Fig. 4.19), which consists of a triangular body and six limbs with ordinary wheels, having six degrees of freedom. Motors and PID controllers are installed on the three lower wheels to realize motion control. The three upper wheels are used to increase its stability, and six springs are used to keep the wheels in contact with the rods. Three basic movements can be realized: vertical climbing along the rod, straightening and rotating around the rod.

The cable inspection equipment developed by Professor Wang Xingsong of Southeast University, China, also adopted a wheeled structure [10, 21]. The equipment consists of two face-to-face equidistant modules (Fig. 4.20), forming a closed hexagon to buckle the stay cable. The upper and lower spokes are connected by springs to provide fastening force for the equipment, and the wheel limbs are hinged with the module so that any small obstacles on the cable surface can passed through the spring's deformation. In order to prevent deviation, besides the V-shaped wheel design, the inspection equipment is also equipped with an anti-deviation device. Four anti-deflection gimbal balls (C1, C2, C3, C4 shown in Fig. 4.20) are in contact with the cable body to prevent the inspection equipment from deviating from the ropeway.

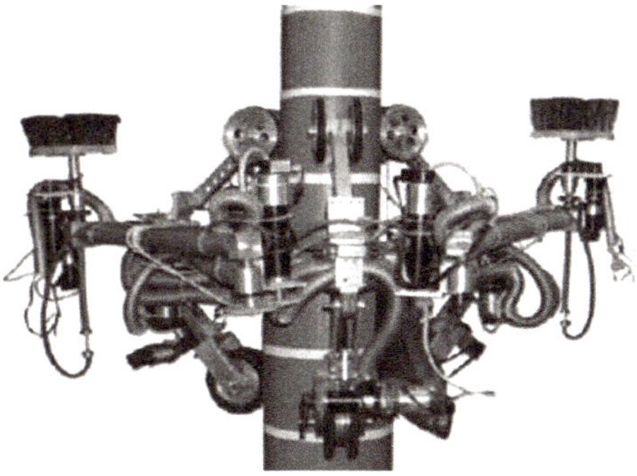

Fig. 4.19 UT-PCR climbing equipment

Fig. 4.20 The climbing equipment developed by Professor Wang Xingsong's team (second generation) [10]

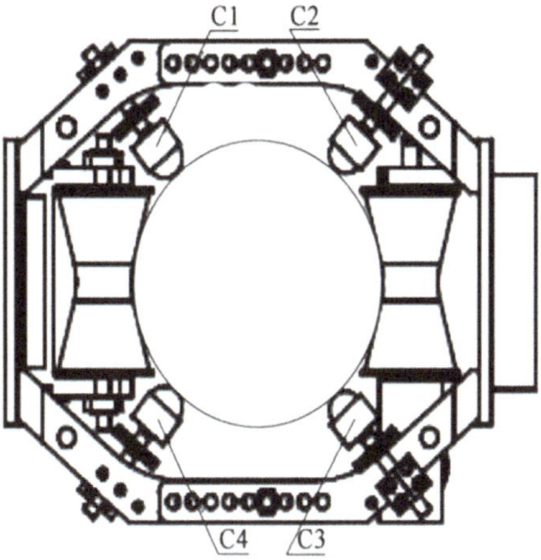

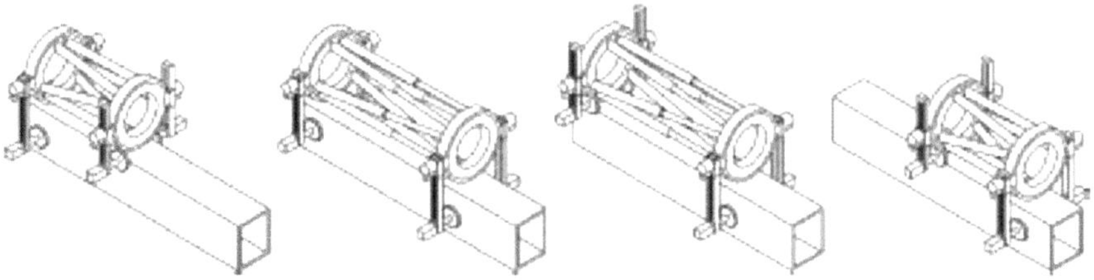

Fig. 4.21 Motion Process of S-G-CPR [22]

(2) Multi-legged Climbing Equipment

Multi-legged climbing equipment uses mechanical jaws to alternately clamp the stay cables when climbin g. Compared with the continuous contact between wheeled equipment and cable surface, multi-legged equipment is based on discrete contact mechanism.

S-G-CPR (Climbing Parallel Robot) is a kind of inspection equipment that can crawl at structural frame joints [22] and perform tasks on pipes and tubular structures. With its working principle, it can also be used for crawling on cables. As shown in Fig. 4.21, the testing equipment consists of two cylindrical bases (lower ring and upper ring), each of which is equipped with two mechanical feet for clamping. The inspection equipment can move in a straight-line by alternately grasping the structure using the front and the rear mechanical feet, and periodically expanding and contracting the linear actuator (Fig. 4.21).

In addition, Shanghai Jiaotong University [23] and Beihua University [24] have also developed their own clamping pneumatic peristaltic cable maintenance robots [23, 24] (Fig. 4.22). Their principle is similar to that of S-G-CPRs, but the difference is that the structure of the inspection equipment developed by Shanghai Jiaotong University [23] and Beihua University [24] encircles the cable body to enhance stability of the inspection equipment on the cable body. Compared with the wheeled equipment, the multi-legged equipment improves its obstacle-crossing ability through discrete contact with the cable surfaces, but its moving speed is limited, and its safety and load-carrying capacity are low.

4.3 Bridge Contact Inspection Robots

Section 4.2 introduces the bridge automatic and semi-automatic inspection equipment which are mature with respect to research and application. The use of these equipment greatly liberated the workforce, expanded the inspection space and improved the inspection efficiency and accuracy. However, with the continuous development of bridge construction at home and abroad, larger bridge volume, more complex bridge structures and more changeable environmental factors require further improvements on the bridge inspection efficiency and quality. Therefore, in order to

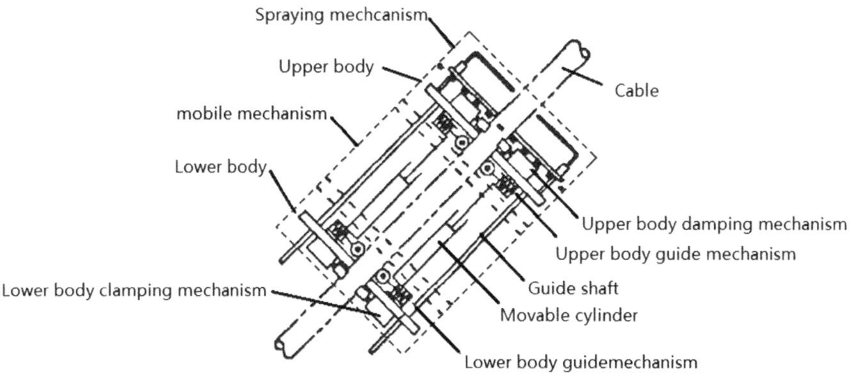

Fig. 4.22 The pneumatic peristaltic cable maintenance robot developed by Beihua University [24]

achieve stronger maneuverability, higher stability, simpler equipment, more optimized control, higher precision and more comprehensive detection content, research of bridge detection equipment is approaching automation, modularization and intelligentization. Many attempts have achieved good results in this research direction. However, due to requirements of technology maturity, R&D cost and operation professionalism, these new attempts only stayed in the stages of model design or prototype testing, and have yet to be put into field tests and application.

4.3.1 *Crawling Inspection Robot for Bridge Deck and Beam Bottom*

At present, most of the apparent defects on the small and medium-sized bridges are inspected by the vehicle-mounted inspection platforms introduced in Sect. 4.2.1. For those bridges with many lanes and wide width, if the above mentioned inspection platforms want to meet the requirements of full detection range coverage, they need to solve the problems of extension length, stability control and lightweight of mechanical arm, which can hardly be addressed. To this end, researchers have come up with another method, that is, to develop an inspection robot that can walk independently on the bridge surface. The advantages of this kind of detection robot lie in its miniaturization, convenient operation and strong maneuverability. At the same time, because it works directly on the bridge surface, it avoids the occupation of driveway on the bridge by the vehicle-mounted inspection platform.

(1) Wheeled Inspection Robots

For steel structures, considering actual inspection requirements and economy, the robots should be attached to the steel structures without consuming electric energy, so magnet should be utilized as an important part of steel bridge inspection robots.

In order to enable the robots with magnet to move freely on the surfaces of the steel bridges, wheeled equipment is naturally adopted.

The robot shown in Fig. 4.23 is called Flexible-based Mobile Sensing Node (FMSN) [25]. The robot is designed for acceleration measurement and can carry accelerometers to designated positions to conduct fixing and measurement of the instruments. FMSN consists of three substructures: two two-wheeled vehicles and a flexible connecting beam. The wheels are wrapped in thin magnets to provide suction, and infrared sensors are placed on both sides of the front and the rear wheels to correct the direction of the movement. The flexible connecting beam is made of spring steel. Compared with the conventional rigid body design, it can firmly attach the accelerometer to the structural surface and help to realize more accurate measurements.

FMSN can move in both front and back directions efficiently, but not in left or right direction, as it is difficult to change its walking route because of the large grounding area of the magnets wrapped on its wheels. A robot named BIREM (Fig. 4.24) was developed to further improve the robot's motion ability [26]. Different from FMSN, BIREM uses frameless wheels with additional magnets. Because the grounding area of the magnet is small, the steering mechanism of BIREM can help to change the left and right motion directions of the robot.

(2) Magnetic Foot Inspection Robot

As mentioned above, wheel design has difficulty in overcoming obstacles, gaps and discontinuous changes in surface gradients. In 2009, Massachusetts Institute of Technology developed Mag-Foot, a steel bridge magnetic foot inspection robot (Fig. 4.25) which can realize long-term adsorption (days or weeks), free walking and basic obstacle passing on bridge surfaces. The robot can be used as both inspection and long-term monitoring equipment, and can change its position on the bridge to acquire more comprehensive bridge information [27].

In order to enable the robot adapts to discontinuous surface gradient changes, researchers designed multi-link legs with magnetic feet. Air gap is used to adjust the

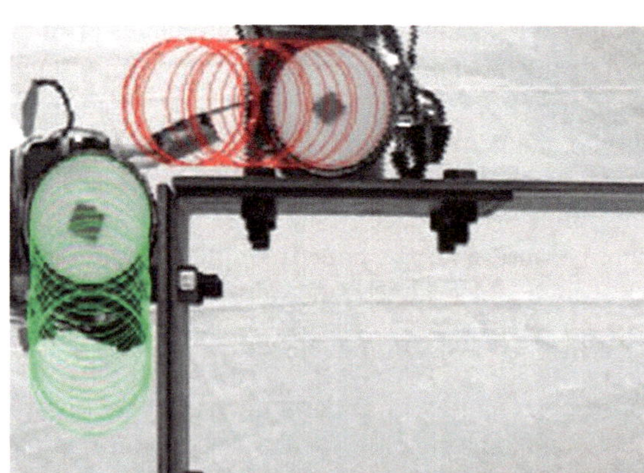

Fig. 4.23 Schematic diagram of FMSN passing through beam-column rotation angle

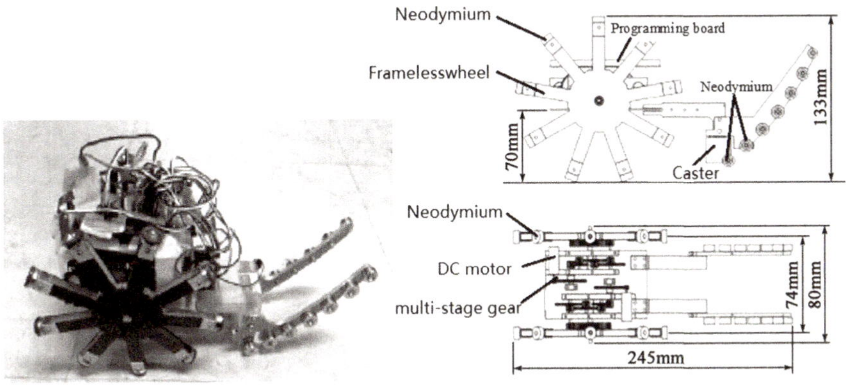

Fig. 4.24 Schematic diagram of BIREM [26]

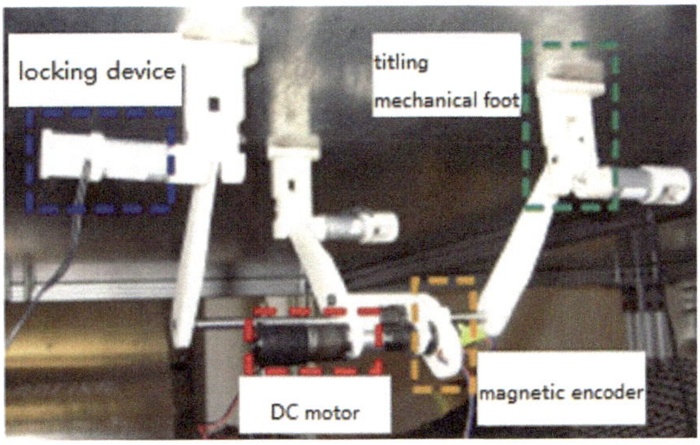

Fig. 4.25 Schematic diagram of Mag-Foot steel bridge magnetic foot inspection robot

force of the permanent magnet foot so that the magnetic foot can be pulled out of the surface. The design of the inclined mechanical foot also makes it easier to pull away or slide along the surface. However, there are many limitations in the current design, for example, the robot can only move on a single relatively flat plane. Additionally, the current design cannot solve the issues of moving between different planes or crossing large obstacles. Even so, the robot already met most of the present needs of bridge surface inspection.

(3) Track-Mounted Inspection Robots

The adsorption function of the robot can also be realized by the track. The water wall climbing robot (Fig. 4.26) developed by Harbin Institute of Technology has two tracks, and some magnetic blocks on each track are in good contact with the wall. The motor drives the harmonic reducer to control the sprocket, which in turn drives the track to realize the robot's movement [28]. This robot can move efficiently on the plane and cross small obstacles, but it is still difficult for the robot to address the issue of discontinuous surface gradient changes.

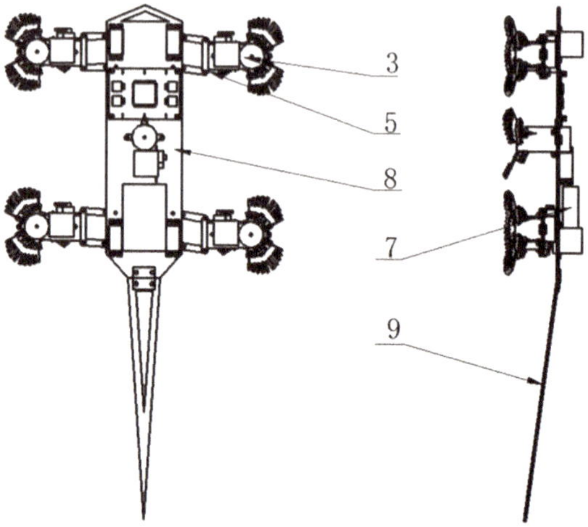

Fig. 4.26 Schematic diagram of water wall climbing robot

Fig. 4.27 Seekur Jr bridge inspection robot

(4) Bridge Deck Inspection Robots

Seekur Jr Bridge Inspection Robot (Fig. 4.27) is equipped with a crack visual inspection camera for detecting surface cracks and other bridge deck defects. It is also equipped with radar and resistive sensors that can penetrate the ground to accurately locate defective parts such as corroded steel bar or degraded concrete in the bridge. The team has tested the robot on four roads and bridges in Nevada, New Hampshire, Maine, and Montana, respectively, and their results showed that its detection accuracy can reach 96%.

4.3.2 Wall-Climbing Robots for High Piers and Towers

To realize fast and economical inspection of high piers and towers of bridges, researchers use wall-climbing robots equipped with visual inspection devices to perform online inspection of apparent defects. This inspection method can flexibly and accurately detect bridge cracks without affecting the traffic on the bridge and therefore has great application value.

Wall-climbing robots are mostly designed based on adsorption mechanism and movement mechanism. The robots can be classified into magnetic adsorption robots, bionic adsorption robots, electrostatic adsorption robots and negative pressure adsorption robots according to their adsorption modes. The magnetic adsorption robots have been introduced in the previous chapter, and the other three types of robots will be introduced below.

(1) Bionic Adsorption Robots

The main working principle of bionic adsorption refers to the microscopic adsorption mechanism of the soles of reptiles such as geckos. The adsorption material used by the robot is made artificially to mimic the fine bristle structure of the gecko's foot so as to uses the intermolecular force to generate the adsorption force. This material can be applied on the walls made of various common materials [4].

On this basis, Carnegie Mellon University has developed a robot prototype, as shown in Fig. 4.28. The robot has a track-mounted structure, in which the track is made manually of bionic adsorption materials, that can be attached to most common smooth wall materials. In addition, the researchers also put forward the scheme of Waalbot robot (Fig. 4.29). Unlike the former, Waalbot uses wheel-leg structure, but similarly, its three soles are adsorbed by using bionic adsorption materials [5].

(2) Electrostatic Adsorption Robots

In recent years, electrostatic adsorption has emerged as a new adsorption technology. Its main principle is to use the adsorption effect between electrostatic charges to generate adsorption forces. More precisely, many electrostatic charges are generated on the material surfaces so as to create opposite charges during contact, thus realizing adsorption.

In 2007, Tokyo University in Japan developed a wall-climbing robot that used this principle [29]. The device for generating electrostatic adsorption force is a flexible electrode plate, which is composed of a plastic film and a conductive metal foil. For the surface of conductive materials, the researchers designed the first prototype

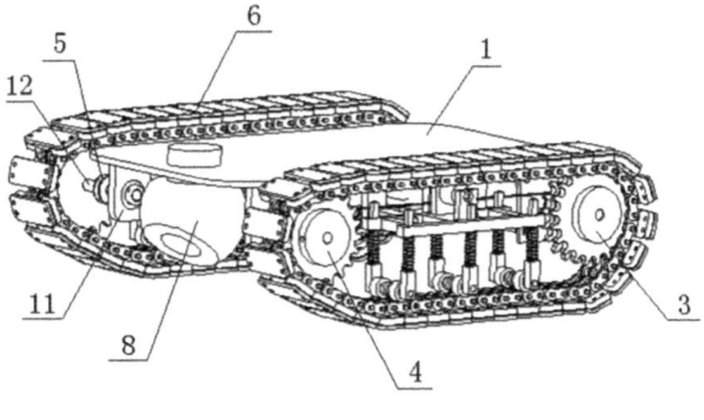

Fig. 4.28 Track-mounted bionic adsorption robot [4]

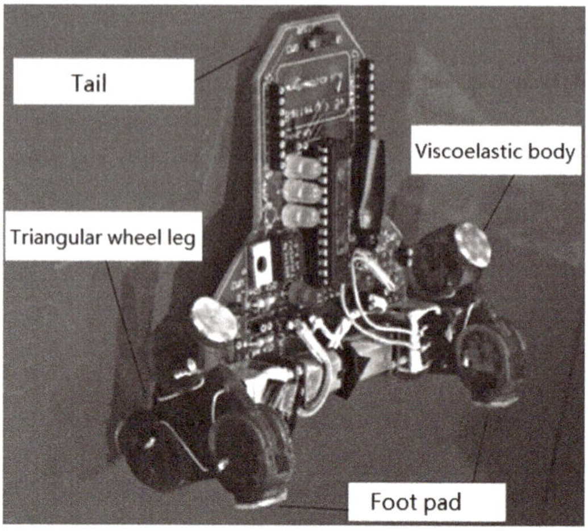

Fig. 4.29 Wheel-leg bionic adsorption robot [5]

(Fig. 4.30) robot that can crawl. Its main body consists of two aluminum frames, and the two electrode plates alternately adsorb to realize crawling movement. For the surface of non-conductive materials, the researchers designed the second prototype (Fig. 4.30). Before using this robot, the measured object's surface had to be pre-charged. The robot is designed as a track-mounted robot which uses a "tail" to maintain its stability in the vertical plane.

Electrostatic adsorption robots are relatively new category of robots that realize lightweight of robots to the maximum extent. However, due to the relationship between materials and mechanism, its adsorption force is inferior to those of other methods, plus its manufacturing process is comparatively complex, so its practicality is presently not strong.

(3) Negative Pressure Adsorption Robots

Negative pressure adsorption is a type of adsorption technique that produces adsorption force by creating a pressure difference between the interior and the exterior

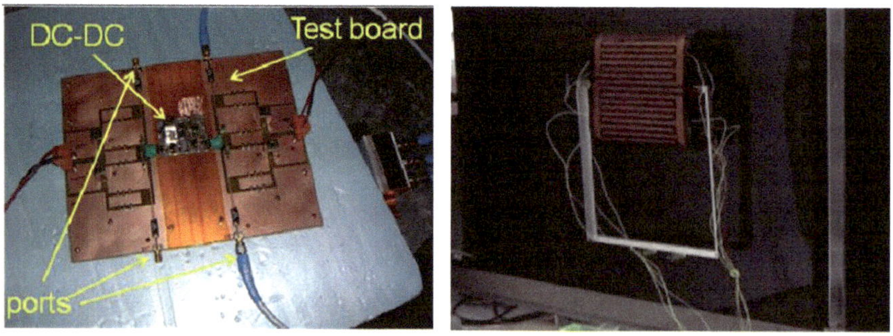

Fig. 4.30 First generation (left) and second generation (right) electrostatic adsorption robots [29]

of the suction cup. The pressure difference created depends mainly on the exhaust device or other ways to create a vacuum state in the suction cup. The commonly used exhaust equipment includes vacuum pump, centrifugal fan, venturi tube, etc. In addition, the vacuum in the suction cup can also be realized by employing propeller to generate a negative pressure inside.

In 2007, Kanagawa University in Japan developed a window-cleaning and wall-climbing robot named WallWalker [30], which consists of a suction cup and two driving wheels (Fig. 4.31). The robot adopts a wet attachment system, which injects liquid (silicone oil) between the sealing ring of the suction cup and the surface of the object to be tested, and uses the sealing function of liquid to improve the sealing effect in the suction cup so as to enhance attachment. On the contrary, the LARVA single suction cup wall-climbing robot [8] (Fig. 4.32) developed by Sungkyunkwan University in South Korea adopts dry attachment mechanism and uses impeller to rotate at high speed to generate a vacuum (Fig. 4.33).

The single suction cup robot has disadvantages in adsorption capacity, safety and wall adaptability, so it cannot ensure smooth wall-climbing operation in all cases. In 2007, Beihang University developed a wall-climbing robot [31], which adopted the Vibration Suction Method (VSM). The negative pressure inside the suction cup is

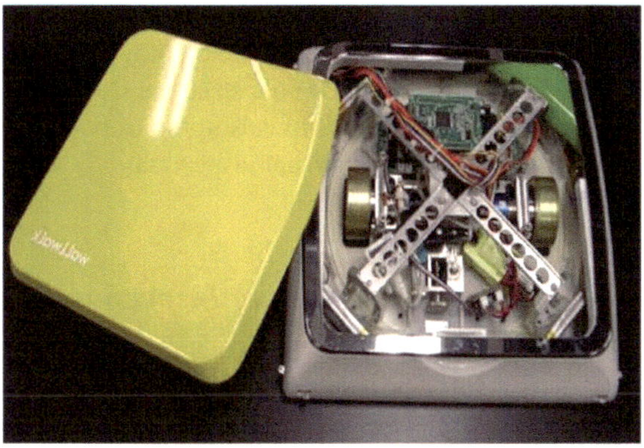

Fig. 4.31 WallWalker robot [30]

Fig. 4.32 LARVA robot [8]

Fig. 4.33 The adsorption robot developed by Beihang University [31]

generated by vibration of the suction cup, which has the advantages of low energy consumption, low noise and high suction stability. The adsorption system consists of a vibration mechanism and an air release mechanism. When the robot needs to be attached, the air release mechanism keeps closed, and when the robot needs to walk, the mechanism starts to release the negative pressure from the suction cup.

It should be noted that the current adsorption robots have high stability and strong motion ability. However, due to restriction of its inspection range, it usually takes a long time to inspect the whole structure using adsorption robots, so it is difficult to inspect piers and pylons by adsorption robots alone, and they should be used in conjunction with manual visual inspection.

4.3.3 Main Cable Walking Inspection Robots

Regarding main cable inspection, there is still no mature automatic inspection equipment at home or abroad. Therefore, developing an effective automatic inspection robot for the main cables is extremely promising. The main cables of a suspension bridge often have long span, steep slope and large diameter, so the cable clamps and suspenders distributed along the main cables plus the crosswind at high altitudes will become obstacles for robots to work on the main cables. To overcome these problems, the authors' team proposed a main cable inspection robot system to realize cable climbing and obstacle crossing, stable movement and accurate measurement of the main cable surface crack and cable clamp slip. Figure 4.34 depicts the design of the proposed main cable inspection robot.

The robot is designed to be wheeled along the main cable's upper surface to avoid obstacles of the suspenders, so it only needs to solve the issue of smoothly passing the cable clamp step obstacles. Secondly, adopting the modular four-axis and eight-wheel independent suspension mechanism increases the number of contact points between the robot and the main cable surface when crossing the step, thus improving

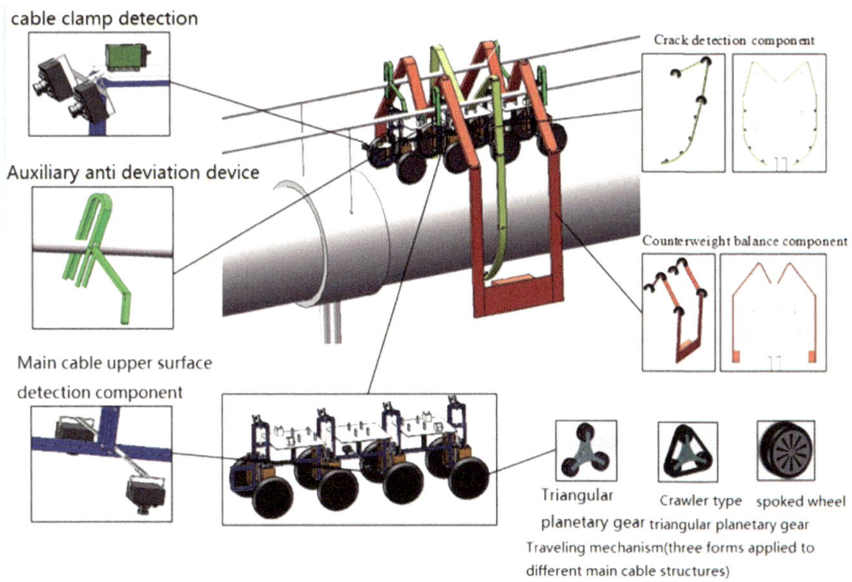

Fig. 4.34 Schematic diagram of the authors' main cable inspection robot

Fig. 4.35 Schematic diagram of snake-like robot climbing [32]

the robot's motion stability. In addition, to realize 360-degree inspection of the main cable surface, the robot is equipped with a three-degree-of-freedom passive lockable robot arm, and the third joint-arm is used as the camera-bearing platform to realize inspection coverage of the main cable's lower surface.

4.3.4 *Pulling (Hanging) Cable Crawling Inspection Robots*

Generally, the cable inspection robots developed have been relatively mature and have been applied widely, thus it is difficult to innovatively make subtle improvements to their shortcomings. This section will introduce a non-mainstream robot structure, besides the wheeled and the multi-legged robots, and methods to solve the wheeled robots' anti-sticking problem.

(1) Snake-Like Climbing Robots

In addition to the two most common robot structures, the wheeled and the multi-legged robots, the snake-like robot is commonly used for cable inspection. This idea was first proposed by Professor Hirese from the Tokyo Institute of Technology, Japan, who developed the ACM-I snake-like robot. On this basis, the snake-like robot has since been further improved [32].

Unlike the conventional motion mechanism adopted by the wheeled and the multi-legged robots, which climb by encircling the cable body, snake-like robots adopt the winding method which generates driving forces by lateral rolling and controls the climbing speed according to the different friction forces generated by changes in the winding tightness. In this way, installation of the robot is relatively easy, and applicable range of the cable diameter is no longer limited by the inherent designed size of the robot. Only the standardized modules need to be continuously added. However, the complexity of this structure is significantly higher than other schemes, and its motion control is much more difficult. Besides, its relatively low adsorption and motion reliability make this kind of robot rarely used in practice.

(2) New Wheeled Climbing Robots

The design of cable-wheeled climbing inspection robot has been relatively mature, and the motion stability, obstacle-crossing ability and anti-deviation ability of the robot have been enhanced. On this basis, Professor Wang Xingsong's team in Southeast University, China, further optimized the robot for its anti-sticking problem.

After long-term use, many defects will appear on the outer surface of the cable, including large area of warped skin. For the low-height warped skin, the robot can directly roll over, but for the high-height warped skin, the robot may be stuck when passing and cannot automatically retry. To overcome this problem, researchers have designed a flexible feedback device (Fig. 4.36). When the retaining ring in the front of the device collides with the warped skin, it moves downward, and the compression spring touches the travel switch, thus providing a feedback signal to the robot's processing system so that it can stop the robot from moving to avoid sticking. Moreover, the diameter of the retaining ring can be adjusted to meet different obstacle-crossing requirements.

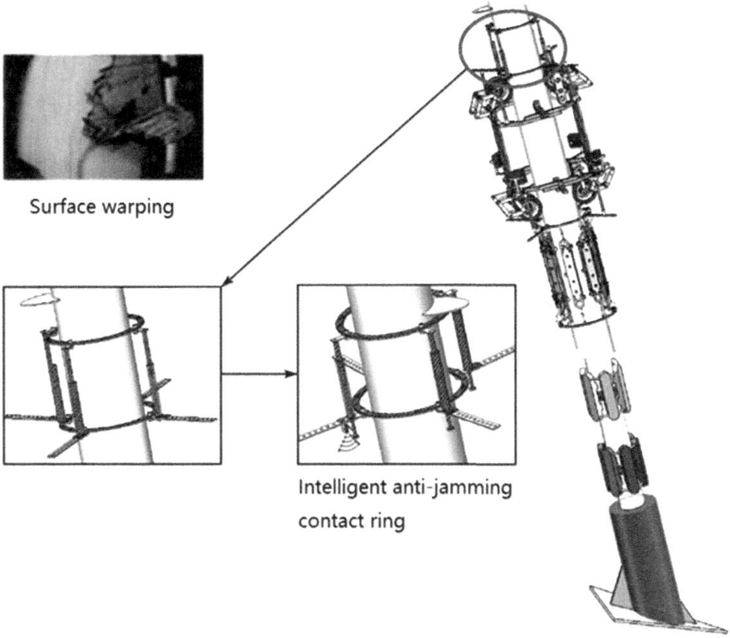

Fig. 4.36 Schematic diagram of flexible feedback device

4.4 Automatic Inspection Equipment for Bridge Underwater Structures

As introduced in Sect. 4.1.2, the current practice of underwater bridge inspection is mainly manual visual based, which requires trained divers to carry lighting and imaging equipment (searchlights, magnifying glasses, digital cameras, etc.) to perform periodic "visual" inspections on the structures. The main advantage of manual visual inspection is that trained divers can carefully inspect the underwater structures on condition of sufficient safety protections. Therefore, this method has good applicability in shallow waters with good visibility. However, in deep, turbid or fast-flowing waters, manual visual inspection is extremely challenging and even life-threatening for the divers. Therefore, in order to solve this problem, researchers have developed many underwater inspection equipment to replace or partially replace the manual work.

4.4.1 Inspection Techniques for Bridge Underwater Structures

This section focuses on the sonar-based bridge underwater automatic inspection technique and the array of camera-based bridge underwater automatic inspection technique.

(1) Sonar-Based Bridge Underwater Automatic Inspection Technique

Sonar (Sound Navigation and Ranging) is an underwater acoustic equipment that uses sound waves to detect, locate, track and identify underwater targets. According to different working principles, sonar can be classified into: (1) passive sonar, which passively receives and processes the radiated noise information or sonar signals emitted by underwater targets; and (2) active sonar, which detects underwater targets by transmitting ultrasonic signals with certain characteristics and detecting the echoes reflected by any obstacles. The multi-beam sonar, commonly used in underwater inspection, belongs to the latter type.

Multi-beam sonar was first put forward in the mid-1960s at an academic exchange seminar held by Woods Hole Oceanographic Institution in the United States. It was mainly used for sounding [33], but in recent years, with continuous improvement of the strategic position of the ocean, human activities in exploring the ocean have been increasing and thus research on sonar inspection technique has gradually attracted the attention of scholars and engineers from all over the world. At present, there are two ways to use multi-beam sonar for underwater inspection: (1) to generate a series of distance and angle measurement information, that is, sounding sonar; and (2) to generate sonar images corresponding to the inspection environment, that is, imaging sonar. Multi-beam sonar has been widely used for seabed topography and water body imaging [34].

Many sonar devices are used in underwater structure inspection, such as sonar detectors, fan sonar scanners, multi-band sonar instruments, etc. Sector scanning sonar technique is an advanced sonar detection technique which has wide range of applications as it is less affected by underwater visibility and has high imaging clarity. Therefore, the equipment can be used to measure and identify erosion subsidence, filling area, exposed pier foundation, debris accumulation and other defects of underwater structures. Sector scanning sonar equipment can also guide the divers to find any possible structural distresses before or during inspection, thus helping them to bypass the potentially dangerous areas. Underwater sonar scanning technique can detect depth change of the underwater riverbed and its influence on the bridge structure foundation. After sonar scanning, detailed images can be obtained by analyzing the sonar signals. Through the images, inspectors can analyze the distress status of the bridges and make correct scientific judgments on arrangements for further inspections so that severity of the defects can be more effectively detected.

In addition, many underwater structures may be affected by the external environment or sudden disasters during service [35], so it is necessary to evaluate the safety and reliability of these structures. Underwater non-contact imaging technique

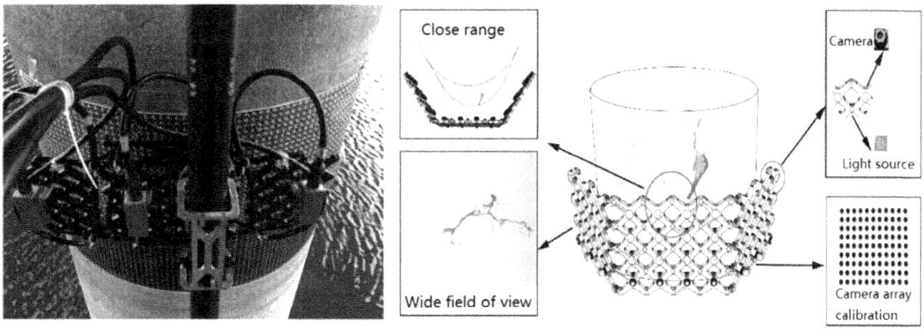

Fig. 4.37 Non-contact camera array detection device with a large field of view

has become an important means of non-destructive testing of underwater structures because of its advantages of fast inspection and intuitive results [36]. However, whether the collected images are directly processed or reconstructed by computer vision technique, clear original structural surface images are the key to subsequent processing. Therefore, obtaining high-definition structural surface images is the basis of structural safety and reliability evaluation under low visibility, especially in turbid water. To this end, He Xiaoyuan and Liu Cong from Southeast University, China, proposed a method for dynamic surface testing of large structures at close range, which is mainly based on camera array devices. The camera array device uses multiple cameras arranged in an array instead of a single camera to acquire images. Because of the use of multiple cameras, many functions that a single camera does not have can be realized. Figure 4.37 shows the non-contact camera array inspection device with a large field of view used in this system.

4.4.2 Unmanned Submersibles for Bridge Underwater Structure Inspection

Unmanned submersible is a device that can perform some underwater bridge inspection tasks. They look more like miniature submarines with shapes designed to adapt according to their specific underwater tasks. Currently, most unmanned submersibles are frame structures with rotating slender bodies similar to submarines. For the unmanned submersibles to work in various complex marine environments, unknowns and challenges such as severe interference with the submersibles' movement and control, communication between the upper and lower computers, and navigation and positioning of the unmanned submersibles need to be solved. This is the main issue hindering the development of unmanned submersibles [37].

European and American countries began to develop marine technologies very early, and the US Navy devoted itself to the research and development of underwater vehicles in the 1950s and 1960s. In 1953, the first submersible, "Poodle," was launched. By 1956, the CURV system developed by the United States successfully recovered a hydrogen bomb in the ocean, which attracted worldwide attention. Manned submersibles and unmanned remotely operated vehicles were developed, including cable-operated submersibles, underwater crawling submersibles, towed submersibles and cable-free submersibles [38]. In 1980, France built an unmanned cable-free submersible, named "Reverse Whale", which carried out more than 100 deep-diving operations in the Pacific Ocean and completed major tasks such as the Pacific Ocean submarine canyon survey and the mid-ocean ridge survey [39]. The research and development of unmanned submersibles in China began in December 1985, when the first unmanned submersible, "Haiyang-1," made its maiden voyage successfully, thus opening the prelude to the development of unmanned submersibles in China. In 1994, China's first cableless autonomous unmanned submersible, "Explorer," was successfully developed, fulfilling the purposes of search and investigation (Fig. 4.38). The successful development of Explorer greatly narrowed the gap between China and the developed countries in this field, marked the maturity of unmanned submersible technology in China, and made important contributions to the country's exploration and development of marine resources. In 2009, the Institute of Underwater Engineering in Shanghai Jiaotong University developed the unmanned remotely operated vehicle "Hailong" (Fig. 4.39) with the strongest functions and the deepest diving capacity in China at that time, and successfully applied it to deep-sea hydrothermal scientific research task of the third leg of "Dayang No.1". Although the research and development of unmanned submersibles in China started late, with the country's strong supports, the technology of unmanned submersibles in China is becoming more and more mature.

Fig. 4.38 "Explorer" cableless autonomous unmanned submersible

Fig. 4.39 Unmanned remotely operated submersible "Hailong"

4.4.3 Bridge Underwater Structure Inspection Robots

The defects of bridge underwater structures are often concealed, directly affecting the structure's overall safety. However, with the gradual development of bridge construction towards "super-long", "super-large" and "super-deep", the conventional underwater inspection technology can no longer meet the needs of inspection of large-scale bridge underwater structures. In the face of complex environments such as turbid waters and rapids, the existing detection methods have the following three main shortcomings:

(1) In turbid water environment, only the outline information of apparent structural defects can be obtained, but the whole and the local underwater distresses cannot be accurately measured;
(2) Facing rapid flow environment, the existing underwater vehicles still have great difficulty in stability and control; and
(3) Because underwater distresses usually have been neglected for a long time, the number of existing image samples is small, the image features are fuzzy and the resolution is low, so the accuracy of defect recognition in the actual detection process still needs to be improved.

Given the above problems, Professors Wu Gang and Song Aiguo of Southeast University, China, designed a new underwater robot for bridge substructure inspection based on human–computer interactive teleoperation, which focuses on solving the above-mentioned three major problems: difficulty in identifying apparent defects of underwater structure due to deep turbid waters and rapids, poor stability and control of the inspection equipment, and low accuracy of defect measurement. Adopting the non-contact optical positioning technique and the contact visual sensing technique, the whole large field of view (24 cm × 100 cm) and the local high precision (0.1 mm) defect detection in turbid water can be realized. With the help of shared control method combining autonomous and bilateral teleoperation (control accuracy of the manipulator reaches 0.5 mm), a high-stability and heavy-load underwater robot platform (control accuracy is 10cm) is developed. After underwater defect detection, accurate segmentation and information synchronous recognition model was independently established and utilized by the team, and the accuracy of defect image recognition with small samples and low resolution was improved (better than 90%). Therefore, efficient and reliable underwater structure detection can be realized, effectively improving the overall efficiency of bridge maintenance and greatly lowering the pressure of underwater structure maintenance in China.

The overall structure of the developed device is shown in Fig. 4.40. The equipment can be divided into three main parts. The first part is for multi-level detection of bridge distresses, including the optical array cameras and the contact vision sensors, which can realize non-contact measurement of the overall distresses of underwater components and contact measurement of the local defects of the components. The second part is the underwater multifunctional detection robot platform, including the robot body, the multifunctional mechanical arm and the force sensing and cleaning devices. The third part is the intelligent distress analysis system, with functions such as data enhancement and distress analysis.

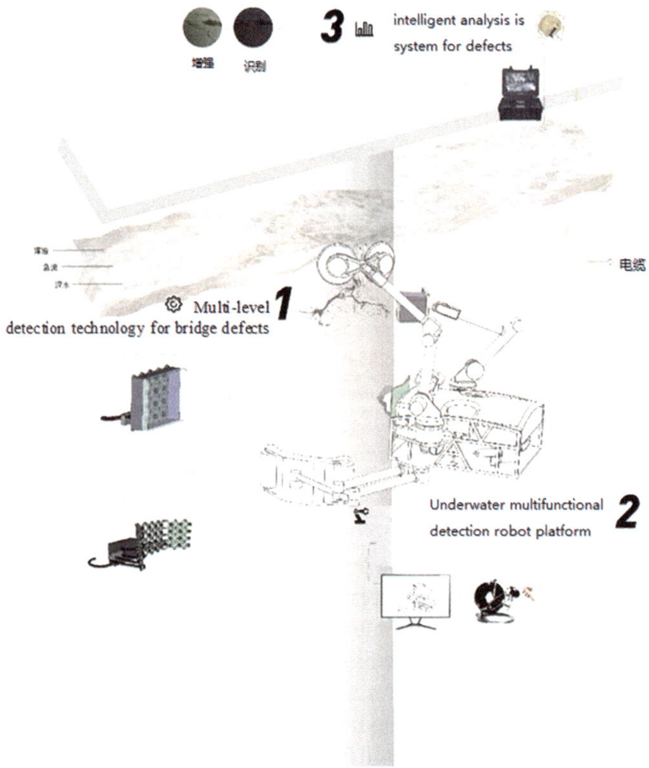

Fig. 4.40 A new underwater inspection robot for bridge substructure based on man–machine interactive teleoperation

The goal of the developed underwater inspection robot is to realize inspection of underwater structures within 100m water depth, with defect inspection accuracy better than 0.1 mm. Operation of the equipment is as follows: the underwater inspection robot dives to the vicinity of the structural surface with a connected cable, ensures its operation stability through the overall power device, and realizes "visual inspection" and "exploration" of underwater distresses by using the optical array cameras and the contact vision sensors respectively. The image, video, force feedback and other information generated in the whole inspection process can be transmitted back to the host computer through the connected cable, and the specific information of underwater structure defects can be output in time by using the distress analysis software.

Specific contents of each component of the underwater inspection robot are:

(1) Multi-level Inspection Technique for Bridge Defects

The multi-level inspection technique included in this instrument pertains to heterogeneous and homomorphic multi-level distress perception, where non-contact and contact (heterogeneous) multi-level perception technique is adopted to realize a large field of view and high-precision description of distressed parts (homomorphic), overall sketchiness and fine local identification (multi-level).

Whole field perception of non-contact camera array: Under turbid water condition, the clarity of optical imaging will be affected by the test distance. In order to enlarge the detection field of view while ensuring clarity, the camera array equipment for direct optical imaging in the complex underwater environment is developed based on integration of array cameras. According to analyses and experiments, to keep the same field of view, the test distance of a 4×4 array pinhole imaging camera will be shortened by four times compared with that of a single camera, which can significantly improve imaging clarity. However, it will lead to a significant increase in the amount of image data generated. Therefore, a high-speed disk array is directly connected with the bus to realize fast and real-time transmission and storage of image data generated by the large array cameras. An additional array illumination source is also designed to improve the underwater ambient illumination and ensure uniform brightness of each camera image. The full-parameter calibration method proposed for camera array is studied, via which the large-field non-contact detection of underwater structure surface is finally realized.

Detailed contact identification of local distresses: local and subtle distresses, such as cracks, damages and pits, which are commonly observed in underwater piers, often form easily; but due to their small sizes and the harsh surrounding environments, these distresses usually cannot be detected in time and accurately. Contact vision sensors use contact vision technique to realize detailed descriptions of the distresses, i.e., to actively contact the concrete through the self-developed high-performance flexible transparent hydrophobic elastomer and to use the elastomer's softness and transparency to reproduce the distressed surface. Then, with the help of high-resolution imaging equipment and an excellent imaging environment created in advance, the texture status in its field of view is collected by the contact vision sensor in the form of high-definition images. Because the elastomer itself has good chemical stability and deformation recovery ability, recovery and reuse of the elastomer after reproducing the distresses can be achieved by optimizing the mixture ratio of raw materials and controlling the interface contact force. Many laboratory tests with real environment simulation suggested that the sensors designed by this technique can realize rapid identification and quantitative analysis of concrete surface distresses (with cracks accurate reaching 0.1 mm) in underwater turbidity or low visibility environment (turbidity above 400NTU). Therefore, it effectively overcomes the technical problems of difficult imaging and low precision encountered by the conventional manual detection or other imaging equipment (sonar, array camera, etc.) in the underwater environment and provides a brand-new idea and technical means for underwater pier distress detection. In addition, with the idea of "array integration," a set of sensing equipment for multi-local distress detection of underwater structures can be developed. The equipment can consider the influences of pier size and shape on the sensor array layout, and realize the detailed local distress detection of various underwater piers by rationally designing deployable mechanical devices and optimizing the size and weight of the single sensors.

Conceptual design of the contact vision sensor is shown in Fig. 4.41, which mainly includes a PDMS elastomer, shell, high-resolution camera, LED strip and various circuit designs and image storage devices. The way to realize contact vision sensing is to actively contact the surface of the underwater structure with an elastomer and to reproduce the texture of its surface by using the softness and transparency of the elastomer itself. Elastomer is a high-performance, transparent, hydrophobic organosilicon compound independently developed by the authors' team. By exploring the deformation sensitivity of elastomer materials with different proportions and thicknesses to the same kind of distresses, the optimal proportions and sizes suitable for various distresses can be obtained. In the shell design, lightweight and high-strength titanium alloy and transparent acrylic plate are used, an elastomer adheres to the outside of the shell, and the inside is composed of a camera and a parallel light source imaging environment. In order to carry out all-around detailed local distress detection, the whole equipment is designed to adopt an array layout. At the same time, it is equipped with a terminal display for real-time image transmission and storage interface of an external memory card, which is convenient for qualitative and quantitative direct judgment of detection results. Macro range cameras with high precision, low power consumption and high frequency are adopted. The camera should also have excellent automatic adjustment and focusing functions to ensure the quality of captured images. In the design of the internal lighting circuit, it is proposed to use an LED strip on four sides oblique at 45 degrees to form a parallel light source environment to ensure that there will be no obvious bright spots in imaging.

In addition, the equipment will also be equipped with a display terminal for real-time dynamic image transmission and an image storage module for data storage. The display terminal is intended for the operators to know, in real time, the underwater position of the equipment and the texture status of the structural surface in the area. The image storage module is intended for saving the collected images in the memory card without loss to facilitate later image processing and to form objective vouchers.

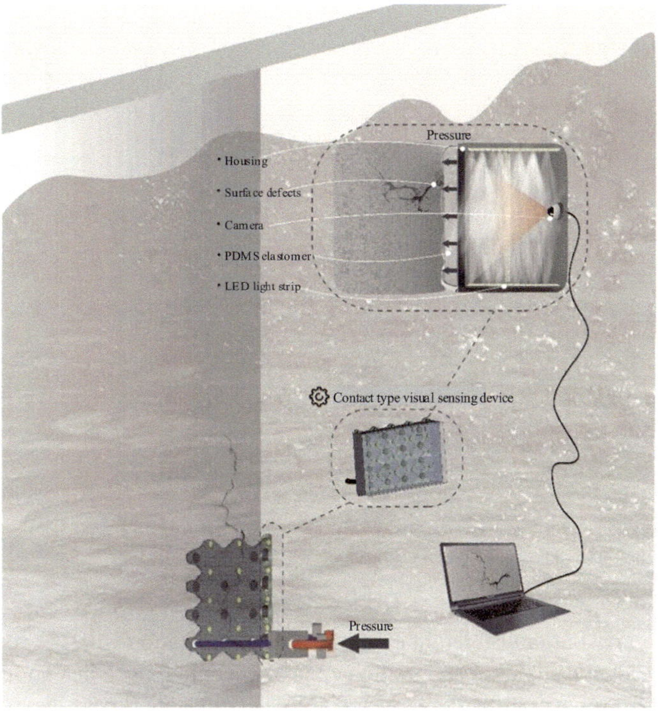

Fig. 4.41 Conceptual design of a contact vision sensor

(2) Underwater Multifunctional Detection Robot Platform

Existing underwater vehicles are mostly used for searching and detecting in open waters. A few of these vehicles that detect underwater structures can complete their detection tasks in simple water flow environments, and most of them adopt the non-contact image acquisition techniques. However, it is difficult for the conventional underwater vehicles to collect effective image information for underwater structure detection in complex water flow environments. Therefore, aiming at improving the performance of underwater vehicles, this project designed a new mechanism that can coordinate control of the robot for use in actual detection.

Design principle of the underwater robot is shown in Fig. 4.42, which includes the following main parts: underwater robot body, clamping device, contact array camera manipulator, non-contact array camera, underwater structure surface attachment removal manipulator and pan-tilt equipped with high-power spotlight and high-definition camera.

The underwater vehicle is equipped with six chinese high-thrust underwater thrusters, four in the horizontal plane and two in the vertical direction. Redundancy design is adopted for the horizontal plane, with a fault diagnosis and fault-tolerant control system. Hence, when a single thruster fails partially or completely, it can

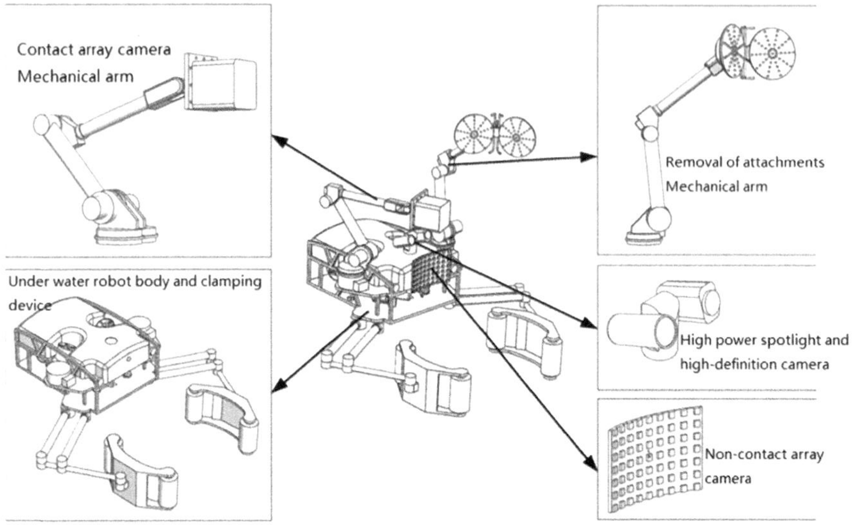

Fig. 4.42 schematic diagram of underwater vehicle

adaptively carry out fault-tolerant control to ensure operation with the conventional sensors for robot navigation and motion control, including the conventional sensing equipment for robot control, depth meter, altimeter, ultra short base line (usbl), electronic compass, doppler velocity log (dvl), inertial navigation system (integrated fiber optic gyroscope, accelerometer, gps, etc.), sonar, etc. The clamping device is used for anchoring the robot in operation, and includes a three-degree-of-freedom parallel manipulator and a roller clamping module. Using a parallel structure can effectively improve rigidity of the manipulator, making the anchorage more reliable. Using a roller clamping module can make the robot move around the pier under the driving force, reduce the anchoring time on the same running horizontal plane, and improve work efficiency. The mechanical arm and the roller clamping module are connected by elastic parts to provide certain adaptability for the clamping device. The operation manipulators, i.e., the contact array camera manipulators and attachment removal manipulators shown in Fig. 4.42, are equipped with end operation tools to complete contact inspection and cleaning as required. It includes two six-degree-of-freedom manipulators distributed on both sides of the robot. A six-degree-of-freedom manipulator can ensure a larger working space to improve work efficiency. The tactile force feedback of the manipulator is mainly provided by a six-dimensional force sensor installed at the end. The equipment for removing attachments on the structural surface consists of an air jet cleaning part and a rotating brush head. Air jet parts are mainly used to clean up dust, aquatic plants and other superficial stains, and a ground air pump provides the gas needed. The rotating brush head cleans stubbornly attached stains such as shellfish. Materials of the two rotating brush heads are stainless steel wires, and the rotating planes are arranged at a certain angle to adapt to the circular arc surface of pier columns. The two brush heads have opposite rotation directions, and the reaction force and torque generated during operation can cancel each other. The pan-tilt is equipped with a high-power spotlight, and the high-definition camera is installed above the front of the robot, facing the workspace of the mechanical arm, in which the high-definition camera provides visual feedback for teleoperation. In contrast, the high-power spotlight ensures a good visual field in the turbid and complex underwater environment.

(3) intelligent Distress Analysis System

Overall block diagram of the intelligent underwater distress analysis system is shown in Fig. 4.43, which includes three main parts: image acquisition, data enhancement and intelligent distress analysis. Image acquisition uses an array camera and contact vision to obtain the image information. Data enhancement can be divided into two parts: image enhancement and image mosaic. For lens distortion and perspective distortion caused by the lens and the scene, camera calibration and image features are combined to eliminate them. Considering that poor water quality and complex water flow environment will corrupt the camera array images with noise, shadow, light spot, etc., it is necessary to reduce the influence of noise by time domain averaging image denoising method and use multi-scale Retinex color constant image enhancement method to improve image clarity. Based on the calibration parameters of the array camera, a multi-refractive index underwater pinhole imaging model is constructed, and the array camera image registration is realized based on image features. The multi-resolution spline method is used to complete the fusion and mosaic of array images, which finally provides the data sources for distress identification and analysis of high-definition array images.

Distress recognition and segmentation include two parts: semantic segmentation based on a semantic segmentation network, and post-processing of recognition and segmentation. Aiming at the problem of high resolution but few samples after image enhancement, a deep transfer learning model is adopted to reduce the dependence of training on the number of samples. The deep transfer learning model utilizes the features of underwater distresses similar to those of aquatic distresses. The pre-trained aquatic distress recognition model is used as the feature extractor of underwater distresses, which can effectively extract distress information. In view of the poor definition of distress boundary and the unstable accuracy of contour segmentation caused by complex underwater environment, the method of combining image enhancement with conditional random field post-processing is used to optimize the result of distress recognition and improve the accuracy of contour segmentation. Considering that the distress boundary is unclear, the image enhancement method can be used to improve clarity of the input distress. The Gibbs binary potential function is used to predict and extract the boundary between the distress pixel and background pixel to segment the distress accurately. By combining the coordinate information of the distress relative to the camera, and the water depth information provided by the underwater robot, distress reconstruction can be realized, and an accurate information source can be provided for distress analysis.

Compared with the conventional manual detection and the local automatic detection equipment (sonar equipment, array camera, etc.), the intelligent underwater

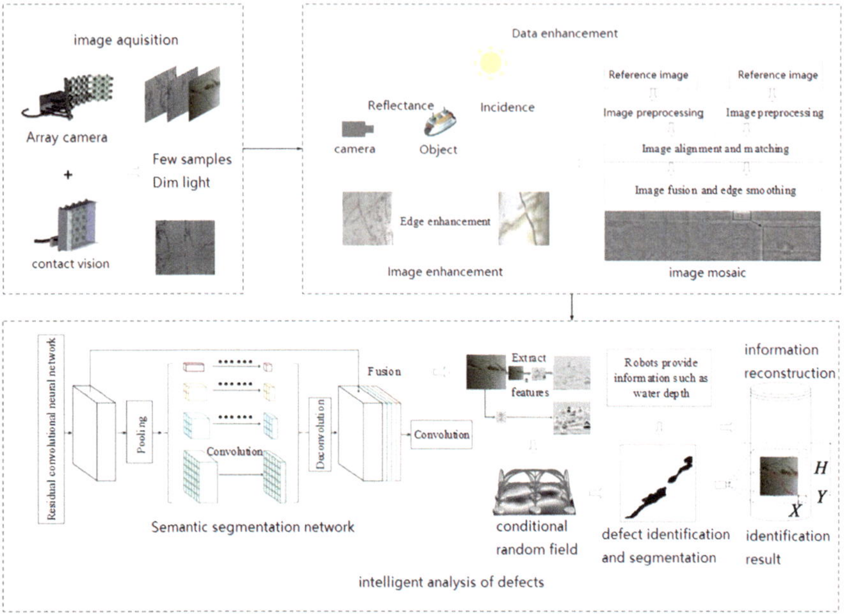

Fig. 4.43 Block diagram of an intelligent analysis system for underwater distresses

distress analysis system integrates the advantages of various detection equipment, adopts the concept of machine intelligence, overcomes the influence of extreme underwater measurement environments, and can realize underwater detection with a large field of view, close distance, refinement, and intelligence.

4.5 Automatic Inspection by Unmanned Aerial Vehicles

UAVs (Unmanned Aerial Vehicles) have been widely used in agriculture, power supply, transportation and other industrial applications. In agriculture, UAVs have been used in crop growth and health assessment, crop yield forecasting, natural disaster assessment and other aspects [40]. In power supply, UAVs have been used to monitor the running status of poles and towers, ground wires, insulators and other components [41]. In transportation, UAVs have been used for traffic monitoring and accident detection at road intersections [42]. For bridges, applications of UAVs greatly improve their safety and detection efficiency [43]. The semi-automatic inspection equipment and robot inspection equipment mentioned above in this chapter are mostly used in specific bridge applications. Use of UAVs for detection is a comprehensive automatic inspection technique for bridge infrastructures, hence this section introduces UAVs for bridge inspection including UAV navigation technique and applications in bridge detection.

4.5.1 *Overview of Bridge UAV Inspection*

(1) Characteristics of Bridge UAV Inspection

Regular inspection of bridges is crucial to ensure its safe operation. Regular inspection refers to daily inspection of technical conditions of the bridges and their ancillary facilities, finding structural defects in time and undertaking maintenance. UAVs can replace human workforce to inspect those parts of bridges that are difficult to reach by manual inspection, and can repeatedly and efficiently sample the details. Furthermore, UAV inspection can be carried out without closing the roads, thus ensuring normal operation of traffic [44].

The authors' team [45] verified the feasibility of using UAVs for bridge inspection through on-site inspection of 13 bridges in Japan. Results of the UAV inspection are compared with results of the reported manual inspections, and shown in Fig. 4.44. It can be seen from the figure that the bridges managed by Japan Ministry of Transport, with sufficient financial support, large inspection machines and experienced staff, reported a large number of damages through a more detailed inspection. However, the number of distresses detected by the local government inspection teams is relatively small. Although UAVs only found 30% of the total bridge distresses, for beams, columns and piers, they detected more distresses than the local inspection teams. At the same time, the study also suggested that generally more damages are more likely to be found by UAVs on bridges with larger spans because these bridges have more space for UAVs to fly under them; where else manual inspection is often difficult to cover all these places. Overall, the results show that although there are some limitations in using UAVs for bridge inspection, generally UAVs shows great potential for automatic bridge detection.

(2) Bridge UAV Detection-Related Technology

The workflow of bridge UAV inspection mainly includes establishing user-oriented comprehensive three-dimensional model of the bridge, UAV automatic inspection route setting, data and image acquisition, automatic computer analysis, statistical data management, display, etc. It involves UAV navigation and positioning,

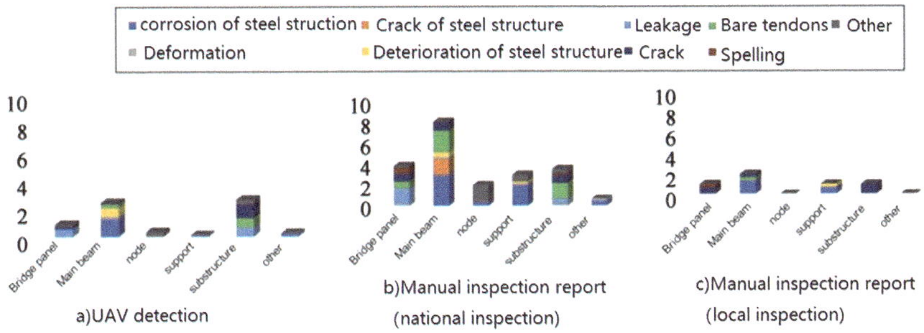

Fig. 4.44 Number of bridge distresses detected by manual detection and by UAV [45]

UAV high-definition photography, computer deep-learning image recognition, data visualization and management, and other related technologies.

(1) Navigation and Positioning of UAV

Due to adverse effects concerning the bridge structure or the surrounding environment, the GPS signal is weak at the bottom or sides of the bridge beams. Therefore, route planning of UAVs based merely on GPS greatly restricts the inspection of some bridge parts, especially for inspecting the bottom of the bridge beams. Application of computer vision in UAV flight directions is mainly intended to obtain peripheral target images through vision sensors mounted in all directions of the aircraft. After digital image processing and feature point extraction, the image coordinates of the target are obtained, and then the combination of spatial geometric parameters and position information of the measured object is computed to realize the aerial positioning of the UAV.

(2) UAV high-definition photography

In bridge data collection, UAVs can obtain two forms of information: dynamic video and static photos. To meet the requirements of later computer vision processing, there are certain requirements of the signal–noise ratio, sensitivity, pixel, and dynamic range of PTZ cameras.

(3) Computer Based Deep Learning Image Recognition

After UAV data acquisition and shooting of the bridge details, it can manually view the acquired image data one by one. However, this cannot fully realize stable, rapid and accurate defect identification. Therefore, with the help of computer based image recognition technique, the types of cracks and distresses in the collected images can be automatically identified. Autonomous recognition based on deep learning is intended to collect a large number of defect images and extract the samples. After training the neural network, the computer provides the defect feature model independently, which is used for defect recognition of the newly collected images. This technique requires a large number of defect samples and needs a database, so investment in the early stage is large. However, with the increase in sample size, the recognition accuracy will continuously improve and finally meet the required detection accuracy.

(4) Visualization and Management of Data

In the past, bridge models were design-oriented, so it is generally impossible to reflect the bridges' detection, monitoring data or performance evaluation results to the user-oriented interface. Combined with three-dimensional real-life BIM model of the bridge built in the early stage, defect locations on the bridge can be accurately marked and correlated [46]. The location and severity of the bridge distresses can be quantified with unified standard data to scientifically evaluate the bridge health status. According to the historical detection data stored in the background database, the specific distribution, severity and other detailed information of the bridge distresses can be managed, maintained and tracked, then the existing historical data can be summarized, and finally the bridge detection comprehensive report can be formulated so that the scientific nature and accuracy of the bridge health management are thus be improved.

4.5.2 UAV for Bridge Inspection

The conventional UAVs face many problems when directly inspecting bridges, such as insufficient accuracy in collecting distress images in the narrow space at close range, and its obstacle avoidance system makes it impossible to shoot when closing on the bridge surface. For the development of bridge inspection, customized improvements of UAVs have become a key link in the application of UAVs for bridge inspection.

(1) Contact UAV

Contact UAV inspection system consists of two parts: (1) UAV's main body, including the UAV fuselage and camera, and (2) ground station, including the wireless video transmission equipment, remote controller and smartphone [47]. The UAV main body adopts the layout of four rotors. The composite frame, brushless motor, flight controller, battery and other components work together to realize the UAV normal flight function. To realize the adsorption function, the UAV is equipped with light wheels extended out of the fuselage at the top and the front to contact the structural surface so that the UAV can stick to the bottom of the bridge or pier to detect cracks at close range. Two horizontal motors are also arranged on the front of the fuselage to provide horizontal thrust. Once the UAV touches the structure surface, such as the pier, the horizontal motors are triggered and pressed the UAV against the structure surface. This special adsorption mechanism combines the advantages of UAVs and wall-climbing robots, which can help to fly to any position quickly and carry out detailed inspections close to the bridge surface. This adsorption detection practice not only realizes high-precision crack detection but also reduces operation difficulty and danger during the inspection process. The ground station of the contact UAV system utilizes a lightweight end-to-end deep learning model and mobile platform transplantation technology. An Android application integrating crack identification and width measurement functions has been developed to provide real-time transmission of the UAV detection video via the smartphone, real-time crack identification and rapid measurement of the designated crack width, and greatly improves the crack detection efficiency. Figure 4.45 is a schematic diagram of the contact UAV.

(2) Crawler UAV

The authors' team developed the crawler UAV [48] shown in Fig. 4.46, which consists of a fuselage, image acquisition module, crawler-type bracket and connecting device. The crawler-type bracket comprises the crawler and the bracket. The bracket comprises two parallel frames, and a plurality of rotating rods are connected between the two frames. The crawler sleeve is arranged on the rotating rods, and the body

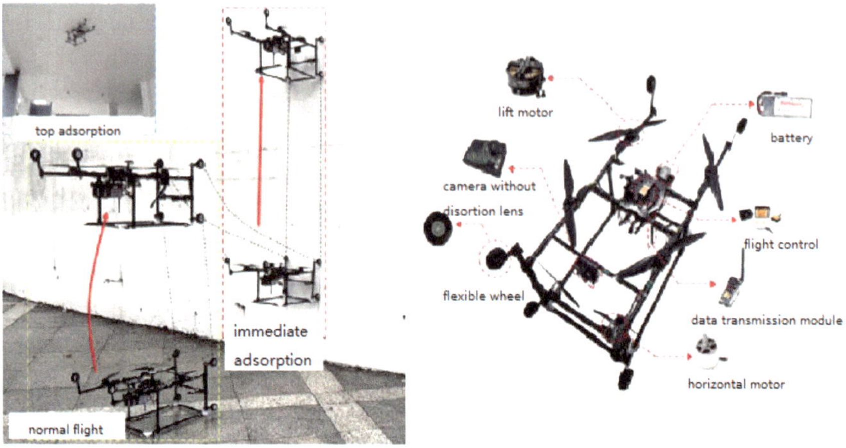

Fig. 4.45 Contact UAV [47]

Fig. 4.46 Schematic of the Crawler UAV [48]

is connected with the frame through the connecting device. The image acquisition module carried by the crawler UAV is responsible for acquiring images of the bridge inspected. The crawler can stick to the bottom of the beam and move at the same time, and the UAV is always in a stable state through rotation process of the rotating rods. Stability of the UAV is thus improved when moving at the bottom of the beam. This type of UAV is suitable for shooting components (such as bearings) and realizing image acquisition in narrow spaces under the bridge.

4.5.3 UAV Navigation Technique for Bridge Inspection

Bridge inspection is a periodic work that has a relatively fixed patrol path. When carrying out bridge inspection, flight path of the UAVs can remain unchanged, so it is of great practical significance to study the navigation techniques suitable for bridge inspection. GPS navigation technique is widely used in UAV positioning and navigation. However, some issues such as low precision and ease of losing GPS signals under the bridge, make it difficult to be used in bridge inspection. In dealing with the case of missing GPS signals, the positioning and navigation techniques based on distance measurements, such as visual and ultrasonic navigation, are generally employed.

(1) Visual Navigation

The UAV navigation scheme based on bridge point cloud model registration makes use of UAVs to fly around the bridge to obtain the overall point cloud model of the bridge. Then it registers the overall point cloud of the bridge with the local point cloud reported by the UAV in real-time to compute the position of the UAV relative to the bridge [49]. Specifically, the scheme uses binocular stereo vision as the bridge point cloud generation technique. It computes the position of the UAV relative to the bridge model through the point cloud registration algorithm. Schematic diagram of the scheme is shown in Fig. 4.47, including the acquisition of three-dimensional point cloud data, establishment of bridge point cloud model, and computation of the UAV attitude. The binocular camera mounted on the UAV is used to obtain the three-dimensional point cloud information of the bridge in the current camera's field of view. Then, the known three-dimensional model of the bridge is registered. The attitude coordinates of the camera are finally obtained by solving the registration algorithm, that is, the position of the UAV.

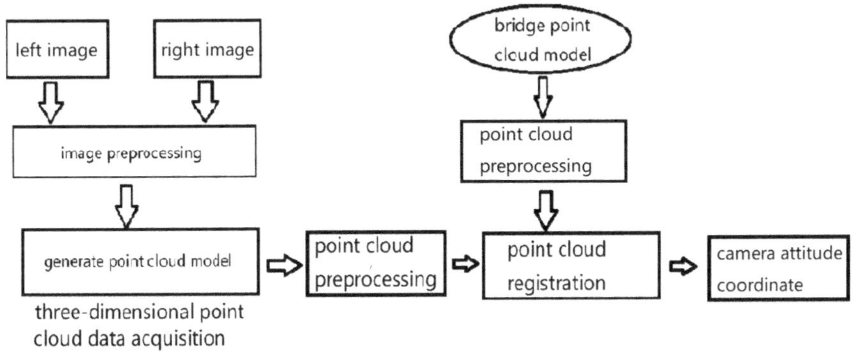

Fig. 4.47 Schematic of UAV navigation scheme based on bridge point cloud model registration [49]

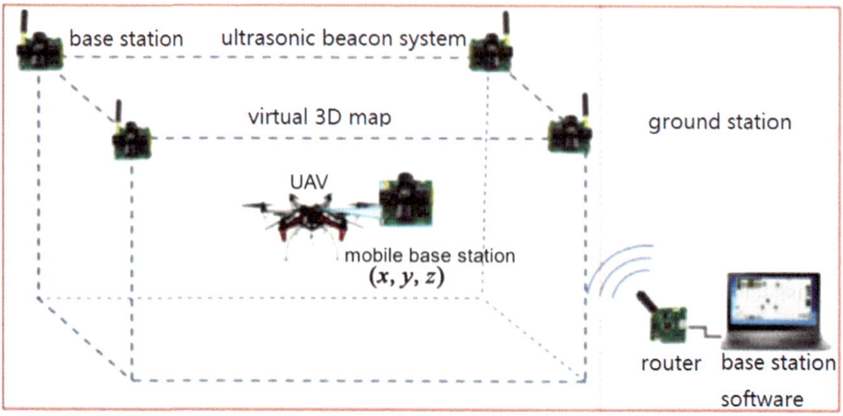

Fig. 4.48 Schematic of ultrasonic navigation of UAV [51]

(2) Ultrasonic Navigation

Ultrasonic positioning technique is often combined with radio frequency technology for indoor positioning systems because of its advantages in accuracy, cost and sensitivity [50]. To deal with the problem of missing GPS signals in UAV bridge inspection, ultrasonic signal of the peripheral beacon is employed to determine the relative position of the UAV. The ultrasonic transmitting module with positioning first sends instructions to each fixed beacon via radio frequency signals. At the same time, it sends ultrasonic waves to each label and determines the relative position of the ultrasonic transmitting module by using the relative distance between different beacons. Ultrasonic positioning technique has a good application prospect in the field of UAV inspection of bridges [51]. The specific positioning process includes ① obtaining GPS coordinate information of the fixed ultrasonic beacon and the UAV; ② acquiring the relative distance dynamic information between the fixed ultrasonic beacon and the UAV; ③ adopting various filtering methods to correct the distance information; and ④ calculating the real coordinates according to the GPS and the relative dynamic distance information. Figure 4.48 is a schematic diagram of an UAV ultrasonic navigation based on an Ad-Hoc network.

4.5.4 Application of UAV in Bridge Detection

UAVs have been used in bridge structure inspection to obtain apparent information, three-dimensional information and dynamic structural response of the bridge through on-site shooting and post-image processing. They have also been used as auxiliary measuring equipment to obtain various structural information.

(1) Obtain Apparent Information of Bridge Structures

Apparent distress detection is an important part of bridge inspection, and UAVs can flexibly and efficiently collect such information. With subsequent image data processing, it can greatly improve the efficiency of bridge inspection. With the rapid growth of bridge safety inspection demand, inspection-oriented UAVs have become a hotspot of researches concerning bridge apparent inspection techniques at home and abroad.

The authors' team [13] developed an intelligent operating system to control the basic UAV flights and to obtain images with the camera. Through a series of physical tests, the payloads, flight speeds, working times and wind resistances of some UAVs were measured and compared, and the feasibility of applying small aerial UAVs to bridge routine and post-earthquake emergency inspections was verified. According to the test results, suitable UAV equipment and optimized flight plans were selected to inspect a steel arch bridge damaged in the Kumamoto earthquake in Japan in 2016. The bridge has a height of 143m and length of 400m, and steel beams with rubber bearings.

For bridge maintenance inspections, small UAVs can be used to shoot 4K video at very close range (3–7 m). The Qingyun Bridge, shown in Fig. 4.49, in Miyazaki City is 143 m high, whose sub-structural members are very difficult for people to reach. Even the largest basket truck can only transport engineers to about 10 m below the bridge floor. Using UAVs, detailed and high-resolution visual information were successfully obtained. From Fig. 4.49, the paint peeling on the surface of the steel bridge and the corrosion of internal steel can be observed.

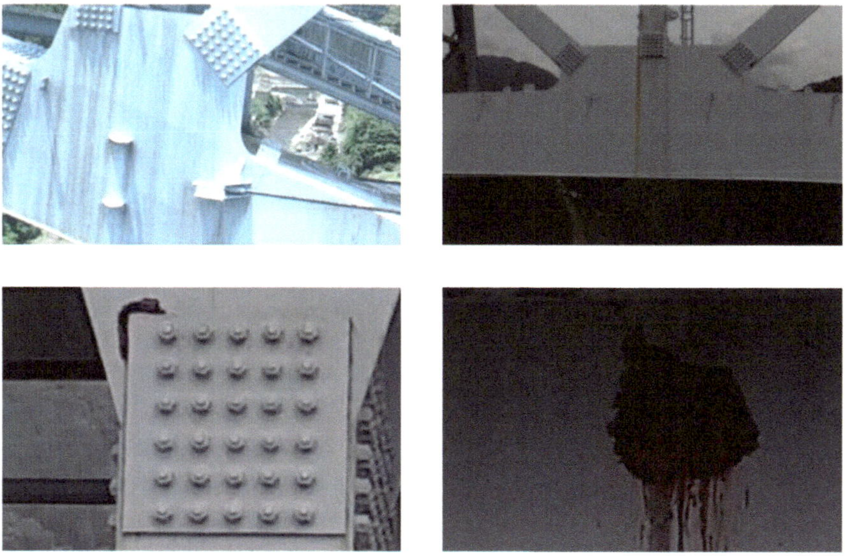

Fig. 4.49 Detailed aerial photos of close flight [13]

(2) Obtain Three-Dimensional Information of Structures

With the maturity of UAV 3D modeling technology, its application in civil engineering shows great potential. In 2015, Freimuth et al. [14] proposed a semi-automatic UAV system based on a 4D building model, which is used for monitoring the construction process of construction projects. The system can use the information provided by the 4D building model to plan the flight mission of a UAV, obtain more comprehensive building images while ensuring no collision with the buildings. In 2015, Lin et al. [15] proposed a model-driven image acquisition framework, which used a 4D BIM model to process the point cloud data collected by UAV to generate the construction progress information, and ultimately realize the automation of construction progress monitoring. In the field of bridge 3D modeling, in 2018, Pang Junyi et al. [52] studied the application of 3D real-life models in project progress and quality management of road and bridge engineering based on BIM 3D real-life modeling method combined with UAV data acquisition technique. Liu et al. [53] proposed an evaluation method of bridge pier cracks based on UAVs and 3D scene reconstruction. This method can realize three-dimensional reconstruction of bridge piers, and accurately identify the location and width of the cracks. They also verified the feasibility of the method on a real bridge. Figure 4.50 depicts a schematic diagram of this three-dimensional reconstruction of pier column with the crack location marked.

(3) Obtain Dynamic Responses of Structures

Vibration-based structural health monitoring (SHM) is one of the most important ways to evaluate the safety of civil infrastructures. The SHM practice is based on measuring the dynamic responses of the structures. The conventional health monitoring methods mostly use wired or wireless sensors to measure the accelerations of

Fig. 4.50 Three-dimensional reconstruction of pier column with crack location marked [14]

the structures, and then obtain the displacements of the structures through integration. However, the acceleration measurement results usually lack accuracies when the structures vibrate at low frequencies. Besides, direct displacement measurement methods such as the approach using laser scanners are always of limited capability due to the need for fixed reference point, high cost and low accuracy. Using the concept of "divide and conquer", Spencer proposed a method to splice together different parts of a modal shape extracted from the vibrating structure's close-up video to generate the structure's overall modal shape. The first six natural frequencies and modal shapes of the structure were measured in the laboratory's shaking table test of a six-story frame. Then, it was tested on a full-scale pedestrian bridge at Lake Woods in Maholm, Illinois. This approach directly addresses many difficulties mentioned earlier that are associated with using vision-based approaches for full-scale infrastructure modal analysis. The tested bridge is Little Golden Gate Bridge on Mahomet Forest Lake, Illinois, USA. The dimensions, the locations of markers and sensors, and the flight path of UAVs are shown in Fig. 4.51. The bridge is a steel pedestrian suspension bridge with plate girders located approximately 18km northwest of Champaign, Illinois. The footbridge has a span of 67m and two 10m high towers. Both sides of the bridge beam body are connected with slings, which are arranged one in every 1m and suspended below the main cables.

To measure the bridge vibration, 13 reference marks were affixed on the south side of the bridge at equal intervals along the longitudinal direction, and four accelerometers were uniformly installed on the front half span of the bridge for comparison. To excite the bridge, three pedestrians jumped at about 1/3 position from the center in the second half of the bridge. The video clip was taken by the commercial UAV DJI Phantom 4 Pro (DJI 2018), with a resolution of 3840 pixels × 2160 pixels and

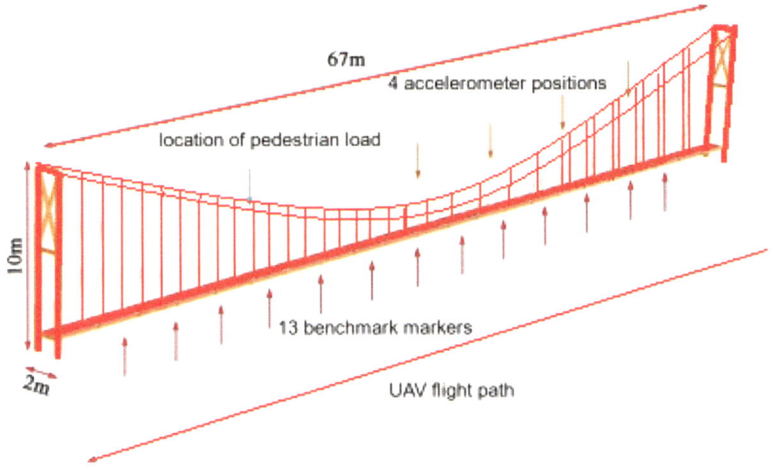

Fig. 4.51 Bridge dimensions, marked positions, sensors, and UAV flight path [50]

a frequency of 30Hz. Comparing the computed modal characteristics with the reference data calculated using samples provided by an accelerometer, all MAC values were found to be above 0.925, and the natural frequency differences of the three comparative modes were all less than 1.6%.

(4) Assist Other Equipment in Measurements

In bridge inspection, UAVs are generally used as the working platform or the carrier of cameras or other inspection equipment to detect apparent bridge defects. UAVs can also be used as auxiliary equipment to obtain structural displacements. The Vision and Control Team of Seville University in Spain innovatively developed a contact multi-rotor UAV based on the ceiling effect, taking the UAV as the target of image measured using a total station, and developing a bridge deflection measurement system [54]. The system consists of an aerial UAV platform, reflector prism, ground control station and laser tracking station.

To meet the demand of bridge deflection measurements, a reflective prism is installed on the upper part of the UAV fairing to provide signal source for the total station. An elastic cushion block is added to adapt to different bridge deck conditions. After the UAV flew to the designated position of the bridge and fixed its flight attitude, the researchers can use a total station to track the reflecting prism module of the UAV by laser, and the ground station to receive all the data and processes them. The laser tracking results are compared with the local three-dimensional laser point cloud scanning data to determine the position change of the UAV reflecting prism relative to the original position so as to measure the bridge deflection. Figure 4.52 shows a schematic diagram of obtaining the UAV position by laser tracking using a total station. Researchers have applied it on the Katuya Bridge and achieved good results, which fully proved the applicability of UAVs as a movable measurement target in bridge inspection.

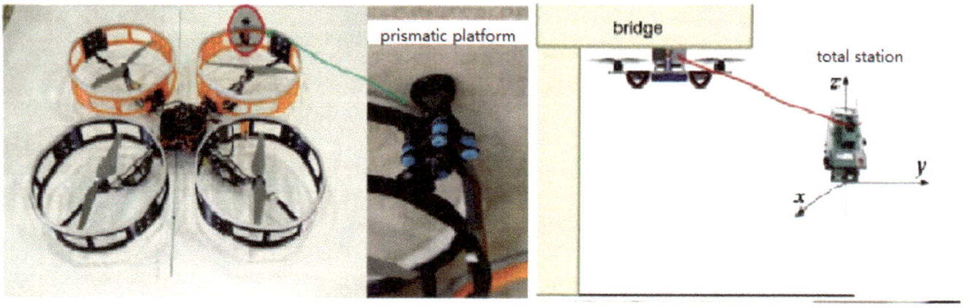

Fig. 4.52 Schematic of UAV positioning by total station laser tracking [54]

4.6 Summary

This chapter introduces different kinds of bridge inspection equipment, divided into bridge deck system inspection equipment, pier tower inspection equipment, cable inspection equipment, main cable inspection equipment, underwater structure inspection equipment and comprehensive inspection equipment according to different inspection objectives. According to the degree of automation, they can be divided into semi-automatic equipment and automatic equipment. It should be noted that the current semi-automation and automation technology and equipment mostly focus on realizing autonomous locomotion. It must be combined with artificial intelligence and navigation system to realize autonomous operation. Presently, navigation is one of the cores of underwater and aerial UAV and robot technology, which is different from contact equipment in realizing autonomous navigation. Based on this, according to its automatic operation mode, this chapter divides bridge inspection equipment into semi-automatic contact equipment, automatic equipment, underwater robot and equipment, and aerial UAV comprehensive inspection equipment.

On the whole, most of the applied bridge inspection robots and UAVs are still in the stand-alone working mode and semi-automatic working state under manual intervention, which still have a certain distance from complete automation. The functions of robot inspection operation efficiency evaluation, refined control method, multi-source inspection data fusion analysis, inspection task planning, etc., need to be developed. Special inspection equipment, autonomous navigation technology and supporting image processing algorithms for bridge inspection must be further studied and improved. In addition, from the perspective of environmental adaptability, the existing inspection equipment design still needs to completely overcome the interference caused by complex environments, such as strong wind and strong vibration, climbing obstacles, turbid turbulence, and so on, which leads to inaccurate inspection results.

Based on the above considerations, automation, modularization and intelligence will be the future development direction of this field. In robot inspection, it is to realize autonomous robot walking, data autonomous acquisition and real-time processing, standardized and quantitative production of robot parts, automatic operation of the robot, in-depth mining and analysis of inspection data, and even form a decision-making scheme. In the application of UAVs, it is to realize special UAVs suitable for various bridge inspection environments, realize automatic planning and autonomous navigation of inspection paths, improve inspection, navigation and route planning based on artificial intelligence, and integrate related technologies to form a fully automatic integrated platform for bridge inspection.

References

1. Hu ZL, Qi YJ, Bai L (2016) Overview of the development of foreign highway pavement and tunnel inspection vehicles. Forestry Mach Woodworking Equipment 5:4–7 (in Chinese)
2. Gucunski N, Kee S, La H et al (2015) Delamination and concrete quality assessment of concrete bridge decks using a fully autonomous RABIT platform. Structural Monit. Maintenance 2(1):19–34
3. Jiang Z (2017) Research on key technologies of self-propelled main cable maintenance vehicle for suspension bridges. Southwest Jiaotong University, Chengdu (in Chinese)
4. Sitti M, Fearing RS (2003) Synthetic gecko foot-hair micro/nano-structures as dry adhesives. J Adhes Sci Technol 17(8):1055–1073
5. Mlirphy MP, Sitti MW (2007) An agile small-scale wall-climbing robot utilizing dry elastomer adhesives. IEEE/ASME transactions on mechatronics: a joint publication of the IEEE industrial electronics society and the ASME dynamic systems and control division 12(3):330–338
6. Zhang L, Jiang Z, Wang SH (2017) Design of self walking main cable maintenance vehicle for suspension bridge. Mech Eng 8:26–28 (in Chinese)
7. Shen P (2011) Inspection and repair of M48 severn bridge main cable in the UK. World Bridges 5:70–73 (in Chinese)
8. Song YK, Lee CM, Koo I et al (2008) Development of wall climbing robotic system for inspection purpose. In: IEEE/RSJ international conference on intelligent robots and systems. IEEE
9. Ye JM, Gong ZG, Li RQ (2014) Review of the main maintenance and repair projects of the fox highway bridge. World Bridges 2:82–88 (in Chinese)
10. Xu FY, Wang XS, Cao PP (2011) Design and application of a new wheel-based cable inspection robot. In: IEEE international conference on robotics and automation. IEEE
11. Zhang GZ, Ma XG (2016) Underwater Inspection Technology for super deep foundation of huge caisson of Hutong Yangtze River Bridge. Bridge Construct 46(6):7–12 (in Chinese)
12. Cho H, Kim B, Yu SC (2018) AUV-based underwater 3-D point cloud generation using acoustic lens-based multibeam sonar. IEEE J Oceanic Eng 43(4):856–872
13. Dang J, Endo K, Matsunaga S et al (2017) Performance evaluation of unmanned aerial vehicle for bridge inspection inspection and application in 2016 kumamoto earthquake. Huixian International Forum on Earthquake Engineering for Young Researchers. United States
14. Freimuth H, König M (2015) Generation of waypoints for UAV-assisted progress monitoring and acceptance of construction work. In: Conference on construction applications of virtual reality
15. Lin JJ, Han KK, Golparvar-Fard M (2015) A framework for model-driven acquisition and analytics of visual data using UAVs for automated construction progress monitoring. In: International Workshop Computing in Civil Engineering
16. Xia J, Zhang L (2019) Current situation and development trend of bridge inspection vehicles. Practical Automot Technol 2:205–207 (in Chinese)
17. Oh JK, Lee AY, Oh SM et al (2007) Design and control of bridge inspection robot system. In: International conference on mechatronics and automation. IEEE
18. Oh JK, Jang G, Oh S et al (2009) Bridge inspection robot system with machine vision. Autom Constr 18(7):875–880
19. Lee BJ, Shin DH, Seo JW et al (2011) Intelligent bridge inspection using remote controlled robot and image processing technique. In: 28th international symposium on automation and robtics in Construction, IAARC (2011)
20. Mahdavi S, Noohi E, Ahmadabadi MN (2007) Basic movements of a nonholonomic wheel-based pole climbing robot. IEEE/ASME International Conference on Advanced Intelligent Mechatronics. IEEE
21. WANG X, XU F. Design and experiments on cable inspection robot. conference of the IEEE industrial electronics society. IEEE (2007)

22. Saltaren R, Aracil R, Reinoso O et al (2005) Climbing parallel robot: a computational and experimental study of its performance around structural nodes. IEEE Trans Rob 21(6):1056–1066
23. Jiang SY, Li RL, Li JY et al (2008) Development of spraying mechanism for pneumatic creeping cable maintenance robot. Mech Drive 32(3):92–93, 96. (in Chinese)
24. Zhang JL, Lv TS, Luo J et al (2000) Development of a pneumatic peristaltic cable maintenance robot. Robots 22(5):397–401 (in Chinese)
25. Zhu D (2012) Wireless mobile sensor network for the system identification of a space frame bridge. IEEE/ASME Trans Mechatron 17(3):499–507
26. Takada Y, Ito S, Imajo N (2017) Development of a bridge inspection robot capable of traveling on splicing parts. Inventions 2(3):22
27. Mazumdar A, Asada HH, Mag-Foot: a steel bridge inspection robot. In: IEEE/RSJ international conference on intelligent robots and systems. IEEE, (2009)
28. LV XQ, Liu G, Zhang XX et al (2008) High temperature corrosion mechanism of water cooled walls and robot detection technology . Journal of Shanghai Jiaotong University, (51):4. (in Chinese)
29. Yamamoto A, Nakashima T, Higuchi T (2007) Wall climbing mechanisms using electrostatic attraction generated by flexible electrodes. In: International symposium on micro-nano mechatronics and human science, IEEE (2007)
30. Miyake T, Ishihara H, Yoshimura M (2007) Basic studies on wet adhesion system for wall climbing robots. In: IEEE/RSJ international conference on intelligent robots and systems. IEEE
31. Haoyang RL, Hong Q et al (2008) A miniature multi-joint wall-climbing robot based on new vibration suction robotic foot. In: IEEE international conference on automation and logistics. IEEE
32. Jiang XJ, Wei W (2011) Research on non-destructive testing system for bridge cables based on snake shaped robots. Transducer Microsystem Technol 30(4):82–84,95. (in Chinese)
33. Chen BW (2012) Research and implementation of ultra wide coverage multi beam sounding technology. Harbin Engineering University, Harbin (in Chinese)
34. Xu YJ (2011) Research on key technologies for underwater target detection. Dalian Maritime University, Dalian (in Chinese)
35. Lu L, Chen GM, Dai Y (2004) Automatic detection and recognition of surface defects in underwater structures. Comput Measurement Cont 12(12):1125–1127, 1142. (in Chinese)
36. Zhang AE (2008) Imaging detection technology in muddy water environment. In: Collection of technical papers on oil and gas exploration and development engineering in Bohai Bay, (in Chinese)
37. Shen XX, Wang YH, Yang SQ et al (2018) Current status and prospects of underwater glider technology development. J Underwater Unmanned Syst 26(2):89–106 (in Chinese)
38. Sun GX, Li CJ, Mu ZL (2017) Application of ROV underwater robot in Changheba hydropower station. Water conservancy Hydropower Const 4:85–87 (in Chinese)
39. Ping W, Ma XF, Zhang JH et al (2017) The unmanned remote-controlled submersible 'Haima' . Ship Sci Technol 39(15):138–141, 145. (in Chinese)
40. Zhou C, Xiang XY, Zhong X et al (2017) The application and prospects of drones in agriculture. Hunan Agricul Sci 2017(11):80–82, 86. (in Chinese)
41. Lv JW (2019) Application of image processing technology in UAV power line inspection. Telecom Power Technol 36(6):84–85 (in Chinese)
42. Zhao JH, Xu BK, Wang BX et al (2017) Current status and prospects of traffic monitoring research based on drones. Comput Prod Circul 9:125 (in Chinese)
43. Chen JQ, Li JY, Li HL et al (2018) Preliminary study on the application of drones in bridge inspection. Traffic World 32:103–106 (in Chinese)
44. Lu YT, Han CH, Zeng P (2017) Research on implementation scheme of UAV bridge detection based on BIM. Civil Const Eng Inf Technol 9(2):73–77 (in Chinese)
45. Dang J, Shrestha A, Chun PJ (2018) Bridge damage detection using unmanned Arial vehicle and deep convolutional neural network. In: Asia conference on earthquake engineering. Thailand

46. Cao ZX (2017) Research on the application of BIM technology in bridge management and maintenance. Chongqing Jiaotong University, Chongqing (in Chinese)
47. Jiang S, Zhang JJCAC, Engineering I (2020) Real-time crack assessment using deep neural networks with wall-climbing unmanned aerial system. Comput-Aided Civ Infrastruct Eng 35(6):549–564
48. Cui MD, Wu G (2019) A tracked unmanned aerial vehicle for automatic positioning of bridge supports: CN109606678A [P/OL]. 2019–04–12. (in Chinese)
49. Zhu PM (2017) Research on autonomous navigation technology of unmanned aerial vehicles for bridge detection. Chongqing University, Chongqing (in Chinese)
50. Hoskere V, Park JW, Yoon H et al (2019) Vision-based modal survey of civil infrastructure using unmanned aerial vehicles. J Struct Eng 145(7):4019062.1–4019062.14
51. Kang D, Cha YJJCAC, Engineering I (2018) Autonomous UAVs for structural health monitoring using deep learning and an ultrasonic beacon system with geo-tagging. Comput-Aided Civ Infrastruct Eng 33(10):885–902
52. Pang JY (2018) Research on 3D realistic modeling and application based on BIM in road and bridge engineering. Chongqing Jiaotong University, Chongqing (in Chinese)
53. Liu YF, Nie X, Fan JS et al () Image-based crack assessment of bridge piers using unmanned aerial vehicles and three-dimensional scene reconstruction. Computer-aided civil and infrastructure engineering, 2020, 35(5):511–529.
54. Sanchez-Cuevas PJ, Ramon-Soria P, Arrue B et al (2019) Robotic system for inspection by contact of bridge beams using UAVs .Nature reviews cancer 19(2):305

Chapter 5
Big Data Enabled Computing

Abstract Big data is data set whose volume greatly exceeds the capabilities of the conventional database softwares in terms of data acquisition, storage, management and analysis. Big data is usually generated in the process of bridge operation and maintenance. Therefore, the storage, processing and utilization of big data is an important research topic in the field of intelligent bridge operation and maintenance. This chapter introduces the basic concept of big data as well as concepts of the common big data computing methods, and the information physical system used to associate the physical space with the information space, with emphasis on the application of big data computing methods in civil engineering, especially in bridge engineering.

5.1 Overview of Big Data Computing

5.1.1 History and Definition of Big Data

Big data refers to data sets that cannot be captured, managed and processed using the conventional softwares within a certain time frame. It is a massive and diversified information asset with high growth rate that requires new processing mode with stronger decision-making power, greater insight and discoverability, and better process optimization capabilities. The overall developing trend of global data volume is shown in Fig. 5.1 [1].

The big data technology commonly referred to today originated from three heavyweight papers published by Google: "*Google File System*" published in October 2003, "*MapReduce: Simple Data Processing for Super Large Clusters*" published in December 2004, and "*BigTable: Distributed Storage System for Structured Data*" published in November 2006. These three articles discussed the "troika" of big data technology: the distributed file system GFS (Google File System), the big data distributed computing framework MapReduce and NoSQL (Not Only SQL), and the database system BigTable. In 2006, Doug Cutting launched an independent project dedicated to the development and maintenance of the big data technology, which gave

G. Wu et al., *Intelligent Bridge Maintenance and Management*, Springer Tracts in Civil Engineering, https://doi.org/10.1007/978-981-97-3827-4_5

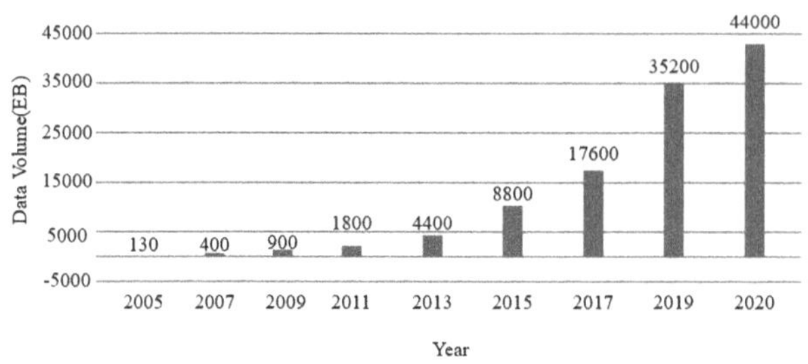

Unit	Quantity
KB	1000
MB	1000^2
GB	1000^3
TB	1000^4
PB	1000^5
EB	1000^6
ZB	1000^7
YB	1000^8

Fig. 5.1 Development trend of global data volume

birth to the classic framework of big data technology—Hadoop, mainly includes the distributed file system HDFS (Hadoop Distributed File System) and the big data computing engine MapReduce.

In 2016, IBM proposed a descriptive definition of big data, i.e., the big data should have some or all of the following 5 V characteristics: volume, velocity, variety, value, and veracity. Accordingly, as shown in Table 5.1, the differences between big data and conventional data are noticeable in many respects.

Big data technology does not rely on the statistical sampling method; it just observes and tracks what has happened. Big data must be statistically analyzed and compared by computers to obtain objective results, and the volume of data contained in big data usually exceeds the processing capacity of the conventional software within an acceptable processing time. It is almost impossible to use most of the database management systems to process big data, and "software running in parallel on dozens, hundreds or even thousands of servers" (clusters) must be utilized. In many fields, scientists often encounter limitations and obstacles in analyzing and processing the big data due to the huge data set.

Based on data discovery and data science, Jim Gray, winner of Turing Prize, puts forward the fourth paradigm of scientific research—the data intensive scientific discovery in 2007. The main feature of the fourth paradigm is that the work of scientific researchers is mainly to find and mine the information and the knowledge required from the big data. Supercomputing capabilities are needed to directly

Table 5.1 Differences between big data and conventional data

Dimension	Conventional data	Big data
Data volume	GB	PB (current)
Generation rate	Every hour, every day	Even faster per second
Data structure	Structured	Semi structured, unstructured
Data sources	Centralization	Distributed
Data consolidation	Easily	Difficulty
Data storage	RDBMS	NoSQL, HDFS
Data processing	Interactive	Batch or real-time

analyze the massive data to find their relevant relationships and then to derive the causalities to obtain new knowledge. Direct facing of the physical objects Models and assumptions are no longer required for the fourth paradigm. The fourth paradigm is a kind of data-driven based scientific research paradigm and can be considered as the criterion of new scientific research represented by big data science.

The technological system of big data science mainly composes of three parts, namely the storage technology, the computing technology and the analysis technology. These are described in the following sections.

5.1.2 Storage Technology: Database and Data Warehouse

Database is a tool for storing and managing data, which is usually divided into relational database and non-relational database. Among them, the Relational Database Management System (RDBMS) is used to operate databases based on the relational model. Currently, there are many RDBMSs, such as MySQL, SQL Server, PostgreSQL, Oracle, etc. Technically, the non-relational database (NoSQL) is not a database, but a collection of data structured storage methods. It has a free mode, and supports partition replication and a large amount of data. It has gradually become the standard for dealing with big data. NoSQL is mainly divided into three categories according to different data models: the key-value storage databases, the column storage databases and the document databases. For key-value storage database, the data is stored in the form of a key and a value. The key values are unique, which are usually used in lightweight data applications, and generally relying on the memory such as Redis. The column storage database is based on the processing of data in columns, rather than the conventional processing of data in rows. It can be divided into multiple nodes for expansion by rows and columns at the same time. The most famous column storage database is Hadoop based Hbase. The document database can support more complex data structures without limiting the storage content and format. Its storage device is usually the hard disk, such as MongoDB.

Data warehouse is a subject-oriented, integrated, non-volatile (non-updatable), time-varying data set, which aims to filter and integrate various business data for the enterprises. It provides the enterprises with certain BI (Business Intelligence) capabilities, helps them improve business processes, and helps them monitor and control time, cost and quality. The inputs of the data warehouse are various data sources (or databases), and the final outputs are used for the enterprise data analysis, data mining, data reports, etc. The data warehouse relies on ETL [extract, transform, load] to integrate data from different data sources, which is also the most time-consuming and resource-consuming link. The four core components of the data warehouse are the various database sources, the ETL, the data warehouse, and the front-end applications, as shown in Fig. 5.2. At present, the mainstream data warehouse tools include Hive based on the Hadoop framework and AnalyticDB (PB level real-time data warehouse) independently developed by Alibaba.

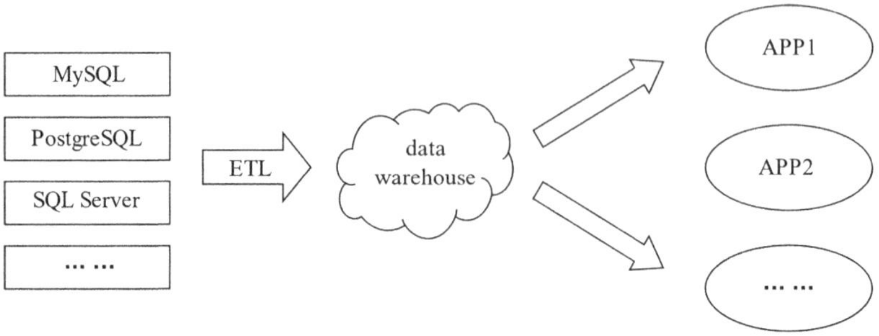

Fig. 5.2 Data warehouse core components

The differences between database and data warehouse are as follows. The database operations, generally online transaction processing (OLTP), are conducted online in the database for specific businesses. They feature small amount of data, and they usually query and modify a small number of data records. The operations of data warehouse are generally online analytical processing (OLAP), which are used to analyze the historical data of certain topics (comprehensive data) and support management decisions.

5.1.3 Big Data Computing Technology

The computing framework for big data can be divided into four main categories, namely the batch processing, the stream computing, the interactive analysis and the incremental computing. It can be seen from Fig. 5.3 that there are specific frameworks under each category, and some of them are cross associated. This "prosperous scene" shows that each specific framework has its own advantages and disadvantages. They complement each other, and are suitable for different scenarios.

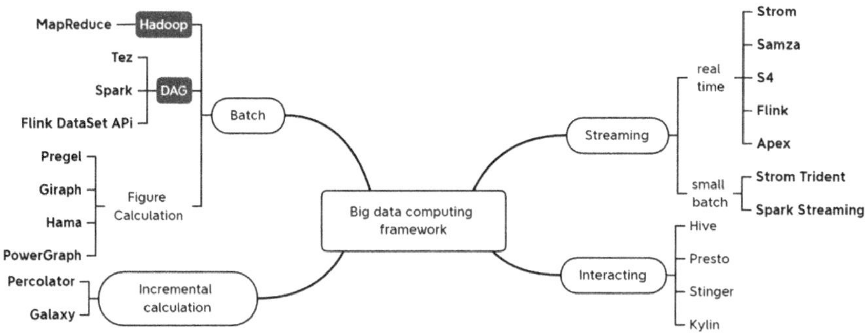

Fig. 5.3 Classification of big data computing frameworks (DAG—Directed Acyclic Graph)

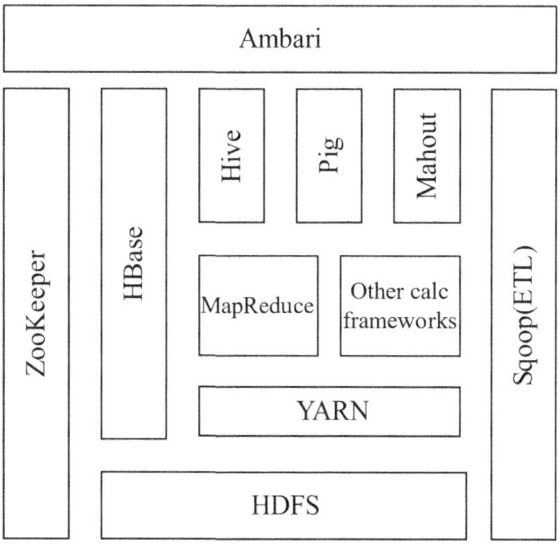

Fig. 5.4 Hadoop ecosystem

Hadoop is a pure batch processing big data computing framework. It is a reliable and extensible open source software supporting distributed computing developed by Apache. Because of its birth more than 10 years ago, big data distributed computing technology developed rapidly. Hadoop can be divided into three major parts: YARN (Yet Another Resource Negotiator) which is responsible for resource and task management, HDFS (Distributed File System) which is responsible for distributed storage, and MapReduce (Computing Framework) which is responsible for distributed computing, making it possible to process large data sets in a distributed way using simple programming models on cluster servers. Hadoop's ecosystem and distributed computing standards are shown in Fig. 5.4.

MapReduce, as the computing core of Hadoop framework, is a simple programming model of high expression ability. It can automatically assign large-scale computing tasks to ordinary PC clusters for parallel and distributed processing. The programming model includes two main user-defined functions—Map function and Reduce function. The Map function converts the original data into intermediate results marked with variables according to certain rules. After that, the computation framework will give the intermediate results with the same variables to the Reduce function to obtain the summary results. However, in face of increasingly complex big data computing tasks, MapReduce is not suitable for cases with complex logical relationships because of its low level of abstraction and limited expression capacity. In order to overcome this problem, a DAG (Directed Acyclic Graph) computing model has been proposed in the industry. Its core idea is to internally decompose the task into several sequential subtasks, so that it can more flexibly express various complex dependencies. Another disadvantage of MapReduce is that it uses disks to store intermediate results, which seriously affects the system performance. Spark improves the DAG model proposed earlier by selectively loading and storing intermediate data in memory to reduce disk IO tasks. Compared with Hadoop, Spark's memory based operations are more than 100 times faster, and its disk based operations more than 10 times faster.

In the age of big data, data is usually generated continuously and dynamically, which requires fast data processing in a very short period of time. In addition, fault tolerance, congestion control and other issues should be considered to avoid missing data or repeated calculation. Therefore, a stream computing framework was born. At present, the most popular and widely used stream computing framework is Storm. It has a simple programming model, supports mainstream developing languages such as Java and Python, and has good performance and fault tolerance. It can process millions of messages per second on a multi-node cluster.

After efficient computation of big data, the next step is to advance data interaction. Hive is the earliest large-scale data warehouse based on Hadoop. Its basic idea is to organize the files in HDFS as a storage system similar to the conventional databases by defining schema information, and to maintain their scalability and flexibility provided by Hadoop. Hive supports the familiar relational database concepts such as tables, columns and partitions, the unstructured data for SQL supports to a certain extent, and the queries expressed in HiveQL, a declarative language similar to SQL. HiveQL is compiled into MapReduce procedure for execution. Hive's main disadvantage is that it needs to be based on MapReduce, and its performance is therefore limited. Some interactive analysis platforms have made corresponding improvements on and extensions to Hive, such as Presto, Styler, Kylin, etc.

5.1.4 Big Data Analysis Technology

Data analysis refers to the process of analyzing a large amount of collected data using appropriate statistical analysis methods and extracting useful information and drawing conclusions by conducting detailed researches and summaries on the data. Data mining is a process of automatically discovering useful information in large data repositories. It is an indispensable part of KDD (Knowledge Discovery in Database). The modeling process of data mining usually consists of five steps: data selection (including training set and data set), model training, model verification, model updating, and model usage. The framework of data analysis and data mining is shown in Fig. 5.5.

Exploring data aims at describing the data according to the centralized trend (mean), the off center trend (standard deviation), the correlation (covariance), etc. Prediction and classification are the two ways of using data to predict future results. They belong to "supervised learning". Clustering and dimension reduction belong to "unsupervised learning". Clustering is the process of dividing a data object set into several subsets according to their similarity. The difference between clustering and classification is that classification refers to a kind of guided learning for which a category of things is known, and the rules of classification should be learnt from the samples; while clustering is a kind of unsupervised learning, which directly obtains several categories based on similarity.

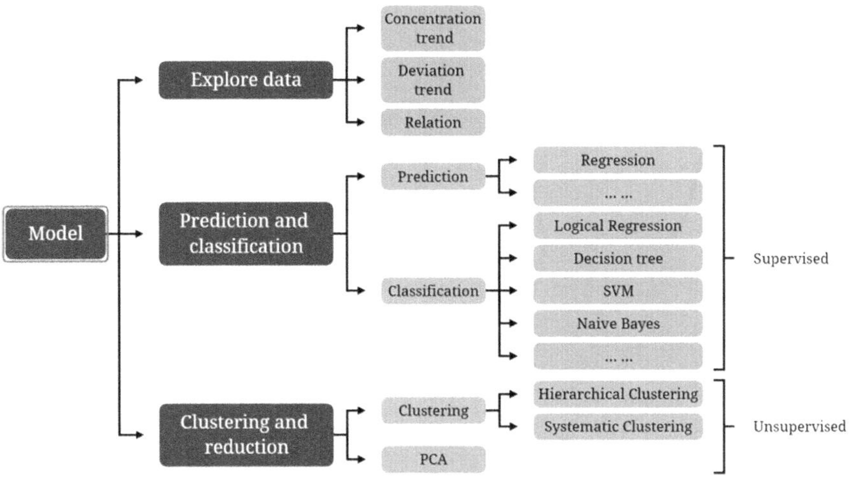

Fig. 5.5 Framework of data analysis and data mining

When analyzing and mining a large amount of data, dimensionality reduction and visualization are generally required. Dimension reduction refers to using linear or nonlinear transformation to reduce some closely related variables to unrelated new variables as much as possible, projecting the data in a high-dimensional space into a low dimensional space with the observational significance, and then realizing the visual representation of the data's internal characteristics in low dimension. At the same time, dimensionality reduction reduces the computational workload and the complexity of the model, as well as the impact of noise variables, and the data storage. The visual presentation after the dimension reduction can help to explore the rules hidden behind the original image set in a human perceptible way. The data dimension reduction and classification, and the model are shown in Fig. 5.6.

In the process of data analysis and mining, anomaly detection is also an important issue. Anomaly values are also called outliers. It is very dangerous to ignore the existence of anomaly values. Including anomaly values in the computation and the analysis of data will have adverse effects on the results. Anomaly detection methods range from the early statistical methods to the later detection methods based on neural networks, to the current methods using decision tree classification, distance-based clustering, density based clustering and other models.

As the monitoring system will encounter noise interference, sensor failure or transmission network failure in the process of data acquisition, the massive raw data obtained will contain a lot of noise, and even serious faulty information. The existence of these faulty data will directly affect accurate assessment and timely warning of the bridge's health in the later stage. Typical sensor data anomalies [2] are: data loss, data drift (transition), long-cycle interference, inconsistent anomaly (jump point), weak noise interference, strong noise that cannot be eliminated, frequency domain signal anomalies, etc.

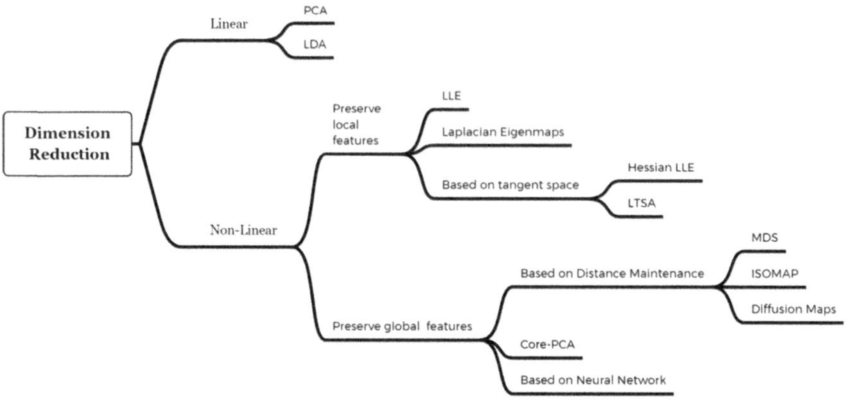

Fig. 5.6 Data dimension reduction classification and model

Therefore, data pre-processing, secondary data pre-processing and post-processing are required manually or using data processing softwares. Among them, data pre-processing includes filtering, denoising, trend item removal, interception and outlier processing. Secondary data pre-processing is mainly basic statistical operations to generate statistical characteristics, such as the maximum, the minimum, the mean, the variance, the standard deviation, the covariance, the kurtosis, the skewness, etc., and the computed results are typically used as the basis for primary early warning. Post processing mainly involves specific analysis of the data, such as modal analysis, correlation analysis between bridge characteristics and environmental factors, nonlinear regression analysis, etc.

5.1.5 Current Status of Bridges

With regard to bridge maintenance and management processes and characteristics of their data, the "high speed" and the "diversity" among the 5 V characteristics of big data are noticeable, while the "volume" is not prominent. Taking a bridge with a deck of 1000 m length and 30 m width as an example, when each pixel corresponds to 1 mm, the total area of the whole deck corresponds to 3.0×10^{10} pixels. It is estimated that the total data obtained on the bridge in one year of operation is about 4.67×10^3 TB; see Table 5.2 for details.

At present, the total volume of field data obtained in the bridge has not reached the narrow big data level. However, with respect to the other four "V" characteristics, the data generated in real time and accumulated in the process of bridge design, construction, monitoring, detection, maintenance and management can be classified as big data. The detection and monitoring process involves a large amount of diverse data, and the emerging big data technology is required to process them. In addition,

Table 5.2 Estimation of total data of single bridge in one year

Sensor type	Quantity	Bytes	Times (day)	Times (year)	Sampling rate (Hz)	Total	Bits
Temperature and other environmental sensors	300	8	24	365	1	1.96×10^{-2}	GB
Crowdsourcing mobile computing	10,000	24300 KB	1	1	1	2.32×10^{2}	GB
Strain gauge	1000	8	1800	365	100	4.90×10^{2}	GB
Accelerometer	1000	8	1800	365	200	9.79×10^{2}	GB
Bridge deck scanning	1	9.0×10^{10}	1	12	1	1.01×10^{3}	GB
UAV	3	24300 KB	3600	24	1	6.01×10^{3}	GB
Traffic camera	10	3 MB	86,400	365	1	9.24×10^{5}	GB
High speed camera	10	3 MB	1800	365	200	3.85×10^{6}	GB
Total data						4.67×10^{3}	TB

the development of more clustered and refined intelligent maintenance and management, coupled with the integration of structural detection and monitoring technology, and big data science and technology are important developing trends of future bridge management, maintenance and operation, especially for regional structure groups.

1. In Terms of Bridge Monitoring

   (1) Deep mining based on monitoring data requires big data technology to support the operation and maintenance of long-span bridge structures.
   (2) Through deep learning technology for image data and big data, monitoring of structural distress can be carried out.
   (3) Through collaborative analysis of temperature data monitored from similar bridges in the same region, the temperature field on bridges in the existing environment can be obtained.
   (4) Based on video technology, rapid identification method of vehicle loads on bridges in the operation stage is developed, which can realize rapid identification and feature statistics extraction of vehicle loads within the collaborative monitoring range of the structure group.
   (5) Through collaborative analysis of monitoring data obtained in the structure group of various bridge structures, the inherent mechanism of the structure in

operation is revealed, the inherent mode of structural change is established, and the control indicators and evaluation methods are formulated.

(6) Data mining, machine learning and other big data application layer related technologies can be used to automatically detect any data anomaly and improve data accuracy.

2. In Terms of Bridge Detection

(1) Image acquisition and artificial intelligence recognition processing;
(2) 3D laser point cloud and reconstruction technology;
(3) Damage and crack detection (acoustic emission, etc.);
(4) Structural geometry evaluation based on spatial data.

In terms of urban scale assessment, the volume of big data has been considerable, for example, the use of UAV photography, image processing and cloud computing technology to quickly assess the condition of a city after an earthquake.

When using database or data warehouse for background management of bridge data, we should consider its scalability, maintainability, security, ease of use, and stability, and make a good preliminary plan, including the selection of database software, database and table, field design, index sorting, read/write separation, master slave replication, load balancing, and other aspects. For example, static structured data from sensors on bridges can be stored in MySQL, a small amount of information such as monitoring data statistics, analysis and evaluation results can be stored in MySQL tables by year/month, and a large number of dynamic data monitored on bridges can be stored in MongoDB or time series data inventory. For another example, the data can be stored in zones according to the call frequency to provide different access speeds. Hot data are within one year, moderate data within one to three years, and cold data over three years.

At present, there are few mature large bridge databases in China and abroad, and the famous ones are listed below:

SuJiaoKe Group has undertaken the design and implementation of structural health monitoring systems for more than 70 bridges in more than 10 years, including 11 Yangtze River Highway Bridges, and is currently the company with the best performance in bridge structural health monitoring of kilometre scale in the world [3]. Jiangsu Long and Large Bridge Health Monitoring Data Center, established in 2010, is the first regional bridge health monitoring system data center in China. Up to 2018, seven bridges under the control of Jiangsu Long and Large Bridge Health Monitoring Data Center had a cumulative data volume of 15 TB, which would continuously accumulate at an annual increment of 3.5 TB (including 1.05 trillion data records). At the same time, a special fault inspection and commissioning system was developed to improve the integrity of the system from 78.8% in 2010 to 85.5% in 2018. 850 billion pieces of data were available for analysis each year.

The NCHRP Bridge Inventory Database (NBI) of the United States, as a part of the Long Term Bridge Performance Program (LTBP), was established by NCHRP. LTBP [4, 5] was launched in 2008 with a term of validity of 20 years. The purpose is to select typical bridges in the United States, collect data on their performance changes, establish a detailed and timely bridge health database, carry out comprehensive researches on the theory of the technologies applicable to bridge structural performance, use big data to improve the management of traffic accidents, and ultimately improve the safety, reliability and long-term life of highway bridges in the United States. The database to be built will integrate multiple data sources, including NBI, PONTIS bridge management database, weather traffic and other data. At present, people can query data in the database through the Federal Highway Administration (FHWA) website, such as bridge type, age, weather statistics, traffic information, geographical location, and evaluation results in the past 20 years.

5.2 Cloud Computing

Conventionally, to establish an IT system, the enterprises need to purchase not only hardware and other infrastructure, but also software licenses, and need special personnel to maintain the IT system. When the enterprises expand, they will need to upgrade the software and hardware supporting facilities to meet the increased needs. For the enterprises, computers and other hardware and software are not what they really need as these are just tools for the enterprises to complete their jobs efficiently. For individuals, we need to install software when using computers, but for some users who do not often use the commercial software, it is not cost-effective to buy the software. So people began to seek services to rent such commercial software and hardware services from centralized sites. In this context, cloud computing came into being. This section introduces the basic concepts of cloud computing, its characteristics, service models, deployment models, commercial products of cloud computing and application scenarios related to bridge engineering.

5.2.1 Overview of Cloud Computing

1. Definition and Characteristics of Cloud Computing

Cloud computing is a new computing business model developed from distributed computing, parallel computing and grid computing. At present, common understanding of cloud computing is constantly changing. There are many definitions of cloud computing in the research community, one of which is given by the National Institute of Standards and Technology (NIST), that is, cloud computing is a model that provides universal, convenient, on-demand network access to configurable computing resources (such as networks, servers, storage, applications and services) in a shared pool. These resources can be quickly configured and released through little management work or little interaction with service providers. Through this technology, network service providers can process tens of millions or even billions of information in seconds, and provide network services with the same powerful efficiency as "supercomputers". Cloud computing has the following characteristics [6]:

(1) Self service on demand. Users can request for one or more services at any time, and can use the "pay as you go" approach to pay, without using the online control panel to interact with people.
(2) Extensive network access. Resources and services in different vendor regions in the cloud can be obtained from a wide range of locations, or provided through the standard mechanism of uncoordinated thin clients or thick clients. The terms "easy to access standardized mechanisms" and "global reach capability" also refer to this feature.
(3) Resource pooling. In cloud computing, infrastructure providers can dynamically allocate computing resource pools to different users. In this sense, users do not know and need not know the actual location of the provided resources.
(4) Fast elasticity. Fundamentally, elasticity is another name for scalability. It refers to the ability to increase (or decrease) resources proportionally when needed. Users can request different services and resources at any time according to their needs. This is a very important feature of cloud computing. Amazon, a famous cloud service provider, named one of its most popular and commonly used services Elastic Compute Cloud (EC2).
(5) Utility based pricing. Cloud computing provides a "pay as you use" billing model, allowing each client to pay according to use of cloud services; i.e., users buy the cloud services on demand and pay in the same way as paying electricity, water and gas supplies.
(6) Multi resilience. This is the fifth cloud feature proposed by the Cloud Security Alliance (CSA). Multi resilience means that there must be policy driven implementation, segmentation, isolation, governance, service levels, and charging/billing models for different consumer categories.
(7) Auditability and verifiability. For services, it is equally important to prepare logs and plan paths to assess their compliance with regulations and policies.

As shown in Fig. 5.7, cloud platform has the functions of data processing and data storage. Users can view cloud data and send instructions to make decisions through the cloud. Unlike edge computing, the data collection end in cloud computing does not perform computing tasks, and all data flows in and out through the cloud.

Cloud computing makes users' computers very simple in function. In addition to sending instructions to and receiving data from the "cloud" through the browser, users' computers can essentially use the computing resources, the storage space and the various applications provided by cloud service providers without doing anything. When the wire connecting the "monitor" and the "host" is infinitely long, the monitor can be placed in front of the computer user and the host can be placed as far away as the user does not know. Cloud computing turns the wires connecting "monitors" and "hosts" into networks, and turns "hosts" into server clusters of cloud service providers. At the same time, users' using concepts have also changed from "purchasing products" to "purchasing services".

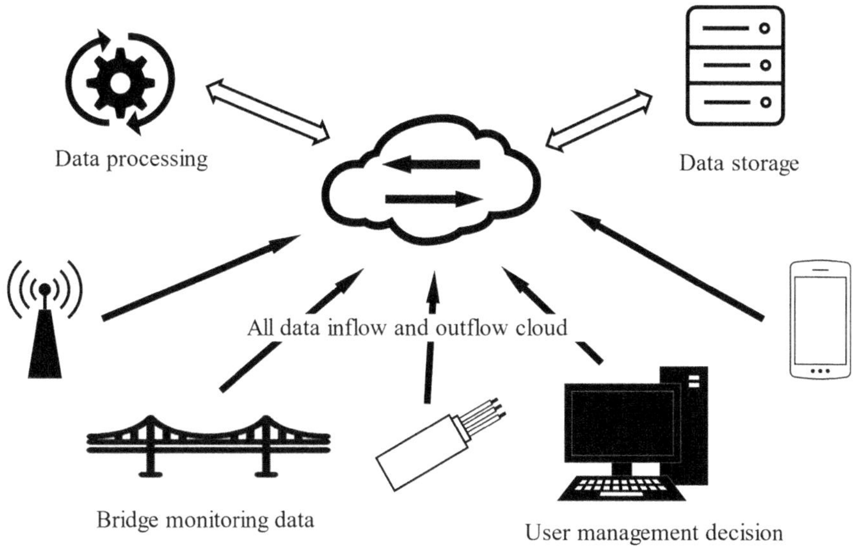

Fig. 5.7 Framework of cloud computing

2. Service Model of Cloud Computing

At present, the main service models of cloud computing include SaaS (Software as a Service), PaaS (Platform as a Service), and IaaS (Infrastructure as a Service).

(1) SaaS—Software or applications executed on the supplier's infrastructure are recognized as services, but consumers have limited access rights. Services are provided through thin clients, such as Web browsers, or programmatic interfaces for sending data and receiving results. Consumers do not know the infrastructure of the application provider, nor do they know that they have limited permission to configure certain settings. For small businesses, SaaS is the best way to adopt advanced technologies. However, some other enterprises, especially large enterprises, will choose not to use SaaS because of security issues to protect their core data. At present, SaaS applications mainly include email services, Google Docs, Microsoft Office Live, online anti-virus software, online conferences, games, videos and other online entertainment applications. Examples of cloud service providers include salesforce.com (SFDC), Google, Microsoft, Oracles, NetSuite, and IBM.
(2) PaaS—In PaaS, service providers provide appropriate basic services, including operating systems, networks and servers, as well as development tools that allow users to develop applications or software acquired and to manage their configuration settings. The main examples are GAE (Google App Engine), force.com (from salesforce.com), and Microsoft Azure. Compared with the software development platform based on data center, PaaS has a very low cost.
(3) IaaS—Users have developed the required applications and only need a basic infrastructure. In this case, the processor, network and storage can be provided

to consumers by the supplier as a service. The most famous providers of IaaS are Amazon EC2 (Amazon Elastic Compute Cloud) and Amazon S3 (Simple Storage Service), Rackspace ECS, Sun Microsystems cloud service, Flexiscale, Terremark and Dropbox.

Classification of the above service models has been widely recognized in the industry. However, there are more detailed classifications according to the services provided. For example, X as a Service (XaaS), where X can be software, platform, infrastructure, hardware, framework, business or organization.

3. Deployment Model of Cloud Computing

(1) Public cloud—This is the main deployment model of cloud computing where the cloud owners provide public services in most cases of the Internet according to predefined rules, policies and pricing models. Because it has a large number of widely distributed world resources managed and operated by public cloud service providers, the providers can offer their consumers with different appropriate resources while ensuring service quality. The cloud may be deployed locally or elsewhere.

(2) Private cloud—Private cloud is designed and established to provide internal public cloud for special organizations or institutions. Similar to the intranet function, all cloud resources and applications in the private cloud are managed by the organization itself. Only the organization and its designated stakeholders have the right to operate a specific private cloud, so as to achieve effective control of data, security, compliance and quality of service under the control of the organization. Due to the organization's firewall settings, such systems have high security. However, since the organization implementing the private cloud needs to be responsible for all transactions in the system, it is relatively more expensive to establish a private cloud. The cloud is located within the organization and may be hosted elsewhere.

(3) Community cloud—For similar interests and needs, some organizations will establish a community to share cloud computing for community members' use. Third party service providers or community members can be responsible for providing the infrastructure required for cloud computing. Reducing costs, allocating expenses among the community members, and supporting high security are the most important advantages of community cloud. The cloud may be deployed locally or elsewhere.

(4) Hybrid cloud—A different cloud model created by combining two or more different public clouds, private clouds or community clouds. The infrastructures that support them not only maintains their own specific properties, but also have standardized or agreed functions to enable interoperability and portability of applications and data. Combination of multiple clouds of the same type belongs to the category of multi cloud, for example, the hybrid cloud formed by combination of two private clouds belongs to the category of multi cloud. Presently, the hybrid private cloud and public cloud is the most popular because when the temporary demand for private cloud

resources is too large, the public cloud resources are automatically leased to stabilize the peak demand for private cloud resources. For example, the website has a huge number of hits during holidays, and during this time it will temporarily use public cloud resources for emergency response.

5.2.2 *Cloud Computing Platform*

1. Overview of Cloud Computing Platform

Cloud platform is a platform for massive data access and data bearing based on hardware resources and software resources to provide the necessary computing, network and storage capabilities. Data is usually stored in the hard disk of the ECS in the form of files, and can be read, written, and managed using database management software. When a user needs a certain functional service, as long as he sends a network request to the cloud platform, the computed results will be presented to him in a short time.

The emergence of cloud platforms is inevitable because the Internet industry is developing rapidly, product requirements are constantly changing, and higher requirements are put forward for rapid development of applications, simplification and automation of management, and reliability and availability of applications. Technically, with the construction of infrastructure and the wide application of virtualization technology and container technology, the emergence of centralized and unified application platforms also became possible. Each cloud platform also has a unique SDK (software development kit) and a rich API (application program interface), which enable it to cooperate with the development code for programmatic operation and realize some common or unique powerful functions of businesses. Cloud platform provides the developers with the basic tools for the entire development cycle, ensures the stability of operation and convenience of management, reduces the requirements for knowledge system, greatly improves the agility of development, and speeds up the production of products. At the same time, it attracts more users and provides more data support for itself while providing richer service content. Unconsciously, modern society has entered the era of cloud platform. Gartner's 2014 survey showed that 72% of server workloads had been virtualized at that time, nearly 1/3 of enterprises had the ability to establish private clouds, and the server virtualization ratio would reach 100% in the future.

Today, cloud computing is in an era of rapid development. There are many kinds of general cloud platform functions, and each has its own characteristics and strengths. Table 5.3 lists some well-known open source and commercial cloud platform products.

There are various companies providing cloud computing services. Companies like Amazon, Microsoft, Salesforce, IBM and Yahoo are some of the pioneers. Many companies are trying to occupy the cloud computing market, while Chinese companies like Alibaba are also growing rapidly. Goldman Sachs Research News pointed out that AWS's real competitors are not in the United States, but Alibaba

Table 5.3 List of general cloud platforms

Type	Main functions	Cloud platform	Company
Open Source	Cloud computing platform	AbiCloud	Abiquo
		Eucalyptus cloud	Huiyu
		Nim-bus	Nim-bus
	Big data platform	Hadoop	Apache Foundation
	Database storage platform	MongoDB	10gen
Business	Comprehensive	Azure	Microsoft
		AWS	Amazon
		Blue Cloud	IBM
		Alibaba Cloud	Alibaba
		Tencent Cloud	Tencent
		Hua Weiyun	Huawei

from China. AWS and Alibaba Cloud will become the world's largest two infrastructure technology companies. The following briefly introduces some industry-leading comprehensive business service providers and their cloud platforms.

2. Cloud Platform Service Companies and Products

   (1) Amazon—In the IaaS field, Amazon is undoubtedly a pioneer and market leader among the cloud providers. Amazon has a set of cloud services called Amazon Web Services. These services include Amazon EC2, Amazon S3 and Amazon SimpleDB. EC2 provides computing servers based on rent; S3 uses a simple Web interface to provide data storage based on rent; and SimpleDB is a non relational database that provides Web based services for real-time query of structured data.
   (2) IBM—The "Blue Cloud" solution is an enterprise level cloud computing solution developed by IBM Cloud Computing Center. This solution can integrate the existing infrastructure of the enterprise, build the enterprise's own cloud computing center through virtualization technology and automation technology, realize the unified management, unified allocation, unified deployment, unified monitoring and unified backup of enterprise hardware resources and software resources, break the monopoly of applications on resources, and thus help enterprises realize the concept of cloud computing.
   (3) Microsoft—In the field of SaaS, Microsoft has launched the Business Productivity Online Suite (BPOS), a software service product for enterprise level communication, which includes services such as Exchange Online, SharePoint Online, Office Communications Online and Live-Meeting Online. These online services are designed to provide the enterprises with simplified communications through high availability, comprehensive security and simplified IT management. In the PaaS field, Microsoft provides a platform called Microsoft Azure, which supports a large number

of open source applications, frameworks and languages, and the number is still growing. Its main goal is to provide a platform for the developers to help them develop applications that can run on cloud servers, data centers, Web and PCs. Cloud computing developers can use the storage, computing power and network infrastructure services of Microsoft's global data center.

(4) Alibaba—The most representative commercial general cloud platform in China is Alibaba Cloud. Thanks to its brand effect and technical capabilities, 40% of China's top 500 enterprises currently use Alibaba Cloud, and about half of China's listed companies use Alibaba Cloud. Alibaba Cloud products fall into six categories: cloud computing foundation, security, big data, artificial intelligence, enterprise applications, and the Internet of Things. The integrity of Cloud Essentials product system is leading the world. The basic products and functions have been continuously upgraded, and the enterprise cloud computing, operation and maintenance, development and management capabilities have been continuously improved through emerging technologies. Alibaba Cloud Realtime Compute (Powered by Ververica) is an enterprise level big data computing platform built on Apache Flink provided by Alibaba Cloud. It can support sub second level processing delay on petabyte level data sets, enabling users' standard real-time data processing processes and industry solutions; support the development of Datastream API jobs, provide unified Flink SQL for batch streams, and simplify the development in BI scenarios. It can be seamlessly connected with the big data components that users have used, and its more value-added features can help enterprises transform in real time.

5.2.3 Application of Cloud Computing in Bridge Engineering

Cloud computing relies on the Internet for on-demand delivery of computing resources. It has the advantages of super scale, high reliability, high scalability, on-demand services, etc. It can manage the drawings of construction sites, track and update information in real time by using Internet resources. The cloud platform enables all parties involved in cloud computing in the project to work together, including the owner, design unit, construction unit, management unit and consulting unit. During the whole project life cycle, all information can be directly traced on the cloud.

In recent years, how to deal with the massive data obtained during bridge construction, operation and maintenance has become an important research direction. Some scholars have introduced cloud computing into bridge management. Jeong et al. [7] presented a cloud-based cyber-infrastructure platform designed to support bridge monitoring in their paper. This cyber-infrastructure platform enables scalable management of SHM data, and facilitates effective information sharing and data utilisation. A cloud-based platform comprises of virtual machines, distributed database and web servers. The peer-to-peer distributed database architecture provides a scalable and fault-tolerant data management system. Platform-neutral web services designed in compliant with the Representational State Transfer (REST) standard enables easy access to the cloud resources and SHM data. For data interoperability, a bridge information model for bridge monitoring applications is adopted. The scalable cloud-based platform was implemented for the monitoring of bridges along the I-275 corridor in the State of Michigan.

Dang et al. [8] proposed a cloud-based digital twinning framework for SHM (cDTSHM) for real-time monitoring and proactive maintenance of civil structures. This method facilitates two-way mapping between physical structure and digital counterpart, as well as interaction among structure, machine and human, paving the way towards real-time intelligent monitoring. A layer of fog computing was implemented prior to the cloud layer to reduce the data volume and the computational demand for the digital model. Case studies of structural damage detection in bridges using DL algorithms were conducted and achieved accuracy of 92%. Deng et al. [9] detailed a case study of commercial cloud computing for connected vehicle (CV) applications in transportation cyber-physical systems with Amazon Web Services (AWS). This study revealed the feasibility of developing a CV application using a serverless cloud architecture that can meet the delay requirements of CV applications without the burden of provisioning, configuring, operating and maintaining server instances. Moreover, serverless cloud architecture is more cost-efficient and provides the scalability and flexibility needed for robust CV applications (Fig. 5.8).

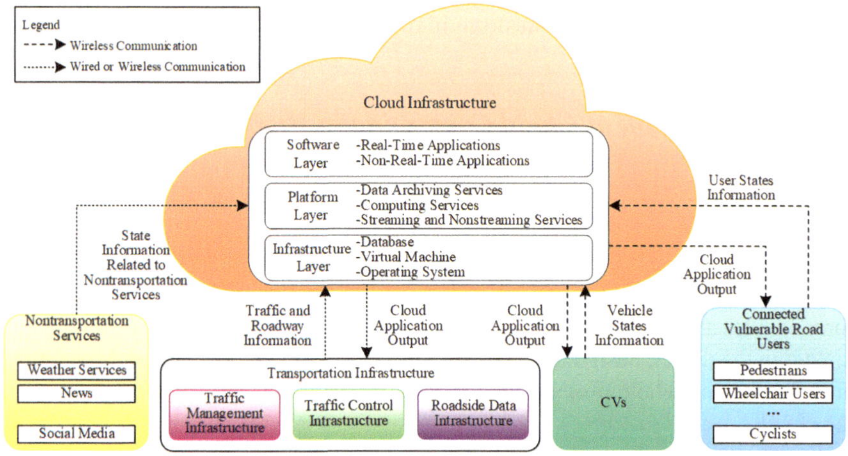

Fig. 5.8 Cloud-based transportation cyber-physical system for connected vehicles

5.3 Edge Computing

With the advent of the Internet of Everything era, the network will generate a large amount of data. If all the data are sent to the cloud for processing, it will pose a great challenge to the bandwidth and the reliability of the network. Edge computing, as a computing model for data processing at the edge of the network, can effectively relieve the pressure of the network and shorten the response time, so researchers gradually attach importance to it. This section focuses on the difference between edge computing and cloud computing, the development history of edge computing and its application in the engineering field.

5.3.1 Concept of Edge Computing and Its Difference from Cloud Computing

Edge computing refers to a new computing model that performs computing at the edge of the network. The objects of edge computing operations include downlink data from cloud services and uplink data from Internet of Things services. The edge in edge computing refers to any computing and network resources on the path between the data source and the cloud computing center [10]. For example, smart phones are the edges between the users and the cloud, gateways in smart homes are the edges between the homes and the cloud, and micro data centers and cloudlet are the edges between the mobile devices and the cloud. Figure 5.9 shows the schematic diagram of edge computing. The request transmission between the terminal device and the cloud center is bidirectional. The edge device can not only request content and services from the cloud center, but also perform some computing tasks, including data storage, data processing, data caching, device management, privacy protection, etc.

Edge computing is a decentralized computing architecture which divides the computing tasks of applications, data and services from the network center node into smaller and more manageable parts, and passes them to the logical edge node of the network for processing. Figure 5.10 shows the application of edge computing in bridge monitoring. The arranged system is used to monitor the bridge displacement. The strain sensor completes the computation at the edge node, and then transmits the computed results to the cloud server. As the bridge edge node is closer to the terminal sensor, the data processing and transmission is speeded up, and the delay is reduced. Under this framework, data analysis and knowledge generation are closer to the source of data, so the edge computing is more suitable for processing big data. Figure 5.11 compares cloud computing with edge computing.

Compared with cloud computing, edge computing has the following advantages: security, cognition, agility, low latency and efficiency. Cloud computing uploads all data to the cloud for processing and storage, while edge computing uses local servers to process the data first, and then transfer it to the cloud for storage.

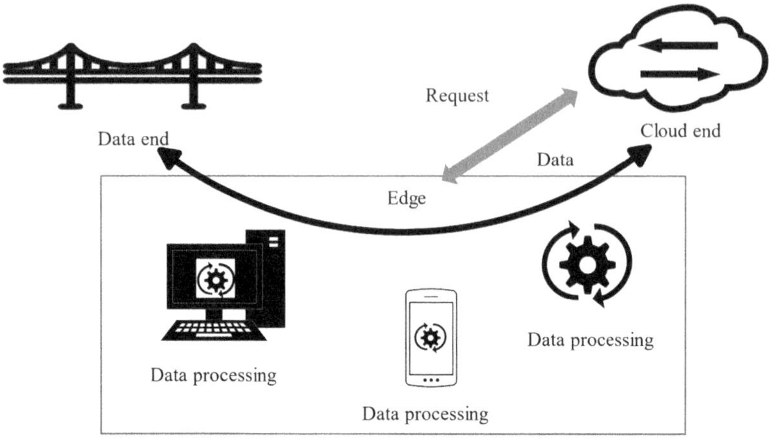

Fig. 5.9 Schematic diagram of edge computing

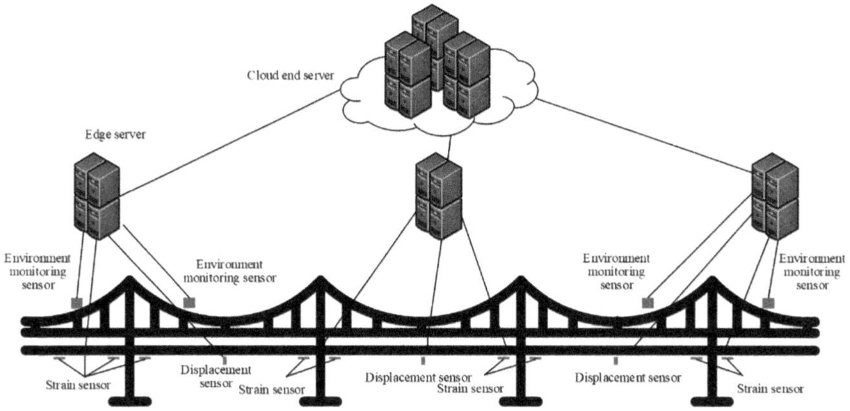

Fig. 5.10 Schematic diagram of edge computing applied in bridge monitoring

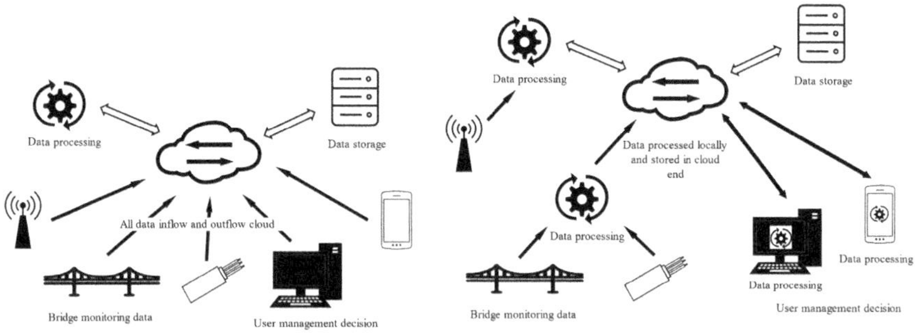

Fig. 5.11 Comparison between cloud computing and edge computing

(1) Edge computing provides additional security to IoT devices to ensure trustworthiness of their transactions. For example, the wireless sensors deployed in outdoor environments now often need to update their remote wireless source codes to solve security related problems. However, due to the dynamic environment factors, such as unstable signal strength, signal interruption, bandwidth constraints, etc., remote central back-end servers may face the challenge of quickly executing updates, thus increasing the risk of network being attacked. In addition, if edge computing devices are available, the back-end can configure the best routing path in the entire network through various edge computing nodes, so as to quickly perform software security updates on wireless sensors.
(2) Edge computing enables customers to realize when and where to deploy computing, storage, and control capabilities to support autonomous decision-making. Essentially, the awareness of edge computing involves a large number of mechanisms such as self adaptation, self-organization, self-healing, self-expression, etc. It transforms the role of IoT devices from passive intelligent devices to active intelligent devices, enabling them to operate continuously and respond to customers' needs without relying on decisions from the remote cloud.
(3) Edge computing enhances the agility of large-scale Internet of Things system deployment. Compared with the existing public cloud service business model that relies on large business holders to establish, deploy and manage infrastructure, edge computing provides the individuals and small enterprises with the opportunities to use the general open software interface or open software development kit (SDK) to obtain edge computing services. For example, the MEC (Multi access Edge Computing) standard of the European Telecommunications Standards Institute (ETSI) and the Indie Fog business model will accelerate large-scale deployment of IoT infrastructure.
(4) Edge computing can provide fast response for applications requiring ultra-low latency. Specifically, in many ubiquitous applications and industrial automation, the system needs to continuously collect and process sensor data in the form of data flow, so as to identify events and perform operations in a timely manner. Obviously, by applying edge computing, these systems can support the time sensitive functions. In addition, edge computing has the characteristics of software, in which the behavior of physical devices can be completely configured by the remote central server using software abstraction, which provides a highly flexible platform for the rapid reconfiguration of IoT devices.
(5) By improving performance and reducing unnecessary costs, edge computing can improve the efficiency of C-IoT. For example, by applying edge computing, a healthcare or elderly care system can assign many tasks to the Internet gateway devices of healthcare sensors, and use these gateway devices to perform sensing data analysis tasks. Ideally, because this process takes place near the data source, the system can generate results faster. In addition, the system uses gateway devices to perform most tasks, greatly reducing the unnecessary cost of output communication bandwidth.

5.3.2 Application of Edge Computing

1. Smartphone Monitoring

Modern smart phones are equipped with accelerometers, gyroscopes and GPS, which can be used effectively for structural state assessment. Additionally, the popularity of cheap smart phones, their mobility, huge storage capacity, powerful computing power and easily modified software make them potential edge computing devices. It has become very common to use them to monitor and transform building structures. Furthermore, due to the attractive features of some smart phones, they have great potential for SHM applications.

A bridge seismic monitoring system based on smart phones developed by the authors' team [11] uses smart phones as smart sensors to process monitored data at the data acquisition end, and remotely control and access data processed at the smart phone end of the cloud. The monitoring system was applied in the long-term monitoring of Takamatsu Bridge in Japan, and successfully captured some seismic information.

The monitoring system obtains, analyzes, stores and transmits data to the cloud through self-developed measurement applications and interaction between the smartphone hardware and operating system functions. As shown in Fig. 5.12, the monitoring system includes smart phones that measure data independently. These devices are connected to a public network to achieve mutual communication. In addition, the process of data in real time by using the smart phone's computing power can quickly diagnose the structural damage and generate damage index report or emergency alarm output. By comparing the waveform and Fourier spectrum of acceleration records with the acceleration sensor (Hakusan Industrial SU501), the effectiveness of using smart phones to monitor earthquakes is verified. See Chap. 3 for examples of detailed comparison results.

2. Intelligent Camera Monitoring

Intelligent camera combines video sensing, advanced video processing and communication with the embedded devices. This camera is a key component of a new type of

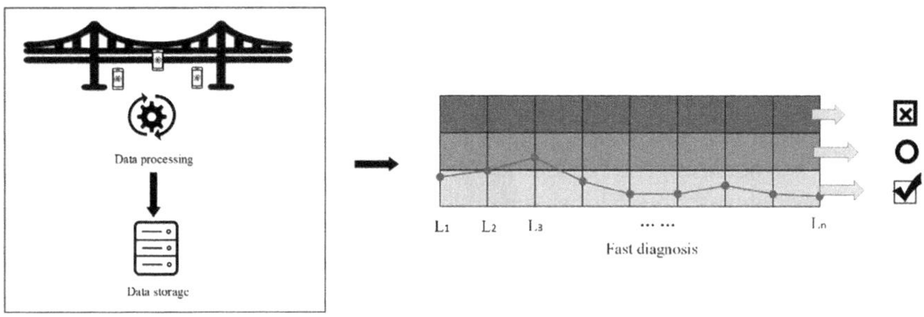

Fig. 5.12 Smart phone monitoring system

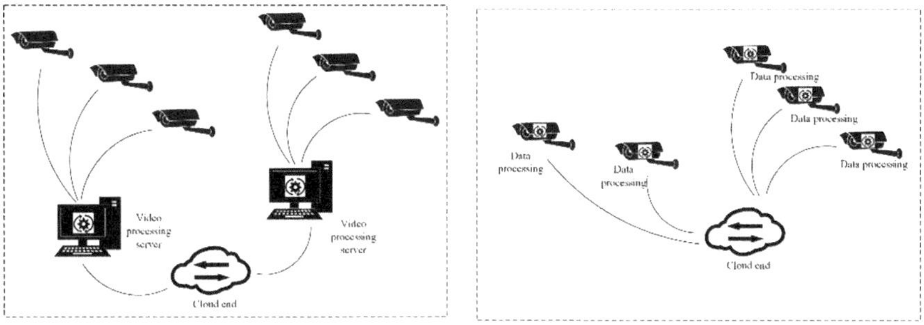

Fig. 5.13 Schematic diagram of intelligent camera monitoring with edge computing [12]

surveillance system. With the advent of the Internet of Things era and the popularity of various video capture devices, putting video processing on the edge can reduce the operating and storage pressure of the central processor and reduce the response time.

Michael et al. [12] developed a prototype intelligent camera for traffic monitoring, proposed an extensible architecture composed of CMOS sensors, digital signal processors (DSP) and network processors, and applied the algorithm for intelligent detection of stationary vehicles in the prototype based on MATLAB to the embedded DSP module of the intelligent camera so that the video data can be transmitted to the central monitoring station. The static vehicles in the video are labeled in real time. Figure 5.13 depicts the intelligent camera monitoring with edge computing.

As mentioned above, the intelligent camera composes of three parts: video sensor, processing unit and communication unit. The video sensor undertakes the works of the first stage of the entire data stream. The sensor captures the incident light, converts it into electrical signal, and then transmits it to the processing unit. CMOS sensors can best meet the requirements of video sensors. Due to their logarithmic characteristics, these sensors have significant dynamic characteristics and can provide amplification. The second stage in the entire data stream depends on the processing unit. Because vehicle image and video processing require very high computing performance, the intelligent camera is equipped with a digital signal processor. The final stage of the overall data flow of the intelligent camera relies on the communication unit, which composes of an Intel XScale IXP425 processor and can manage PCI, Ethernet, USB and other interfaces.

A series of images in the figure show the static vehicles were detected after a traffic accident in the tunnel. The left subfigure shows the situation after the accident. About 3 s after the accident vehicle stopped moving, the first part of the accident vehicle was identified as a potential stationary area (middle image). After 3 s, the system detected that the accident vehicle was stationary and gave an alarm (right subfigure). Since the truck stopped moving several seconds after the accident vehicle stopped moving, it was not detected that the truck was stationary. This result confirmed the effectiveness of the monitoring practice.

5.4 Crowd Sourcing

Crowdsourcing refers to the participation of non specific groups in a task with or without payment (mostly free), and achieve completion of the overall task similar to each person completing a small part of it. This concept was put forward in 2006, and since then, it has been more and more widely used with the rapid development of Internet and computer technology. Compared with the conventional model, crowdsourcing has the advantages of cost, efficiency, data sources and various other aspects. In the field of civil engineering, massive detection, monitoring data collection, simple or complicated processing tasks can all be completed in a faster and better way through crowdsourcing.

5.4.1 Concept and Classification of Crowdsourcing

1. Concept of Crowdsourcing

The term "crowdsourcing" first appeared in Wired magazine in 2006. Its original meaning is that a company or institution outsources a task to an unspecified (usually large) crowd. Later, its application scope gradually expanded. In addition to the original form of "crowdsourcing" in which non specific groups jointly complete a task, other forms of practices can be seen as applications of crowdsourcing in the commercial, industrial, scientific and other fields, e.g., groups composed of single users contribute their individually collected/owned data to construct databases, product design competitions held to select excellent design schemes, etc.

2. Classification of Crowdsourcing

Crowdsourcing activities can be divided into two categories: selective crowdsourcing and integrated crowdsourcing [13]. Selective crowdsourcing is mostly used for the comparison and selection of schemes of institutions or companies, which requires high professional skills of the participants. The number of participants shows no direct correlation with the success of crowdsourcing. Integrated crowdsourcing involves a large number of users to produce massive results and achieve their goals. The more participants the crowdsourcing involves, and the more data the participants provide, the more valuable crowdsourcing activities will be.

Integrated crowdsourcing is widely used in civil engineering, so it is necessary to introduce two categories of integrated crowdsourcing—crowdsourcing computing and crowdsourcing awareness, as shown in Fig. 5.14.

Crowdsourcing computing refers to the crowdsourcing tasks perform by the public that involve computing or processing. Users upload their results according to the assigned information to the platform after processing, such as annotated image, translated text, efficiency evaluation result, etc. Crowdsourcing awareness is used for data collection and processing. In typical application scenarios, such as the establishment of databases and information databases, the users use their own mobile devices to

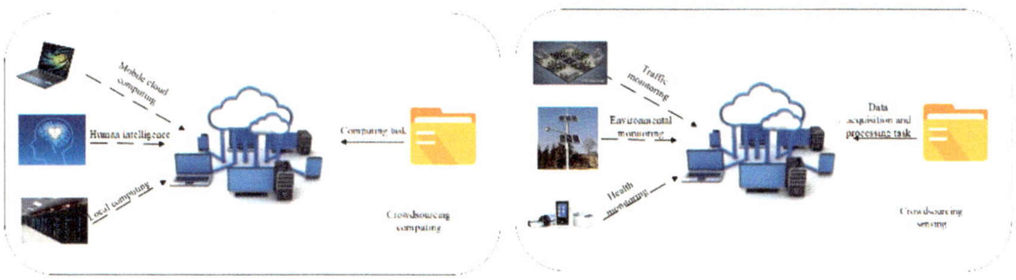

Fig. 5.14 Crowdsourcing computing (left) and crowdsourcing awareness (right)

collect a large amount of data, process and upload these data through local servers or mobile cloud computing.

5.4.2 Implementation Process of Crowdsourcing and Mobile Computing

1. Implementation Process of Crowdsourcing

A crowdsourcing project can be divided into three phases. First, a complex and huge task is divided into several simple and manageable micro tasks. The size of the task should be for a single user to quickly get accurate results, while making the user feel that he/she has made meaningful contributions to the project. Second, the crowdsourcing platform needs to strive to attract and organize sufficient number of users who can contribute, and encourage them to make accurate contributions. Finally, after collecting large number of user feedback results, the platform should distinguish the quality of individual users' contributions through back-end processors, and combine them to produce an integrated reliable solution.

In order to make crowdsourcing activities more efficient and the final project results more accurate, the following three points should be paid special attention to:

First, selection and training of participants. Extensive use of social media and certain rewards provided to recruit more volunteers to participate in the project. Participants should receive some training to ensure that the results they submit are reliable and effective.

Second, construction of crowdsourcing platform. Internet based applications should be developed so that users can log in to the platform with their own accounts to access the project units they are responsible for, and can use the tools provided on the platform to quickly complete their tasks.

Third, evaluation of individual contributions. Due to different knowledge backgrounds and abilities, participants' contributions may have gaps in quality. Based on the quality of the data contributed by each user, the contributions will be marked with credibility to facilitate future data screening, and users with more contributions will be given more rewards.

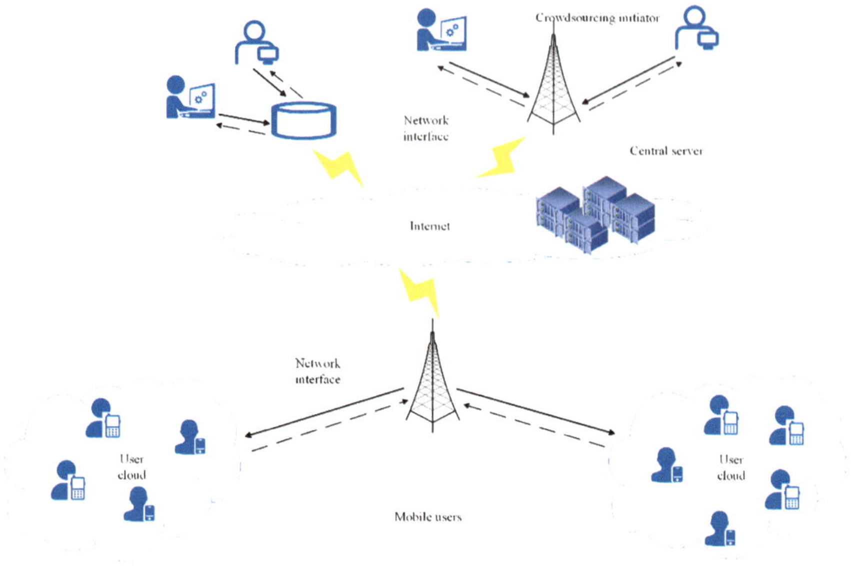

Fig. 5.15 Framework of mobile crowdsourcing [14]

2. Mobile Computing [14]

Mobile computing enables computers and other information devices to transmit data and share resource without being connected to a fixed physical device. It helps to share the computing pressure of the central information system and provides useful, accurate and timely information to any user who needs it at anytime, anywhere.

Based on mobile computing, group participation, processing and uploading of results in crowdsourcing become possible. Individual users use their own mobile devices to upload their contributions to the unified cloud, and can view their tasks in the same cloud, which enables each individual in different space to participate in the same task at different time. Mobile computing and cloud services provide the most economical and convenient implementation solution to crowdsourcing, which is called mobile crowdsourcing (Fig. 5.15).

5.4.3 Application of Crowdsourcing in Bridge Maintenance and Management

1. Crowdsourcing for Bridge Health Monitoring

In conventional bridge monitoring, the sensors are of high installation and maintenance costs, plus with problems concerning remote monitoring and messy wire arrangement. Additionally, the conventional sensors have shortcomings such as high power consumption, low data acquisition stability and limited network stability.

Hence, the development of Internet, smart phones and mobile networks has led to crowdsourcing being increasingly introduced to bridge monitoring.

The sensors embedded in smart phones can be used to measure structural vibration. Bridge inspectors can use their smart phones to transmit data to the online server, and compute the modal information of the bridge. This belongs to the application of crowdsourcing awareness mentioned earlier in Sect. 5.4.1, and its main purpose is data collection.

The crowdsourcing platform consists of three parts: participating citizens, administrators and network platform. Among them, participating citizens refer to those who use smart phones to collect and upload data. The administrator collects the best available vibration data and maximizes the efficiency and accuracy of the subsequent structural system identification; Finally, data integration and processing involve mobile sensing, server acquisition, digital signal processing and database storage, all of which are completed on the network platform [15, 16]. Figure 5.16 depicts the bridge monitoring crowdsourcing platform.

MyShake [17] is an Android based application that uses personal smartphone sensors to collect data and analyze earthquakes, it can detect whether the mobile phone movement is caused by earthquake or other human activities, and send the information back to the processing center, where the detection algorithm will confirm whether the obtained information is a seismic signal. The location, starting time and magnitude of the earthquake can be determined by multiple triggers in the telephone network. This information can be used to estimate the seismic intensity and the remaining time before the seismic wave reaches the target location.

2. Crowdsourcing for Bridge Condition Assessment

(1) Bridge Deck Condition and Performance Evaluation

Based on the data collected by vehicles, the Michigan Department of Transportation and the Automotive Research Center of the United States evaluated the condition

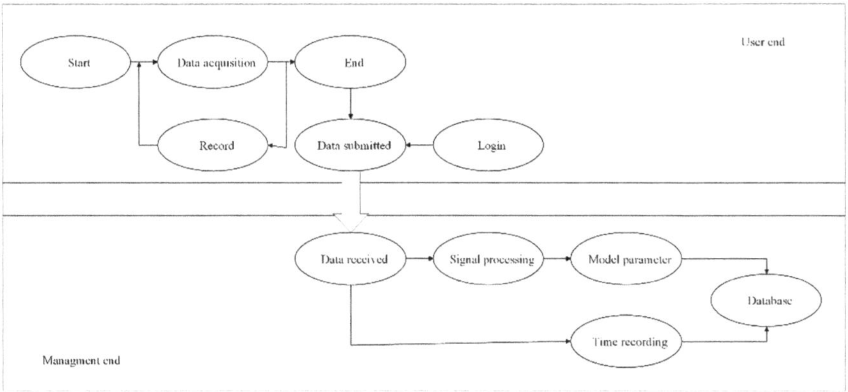

Fig. 5.16 Schematic diagram of bridge monitoring crowdsourcing platform [16]

and performance of a bridge deck [18]. The approach is to measure the acceleration data by the smart phone sensor when the vehicle is running on the bridge. The data obtained by this method is affected by coupling of the vehicles, the bridge deck and the driving conditions, leading to poor repeatability of the experiment and low accuracy of the data collected. Crowdsourcing, on the other hand, uses a large amount of data collected from multiple vehicles at different times to eliminate the limitation of generalization.

(2) Assessment of Bridge Distresses

In today's bridge detection and evaluation, the usual practice is to take the distress images during inspection, while the image analysis and bridge evaluation are carried out in the follow-up work, which can hardly be termed teamwork. The researchers established an inspection platform with the help of Collaborative Mobile Cloud Computing (CMCC). For the inspection to be successful, multiple groups of personnel need to check the target facilities synchronously, using mobile devices to capture the distress images and upload them to the cloud platform, and at the same time, simply mark the location, type and other basic parameters of the distress. As shown in Fig. 5.17, all uploaded images can be automatically classified according to the parameters, facilitating the image analysis and subsequent detection in the next step.

3. Crowdsourcing for Post Disaster Damage Assessment [19–21]

In post disaster emergency processing, the real image of the area provided by remote sensing technology greatly help in the assessments of damage distribution, economic loss and potential area that may need disaster relief within the affected region, while crowdsourcing provides a mechanism to decompose information processing tasks and relief the pressure on the emergency personnel. The Global Observing Disaster

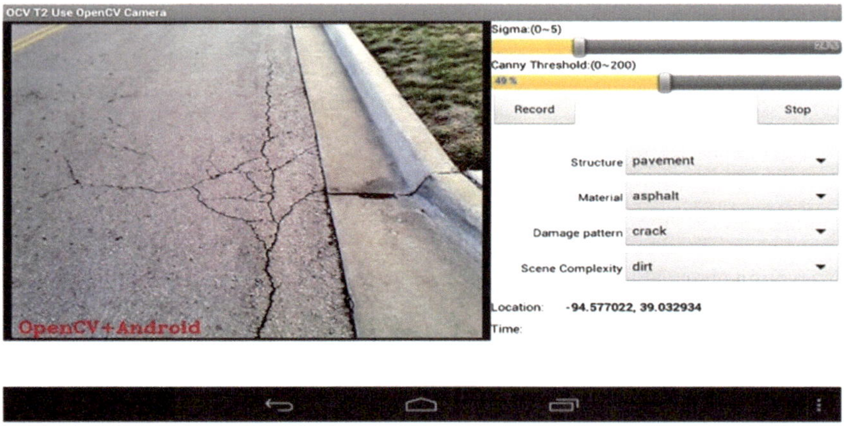

Fig. 5.17 Image upload and parameter input interface

Fig. 5.18 Example of post disaster assessment platform interface and post disaster assessment results [19]

Assessment Network (GEO-CAN) is such a crowdsourcing platform. Each participant will be assigned a certain area. By comparing the images before and after an earthquake in the area, the damage degree will be systematically assessed and submitted to the central database. Figure 5.18 shows a post disaster assessment map provided by the platform. Each polygon on the map is drawn by a participant and marked by the damage level.

After the Haiti earthquake, more than 600 people from 23 countries participated in this activity. Figure 5.19 shows the daily changes in the number of buildings damaged at levels 4 and 5 identified by volunteers in this activity. It can be found that the use of GEO-CAN has greatly accelerated the speed of building marking. After Christchurch earthquake in New Zealand, more than 200 users analyzed an area of more than 77 km^2 and evaluated more than 1400 buildings. At this time, users can use the simple interface on the web end instead of the GIS platform, and quickly get familiar with the task through the platform's training module.

In general, crowdsourcing has been successfully applied in post disaster building damage assessment, hence the same idea can be implemented for bridge damage assessment. This not only refers to the identification of damage level of a bridge after any disaster, but also includes assessment of the overall technical condition of the bridge or the technical condition of a single component in normal times. For the former, the requirements of professional participants are low; while the latter should be undertaken by people with relevant professional skills.

5.5 Collaborative Computing

The concept of collaborative computing originated from Computer Supported Cooperative Work (CSCW), which refers to a geographically dispersed group that cooperates with each other to complete a task with the help of computer and network

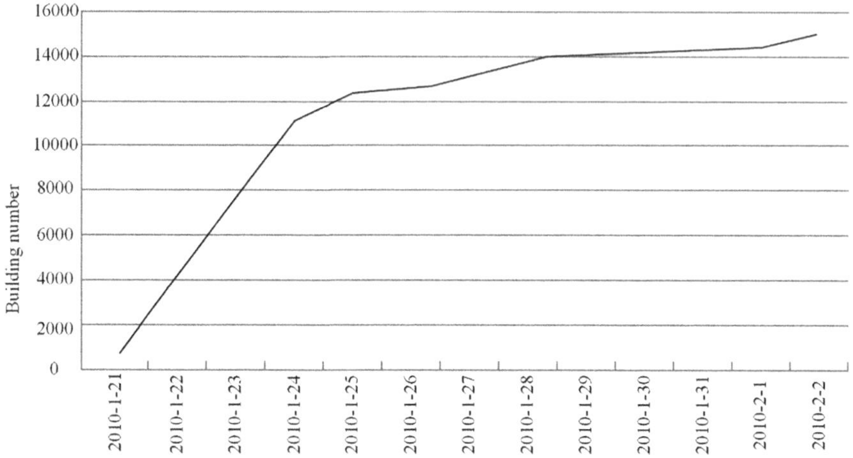

Fig. 5.19 Changes in the number of buildings damaged at levels 4 and 5 identified by volunteers every day [19]

technology. Since the concept was put forward in the mid-1980s, its application has spread in both the academic and industrial circles. Collaborative computing has the characteristics of group, interaction, distribution and collaboration, which can fully take the advantages of various computing resources and improve the quality and efficiency of group work. This section focuses on several collaborative computing technologies widely used in bridge engineering.

5.5.1 Cloud Edge Collaborative Computing

As shown in Fig. 5.20, in practices, close collaboration of edge computing and cloud computing is often needed to better meet the demands of various scenarios, thus maximizing the application values of edge computing and cloud computing. Edge computing is not only close to the execution unit, but also the collection and preliminary processing unit of high-value data required by the cloud, which can better support cloud applications. On the contrary, the business rules or models optimized by cloud computing through big data analysis can be distributed to the edge, and edge computing runs based on the new business rules or models assigned.

Cloud edge collaboration needs to ensure collaborations with regard to the following aspects: resources, security policies, application management, business management, etc.

(1) Resource collaboration includes the collaboration of computing, storage, network, virtualization and other infrastructure resources provided by the edge nodes for value-added network services, as well as the collaboration of life cycle management of the edge node devices. Among them, the collaborations

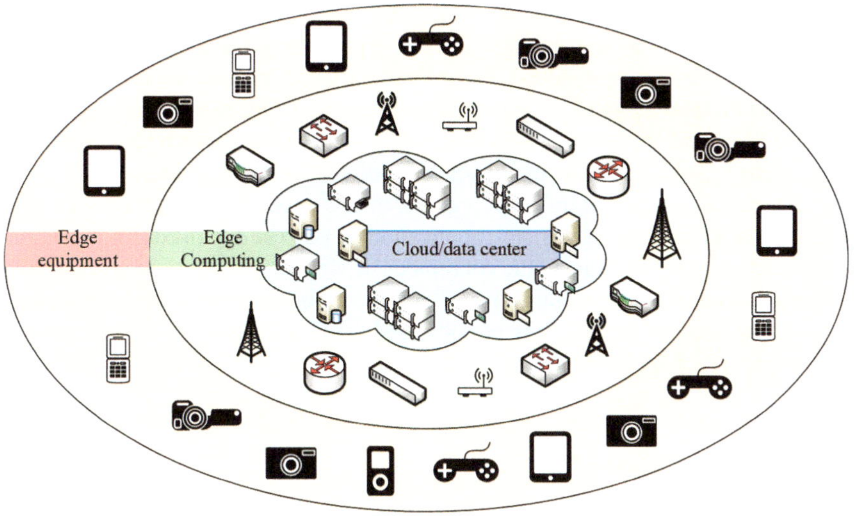

Fig. 5.20 Relationship between edge computing and cloud computing [22, 23]

of computing resources, network resources and storage resources are the major elements.

① Computing resource collaboration means that when the edge cloud resources are insufficient, the resources of the central cloud can be called upon to supplement the insufficiency of the resources of edge applications.

② Network resource collaboration refers to cases where there are multiple links between the edge side and the central cloud connection network. When the nearest network is congested, the network controller can sense the congestion and direct the traffic to the idle network.

③ Storage resource collaboration means that when the edge side storage is insufficient, part of the data will be stored in the central cloud so as to save the edge side storage resources, and these data will be transmitted to the client through the network when the application needs them.

(2) Security policy collaboration: with regard to security policy collaboration, if the central cloud finds a malicious traffic in an edge cloud, it can block it to prevent the malicious traffic from spreading across the entire edge cloud platform.
(3) Application management collaboration: edge nodes provide a deployment and running environment for network value-added applications. The cloud implements the lifecycle management of value-added network applications on edge nodes, including application push, installation, uninstallation, update, monitoring, logs, etc.

(4) Business management collaboration: edge nodes provide value-added network business application examples. The cloud provides unified business orchestration services for value-added network services, and provides the customers with relevant network value-added services on demand. Due to resource shortage on the edge side, the central cloud can process some applications with high priority so as to classify and process businesses with different priorities.
(5) Edge collaboration in different regions: in some specific application scenarios, such as the Internet of Vehicles, applications need to be deployed or hot migrated to different regions due to the continuous driving of vehicles. The central cloud needs to deploy applications in advance according to the regional requirements of applications in different periods, and issue strategies to achieve smooth migration of applications.

5.5.2 *Mobile Cloud Computing*

With the development of science and technology, smart phones have become an indispensable part of people's life and work. By combining mobile devices with cloud computing, resource intensive computing is migrated from mobile devices to clouds or servers of rich resources, enabling smart phones to execute more complex applications and achieve more difficult computing tasks. From this perspective, mobile cloud computing is a special kind of cloud edge collaborative computing. This section focuses on the concept, main architecture and implementation of mobile cloud computing.

As an emerging computing model, mobile cloud computing aims mainly at greatly improving the computing power of mobile devices by taking advantage of cloud resources in computing, storage and other aspects, so as to provide the mobile users with richer applications and better user experiences. A typical mobile cloud computing architecture is shown in Fig. 5.21. Mobile devices are connected to cloud servers on the Internet through wireless network using access methods such as base stations.

The core technology of mobile cloud computing is computing migration technology. As the name implies, computing migration technology is to migrate the storage, computing and other tasks of mobile terminals to the cloud data center or high-performance servers for execution, so that mobile terminals can obtain flexible and efficient services. The flowchart for the basic process of computing migration is shown in Fig. 5.22. The computing migration workflow starts with the execution of application program by checking the migration permissions of users. If the user has granted the permissions, the available (allocatable) cloud computing resources are checked and recorded. The next step is to determine whether the computing migration is beneficial to the user according to the user's expectations. If so, the computing migration will be performed and the relevant data and computing tasks will be migrated to the cloud for computing. Otherwise, the application performs all computations locally. After the mobile terminal is connected to the remote server,

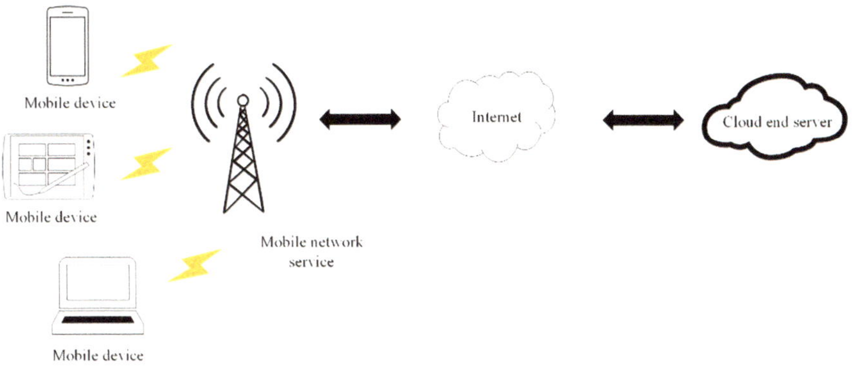

Fig. 5.21 Typical architecture of mobile cloud computing [24]

it can perform computing migration, which is a complex process and affected by various aspects of system hardware and software. Othman et al. [25] summarized the main factors affecting computing migration, as listed in Table 5.4.

However, mobile cloud computing at this stage also faces the following challenges:

(1) Mobility of users: mobility is one of the important characteristics of mobile cloud computing. By accurately and efficiently identifying the users' location, users can obtain cloud data on demand anytime anywhere.
(2) Mobile terminal's battery capacity: limited by the size, weight and battery technology of mobile terminals, it is difficult to significantly improve the available power supply capability of terminals in short time. In order to prolong the available time of mobile terminals, researchers have been paying attention to the development of energy saving technologies for mobile terminals.
(3) Protection of privacy: people have been used to backing up personal information such as the address books and photos in the cloud, or even directly using mobile phones to complete electronic payment. Therefore, data security and privacy protection in the mobile cloud computing environment are becoming more and more important.

5.5.2.1 Other Collaborative Computing

In addition to the relatively mature cloud edge collaborative computing, there are some other collaborative computing models as introduced below.

1. Edge to Edge Collaboration

Edge to edge collaboration is a research hotspot at present, which is mainly intended for solving two problems: (i) limited ability of edge computing—to deal with this issue, multiple edges can be required to carry out collaborative training; and (ii) data islands—the data source of the edge has strong locality, and needs to cooperate with

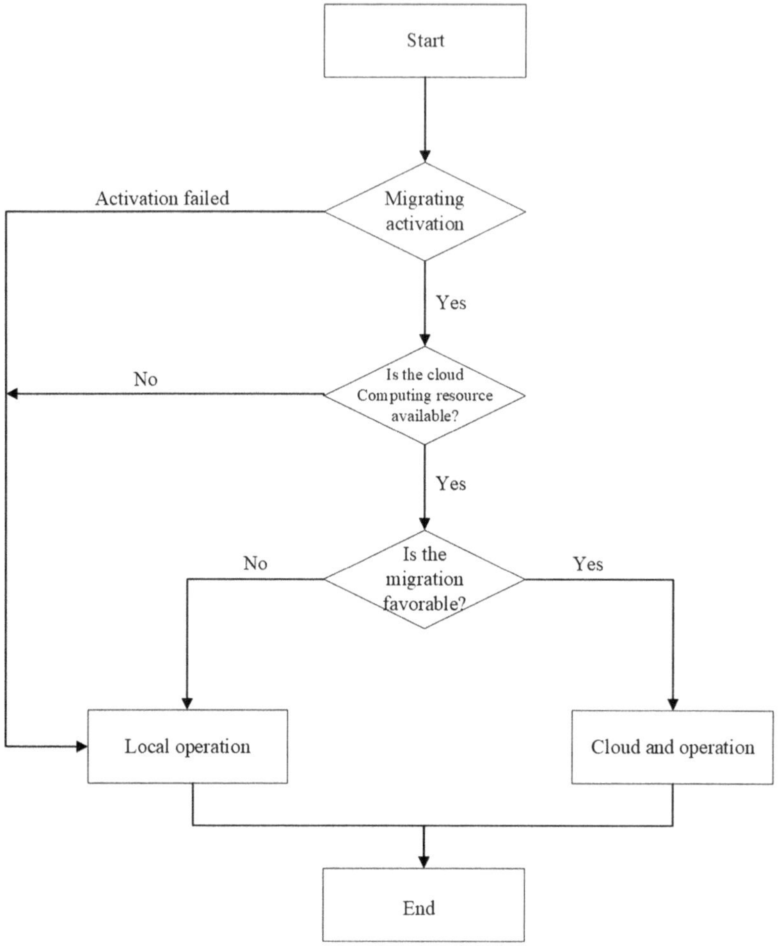

Fig. 5.22 Process of computing migration

Table 5.4 Main influencing factors of mobile cloud computing

Affected entity	Main influencing factors
User	Network data cost, service cost, data privacy, computing speed, application support
Mobile Devices	CPU speed, memory, storage, power
Connectivity	Communication technology (2G/3G/4G/5G, Wi Fi, etc.), communication bandwidth, communication delay
Application model	Context awareness, application partitioning, code availability
Application program	Migratability, data availability, input size, granularity
Cloud server	Computing power, memory, storage, running time, turnaround time

other edges to complete wider range of tasks. There are two main modes of edge to edge collaboration:

(1) Edge to edge distributed collaboration. In this collaborative mode, each edge undertakes computing tasks as a computing node. The edge owns the whole model or part of the model, and the computing data comes from the data generated by the edge itself. After the model is trained to reach a certain stage, the computed results will be summarized to the central node (server), and finally the computed results will be obtained through training.
(2) Learning synergy in the edge to edge federation. This method is proposed for the purpose of data security and privacy protection. An edge node stores the optimal model, and each edge participates in the training of the model as a computing node. Other nodes update the parameters to the node without violating the privacy rules.

2. Edge Thing Collaboration

The "thing" in the edge thing collaboration refers to the end of things equipment, which can also be described as the Internet of Things equipment, mainly including sensor equipment, cameras, factory mechanical equipment, etc. The main purpose of this collaboration is to enhance the ability of edge nodes. This collaboration is widely used in the Internet of Things, especially in smart home and industrial Internet of Things. Under the cooperation of edge and things, the object end is responsible for collecting data and sending it to the edge, and receiving the instructions from the edge for specific operation execution. While the edge is responsible for the centralized computation of multi-channel data, issuing instructions and providing external services.

5.5.3 *Application of Collaborative Computing Technology in Civil Engineering*

With the development of collaborative computing technology, its application has also been paid more attention in civil engineering. In the early engineering applications, data collected by mobile devices (edges) were usually transmitted to and displayed in the cloud, such as Intergraph [26] using mobile cloud computing to assist the infrastructure inspections. However, these applications did not reflect well the advantages and characteristics of collaborative computing.

1. Infrastructure Inspection Based on Collaborative Mobile Cloud Computing

Chen et al. [27] realized the detection of frame infrastructure based on CMCC. Its basic process is shown in Fig. 5.23.

To realize this process, Chen et al. [27] designed two sub-systems, including the front-end mobile application and the back-end cloud computing service platform.

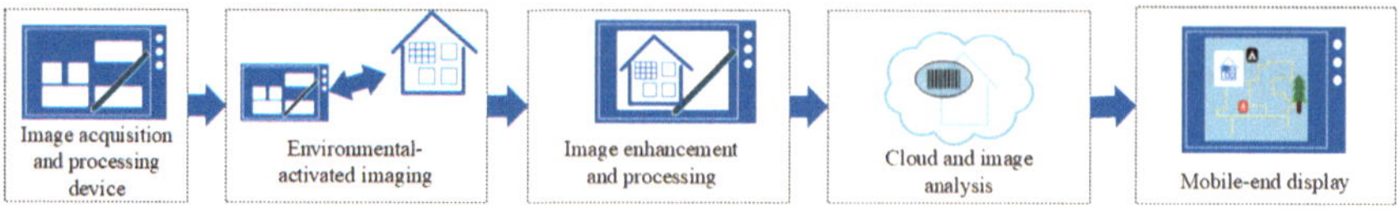

Fig. 5.23 Process of CMCC infrastructure detection [27]

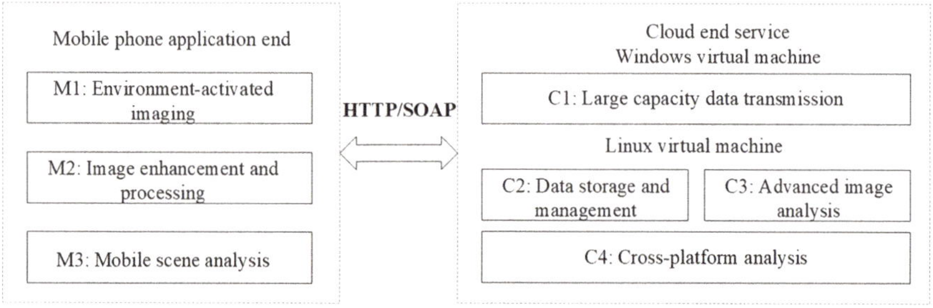

Fig. 5.24 Basic architecture of CMCC system [27]

The basic architectures of the two systems are shown in Fig. 5.24. As can be seen, the two systems exchange information through HTTP or SOAP protocol.

The platform designed by Chen et al. [27] can be used for the analysis of a complex damage scenario. Figure 5.25 shows an example of crack treatment using this platform.

2. Structural Monitoring System Based on Mobile Cloud

Zhao et al. [28–30] developed a smart phone driven structural monitoring system based on mobile cloud, and used it for daily monitoring of bridge structures. Its basic principle is to obtain the essential data such as GPS information, temperature, acceleration, inclination, displacement, etc. through the built-in sensor in the smart

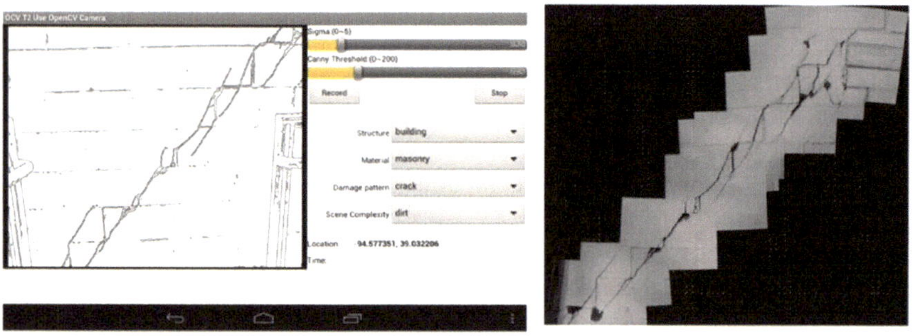

a) Image processing based on interaction b) Image stitching at cloud end

Fig. 5.25 Using mobile cloud computing to deal with crack damage [27]

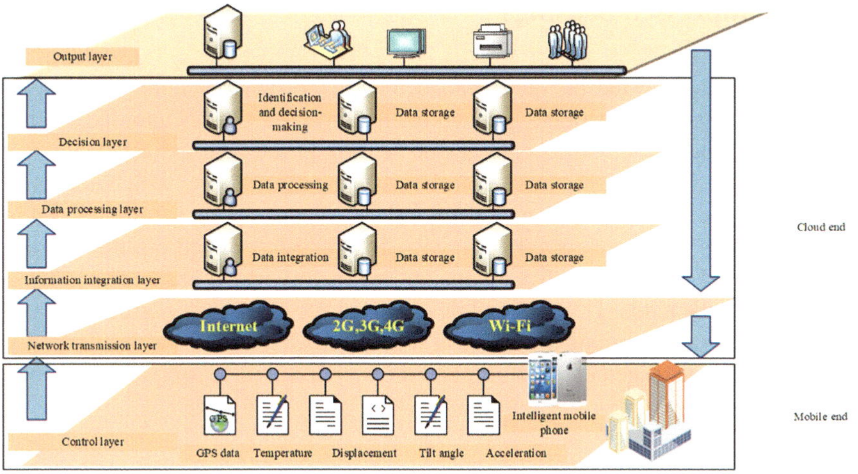

Fig. 5.26 Basic architecture of mobile cloud based structure monitoring system [29]

phone, and input these earthquake intensity information through the interactive interface of the smart phones; then upload the monitored data to the cloud server or the website, so as to integrate them with the structure related information. The cloud server processes the data and uses the relevant algorithms for damage identification and structural evaluation to obtain changes of the structural parameters and evaluate the structural damage. Finally, the relevant information is sent back to the smart phone user, and the user can then take measures corresponding to the computed results. The overall architecture of the system is shown in Fig. 5.26.

In order to make data collection more convenient, the authors also developed the mobile phone App-Orion for data collection and interactive analysis. The main functions of App-Orion are: ① obtain the location of the monitored items, ② establish monitored projects for a project database in the cloud, ③ perform data analysis: spectrum data is obtained from the accelerated time history sample, and ④ compute the cable force according to the related parameters.

The authors applied this system to monitor the cable forces of XingHai Bay Bridge and demonstrated that application of mobile cloud monitoring system can greatly enhance the convenience of structural monitoring and reduce the monitoring cost. The use of smart phones makes on-site monitoring more convenient, low-cost and efficient. In addition, the friendly interface is easy to operate, which enables ordinary people to actively participate in structural health monitoring and makes it possible to conduct rapid and safe assessment, which is not possible using the conventional monitoring systems.

5.6 Compressed Sensing

In order to alleviate the pressure from demands for signal sampling rate, signal transmission speed and storage space in the big data era, Candès introduced a new signal sampling theory called compressed sensing at the International Mathematical Congress in 2006. He pointed out that using this theory, with samples far less than those required by the conventional Nyquist sampling theorem, the original signal can be reconstructed accurately. It should be pointed out that the core of compressed sensing is still a computational theory, which has been applied in different sensing fields. This section briefly introduces the compressed sensing technology and its applications in structural damage identification, and then proposes a smart bridge management platform based on compressed sensing and Internet of Things technology.

5.6.1 Overview of Compressed Sensing Technology

Compressed Sensing (CS) is also called Compressive Sampling or Sparse Sampling. It can be inferred from the name that CS is a technology for signal sampling. It realizes "compressed sampling" using some algorithms. To be exact, it completes data compression in the sampling process [31]. Nowadays, CS has been used in electronic engineering, especially in signal processing, to acquire and reconstruct sparse or compressible signals [32].

1. Signal Sampling and Nyquist Sampling Theorem

Continuous signal refers to the information expressed by continuously changing physical quantities, such as temperature, humidity, pressure, length, current, voltage, etc. Electrical signals are often used to simulate the changes of these physical quantities. Therefore, continuous signals are also called analog signals, which can have infinite values within a certain time range. Various physical quantities in actual production and life, such as images taken by cameras, sounds recorded by recorders, and pressure, flow rate, speed, humidity recorded in workshop control room, are analog signals. But the computers obviously can do nothing to process these analog signals. To enable the computer to recognize, compute and process these signals, it must convert them into digital signals, that is to digitize the analog signals. Commonly used digital photos, videos, etc. can all be applied to computers after analog signal digitization [33, 34].

As shown in Fig. 5.27, conversion of analog signals to digital signals requires four basic steps: signal sampling, signal holding, signal quantization and signal coding. Among them, signal sampling refers to discreting analog signals in time, i.e. acquiring instantaneous values in the original analog signals point by point based on a specific time interval [33, 34]. A problem with signal sampling is that after sampling, the computer gets discrete points, and using these discrete points to replace continuous lines will inevitably lead to errors.

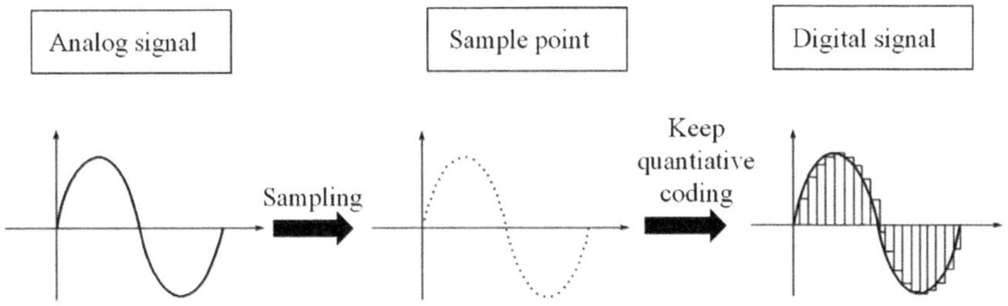

Fig. 5.27 Digitization process of analog signal

In 1928, H. Nyquist proposed a sampling theorem to solve the problem of sampling frequency, which is called Nyquist sampling theorem. Nyquist sampling theorem points out that only when the sampling frequency reaches more than twice the maximum frequency of the original signal, can the original signal be accurately reconstructed from the sampled signal [35]. If the sampling frequency is less than twice the maximum frequency of the original signal, the signal will suffer aliasing after its frequency spectrum shifted in the frequency domain. This theorem provides a standard for the sampling rate, which allows the discrete sampling sequences to capture all information from the continuous time signals with limited bandwidth. In the field of digital signal processing, Nyquist sampling theorem has always been the bridge between analog signals and digital signals, and the basis for continuous signal discretization.

However, with the increasing information demand in the world today, signal bandwidth is becoming wider and wider, and the sampling rate and processing speed are expected to be higher and higher for information acquisition. In order to relieve the pressure from signal transmission speed and storage space, the current common solution is to undertake signal compression, such as JPEG2000 standard based on wavelet transform [36]. Figure 5.28 shows the schematic diagram of conventional signal sampling and compression/decompression process [35]. However, signal compression is actually a serious waste of resources, because a large number of sampled data are discarded in the compression process ($N \gg K$), and they are not important to the signal itself or just redundant information. In this sense, it can be concluded that the bandwidth cannot essentially express the signal information, and the Nyquist sampling mechanism based on the signal bandwidth is redundant or non-informative.

Different from Nyquist signal sampling mechanism, in recent years, Candès et al. [37] proposed a novel sampling theory called compressed sensing or compressive sampling based on signal sparsity and successfully realized simultaneous sampling and compression of signals by using compressed sensing. Although the theory of compressed sensing has only appeared for a few years, it has attracted wide attention soon after it is reported, and has quickly become a hot research topic in the world. Figure 5.29 shows the sampling/reconstruction process of compressed sensing.

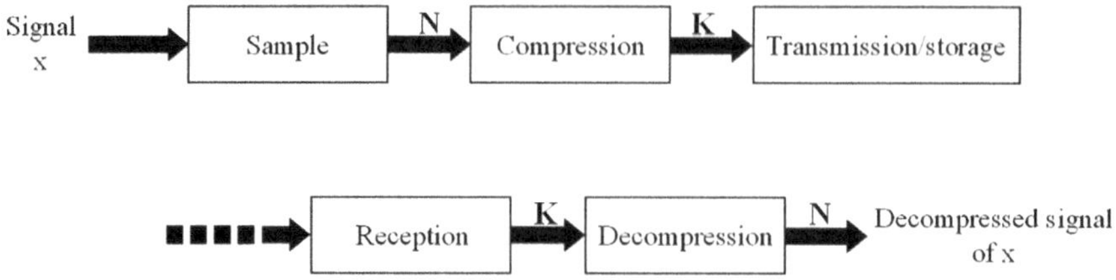

Fig. 5.28 Conventional signal sampling and compression/decompression process

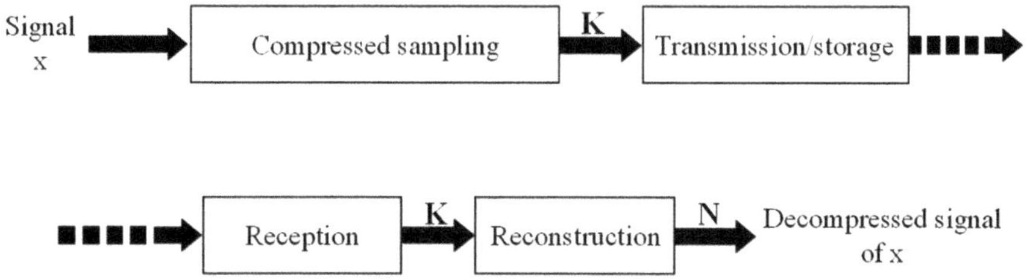

Fig. 5.29 Schematic diagram of compressed sensing sampling/reconstruction process

2. Basic Methods of Compressed Sensing

Based on the above introduction, we can briefly describe the basic idea of compressed sensing as: if a signal is sparse in a certain transformation domain, we can use an observation matrix that is not related to the transformation basis to project the transformed high-dimensional signal into a low dimensional space, and then reconstruct the original signal from these small projections with high probability by solving an optimization problem. In this theoretical framework, the sampling rate will no longer depend on the bandwidth of the signal, but largely depends on two basic criteria, namely sparsity and non-correlation, or sparsity and isometric constraints.

Figure 5.30 shows the basic method of compressed sensing which includes three main parts: ① sparse representation of signals; ② design observation matrix; and ③ design signal recovery algorithm.

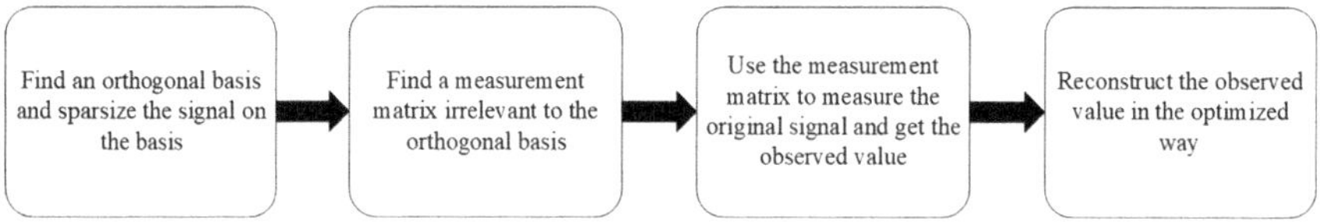

Fig. 5.30 Basic methods of compressed sensing

This section briefly introduces the basic technical route of compressed sensing. For the detailed algorithm, extension, promotion and variation of the basic theory, interested readers can read the relevant references on compressed sensing [38, 39].

5.6.2 Structural Damage Identification Based on Compressed Sensing

Because structural damage identification is an inverse problem and the dynamic structural response usually involves only a few modes, the structural response sample is a narrowband signal in the frequency domain with sparsity. Even with the effects of measurement noise, the structural dynamic response data is still approximately sparse in frequency domain or other transform domains. To this end, compressed sensing provides a new way to solve some key structural damage identification problems. In addition, wireless sensors and sensor networks have been widely used in structural damage identification. The issues of energy consumption, data loss and limited bandwidth with wireless sensors, which have puzzled the academic community for a long time, can also be solved by compressive sampling. This section briefly introduces the research status of application of compressed sensing to structural damage identification.

1. Data Acquisition and Signal Reconstruction

In recent years, more and more scholars have studied compressed sensing and compressive sampling. Bao et al. [40] studied the application of compressed sensing to health monitoring acceleration data acquisition. Their research results showed that the sparsity of structural acceleration data is an important factor affecting the accuracy of data reconstruction. Mascarenas et al. [41] studied the application of compressed sensing technology to structural damage identification. According to their research, compression sensors are first used to collect compression coefficients from measured signals, which are then transmitted to off board processors to reconstruct the data by minimizing the L1 norm. The compressed version of the matched filter is used in the sensor nodes for structural damage identification. O'Connor et al. [42] tried to reduce the energy consumption of wireless sensors used in health monitoring systems by using compressed sensing. They improved the wireless sensor based on compressed sensing so that it randomly sampled the data, and then sent the data to the data computing server outside the station. They reconstructed the data using the CoSaMP matching tracking recovery algorithm, and then further processed the data to extract the modal parameters of the structure.

In this section, several advanced technologies are introduced.

(1) Data Loss Recovery Method for Wireless Sensors Based on Compressed Sensing

In 2013, Bao et al. [43] studied the application of compressed sensing to data loss recovery of wireless sensors for structural health monitoring. In their research, the randomly lost data is equivalent to the compressed data in compressive sampling, and sparse optimization algorithm can be used to recover the lost data with high accuracy. In addition, the compressed sensing algorithm is embedded in the wireless sensor using random demodulation technology. The field test results on Harbin Songpu Bridge showed that the data loss recovery based on the compressed sensing algorithm is feasible and successful.

(2) Multi-sensor Data Fusion Method for Structural Damage Signals Based on Compressed Sensing

In structural health monitoring, damage identification and location are often achieved through the cooperation of multi-sensors. In order to improve the accuracy of damage identification and reduce the amount of data transmitted, multi-sensor data fusion is often used. Wireless sensor nodes often feature high-frequency sampling, the amount of data transmitted is very large, and the conventional direct fusion method will lead to excessive data transmission in the network and increase energy consumption of the entire network, affecting service life of the monitoring system. Therefore, a multi-sensor data fusion method based on compressed sensing is proposed. Its main idea is to linearly project the original collected data from multiple sensor nodes through the measurement matrix to realize compressed transformation of the collected data from high-dimensional sequence to low dimensional linear measurement signal, and then undertake data-level information fusion of the linear measurement signals from multiple sensors, and finally reconstruct the original collected data. The main process is shown in Fig. 5.31.

(3) Compressive Sampling and Reconstruction of Structural Health Monitoring Data Based on Group Sparse Optimization

In an actual health monitoring system, multiple measuring points are often arranged on the structure for multiple sensors to collect data at the same time. These data reflect the dynamic characteristics of the same structure at the same time. The signals have space–time correlation and basically the same sparsity in the frequency domain. Therefore, according to the structural health monitoring compressed sensing data

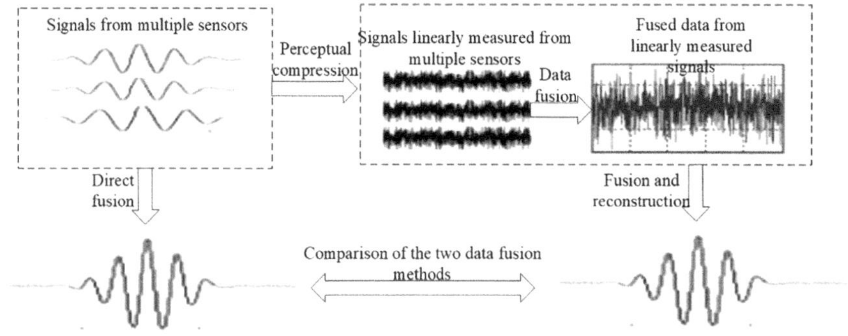

Fig. 5.31 Multi-sensor data fusion method based on compressed sensing

reconstruction algorithm, by introducing the group sparse model and using the correlation between the original signals and the group sparsity in the frequency domain, we can remove the redundant information and improve the availability of the information contained in the compressed sampling data. As shown in Figs. 5.32, 5.33 and 5.34, the test on Xiamen Haicang Bridge proved that this method based on compressed sensing improved the accuracy of data reconstruction [44].

2. Damage Location and Quantification

(1) Sparse Damage Location Method Based on System Model [45].

Based on guided wave system model, in order to locate damages in the structure, the overall wave field is observed from discrete observation positions in space and the inversion problem is then constructed to locate the damages. Due to the fact that spatial distribution of the actual damage is often sparse compared with the overall detected area, and based on compressed sensing, the priori knowledge of spatial sparsity is used as the regularization constraint inversion problem to overcome the

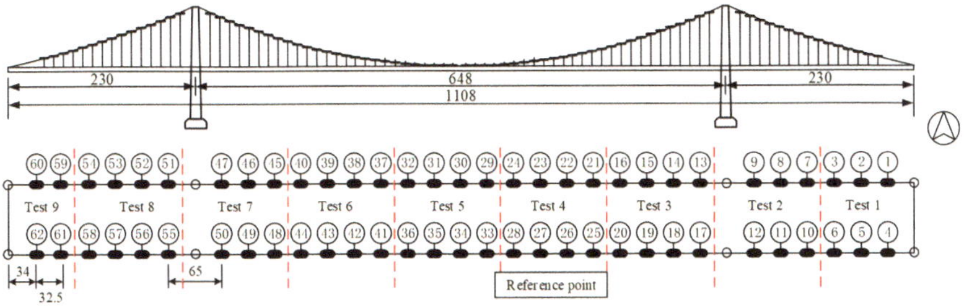

Fig. 5.32 Sensor layout [44]

Fig. 5.33 Sensor for monitoring [44]

Fig. 5.34 Field acquisition module [44]

problem of inverse model being highly uncertain, and cross validation is used to ensure that the sparse damage distribution is obtained. The simulation results showed that this method can effectively locate the damage.

(2) Structural Damage Identification Method Based on Structural Sensitivity Analysis and Sparse Constraint Optimization

Zhou et al. [46] proposed a structural damage identification method based on structural sensitivity analysis and sparse constraint conditions, and verified the effectiveness of the method through numerical simulation. Based on the sparse damage of the structure, the relationship between the damage degree of the structural element and the measured frequency or mode change is established through sensitivity analysis. However, the equations are ill conditioned linear ones sensitive to noise, and it is difficult to obtain accurate solutions by the conventional methods. The sparse constraint is introduced in the compressed sensing to solve the equations, and the numerical simulation of a simply supported truss model verified that the accuracy of damage identification can be significantly increased by using sparse constraint.

(3) Modal Identification Method Based on Output Data

Yang et al. [47] proposed a modal identification method based on output data. combining Blind Source Separation (BSS) with compressed sensing, non-uniform low rate random sampling is directly used for modal parameter identification. This method reduces the potential high sampling rate and requires only a lower sampling rate for modal identification, which provides an alternative to the conventional Nyquist homogeneous sensing.

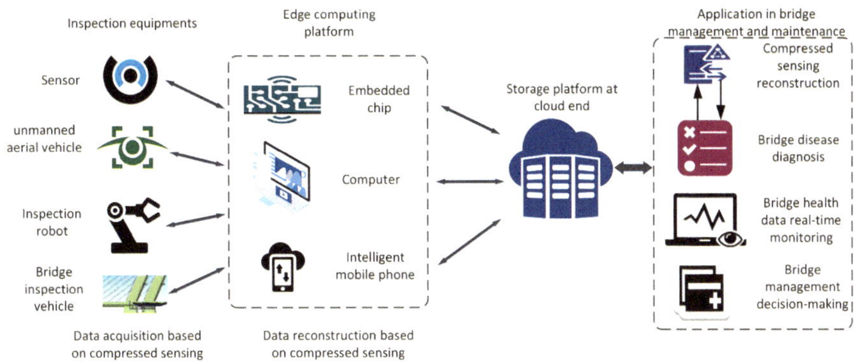

Fig. 5.35 Intelligent bridge maintenance and management platform based on CS and IoT

5.6.3 *Smart Bridge Management Platform Based on Compressed Sensing and Internet of Things*

1. Application Background

Inspired by compressed sensing technology and existing literature [48], the authors proposed a smart bridge management platform which comprehensively uses compressed sensing, edge computing, cloud computing and Internet of Things to minimize the amount of collected data while ensuring data quality.

2. Technical Route

The intelligent bridge maintenance and management platform based on CS and IoT includes four modules, namely, data acquisition module, data reconstruction module, cloud storage module and maintenance and management application module, as shown in Fig. 5.35.

(1) Data Acquisition Module

The data acquisition module composes of intelligent monitoring equipments deployed at different positions on the bridge, including intelligent sensors, intelligent UAVs, intelligent detection robots, intelligent bridge inspection vehicles, etc. The monitoring equipments will keep collecting various types of bridge health data for a long time to satisfy different maintenance and management needs. The data acquisition module is the foundation of the whole bridge maintenance and management platform. However, with the continuous improvement of bridge maintenance and management, the requirements for data acquisition efficiency and transmission speed are also increasing. At the same time, how to minimize power consumption on the premise of ensuring data acquisition efficiency and transmission speed is the key problem to be solved. Compressed sensing can be an effective method to develop an energy-saving data acquisition and transmission scheme.

(2) Data Reconstruction Module

Compressed sensing and edge computing are deployed in the different computing systems (embedded chips, computers, smart phones, etc.) to implement local processing units, so as to build a "CS based edge computing platform" [49]. The module first obtains the bridge health monitoring compressed data from the data acquisition module, then stores and reconstructs the local processing unit to extract the important information, and then sends the reconstructed and filtered data to the cloud storage platform through wireless transmission. The local processing unit acts as a gateway to only route the useful information to the cloud, rather than all the collected data, so as to significantly reduce the network bandwidth.

(3) Cloud Storage Module

The cloud storage module is mainly used to store the effective monitoring data compressed and filtered by the data collection module and the data reconstruction module. In the cloud storage module, CS can also perform operations with other data analysis algorithms, and provide solutions according to the specific tasks delivered by the management application module. Cloud storage can automate and intelligentize management of the bridge monitoring data, and integrate all the storage resources to improve storage efficiency.

(4) Maintenance and Management Application Module

The main function of the maintenance and management application module is to assist the bridge maintenance and management engineers in bridge health monitoring and maintenance oriented decision-making by calling the data stored in the cloud storage module, including bridge distress diagnosis, real-time health data monitoring, bridge maintenance and management decision-making, etc. Although compressed sensing is not used for signal processing and data acquisition, the essence of signal reconstruction in compressed sensing is to find the thinnest solution of underdetermined equations. Compressed sensing also promotes solving sparse optimization problems, providing new ideas for solving several inverse problems with sparse representation in time domain and space domain. Therefore, compressed sensing can also be used to solve some problems in bridge distress diagnosis and management decision-making.

5.7 Summary

This chapter started with the basic concept of big data, introduced the main characteristics of big data, and expounded the necessity of applying big data in intelligent operation and maintenance of bridges. It also introduced several common computing methods in big data computing, including cloud computing, edge computing, crowdsourcing and collaborative computing. In view of the large amount of data in big data computing, the compressed sensing theory for data compression dimension reduction is introduced. Finally, information physical model is used to unify digital model with big data as the core and the physical entity space.

References

1. Zhang J, Wu G (2019) Health monitoring and big data analysis of long-span bridges—methods and applications. China Construction Industry Press, Beijing (in Chinese)
2. Zhang YF, Li XQ (2018) Bridge structure health monitoring and condition evaluation. Shanghai Science and Technology Press, Shanghai (in Chinese)
3. Jiangsu Province, Long Bridge Health Monitoring Data Center. Jiangsu Transp Res 2012(4):1 (in Chinese)
4. Han YX, Zhang YF, Zhao L et al (2015) Introduction and consideration of the latest progress in long-term performance research of bridges abroad. J China Foreign Highway 35(4):217–221 (in Chinese)
5. Zhu CM, Zhang YF, Dai YF (2012) The long-term performance research program of American bridges and its enlightenment. Modern Transp Technol 9(2):18–21 (in Chinese)
6. Jula A, Sundararajan E, Othman Z (2014) Cloud computing service composition: a systematic literature review. Expert Syst Appl 41(8):3809–3824
7. Jeong S, Hou R, Lynch JP et al (2019) A scalable cloud-based cyber infrastructure platform for bridge monitoring. Struct Infrastruct Eng 15:1, 82–102
8. Dang HV, Tatipamula M, Nguyen HX (2022) Cloud-based digital twinning for structural health monitoring using deep learning. IEEE Trans Industr Inf 18(6):3820–3830
9. Deng H-W, Rahman M, Chowdhury M, Salek MS, Shue M (2021) Commercial cloud computing for connected vehicle applications in transportation cyber physical systems: a case study. IEEE Intell Transp Syst Mag 13(1):6–19
10. Yang HZ (2017) Application of PHM system for maintenance and repair of large bridges on high-speed railway. Railw Eng 3(5):637–646 (in Chinese)
11. Shrestha A, Ji D, Xin W et al (2020) Smartphone-based bridge seismic monitoring system and long-term field application tests. J Struct Eng 146(2):4019208.1–4019208.14
12. Bramberger M, Brunner J, Rinner B et al (2004) Real-time video analysis on an embedded smart camera for traffic surveillance. In: Proceedings RTAS 2004. 10th IEEE real-time and embedded technology and applications symposium. IEEE, pp 174–181
13. Geerts SAM, Weggeman M, Gevers JMP (2009) Discovering crowdsourcing: theory, classification and directions for use. Eindhoven: Technische Universiteit Eindhoven
14. Zimmerman JB (1999) Mobile computing: characteristics, business benefits, and the mobile framework. University of Maryland European
15. Feng M, Fukuda Y, Mizuta M et al (2015) Citizen sensors for SHM: use of accelerometer data from smartphones. Sensors 15(2):2980–2998
16. Ozer E, Feng MQ, Feng DM (2015) Citizen sensors for SHM: towards a crowdsourcing platform. Sensors 15(6):14591–14614
17. Kong Q, Allen RM, Schreier L et al (2016) MyShake: a smartphone seismic network for earthquake early warning and beyond. Sci Adv 2(2):e1501055
18. Dennis EP, Hong Q, Wallace R et al (2014) Pavement condition monitoring with crowdsourced connected vehicle data. Transp Res Record 2460(1):31–38
19. Barrington L, Ghosh S, Greene M et al (2012) Crowdsourcing earthquake damage assessment using remote sensing imagery. Ann Geophys 54(6):680–687
20. Ghosh S, Huyck CK, Greene M et al (2011) Crowdsourcing for rapid damage assessment: the global earth observation Catastrophe Assessment Network (GEO-CAN). Earthquake Spectra 27(s1):179–198

21. Minson SE, Brooks BA, Glennie CL et al (2015) Crowdsourced earthquake early warning. Sci Adv 1(3):e1500036
22. Zhang XZ, Lu SD, Shi WS (2019) Research on collaborative computing technology in edge intelligence. Artif Intell 5:55–67 (in Chinese)
23. Cui Y, Song J, Miao CC et al (2017) Research progress and trend of mobile cloud computing. Chin J Comput 40(2):273–295 (in Chinese)
24. Khan AN, Kiah MLM, Khan SU et al (2013) Towards secure mobile cloud computing: a survey. Futur Gener Comput Syst 29(5):1278–1299
25. Othman M, Madani SA, Khan SU et al (2014) A survey of mobile cloud computing application models. Commun Surveys Tutor 16(1):393–413
26. Intergraph. Intergraph mobile alert. 1 Nov 2013 (4 Sept 2021). https://geospatial.com/intergraph.com/products/aspx/mobile
27. Chen ZQ, Chen J, Shen FC et al (2015) Collaborative mobile-cloud computing for civil infrastructure condition inspection. J Comput Civ Eng 29(5):04014066
28. Zhao XF, Yu Y, Li MC et al (2015) Cloud-structural health monitoring based on smartphone. Vibroeng Procedia 5:241–246
29. Zhao XF, Han R, Ding YB et al (2015) Portable and convenient cable force measurement using smartphone. J Civ Struct Heal Monit 5(4):481–491
30. Han RC (2018) Research on multi-parameter monitoring technology and earthquake emergency response based on smart phone. Dalian University of Technology, Dalian (in Chinese)
31. Jiao LC, Yang SY, Liu F et al (2011) Review and prospect of compression sensing. Acta Electron Sin 39(7):1651–1662 (in Chinese)
32. Shi GM, Liu DH, Gao DH et al (2009) Compressed sensing theory and its research progress. Acta Electron Sin 37(5):1070–1081 (in Chinese)
33. Wang JP (2004) Analog and digital signals. J Phys Teach 22(5):49–51 (in Chinese)
34. Xu SS, Tan Y, Guo W (2018) Signal and system: theory, method and application, 3rd edn. China University of Science and Technology Press, Hefei (in Chinese)
35. Shao WZ, Wei ZH (2012) Basic theory of compressed sensing: review and prospect. J Image Graph 17(1):1–12 (in Chinese)
36. Liu LB (2004) Research on key technologies of JPEG2000 still image compression and VLSI implementation. Tsinghua University, Beijing (in Chinese)
37. Candes EJ, Wakin MB (2008) An introduction to compressive sampling: a sensing/sampling paradigm that goes against the common knowledge in data acquisition. IEEE Signal Process Mag 25(2):21–30
38. Jin J, Gu YT, Mei SL (2010) An introduction to compressive sampling and its applications. J Electron Inform Technol 32(2):470–475
39. Tsaig Y, Donoho DL (2006) Extensions of compressed sensing. Signal Process 86(3):549–571
40. Bao YQ, Beck JL, Li H (2011) Compressive sampling for accelerometer signals in structural health monitoring. Struct Health Monit 10(3):235–246
41. Mascarenas D, Cattaneo A, Theiler J et al (2013) Compressed sensing techniques for detecting damage in structures. Struct Health Monit 12(4):325–338
42. O'Connor SM, Lynch JP, Gilbert AC (2014) Compressed sensing embedded in an operational wireless sensor network to achieve energy efficiency in long-term monitoring applications. Smart Mater Struct 23(8):085014
43. Bao YQ, Li H, Sun XD et al (2013) Compressive sampling-based data loss recovery for wireless sensor networks used in civil structural health monitoring. Struct Health Monit 12(1):78–95
44. Wang XY (2016) Group sparse optimization algorithm for compressed sampling and reconstruction of structural health monitoring data. Harbin University of Technology, Harbin (in Chinese)
45. Pan P (2015) Research on structural damage identification and location algorithm based on guided wave. Shanghai Jiaotong University, Shanghai (in Chinese)
46. Zhou SM, Bao YQ, Li H (2016) Structural damage identification method based on structural sensitivity analysis and sparse constraint optimization. J Vib Shock 35(9):135–140 (in Chinese)

47. Yang YC, Nagarajaiah S (2015) Output-only modal identification by compressed sensing: non-uniform low-rate random sampling. Mech Syst Signal Process 56:15–34
48. Djelouat H, Amira A, Bensaali F (2018) Compressive sensing-based IoT applications: a review. J Sens Actuator Netw 7(4):45
49. Bao YQ (2009) Data compression sampling and damage identification fusion method for structural health monitoring. Harbin University of Technology, Harbin (in Chinese)
50. China Information Physics System Development Forum. White paper on information physical systems. 2 Mar 2017 (4 Aug 2021. http://www.cesi.cn/201703/2251.html (in Chinese)
51. Li HY, Wei MH, Huang J et al (2019) Overview of information physical system technology. Acta Automatica Sinica 45(1):37–50 (in Chinese)
52. Bhuiyan MZA, Wu J, Wang G et al (2017) Towards cyber-physical systems design for structural health monitoring: hurdles and opportunities. ACM Trans Cyber-Phys Syst 1(4):1–26
53. Yuan X, Anumba CJ, Parfitt MK (2016) Cyber-physical systems for temporary structure monitoring. Autom Constr 66:1–14

Chapter 6
Machine Learning and Artificial Intelligence

Abstract With the advent of big data analytic, Machine Learning and more generally, Artificial Intelligence (AI), has been developing rapidly, especially in the last few years, drawing much worldwide attention among universities, research institutes and industries of all sectors, and even government agencies. Civil engineering, a traditional pillar industry during the period of explosive developments of AI, urgently requires a fresh wave of technological and industrial upgrading. The application of AI technology in civil engineering is thus a critical milestone in the evolution of intelligent engineering.

6.1 Overview

6.1.1 Artificial Intelligence

(1) Basic Conception

Despite a common misunderstanding that AI is unattainable, it has actually infiltrated our daily life, as illustrated by the facts that QR codes are generally replacing wallets, traffic congestion can be avoided using advanced navigation, photographs can be automatically retouched by apps, etc. Hence, it is evident that AI is omnipresent and changing our lifestyles [1].

AI definitions differ along two dimensions. One focuses on the mechanism underlying human thought process [2], whereas the other is concerned with activities in need of human intelligence [3]. Overall, AI programs are designed to imitate human ingenuity, which is capable of thinking and independent event handling.

In accordance with the targets and underlying philosophy, AI can be categorized into three classes [4]:

Weak AI, also known as narrow AI, is AI limited to specified areas, for instance, automatic interceptions of harassments via phone calls and e-mails.

Strong AI, also known as full AI, refers to AI with intellectual capabilities that exactly emulate human brains in all respects, including the abilities to think, plan, and study.

G. Wu et al., *Intelligent Bridge Maintenance and Management*, Springer Tracts in Civil Engineering, https://doi.org/10.1007/978-981-97-3827-4_6

Superintelligence, as introduced by Nick Bostrom (a renowned philosopher specializing in AI), denotes an intellect smarter than the best human brains in virtually every discipline, including scientific innovation, general knowledge and social abilities. It indicates the exploration and imagination toward the ultimate development trend of AI.

Strong AI has not yet been realized due to the absence of technical support, although weak AI has been substantially fulfilled. As shown in Fig. 6.1, computational intelligence, perceptual intelligence and cognitive intelligence are the three tiers of AI technology in the order of progress and difficulty.

The application of computational intelligence expedites the processing of massive data sets. By utilizing the computational capacity of supercomputers, for instance, it is feasible to accomplish some extremely complicated tasks in short order. Perceptual intelligence refers primarily to visual sense, auditory sense, tactile sense, and other perceptual abilities. Face and speech recognitions are popular examples. Comprehending, analyzing, reasoning, and evaluating are components of cognitive intelligence. Its ultimate objective is for AI to be able to assist or substitute human labor, such as via fully automatic systems.

(2) History of AI

Rises and falls of AI followed each other in rapid succession since the emergence of AI.

Birth of AI (1950–1956). The first neural net machine was built in 1950 when Alan Turing, "the father of computer science", devised his famous Turing Test that is the first serious proposal in the philosophy of AI for evaluation. John McCarthy, "the

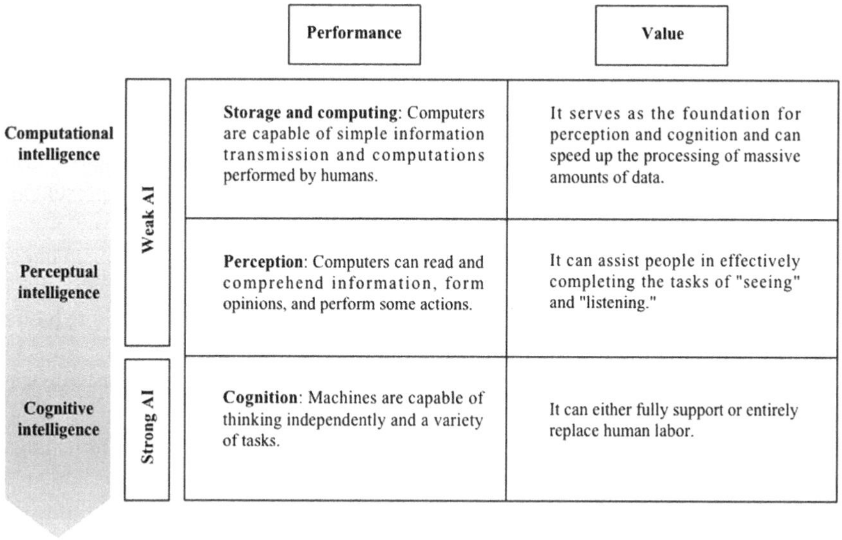

Fig. 6.1 Technological development of AI

father of AI," presented the term "Artificial Intelligence" at the Dartmouth conference in 1956, which is universally regarded as the official birth of AI.

Golden years (1956–1974). The first pinnacle of AI was reached when computers could solve algebra problems, prove theorems in geometry [5], and translate English [6]. Several symbolic events emerged in succession with an intense optimism expressed by the first generation of AI researchers, both in private and in print. In 1956, the Logic Theorist (LT), commonly referred as the first running AI program, was demonstrated by Allen Newell et al. for the first time. Two years later, John McCarthy's LISP language was originally specified as a family of programming languages with a long service life. Margaret Masterman coined the phrase "semantic nets" for machine translation in 1961, and in the same year, Leonard Uhr and Charles Vossler published a seminal study on pattern recognition. This paper was later reprinted in the first collection of AI-related essays, Computers and Thought. It is believed that the software program Dendral started in 1965 by Edward Feigenbaum et al., was the first expert system, capable of automatic decision-making and problem-solving behavior.

First AI winter (1974–1980). In a renowned 1966 report from the US ALPAC committee, machine translation was condemned to failure and deemed expensive, inaccurate, and unpromising. Another scathing report in the UK, now known as the Lighthill report, also criticized the utter failure of AI in 1973. Over the same period, there appeared to be an enormous abandonment of connectionism. AI was thus subjected to extensive critiques and financial setbacks. There were three fundamental limitations on AI programs: insufficient computer power, limited databases, and intractable problems.

Boom (1980–1987). A newly established organization, the American Association for Artificial Intelligence (AAAI), held the first national AI conference in 1980 at Stanford, while the fifth-generation computer project was generously funded by the Japanese Ministry of International Trade and Industry in 1981. Spring follows winter, and so does the second resurgence of AI. CART (Classification and Regression Tree), a revolutionary DT (decision tree) method, was a rising algorithm in 1984. That year saw the first official publication of CART software, the growth of which is still anticipated to this day. After another two years, in 1986, a revival of machine learning came after the rediscovery of back-propagation (BP), an algorithm applied widely in feed-forward ANNs (artificial neural networks) training. Then the multilayer perceptron (MLP), a fully connected class of ANNs, began flourishing naturally in the 1980s.

Second AI winter (1987–1993). Eventually, enthusiasm for expert systems spiraled out of control, and AI suffered a series of financial setbacks again in the late 1980s and early 1990s. These successful expert systems were increasingly shown to be overpriced and replaceable. Following a classic pattern of economic bubble, the LISP machine market suddenly collapsed in 1987, as various benchmarks demonstrated that alternatives such as the workstations outperformed LISP machines. As a result, the Strategic Computing Initiative canceled any future funding for AI, and the aforementioned Fifth Generation Project in Japan also failed to meet the unrealistic expectations by 1991.

Breakthrough (1993–present). In 1997, the IBM Deep Blue (a computer chess-playing system) defeated the reigning world chess champion, Gary Kasparov, for the first time. This milestone indicated the gradual progress in tackling the underlying problem of "raw computing power." Commercialization of AI began in the early twentieth century. A renaissance of interest in deep neural networks was precipitated by the discovery of a simple but efficient method for pre-training deep neural networks in 2006. With remarkably accelerated computation speed and substantially accumulated databases, a brand-new AI renaissance is imminent, owing to the continuous advancement of deep learning techniques.

(3) Industry Structure

Industrial AI is typically classified into an infrastructure layer, a technology layer and an application layer, as depicted in Fig. 6.2. The first and fundamental layer comprises not only the "hard part," hardware and computing systems, but its "raw materials" as well: data. Their developments provide a guarantee for AI's computing power and database requirements.

The technology layer is the kernel of the AI industry aiming at constructing the technological paths to simulate human intelligence, which includes the algorithm theory, development platforms, and application techniques.

As an extension of the AI industry, the application layer integrates one or more application techniques to make software and hardware products or solutions available on demand. The applications shown in Fig. 6.2 spread across all trades and professions, as "X + AI" has already become a future tendency of intelligent industry and scenarios.

(4) Relationship Between AI, Machine Learning and Deep Learning

AI, as a subject, aims to study and develop human-like intelligent systems, involving robots, computer vision and expert systems. During the implementation process, machine learning serves as the most feasible method to deal with AI problems, providing the essential technical supports. Its core idea lies in the algorithms to

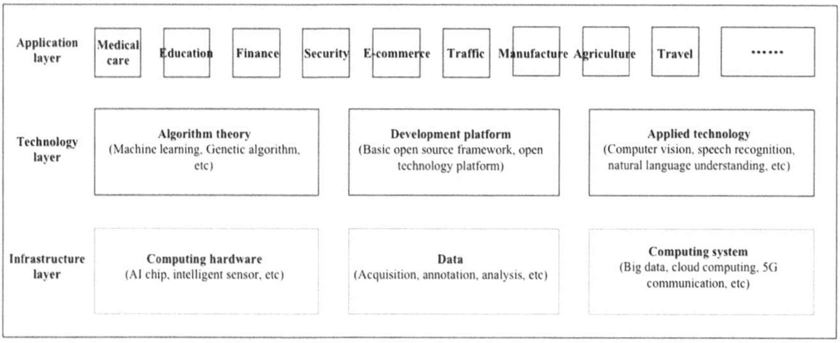

Fig. 6.2 Structure of AI industry

Fig. 6.3 Relationship between AI-related concepts

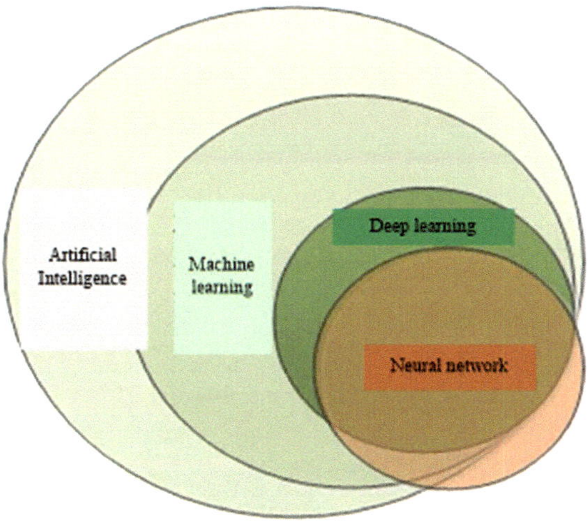

guide computers in search for appropriate mathematical models through data. Long-standing research has been conducted on neural networks since they were originally a biological concept, and later artificial neural networks were developed from their inspiration. By now, neural networks refer to a considerable research field with cross-over and integration among multiple disciplines. The specified "neural networks" in machine learning means "neural network learning," an intersection of machine learning and neural networks among various machine learning algorithms.

Deep learning also belongs to machine learning. It employs Deep Neural Networks (DNN) to generate complicated models based on profound understanding of data. This algorithm is applicable to big data processing rather than other machine learning algorithms, broadening the application fields of machine learning and AI. Deep learning has now dramatically enhanced the performance of AI systems in several crucial areas, including computer vision, speech recognition, natural language processing, etc. The relationship between AI, machine learning and deep learning is depicted in Fig. 6.3.

6.1.2 Machine Learning

Basic Conception

Machine learning is an inexorable outcome of AI development. Before 1980s, the predominant AI problem-solving techniques were logic inference, knowledge engineering, and expert systems. They established rule bases based on human knowledge to provide a foundation for deduction and decision-making. In contrast, these predefined rule bases are unnecessary for machine learning, as it enables the computer to learn the rules and correlations automatically from "empirical data," i.e., empirical

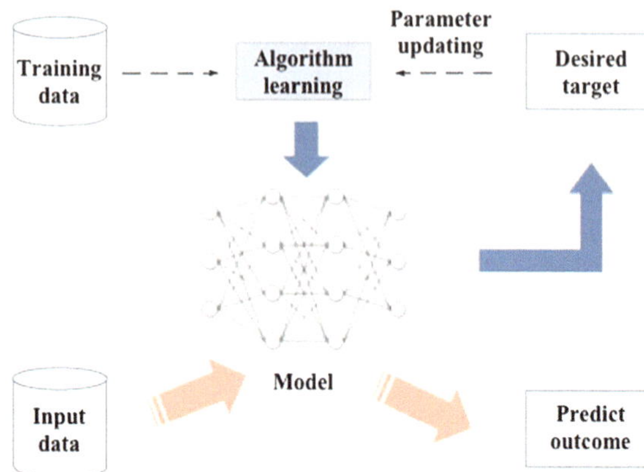

Fig. 6.4 Basic processes of machine learning

knowledge. Learning algorithm becomes, therefore, the focus of machine learning to fulfill this purpose. By utilizing the learning algorithm, computers can establish models after learning rules from the given training data, and then predict the unknown data (Fig. 6.4).

Machine learning can merely be regarded as a simulation of human cognitive processes. For humans to discover the underlying "laws," they will "generalize" their life experiences accumulated from repeated observations and then make "speculations" for unknown problems in advance. People can summarize the characteristics of cats and dogs, for example, through extensive observations in order to distinguish them more effectively in the future. The "training" and "prediction" processes in machine learning correspond to the "generalization" and "speculation" processes for humans, respectively.

The objective of machine learning is to accurately predict future samples rather than to invert the given training data. In general, the difference between prediction outputs (Fig. 6.4) and actual samples is referred to as "error," and "training error" and "generalization error" denote that within the training sets and other samples, respectively. A model's generalization ability refers to its capacity to generate predictions for samples beyond the training set. Apparently, models of the strong generalization ability are the target of machine learning, striving for the fewest generalization errors.

Although the generalization error is the ideal measure to evaluate a model, it cannot be directly computed for new samples in advance. An alternative approach is thus adopted to minimize the training error. However, there always turn out to be two undesirable situations. In one case, the model "overlearns" the training set, even capturing a specific point of a single sample, and then considers it as the general rule for all samples, resulting in the diminished generalization performance. This phenomenon is termed "over-fitting," as the model with an acceptable training error predicts new data inaccurately. Another instance of "under-fitting" occurs when a model's learning capacity is insufficient to acquire the general features of the training set. Therefore, the "training set" is generally established to evaluate the models' ability to predict new samples, with the "training error" roughly equaling the "generalization error."

Performance indices are also utilized to assess the model's generalization capability, indicating specific requirements for tasks. The model's "quality" thus becomes relative as the performance of various models under comparison is reflected by the corresponding performance index applied. Common indices include error rate and accuracy, precision, recall, ROC curve (a figure-based index), cost-sensitive error rate and cost curve, etc.

1. History of Machine Learning

Machine learning truly entered history in 1980, with previous algorithms that basically were scraps without systems still served as a great push forward. Then, numerous algorithms were consistently introduced into the field, inducing an accelerated growth of machine learning with extensive applications. Machine learning, according to its developments in the hierarchical structure of models, has roughly experienced two waves: shallow learning and deep learning [8].

(1) First Wave: Shallow Learning

In late 1980s, the introduction of back-propagation (BP) in neural networks produced a resurrection of hope for machine learning. Researchers of BP demonstrated how it enabled ANN models to learn statistical rules from a large number of training samples and then predict the unknown events. However, due to the difficulty of model training, the ANNs then were mostly with a single hidden layer.

1990s were the time of massive push in machine learning, as various shallow learning models were proposed, such as support vector machines, AdaBoost, random forests, etc. Huge successes were achieved by these models in both theoretical analyses and practical applications. In contrast, the development of multilayer neural networks remains stagnant due to difficulties in theoretical analyses and model training.

(2) Second Wave: Deep Learning

Deep learning can be traced back to a deep neural networks training method developed in 2006 by a group led by professor Hinton. They subsequently invented the AlexNet network model in 2012, successfully solving image categorization issues. It actually sparked a surge of deep learning, which was thereafter extensively employed to address machine vision issues. In comparison with the conventional shallow learning, deep learning is distinguished by: (i) emphasis on the depth of model structure with more than five hidden layers in common; and (ii) focus on characteristic learning. Features learned from big data are superior to artificially produced features for the former's description with enriched internal information. The general convolutional neural network, cyclic neural network and confrontation generation network all belong to the field of deep learning.

The success of cyclic neural network in addressing the problem of time series analysis has a direct impact on the implementation of speech recognition, machine translation, etc. Regarding data generation, the deep neural network represented by confrontation generation network has produced favorable results. It can generate realistic visuals, pleasant music, etc., which enables innovative approaches to deal with data production issues. Deep reinforcement learning has gained progress in meeting strategic and controlling challenges.

(3) Applications of Machine Learning

Machine learning is becoming more prevalent as the age of big data approaches; for instance, with the expansion of Internet of Things and mobile devices, there is a growing amount of unstructured data, such as images, texts, videos, etc., and models of machine learning can thus access a broader variety of data sources. Simultaneously, distributed computing enhances machine learning, and the benefits of machine learning become increasingly apparent.

The extension and deployment of machine learning in diverse fields have facilitated their advancements and benefited people's lives. For example, computers can successfully recognize human speech [9]; machine learning can accurately predict the recovery rate of patients with pneumonia [10]; major search engines can use machine learning to summarize and analyze the user's search records, click habits, etc. and realize regular and targeted advertising, and traffic systems can use image recognition to identify lawbreakers, etc. Likewise, in the area of engineering, research and development have been undertaken on machine learning-based structural safety monitoring systems [11, 12].

Civil engineering has been significantly influenced by the three waves of AI. From 1980s to the dawn of the twenty-first century, combination of AI and civil engineering focused primarily on the application of expert systems in structural design and performance evaluation, and yielded a series of social and economic benefits. Since 2012, as a consequence of the rapid development of deep learning networks, AI applications have become a real success and a hot direction in numerous fields. AI has fostered advancements in the intelligentization, informationization and automation of civil engineering, gradually infiltrating the entire life cycle of civil infrastructures, comprising structural design, construction, management, maintenance, operation, and disaster mitigation.

Despite the extensive application potential of machine learning, it is still a young field of research undergoing rapid progress. Through identifying historical patterns and correlations in data, it has gained the merit of generating reliable and repeatable decisions and outputs, as well as revealing previously undiscovered insights. With the flow of time and explosive increase in digital data, the outputs of machine learning are becoming more accurate and offering more powerful analysis and prediction. In the future, machine learning will contribute substantially to all industries, with particular emphasis on the intelligent development of civil engineering.

6.1.3 Recent Advances in Large AI Models

More spectacular are the recent advances in large AI models (e.g., large language or vision models) and their impact on many engineering and scientific fields. Pioneered by OpenAI, significant advancements in large-scale AI modeling, represented by

ChatGPT and many other large language models (LLMs), have been achieved. These LLMs showcase improved understanding, coherence, and versatility that transform the evolution of AI capabilities. Large vision models (e.g., SAM from Meta) redefine the boundaries of computer vision, integrating multi-modal capabilities to process and interpret both textual and visual data with higher precision and accuracy.

To this end, this chapter focuses on introducing the basic concepts of classic AI and machine learning technologies, rather than on the recent large AI models, as the majority of the engineering problems, particularly in the domain of bridge maintenance and management, do not necessarily rely on these large AI models. Frontier research about the use of large AI models certainly are being discussed by researchers for advancing automation in Civil engineering; however, most are in their infancy stage.

6.2 Machine Learning Algorithms

6.2.1 Classification of Machine Learning Algorithms

Machine learning algorithms have various classifications in accordance to different criteria. They may be classified into statistical and non-statistical approaches based on the learning criteria applied or be subdivided into linear and non-linear models for different functions. According to different types of information feedback, Fig. 6.5 depicts four common categories: supervised learning, unsupervised learning, intensive learning, and weakly supervised learning.

1. Supervised Learning

In supervised learning, a model is trained by a machine with a large number of data samples, and hence it can predict the outcomes based on inputs when new data is introduced [13]. Supervised learning is also known as "instructed learning," and the input learning data is referred to as "labeled data". These input data are denoted

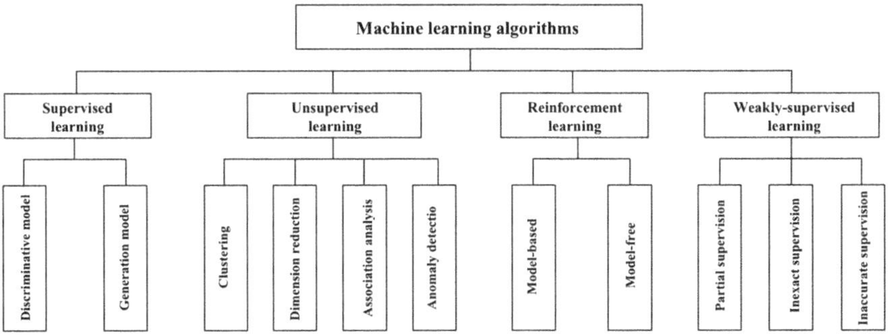

Fig. 6.5 Tree graph of machine learning algorithms

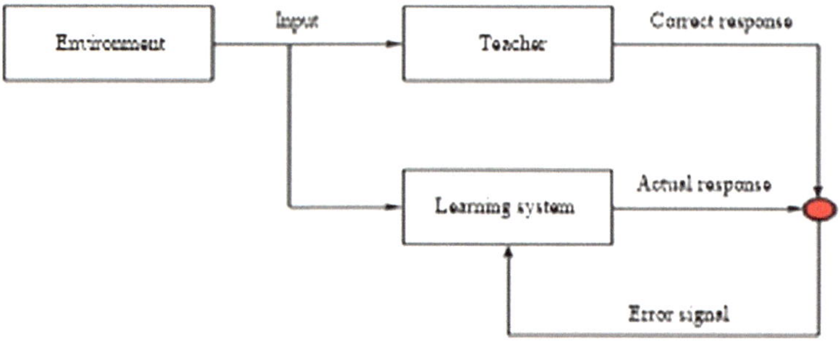

Fig. 6.6 Learning process of supervised learning

by "training data," as each set of training data has a distinct identifier or outcome, giving input x and output y (e.g., in the spam interception system, each email is labeled as "spam" or "non-spam"). The learning process of supervised learning is depicted in Fig. 6.6. During the prediction model training process, supervised learning compares the predicted results with the actual outcomes of the training data, and then continuously adjusts in response to the errors observed until an acceptable level of accuracy is achieved.

Based on whether the distribution of observation variables is modeled, supervised learning methods can be classified as discriminating methods and generation methods, with the corresponding learning models referred to as the discriminating model and the generation model, respectively.

The discriminating model establishes a prediction model based on probability theory by a conditional probability distribution function $P(x|y)$ (i.e., the probability of y occurring when x occurs), and it can also build the decision function $f(x)$ from the training data. This model focuses on predicting the output y for a given input x. While the generation model explains the joint distribution between x and y as the joint probability distribution function $P(x, y)$ (i.e., the probability of the joint occurrence of x and y), a conditional probability distribution function $P(x|y)$ is thus proposed to be the prediction model. This model can learn the data generation mechanism, representing the generation correlation between the input x and the output y. Common discriminating models include logical regression, support vector machine, boosting, neural network, random forest, etc. Typical generation models are the Gaussian mixture model, hidden Markov model, naive Bayes model, etc.

Both models have benefits and drawbacks. The generation model can provide more information with additional samples and calculations. For classification problems in particular, accurate estimation of the distribution of category conditions requires an increasing number of samples, as much information regarding the category conditional probability is not utilized in the classification process. The discrimination model cannot reflect the characteristics of the training data; but by reflecting variations between different data sets, it may be able to determine the optimum classification between categories. Discrimination model learning is generally more precise for direct prediction, which can simplify learning problems and save computing resources. However, generation model has a higher convergence rate for learning, as the learned model can converge to the real model more quickly when increasing the sample size. It remains effective for hidden variables, while the discrimination model loses its function in this case.

In general, discriminant models are more practical. Zhou [7] pointed out that direct estimation of joint probability based on limited samples would induce combination explosion in calculation and sparse samples in data. More severe problems may emerge with the increase of data dimension. In practice, limited samples are available, with potential imbalances between different categories. A joint probability close to the true distribution is thus difficult to obtain, especially for generation models.

Classification and regression are two common application scenarios for supervised learning. Classification refers to mapping the samples in the test set to the known discrete class tags by training the classification model. It has been widely used in bridge structural health monitoring, bridge reinforcement performance evaluation, load pattern recognition, etc. The classification model will output the damage according to the input characteristics [14]. Regression is mainly used to investigate correlations between variables, which can also describe the system characteristics revealed by the inputs and the outputs in the form of time series data. The optimum fitting line is the target of regression, approaching each point in data sets. In bridge structural health monitoring, it is often utilized to analyze the load impact for structural state assessment, or to separate load components in response data by establishing a regression model to depict the relationship between external load and structural response. Regression can also be directly used for damage identification of bridges [15].

Supervised learning has been widely used in various other fields, such as handwritten character recognition, image processing, spam classification and interception, web page retrieval, gene diagnosis, stock forecasting, etc. In civil engineering, supervised learning is also widely used for damage identifications of cable-stayed cables, irregular continuous beam bridges, urban overpasses and buildings [16, 17].

2. Unsupervised Learning

According to an article published in 1996 by Bruno [18], unsupervised learning is the process of identifying patterns in data via training samples of undefined categories. It differs from supervised learning in that the input data has no labels attached; that means the algorithm is provided with merely the input variable x without a corresponding output variable y. Therefore, the computer is incapable of precisely determining which data has which labels, and can only rely on its powerful computing power to analyze the data features to obtain certain findings. Unsupervised learning resembles the self-study of "students" without "teachers" in the learning process.

There is a school of thought that believes unsupervised learning has higher potential, despite the fact that supervised learning has only achieved a few successes. Training restricts the capacity of the supervised learning model, which only learns the tasks after training. In contrast, unsupervised learning represents the realization of general AI, implying that AI can learn any task like humans. The majority of unsupervised learning problems involve clustering [19], dimension reduction [20], association analysis [21], and anomaly detection [22].

(1) Clustering

Clustering problems are typical cases of unsupervised learning to determine the object's category, which is similar to classification problems but without any predefined category. The model needs to independently divide the samples into multiple categories. In accordance with the concept that "birds of same feathers flock together," the clustering algorithm divides the data set into distinct classes or clusters based on characteristics of the data structure, following a specific standard such as the distance between data points. Hence, the optimal similarity for data objects within the same class or cluster is obtained, whereas the data differences between different classes or clusters reach their maximum (Fig. 6.7). The clustering algorithm has no training process.

Due to the lack of a defined category standard, the core issue for clustering algorithms is how to define clusters, with the mere requirement that samples in a cluster should be potentially similar. Typically, it is determined by the distance between samples in the cluster or the sample point density in the data space. Various clustering algorithms can be derived from different cluster definitions.

Clustering allows for the analysis of prestress in the evaluation of bridge structural reinforcement performance. With the clustering results, hidden information in the data set can be extracted, and future data can be classified and predicted [22]. The hierarchical clustering method can be utilized to identify the section damage of a bridge model, and it can also prevent any change of the structural reference state [23].

(2) Dimension Reduction

Dimension reduction is an important concept in machine learning. In vector computing problems, an increase in dimension results in sparse data samples, exponentially increased computation and computational difficulties. This is a problem encountered by all machine learning methods and is known as "dimension disaster." In addition, the linear correlation between high-dimensional features is common, suggesting redundant features.

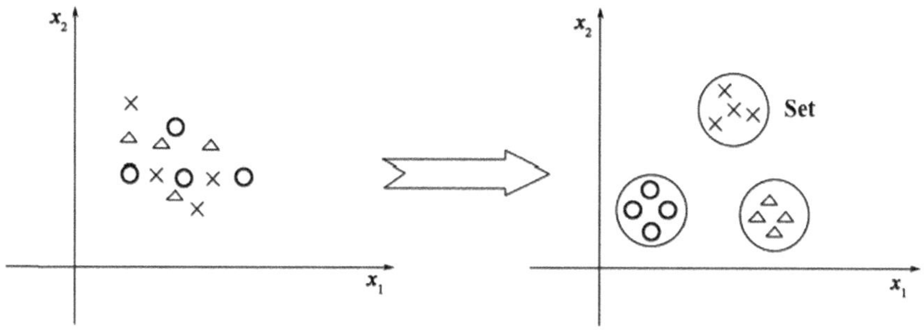

Fig. 6.7 Clustering

Dimension reduction theory is proposed to deal with the aforementioned problems, by mapping the data from the original high-dimensional space to the low-dimensional space via a mapping technique. It thus overcomes the dimension disaster, among other merits such as obtaining the essential characteristics of data, conserving storage space, eliminating unnecessary noise, enhancing training efficiency, etc. Principal component analysis, linear discriminant analysis, local linear embedding and Laplace feature mapping are typical dimension reduction algorithms. By reducing dimensions, researchers [24] were capable of simulating the virtual deformation of large and complicated bridge structures, and proposed an effective bridge surrogate model.

(3) Association Analysis

Association analysis is the process of determining the relationship between elements in a large data set by employing frequent item sets or association rules to describe the rules and patterns by which some attributes of a thing are shown simultaneously. It was initially conducted to determine the relationship between various commodities in the supermarket transaction database, i.e., the famous "beer and diapers" story. The Wal-Mart supermarket conducted a comprehensive analysis of the original transaction data in the database and discovered that 67% of customers would purchase diapers and beer at once. This is due to the fact that in the United States, wives frequently instruct their husbands to purchase diapers for children after work, and most husbands would purchase their favored beer after. With the assistance of correlation analysis, Wal-Mart revealed this hidden phenomenon, rearranged the shelves, and began selling diapers and beer together, thereby substantially increasing the sales of both products.

With association analysis, shelf arrangements can be determined by analyzing the user's credit card bill, which is desirable for future market promotion; and medical researchers can explore superior preventive measures by analyzing the common characteristics of patients suffering from certain diseases from numerous medical records. Association analysis is also employed in bridge operation and maintenance, such as examining the correlation between the defect status of each component and the overall defect status, as well as the correlation between each component's defect status [25].

(4) Anomaly Detection

A typical application of machine learning algorithms is the anomaly detection, also known as outlier detection. Hawkins [26] provided a general definition of anomaly as the distinct data within a data set. These data are not random deviations; rather, they are generated by entirely different mechanisms, such as large credit card transactions in consumption records. This definition of abnormal data suggests their two characteristics: (1) abnormal data differs from the majority of sample data; and (2) the proportion of abnormal data in the overall sample of data is limited. Based on the information in the test data, the anomaly detection algorithm ought to be able to determine whether new data is abnormal compared with a given set of normal data.

As a key aspect of data mining, anomaly detection and analysis identify objects in a data set that are significantly different from the others. Researchers have implemented deep learning techniques for anomaly detection, such as automatic coders based on deep neural networks [27], LSTM (long short-term memory) model for processing time series data [28], and DeepLog [29] for anomaly detection of text log data. Anomaly detection has been extensively utilized for credit card fraud detection, loan approval detection, consumer behavior analysis, weather forecasting, financial customer classification, and network intrusion detection, among other applications.

3. Reinforcement Learning

Reinforcement learning is a unique machine learning technique that utilizes environmental feedback as input and then adapts to it. Research on reinforcement learning can be traced back to Pavlov's experiment on conditioned reflexes, but it was not until the late 1980s that reinforcement learning research and applications began to develop rapidly. Reinforcement learning is fundamentally the transfer of learning from the state of the environment to the action mapping in order to maximize the cumulative reward value of the action from the environment. This technique is distinctively different from supervised learning technologies that use positive and negative examples to determine appropriate behavior. Instead, it uses the "trial and error" practice to discover the optimal behavioral strategy by learning from mistakes. Its primary components include agent, environment, state, action, and reward. Figure 6.8 illustrates the learning process of reinforcement learning, wherein the agent represents the reinforcement learning algorithm model and the environment is the scene where the agent executes actions.

According to the learning process of reinforcement learning, the environment sends the agent a state, and the agent thus takes action based on its knowledge in response to the state. The environment then sends the subsequent status and returns the reward to the agent. This reward is employed by the agent to update its knowledge and evaluate the last action. This cycle will continue until the environment signals the end of the process with an end state.

Reinforcement learning can be divided into model-based reinforcement learning algorithms and model-free reinforcement learning algorithms, depending on whether the algorithm employs the model. Both can obtain data by interacting with the environment, but their utilization of this data is different. The algorithm for model-based

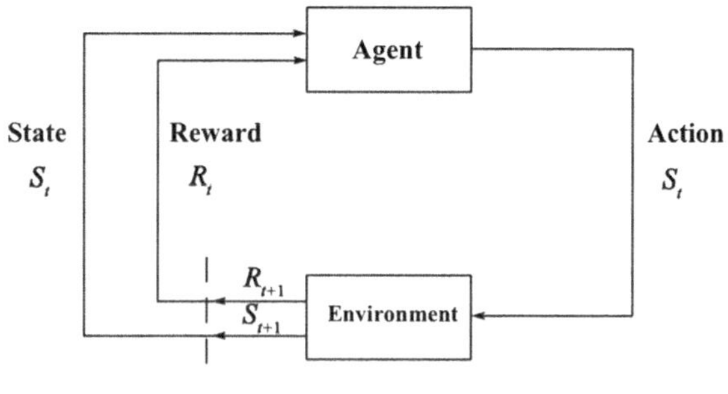

Fig. 6.8 Learning process of reinforcement learning

reinforcement learning employs the data learning system or environment model obtained through interaction with the environment and then makes sequential decisions based on this model. In contrast, model-free reinforcement learning algorithms use the data collected through interactions with the environment to directly improve their own behaviors. Both approaches have merits and drawbacks. In general, model-based reinforcement learning algorithms are more efficient as their agents can utilize model information when exploring the environment. However, for tasks that cannot be modeled at all, model-free reinforcement learning algorithms are the only option. The model-free reinforcement learning algorithm is more applicative, requiring no model.

Reinforcement learning is fundamentally a decision-making process. Supervised learning refers to the process of learning prediction, while unsupervised learning is the process of analyzing data to discover the data structure. It is not sufficient for a system to merely recognize and predict; it must also complete a large number of tasks and engage in a series of actions or behaviors to be capable of making decisions. Reinforcement learning is the process of repeatedly learning from a large amount of data in order to find the optimal solution and then backwardly deducing the necessary model from the final results, which correspond precisely to the decision-making process. Consequently, reinforcement learning is primarily used for manufacturing process control, task scheduling, robot design, and games.

In typical discrete cases, the conventional reinforcement learning is limited to the action space and the sample space. However, increasingly complex tasks, closer to the real-world cases, typically have large state space and continuous action space. When the input data are images and sounds, they usually have high dimensions, making them difficult to process with the conventional reinforcement learning. Deep-reinforcement learning combines the power of deep learning targeted for the high-dimensional data with the reinforcement learning. In recent years, reinforcement learning has made remarkable progress in decision-making, proven by the dominance of AlphaGo and AlphaGo Zero over humans in the popular game. This was possible due to their utilization of deep reinforcement learning techniques. The tremendous success of AlphaGo underscores the capacity of this approach to successfully accomplish assigned tasks without human guidance or prior experience [30].

4. Weakly-Supervised Learning

In the past few years, numerous effective deep learning models have been developed, depending on a significant amount of training datasets that are manually labeled. However, generating these data sets can be a challenging and time-consuming task for multiple applications. To overcome the difficulty in obtaining highly informative supervision information, scientists have incorporated successful research advancements, like active learning and semi-supervised learning. The has led to the development of weakly-supervised learning, which involves the creation of prediction models utilizing weak supervision signals. Weakly supervised learning has three categories: partial supervision, inexact supervision, and inaccurate supervision.

(1) Partial Supervision

Partial supervision involves labeling only a portion of the training data as sometimes the cost of labeling full and strong supervision signals is excessive (i.e., hiring domain experts to manually label extensive datasets). Furthermore, in certain professional fields, some annotators lack sufficient knowledge and experience in data science; as a result, their labeling cannot be used. This results in a significant number of training samples with ineffective annotation, such as the disease contour framed for segmentation tasks. In a partially supervised environment, active learning, semi-supervised learning, and transfer learning are the most prevalent machine learning techniques utilized to overcome these challenges.

Active learning assumes that human experts can be consulted to provide truth labels of unlabeled data, in order to label the most valuable data points [31]. The main objective of active learning is to improve query efficiency and achieve optimal performance with minimal query times when measuring annotation cost solely by query times. In contrast, semi-supervised learning can analyze and utilize unlabeled data without the assistance of human experts. Relevant information on semi-supervised learning will be presented in the following section. Transfer learning, on the other hand, is a concept that aims to enhance data utilization by learning from humans' ability to draw conclusions from a single instance. It has been extensive studied in recent years.

(2) Inexact Supervision

In the context of inexact supervision, training samples only contain coarse-grained labels; for example, when considering an image, it will only identify the category, without providing any supervision information for each entity within it. This limitation becomes apparent when classifying a lung X-ray image, since the presence of pneumonia cannot be verified when an image is only recognized as a lung image from a patient with pneumonia. In order to overcome this obstacle, multi-instance learning has been applied successfully to deal with a variety of tasks, including image classification, retrieval, annotation, text classification, spam detection, medical diagnosis, face detection, target detection, and target tracking, among others.

(3) Inaccurate Supervision

There are several factors that contribute to inaccurate monitoring, such as incompetent labeling personnel, careless labeling processes, and difficulty of labeling. One example of inaccurate supervised learning is learning with noisy labels. In recent years, crowdsourcing has emerged as an important application of the inaccurate supervised learning paradigm, providing a means of collecting training data that addresses some of the challenges related to this type of learning.

5. Semi-Supervised Learning

Machine learning technology typically uses either labeled or unlabeled sample sets for learning. However, in practice, these sets often coexist. To better utilize the value of unlabeled data, semi-supervised learning has been introduced, enabling machines to automatically improve their learning performance by using unlabeled samples without requiring external intervention. Semi-supervised learning is a method that combines supervised and unsupervised learning, where only a portion of the training data is labeled, while the remainder is unlabeled. In the fields of pattern recognition and machine learning, this approach [32] has become increasingly crucial.

Semi-supervised learning has roughly undergone three phases of development [33].

Infancy: In 1990s, researchers began investigating the role of unlabeled samples in the conventional machine learning algorithms [34]. At the same time, most semi-supervised learning algorithms were developed by incorporating unlabeled samples into the conventional supervised learning algorithms, resulting in enhancements.

Maturity: The early 2000s saw the introduction of the concept of "semi-supervised learning", along with numerous algorithms, and the beginning of semi-independent learning algorithms and theoretical systems. Key semi-supervised learning algorithms included hybrid model, pseudo-label or self-training, graph-based semi-supervised learning, and manifold semi-supervised learning.

New height: In recent years, there have been significant advancements in deep learning, which has gained widespread acceptance. However, popularity of deep learning is still hindered by the needs to manually label a large number of samples, which can be challenging to accomplish in many practical situations. As a result, the combination of deep learning and semi-supervised learning has become a popular topic of discussion as a means to address these challenges.

6.2.2 *Classical Algorithms of Machine Learning*

Several commonly used machine learning algorithms, including logical regression, random forest, decision tree, support vector machine (SVM), Bayesian network, k-nearest neighbor (KNN), AdaBoost classifier, k-Means, and artificial neural network (ANN), are now widely utilized in civil engineering. Below is a brief overview of a few of the frequently employed algorithms.

(1) Decision Tree

Decision tree is a case-based inductive learning technique that belongs to the category of supervised learning. It can extract a tree-like classification model from unordered training samples with class names. The algorithm uses a top-down recursive method to compare attribute values at the internal nodes of the decision tree, branching downward based on various attribute values. Each non-leaf node in the tree represents a criterion used to determine a category, while each leaf node represents the final category. A decision tree typically consists of a root node, multiple internal nodes, and several leaf nodes, as depicted in Fig. 6.9. The root node contains the entire set

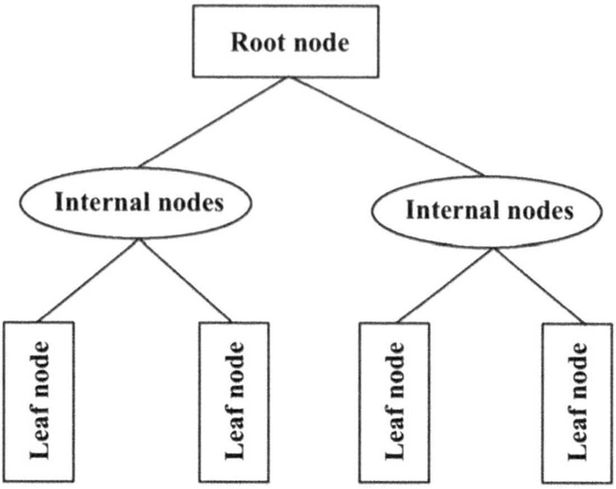

Fig. 6.9 Decision tree structure

of samples, the leaf node represents the decision result, and each additional node represents an attribute test.

The sample sets are split at each node based on the attribute tests. A classified path is formed from the root node to each leaf node. When testing a new sample, simply start at the root node, test at each branch node, recursively enter the sub-tree along the corresponding branch, and then retest until reaching the leaf node. The category represented by the leaf node corresponds to the predicted category of the test sample.

The process of decision tree learning typically involves three steps.

(1) Feature selection: the first step of decision tree learning involves selecting a feature from the features of the training data to serve as the splitting criterion for the current node. Optimal selection of the partition attribute is critical to decision tree learning, and as the partitioning continues, it is expected that the samples contained in the branch nodes of the decision tree belong to the same category; leading to increased "purity" of the nodes. The most commonly used feature selection principles are information gain, information gain rate, and the Gini coefficient, corresponding to the ID3 algorithm, C4.5 algorithm, and CART algorithm, respectively.
(2) Decision tree generation: Starting from the root node, internal nodes are created recursively from top to bottom based on the evaluation criteria of the selected features until the decision tree is terminated for an indivisible data set.
(3) Pruning: during the process of building the decision tree, the node division is repeated multiple times to properly classify the training samples. However, this can potentially result in an excessive number of branches and overfitting. To prevent overfitting, the primary form of the decision tree algorithm is pruning. There are two types of pruning: pre-pruning and post-pruning. Pre-pruning involves immediately stopping the construction of a branch when a node meets

the pruning conditions during the construction process. On the other hand, post-pruning builds a complete decision tree and then prunes it by traversing the tree under certain conditions.

The first decision tree algorithm was proposed in 1966 by CLS (Concept Learning System). In 1986, the ID3 algorithm was introduced and served as the basis for Quinlan's C4.5 algorithm, which was proposed in 1993. Since then, several enhanced algorithms have been proposed to meet the requirements of large-scale data processing (Table 6.1).

Currently, decision tree algorithm has a wide range of applications, including medical diagnosis, weather forecasting, financial analysis, network security, and more. In civil engineering, decision trees have several advantages in addressing multi-classification problems involving non-differentiable models and discrete features. They are commonly used in the construction of discrete element models and complex finite element models for bridges, houses, and other structures. Additionally, decision trees have been used for damage identification of long-span bridges [35].

(2) Support Vector Machine

Support vector machines were first proposed in the 1960s [36] and have undergone rapid development since the 1990s. The primary concept is to locate a hyperplane in the input space to separate distinct types of training set samples. As shown in Fig. 6.10, there may be multiple separation planes, but the hyperplane represented by the red line provides the most effective separation. This hyperplane has the highest "tolerance" and is the most robust to local training sample disturbances. The classification principle behind the support vector machine is to maximize the interval.

In the sample space, the direction of the hyperplane is determined by the normal vector $w = (w_1, w_2, \ldots, w_d)$, and b represents the distance between the hyperplane and the origin. Following is a description of the hyperplane (w, b):

$$w^{T}x + b = 0 \tag{6.1}$$

Table 6.1 Comparison of common decision tree algorithms

Algorithms	Advantages	Disadvantages
ID3	It has transparent theory, straightforward method, and powerful learning ability	It only works for small data sets and is susceptible to noise
C4.5	It generates classification rules that are easily understood and accurate	It is inefficient, and the program will not run if the memory cannot hold the training set
SLIQ	It scales well with increasing number of records and attributes within a certain range	It somewhat restricts the size of the data set that can be processed
SPRINT	It facilitates identification of the optimal splitting criteria for each node	Poor scalability

The distance from any point x in the sample space to the hyperplane can therefore be expressed as

$$\gamma = \frac{|w^{T}x + b|}{\|w\|} \tag{6.2}$$

Assuming that the hyperplane (w, b) is capable of correctly classifying the training samples as

$$\begin{cases} w^{T}x_i + b| + 1, y_i = +1 \\ w^{T}x_i + b| - 1, y_i = -1 \end{cases} \tag{6.3}$$

The decision boundary satisfying this condition creates two parallel hyperplanes as the interval boundary. The distance in between is defined as the interval with size of $2/\|w\|$. The equality relation is valid for the sample points located on the interval boundary. As depicted in Fig. 6.10, they are referred to as support vectors, or the sample points closest to the hyperplane.

In practical scenarios, it is common for there to have no hyperplane in the original sample space that can accurately divide the training samples, which indicates that it is linearly inseparable. In such cases, SVM map points from the low-dimensional space to the high-dimensional space by introducing a kernel function to make them linearly separable. This makes it possible to use the linear SVM approach to locate the partition hyperplane. Though this partition is linear within the high-dimensional mapping space, it is nonlinear within the original sample space.

When training samples are linearly separable in the original space or high-dimensional feature space, a linearly separable hyperplane can be learned by maximizing the interval, which is also known as the hard interval. However, in practical scenarios, it can be challenging to determine the appropriate kernel function that makes the training samples linearly separable. Even if the proper kernel function is

Fig. 6.10 Support vectors and interval

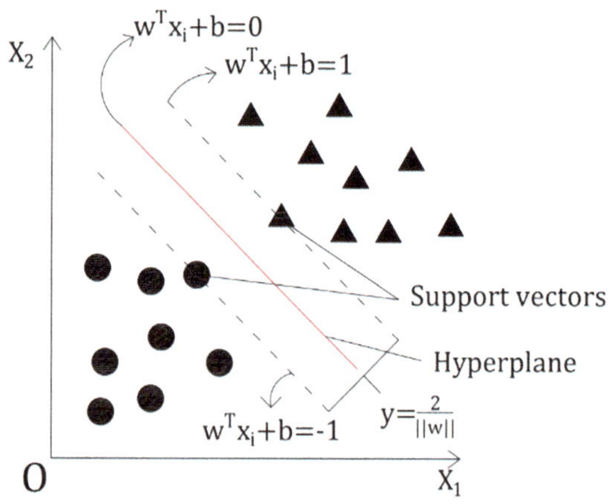

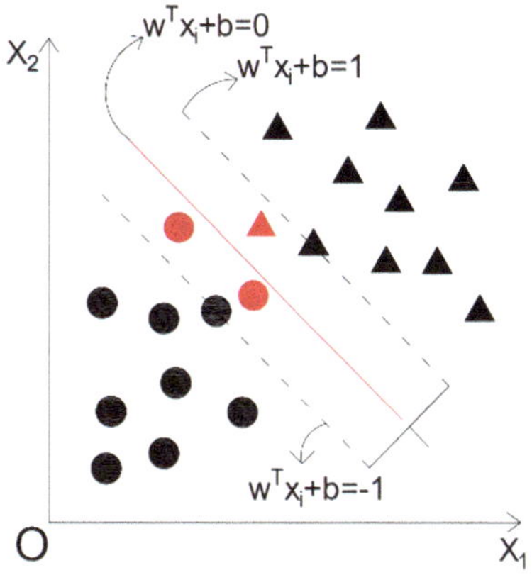

Fig. 6.11 Soft interval

identified, overfitting remains a possibility. To address this issue, the hard interval of linear support is transformed into a soft interval, as shown in Fig. 6.11. This means that some sample points that do not meet the classification criteria are permitted during classification.

The standard SVM algorithm is used to solve binary classification problems, but researchers have expanded its application into two additional types:

(1) SVM Extended from Classification to Regression Tasks

The hyperplane decision boundary used in SVM can be employed as regression model to generate support vector regression (SVR). In such approaches, the standard SVM algorithm is referred to as Support Vector Classification (SVC); for example, in bridge damage identification, SVC is frequently used to identify the types of damage in bridge structures, whereas SVR is commonly used to assess the degree of bridge damage [37].

(2) SVM Expanded from Binary to Multiclass Classification Problems

In such cases, SVM is often used in combination with decision trees. Binary trees are constructed to address a series of two-class problems from the sample set to be classified, and a support vector classifier is constructed for each of the two classes. Furthermore, SVM based on clustering is a generalization of SVM for clustering problems and is frequently applied to multiclass classification problems.

Based on the theory of statistical learning, SVM can quickly and efficiently "deduce" from the training set to the prediction set, thereby improving the accuracy and efficiency of classification and regression for limited samples. By calculating the optimal global solution, SVM theoretically avoids the local extremum problem. The complexity of SVM operations is proportional to the number of support vectors making it vector-dimension independent and suitable for multidimensional or even infinite-dimensional data classification and regression. However, SVM algorithm can be challenging to implement in large-scale training samples due to its sensitivity to missing data, lack of theoretical guidance on the construction and selection of kernel functions and parameters, and limited explanatory power for high-dimensional kernel mapping.

SVM has found a wide range of applications, including fingerprint and face recognitions, text classification, 3D object recognition, and bio-informatics. In building structural health monitoring, SVM-based damage identification technology has reached a higher level of maturity. In a typical application scenario, the parameters of bridge structure dynamics are used as input variables, and SVM is used for training and learning to identify the bridge damages [38].

(3) Bayesian Network

Pearl, a Turing Award winner in 1988, proposed the Bayesian network, which is also known as the confidence network [39]. It is a standard generation model constructed using the Bayesian theorem and graph theory. Besides having a flawless mathematical interpretation of probability theory, Bayesian networks also feature a graphical and intuitive display form. They are one of the most effective theoretical models for expressing and reasoning with uncertain knowledge.

The network structure of a Bayesian network expresses the joint distribution and conditional independence of variables. It is composed of nodes representing variables and directed arcs connecting these nodes, resulting in a directed acyclic graph with multiple nodes. In brief, the random variables involved in a particular research system are plotted on a directed graph based on whether or not conditional independence is met, forming a Bayesian network. In Fig. 6.12, the circle nodes represent random variables, which can be observed variables, hidden variables, unknown parameters, etc. The arrows connecting the nodes represent the causal connection between two random variables. If two nodes are connected by a single arrow, the node that emits the unidirectional arc is the "cause," and the node that points to it is the "effect." When node x_1 directly affects node x_2 in the figure, the arrow from x_1 to x_2 is used to establish the directed arc from node x_1 to node x_2, and the weight value is expressed as the conditional probability P $(x_2|x_1)$. By multiplying their local conditional probability distributions, the joint probability of any random variable can be obtained. For example, in Fig. 6.12, $P(x_1, x_2, x_3) = \mathrm{P}(x_3|x_1, x_2)\mathrm{P}(x_2, x_1)\mathrm{P}(x_1)$ because x_1 influences x_2, and x_1 and x_2 affect x_3.

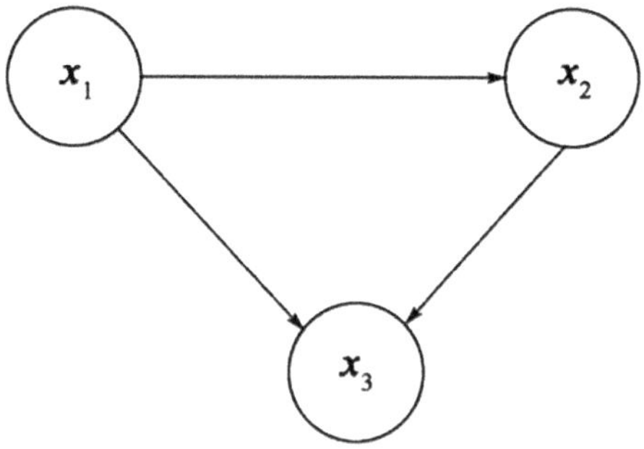

Fig. 6.12 Simple Bayesian network

Figure 6.13 shows a typical Bayesian network structure with three primary components: v-structure, common parent, and cascade. In the v-structure illustrated in Fig. 6.14a), *A* and *B* are independent of one another, just like x_1, x_2, and x_4 in Fig. 6.13. In the common parent structure depicted in Fig. 6.14b, the relationship between *A* and *B* depends on the status of *C*. *A* and *B* are not independent when *C* is unknown, but *A* is independent of *B* when *C* is known. In Fig. 6.13, examples of the common parent structure include x_3, x_4, and x_5 The cascade structure is depicted in Fig. 6.14c, where the relationship between A and B must be divided into two cases based on the status of *C*, as described previously.

Fig. 6.13 Typical Bayesian network structure

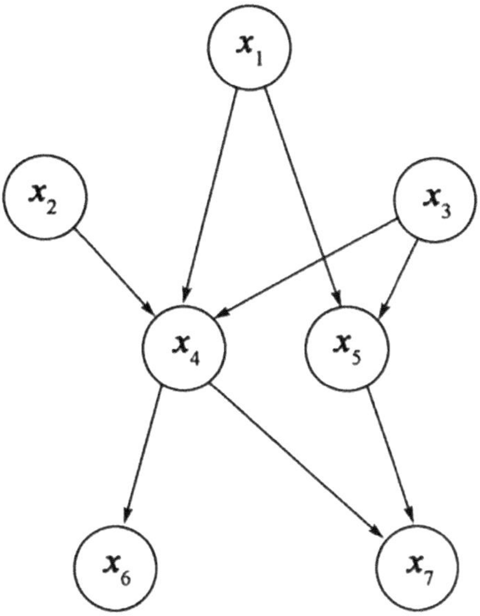

Fig. 6.14 Bayesian network structures

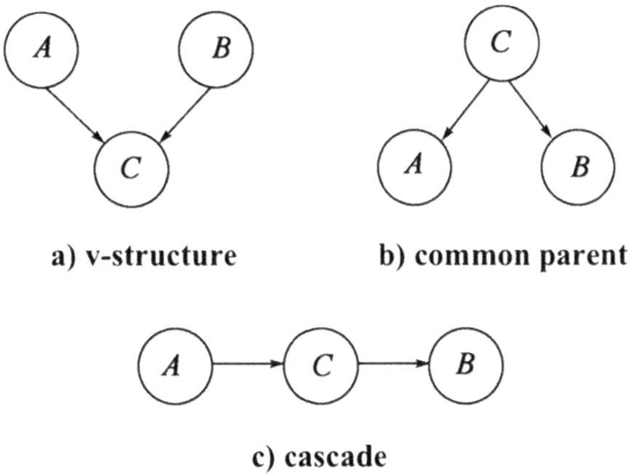

The Bayesian network method has three main components: network construction, network learning, and network reasoning. There are two methods of network construction. One involves manual construction based on experience, while the other involves analyzing huge amounts of information. The manual approach often uses the causal relationships between variables to determine the network structure.

The purpose of network learning is to determine the relationship between events by utilizing a data-training model. Typical network learning algorithms include SGS algorithm, PC algorithm, TPDA algorithm, K2 algorithm, simulated annealing algorithm, and greedy algorithm.

Network reasoning includes both prediction and diagnosis, and it can be categorized into two types. Approximate reasoning methods include sampling, Markov chain Monte Carlo, and other techniques. Precise reasoning techniques involve causal tree algorithm, poly tree algorithm, and other approaches.

Bayesian network provides a simple and intuitive topology structure to visualize the probability relationship between variables. It displays the dependency or independence between events, and enables bi-directional data reasoning. Unlike the conventional probability and statistics theory, Bayesian networks do not require large amount of historical data and can directly use prior knowledge to construct the probability relationship between variables. Unlike neural network and logistic regression models, all nodes in the Bayesian network model are visible, and the outcomes are highly interpretable. Furthermore, Bayesian network can complete network construction with incomplete data and supplement data samples with its unique learn algorithm to complete network model construction, hence expands the network's application.

Bridge structure system is a complex and uncertain system, with structural parameters similarly fraught with uncertainty. Bayesian network is a powerful tool for quantifying uncertainty, with robust uncertainty reasoning and data analysis capabilities. In evaluating bridge structures, Bayesian networks can simultaneously consider the comprehensive effects of multiple variables, effectively determine the uncertain information contained in the monitoring data, and provide a more scientific and reasonable basis for evaluation and early warning [40]. Thus, Bayesian network can effectively assist in identifying potential risks in the bridge structure system and provide a reliable basis for decision-making and risk management.

(4) K-Nearest Neighbor

K-nearest neighbor is a fundamental classification and regression algorithm proposed by Cover and Hart in 1968 [41] as a component of the inert learning method. This algorithm's concept is simple and intuitive, requiring neither parameter estimation nor training. Similar to majority voting in elections, the k-nearest neighbor algorithm classifies data by measuring the distance between different feature values. According to this algorithm, if the majority of the closest k samples in the feature space belong to a specific category, then the sample belongs to this category and possesses its characteristics. In other words, samples belonging to the same category should be grouped together in the feature space.

Figure 6.15 depicts the basic principle of the k-nearest neighbor algorithm. Given a training set as a sample set, each data element within it is labeled with a category.

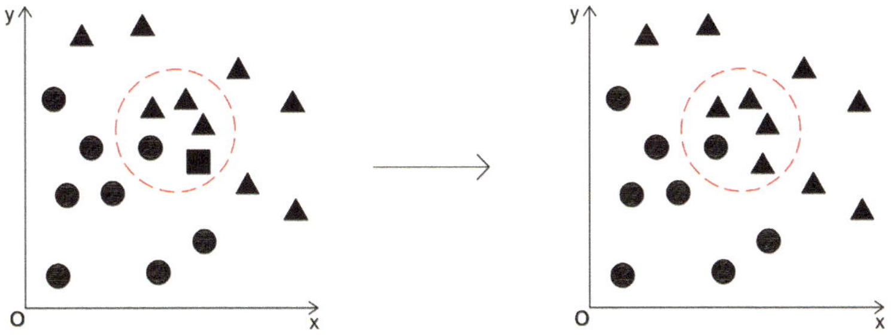

Fig. 6.15 Basic principle of k-nearest neighbor algorithm

When a test object is entered, its characteristics are compared to the sample set's corresponding attributes by calculating the distance between the test data in the feature space and each data point in the sample set. The category labels of *k* data from the sample set that are closest to the test data are then extracted, and the category with the most occurrences among the *k* most similar data is chosen as the category for the new data. The k-nearest neighbor algorithm only selects the first i data points from the sample dataset, where k is typically an integer less than 20.

Important aspects of the k-nearest neighbor algorithm include decision rules, k-values, and distance measures. The majority of classification decision rules in the k-nearest neighbor algorithm are based on majority voting. This means that the category to which the majority of *k* points belong is used to determine the final classification attribution of the new instance points. To measure distances, the k-nearest neighbor algorithm employs the Euclidean distance, with the following computation formula:

$$d(x, y) = \sqrt{(x_1 - y_1)^2 + (x_2 - y_2)^2 + \cdots + (x_n - y_n)^2} = \sqrt{\sum_{i=1}^{n} (x_i - y_i)^2} \tag{6.4}$$

The choice of k-value has a significant impact on the classification outcomes of the k-nearest neighbor algorithm. An excessively large *k* will increase the training error, while a too-small *k* can increase the generalization error, leading to over-fitting. The value of *k* is not predetermined, and must be determined based on the specific problem at hand. In most cases, the cross-validation method is used to determine the optimal *k* value by continuously minimizing the error. This involves dividing the sample data into a training set and a validation set in a proportional manner. We begin with a small *k* value and perform k-nearest neighbor computations on these two sets to compute the variance of the validation set. As the value of *k* increases, the optimal classification result will eventually be determined.

Due to its early introduction and other evolving technologies, the flaws of the k-nearest neighbor algorithm have gradually become evident, resulting in the emergence of improved algorithms. For classification effectiveness, the k-nearest neighbor algorithm is improved by using the weight method, in which the neighbor closest to the sample is given a high weight. For classification efficiency, some representative samples can be selected as new training samples by reducing sample attributes ahead of time, or the center points created by clustering can be used as new training samples. The k-nearest neighbor search process can be accelerated to improve classification efficiency by reducing search time for the k-neighbors. Based on this, researchers have proposed a variety of fast search techniques, such as the k-nearest neighbor search algorithm based on partial distance calculation in the wavelet domain. Additionally, the computation cost of k-nearest neighbors can be reduced by introducing efficient index methods, which has a substantial impact in high-dimensional space. In recent years, the k-nearest neighbor algorithm has frequently been combined with other algorithms, such as SVM, Bayesian classifiers and P-tree [42], to enhance performance and make greater application breakthrough.

The k-nearest neighbor algorithm is straightforward and efficient, as new data can be added directly to the dataset without retraining or making data input assumptions. Currently, the k-nearest neighbor algorithm is extensively used in face recognition, character recognition, medical image processing and other fields, and has achieved significant success. Despite its advantages, the K-nearest neighbor algorithm still has some shortcomings. For highly complex spaces, calculating the k-nearest neighbor algorithm is time-consuming for extremely large data sets. Similarly, if the samples are not evenly distributed, the prediction error of the k-nearest neighbor algorithm can be substantial.

The bridge health monitoring system collects a large amount of monitoring data through long-term and comprehensive monitoring to study the structural response characteristics of the bridge during operation. However, due to various uncertain factors, monitoring data often contain anomalous information, which can impact the evaluation of bridge structure safety. The k-nearest neighbor algorithm, a simple supervised learning classification method, can effectively detect and identify these abnormal data in the monitoring data [43].

(5) Artificial Neural Network

Artificial neural network (ANN) is a mathematical model based on biological neural networks and network topology, simulating the processing mechanism of the neural system in the human brain for complex information. Comparable to the nervous system of the human brain, ANN is a complex network structure consisting of a large number of interconnected simple "neurons" that communicate with one another. At present, the most extensive definition of ANN is a widely parallel interconnect network composed of adaptive simple units. Its structure can mimic the interaction response of biological neural systems to real-world objects [44]. ANN possesses a high degree of fault tolerance, robustness, self-learning, self-organization, adaptability, and parallelism. These features make it a potent instrument for AI, computational intelligence, and data processing.

The "M-P neuron model" (Fig. 6.16) was first proposed by American psychologist McCulloch [45]. This model is a mathematical simplification of the information processing mode of biological neurons and serves as the basic information processing

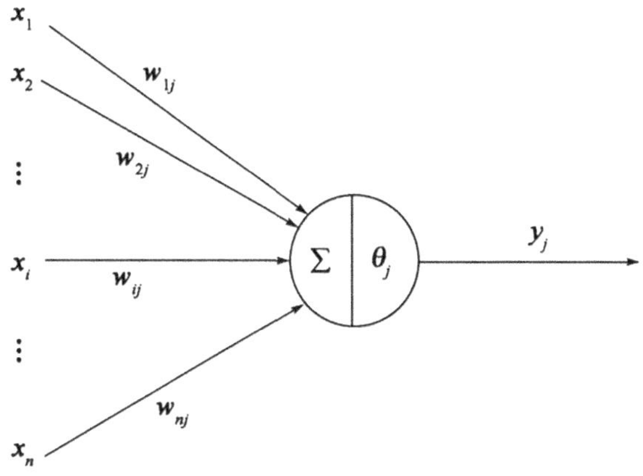

Fig. 6.16 M-P neuron model

unit of neural network structures, and paving the way for theoretical research into neural networks.

As depicted in Fig. 6.16, the neuron receives input signals from n other neurons via weighted connections. It then generates output through activation function processing, based on comparing the received total input value to its own threshold value.

$$y_j = f\left(\sum_{i=1}^{n} w_{ij}x_i - \theta_j\right) \tag{6.5}$$

where x_i $(i = 1, 2, \ldots, n)$ represents the transmission signal of other neurons connected to the neuron; w_{ij} represents the connection weight value from neuron i to neuron j; θ_j represents the threshold of the neuron; and $f(\cdot)$ is the activation function reflecting the mapping relationship between various network inputs and outputs. The activation function can be of different types, such as threshold function, linear function, S-type function, or other non-linear functions.

The neural network is obtained by connecting the aforementioned neurons according to a certain hierarchical structure. In the early stages of development, the network structure consisted of only an input layer and an output layer. The perception model depicted in Fig. 6.17 with two inputs consists of two layers of neurons, including the input layer and the output layer. The input layer receives external input signals and transmits them to the output layer, which is an M-P neuron.

The perceptual function can perform simple logical AND, OR, and NOT operations, but with only one layer of functional neurons. It lacks the learning capacity to solve the problem of nonlinear separability. Therefore, the development of multilayer neural networks was an inevitable consequence of the growth of neural networks. In addition to the input layer and output layer, a multilayer neural network contains an intermediate layer known as the hidden layer, which can consist of one or more layers. As shown in Fig. 6.18, a typical neural network structure is hierarchical, with each neuronal layer fully interconnected with the neuronal layer beneath it. There

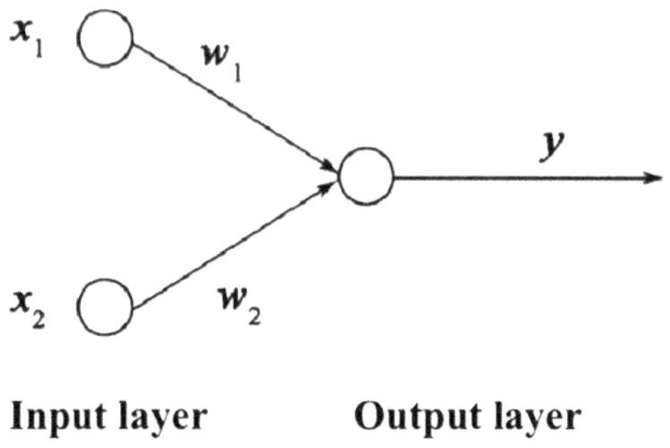

Fig. 6.17 Perceptron model with two inputs

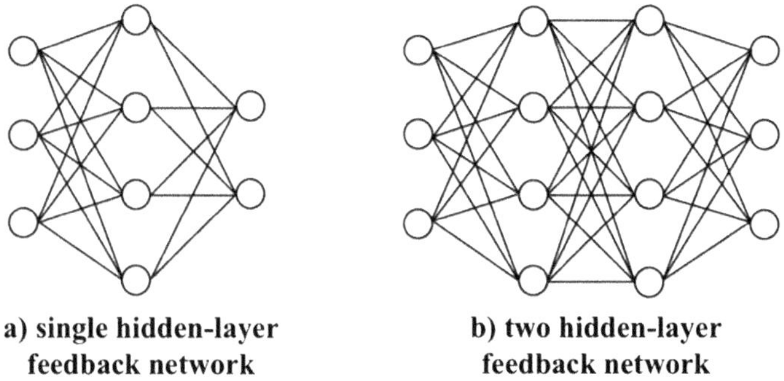

Fig. 6.18 Multilayer feed-forward neural networks

are no connections between neurons either within or between layers. This structure is referred to as a "multilayer feed-forward neural network" [7].

The connection mode, number of network layers, and number of neurons in each layer are typically set in advance as the "super parameters." Learning in neural networks depends on the connection weights between neurons and the threshold of neurons. The most fundamental aspect of learning is the dynamic adjustment of weights, which stores the most crucial network information.

There are numerous models of neural networks, each with distinct characteristics and application functions, so only a few prevalent neural networks are briefly introduced below.

(1) BP Neural Networks

Adding layers can increase the network's flexibility, but the training of parameters is a significant impediment to the advancement of multilayer neural networks. In his 1974 doctoral dissertation, Werbos proposed the now-famous Error Back Propagation (BP) algorithm to solve the learning problem of multi-layer neural networks and significantly accelerate their development. The fundamental concept underlying the BP algorithm is that the learning process is comprised of two processes: forward propagation of signals and backward propagation of errors. When the signal propagates forward, the input samples are transferred from the input layer, processed in order by each hidden layer, and then transferred to the output layer. The error back-propagation phase is initiated if the actual output of the output layer does not match the expected output. During the error back-propagation phase, the output error is back propagated to the input layer via the hidden layers in a specific order, and the error is allocated to all units of each layer to obtain the error signal of each layer, which forms the basis for correcting the unit weight. The weight adjustment process of forward signal propagation and error back-propagation is repeated until the network output error is reduced to an acceptable level or until the predetermined learning times expire.

BP neural network is a multilayer feedforward network that utilizes the BP algorithm. It can learn to store complex mapping relationship without knowing the specific numbers or expressions for input and output in advance. BP neural network adopts the learning mode of information forward propagation and minimum mean square error back propagation. It uses the fastest gradient information to identify the parameter combination that minimizes network error, thereby solving the problem of adjusting the weight and threshold.

(2) RBF Neural Networks

Radial basis function (RBF) was introduced by D. S. Broomhead to simulate the local response characteristics of biological neurons [46]. RBF neural network utilizes the RBF-composed hidden layer space to project the low-dimensional input vector into the high-dimensional space, thereby transforming the original linearly indivisible problem into a linearly separable one. This is its fundamental operating principle. The structure of a RBF neural network is depicted in Fig. 6.19.

RBF is employed as the activation function for the hidden-layer neurons in RBF neural networks. The input layer and hidden layer require no weight connection, and the output layer is a linear combination of neurons from the hidden layer. Typically, RBF is described as a monotone function of the Euclidean distance between the sample x and the data center c_i. The commonly used Gaussian radial basis function is given by

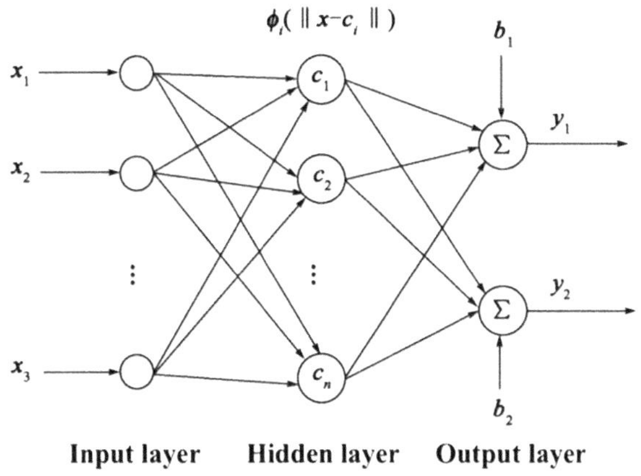

Fig. 6.19 Structure of a RBF neural network

$$\phi(x, c_i) = \mathrm{e}\beta i\|x \cdot c_i\|^2 \tag{6.6}$$

RBF neural network is typically trained in two steps. The first step involves identifying the neural centers from the sample data using unsupervised learning techniques, such as random sampling, clustering, etc. This step initializes the network for the supervised learning stage. In the second step, the weights between the hidden layer and the output layer are modified using supervised learning techniques.

RBF neural network has a rapid operation speed, robust nonlinear mapping capability, and superior approximation performance. It has been demonstrated [4] that an RBF neural network with a sufficient number of neurons in the hidden layer can approximate any continuous function with any level of precision. The network structure and algorithm of RBF neural networks are straightforward, making them easy to use. It has been extensively utilized in applications such as time series prediction, speech recognition, automatic control, signal prediction, and signal processing.

(3) Feedback Neural Networks

BP neural networks and RBF neural networks are examples of feed-forward neural networks, in which each layer of neural nodes receives data input from the previous layer and outputs it to the next layer after processing. The data flows forward without a feedback connection. The output of a feed-forward neural network is determined solely by the current input and the network's weight, while the output of a feedback neural network is also dependent on the network's previous input. Figure 6.20 depicts the structure of a feedback neural network.

Hopfield network, Elman network, CG network model, BSB (Brain State in a Box) model, BAM (Bidirectional Associative Memory), etc. are examples of typical feedback neural networks. The feedback neural network has greater computing capacity than the feed-forward neural network, and exhibits superior associative memory and optimal computing capabilities.

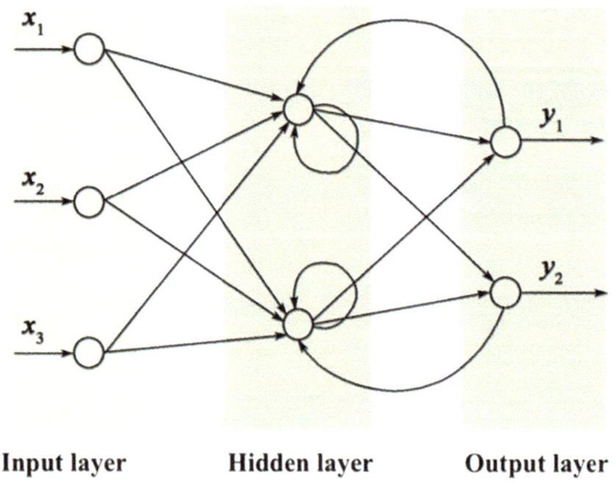

Fig. 6.20 Structure of feedback neural network

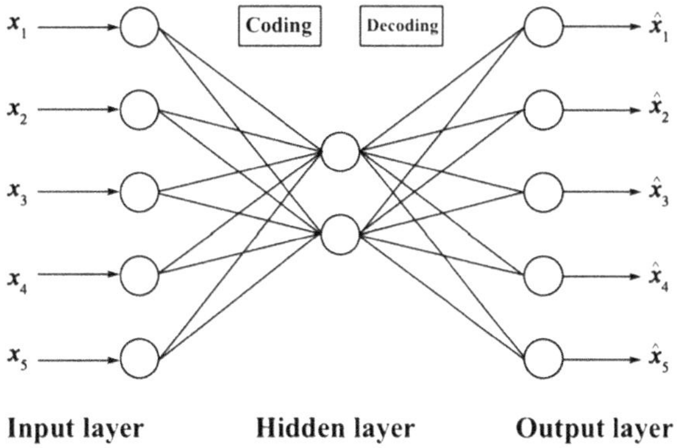

Fig. 6.21 Structure of autocoder

(4) Autocoder

Autocoder is both an unsupervised neural network model and a specialized form of supervised learning, with the supervision signal being the input data itself. Some researchers therefore refer to Autocoder as a "self-supervised" neural network. An autocoder is designed to replicate input into output through training. Figure 6.21 depicts its general composition—coding and decoding. The process from the input layer to the hidden layer that learns the hidden features of the input data is called "coding." The process from the hidden layer to the output layer that reconstructs the original input data using the learned hidden features is called "decoding." The network aims to ensure that the output result $\hat{x}$ generated by the hidden layer, after encoding and decoding the input data x,, is as consistent as possible with the input data x. The network is trained by minimizing reconstruction error between input x and output $\hat{x}$.

Training an autocoder to perform input reconstruction aims to discover more beneficial properties in the hidden layer. Many improved autocoder methods add constraints to the expression of the hidden layer to force it to be copied approximately. If, after adding these constraints, the model can still recreate the input signal, it indicates that the hidden layer representation is an effective feature. Dimensionality reduction autoencoder, denoising autoencoder [47], sparse autoencoder [48], and variational autoencoder [49] are typical examples of autocoder methods.

By restricting the number of hidden layer nodes, the dimensionality reduction autoencoder reduces the number of nodes below that of the input/output nodes. Since the number of hidden layer nodes is less than the number of input nodes, the model can only learn the most essential features of the data and remove the irrelevant features.

A denoising autoencoder, on the other hand, amplifies input noise and produces a noiseless output after training. This prevents the autoencoder from simply duplicating the input to the output. Despite adding noise to the input data, this model can still extract useful features. Gaussian noise and random occlusion noise are the most widely used noise techniques in this model.

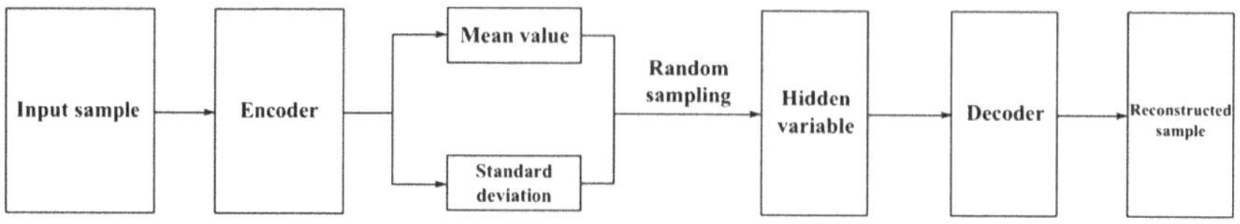

Fig. 6.22 Structure of variational autoencoder

The sparse autoencoder is an extension of the traditional autoencoder that adds sparse constraints to the hidden layer by targeting specific neurons with sparsity. By suppressing the majority of hidden layer neuron outputs, only a small fraction of neurons will be activated each time the network executes, allowing it to achieve a sparse effect.

The variational autoencoder is a popular unsupervised learning method proposed in 2013 [49] for complex probability distributions. It is an improved version of the autoencoder and consists of an encoder and a decoder. Unlike the conventional autoencoder, variational self-coding incorporates constraints to the coding network. It maps the training samples to the mean and standard deviation of the probability distribution that the hidden variables obey, and then obtains the hidden variables through sampling. In addition to the encoding process, the decoder of the variational autoencoder maps the hidden variables back to the sample variables for reconstruction, allowing for the re-creation of the original input data. Figure 6.22 highlights that the key difference between the variational autoencoder and the standard autoencoder is that the variational autoencoder outputs the parameters of the probability distribution instead of the direct hidden variables. The variational autoencoder adds randomness through random sampling during the reconstruction process, allowing for increased flexibility in generating new outputs. The standard autoencoder is limited to only recreating the input sample.

(5) Deep Neural Networks

Depth neural network (DNN) is obtained by extending the hidden layer of a simple single-layer neural network to multiple layers, enabling it to handle complex problems. Theoretically, the capacity of a neural network model to handle complex problems is related to its complexity. Increasing the number of hidden layers or the number of hidden layer neurons is a straightforward way to increase the complexity of a neural network model. Increasing the number of hidden layers in a DNN is more effective than simply increasing the number of hidden layer neurons. This is because adding more hidden layers increases not just the number of neurons with activation functions but also the number of neurons with activation functions nested within the hidden layers [7]. However, a higher number of hidden layers also means an increase in the number of parameters in the model. In addition, training a DNN with multiple hidden layers using classical algorithms like the standard BP algorithm can be challenging. This is due to the problems of "vanishing gradient" or "exploding gradient" that frequently occurs during error back propagation in multiple hidden layers.

Unsupervised layer-by-layer training is an efficient method for DNN training based on the principle of "pre-training + fine tuning." In this method, the network is trained layer by layer using unsupervised algorithms to learn the features of each layer. Once "pre-training" is completed, the entire network is fine-tuned using the supervised learning algorithms [50]. Pre-training and fine-tuning can be thought of as grouping a large number of parameters, determining better local settings for each group, and then combining these local optimal results for global optimization. This approach effectively reduces the training costs and provides a large number of model parameters for increased degrees of freedom [7].

Weight sharing, which involves allowing a group of neurons to share the same connection weight, is another method for reducing training costs. This technique plays a crucial role in the convolutional neural network [51] described in detail in Chap. 7.

(6) Graph Neural Networks

The conventional algorithms for machine learning are primarily designed to handle data in European space, which has a regular spatial structure that can be represented by one-dimensional or two-dimensional matrices. For instance, images can be represented as regular two-dimensional matrices, while voice data can be represented as one-dimensional sequences. These algorithms assume that samples are independent of one another. In practice, however, various non-European spatial data exist, such as abstract maps of electronic transactions and recommendation systems. These forms of data do not have a fixed connection between each node on the map and require different machine learning algorithms to handle them effectively.

A graph neural network (GNN) is a neural network that can handle non-Euclidean data with internal dependencies in an irregular and unordered fashion. The concept of this model was derived from a neural network model created by Gori for processing graph-structured data [52]. As shown in Fig. 6.23, GNN captures the dependency of each node by passing messages between them. In contrast to the conventional neural networks, GNNs can represent information from its neighborhood at any depth. It takes scale, heterogeneity, and deep topology information of the input data into account, and are capable of extracting complex characteristics of the data, realizing rapid processing of massive data, and mining deep and effective topology information [53].

ANN refers to models that are designed to solve problems in systems with a limited understanding of the internal operating mechanism and rules. These models are particularly effective when working with fuzzy or incomplete input data, when it is not possible to establish appropriate mathematical models based on existing knowledge, and when there is a need for fault tolerance. Civil engineering problems often involve nonlinear relationships that can be difficult to capture using exact mathematical and mechanical models. The complex relationships between variables further adds to the challenge. To overcome these difficulties, many researchers in civil engineering have adopted neural networks as a viable tool to solve such problems.

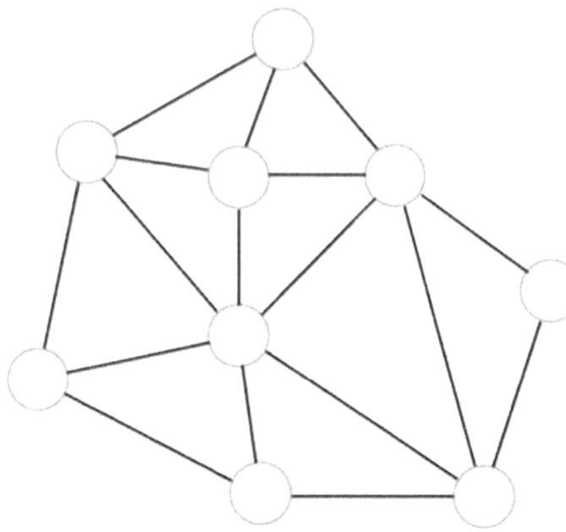

Fig. 6.23 Structure of graph neural network

6.3 Application Scenarios and Challenges of Machine Learning

6.3.1 Application Scenarios

1. Structural Health Monitoring

Structural health monitoring system (SHM) is a system that quantifies and localizes structural damage by monitoring and analyzing structural deformation, displacement, dynamic response, and other parameters. SHM systems are used to make informed maintenance and management decisions based on the structural states. The monitoring process involves collecting data from various sensors attached to the structures, processing the data to extract meaningful information, performing statistical analyses to identify patterns, and using visualization tools to display the results. The incorporation of machine learning technology can significantly enhance the effectiveness and precision of SHM systems, thereby improving the dependability of structural maintenance and management decisions. Machine learning techniques have demonstrated success in various areas, including optimization of sensor layouts, fault detection, pre-processing of monitoring data, and feature analysis.

(1) Optimization of Equipment Layout

Equipment plays a crucial role in the acquisition of data. In the conventional monitoring systems, sensor networks are commonly used to collect data. In recent years, technological advancements have led to the use of unmanned aerial vehicles (UAVs) and underwater robots that come equipped with cameras, radars, lasers, and other sensors. These tools have significantly expanded the means of structural monitoring and have increased the effectiveness of monitoring structural distresses in complex situations. Optimizing the layout of monitoring equipment can significantly improve the monitoring network's precision and sensitivity. Machine learning algorithms have been applied in two areas to achieve this: sensor layout optimization and UAV patrol route optimization.

Optimizing sensor layout can significantly improve the efficiency and effectiveness of SHM systems, leading to more rational evaluation decisions. Machine learning algorithms, like swarm intelligence algorithm and neural network algorithm, can be used for optimizing sensor placements. Swarm intelligence algorithm is a type of bionic optimization algorithm that simulates biological behaviors such as foraging and reproduction. Various algorithms, including genetic algorithm [54], particle swarm algorithm [55], and monkey swarm algorithm [56], have been applied to the optimal layout of sensors.

Presently, among the various optimization techniques used to optimize senor layout, genetic algorithm is the most widely used. Its robustness and capability to seek out the global optimal solution confer a number of benefits to the optimal layout of sensors. Liu et al. [54] utilized a genetic algorithm to investigate the optimal sensor layout for the Runyang long-span cable-stayed bridge. The results showed that the approach is robust to noise, highlighting the reliability and effectiveness of genetic algorithm for sensor layout optimization. Swarm intelligence algorithm is susceptible to local optimization as it must find the optimal solution via the direct relationship between the layout scheme and the evaluation index. To mitigate this issue, Yuan Can [57] used deep learning to establish the mapping relationship between the layout scheme and the evaluation index.

Optimizing UAV patrol routes can be challenging due to the complex historical data of the structural health status. It can be difficult to comprehensively consider all aspects of the complex data situations in manually planned routes, resulting in low efficiency of UAV in structural health monitoring. Machine learning algorithms can optimize the key cruising routes of the UAVs based on the historical evaluation data of structural damage. This optimization allows the UAV to focus on the long-term patrol inspection of important structural components. This automated and optimized approach can effectively monitor the progression of bridge damage and prevent safety accidents.

The sensor used in SHM are subject to deterioration over time due to factors like aging and the environmental conditions. Once a significant number of sensors in a network failed, effectiveness of the monitoring system is compromised, and it becomes unable to monitor the structural health effectively. Reliable functioning of a health monitoring system depends on the timely detection and repair of sensor malfunctions, to ensure that monitoring and early warning functions can be perfomed accurately and reliably. Current research on sensor fault detection includes utilizing neural networks [58], SVM [59], principal component analysis [60], Bayesian network [61], and other algorithms. In mechanical, chemical, and other fields, machine learning technology has been employed to address sensor failures, with processing methods involving compensation [61, 62] and isolation [63]. Zeng et al. [62] proposed a compensation-based approach to address sensor failure in SHM systems. Their high-precision sensor error compensation method utilized an orthogonal basis function neural network to model the relationship between measurement errors and input data. While this approach can improve the accuracy of compensated data, it can still be challenging to compensate for data anomalies caused by faulty sensors. Isolating the faulty sensors is another processing strategy that can be employed to mitigate the impact of faulty sensors. Follosa et al. [63] implemented fault sensor isolation by shielding the fault data input channel using multicore SVM. Application of the above-mentioned machine learning technology is not limited to specific fields, and can also be utilized in structural health monitoring. To ensure effective and reliable monitoring of structures, it is essential to establish a sensor network with a certain level of redundancy. This provides the monitoring system with a backup in case one or more of the sensors failed. Machine learning algorithms can then be employed to detect and isolate the faulty sensors, allowing the monitoring system to continue operating normally without the need for sensor replacements.

(2) Data Preprocessing

During long-term monitoring of large structures, there may be limitations to monitoring equipment, environmental factors and other impacts that can lead to missing data, errors, inconsistencies and other issues in the data collection process. These issues can subsequently affect the reliability and accuracy of structural evaluation and early warning. To address these issues, it is necessary to first clean the monitoring data by repairing the missing data, reducing data noise, and correcting inconsistent or erroneous data. This will enable the provision of high-quality data for subsequent structural evaluation and early warning.

Zheng et al. [64] proposed a method for restraining and guiding the cleaning of abnormal data from dam safety monitoring using association rules. In their research, they found that the method, which takes data association into consideration, is more precise than the conventional method for distinguishing abnormal data resulting from gross error and environmental mutation, which considers single monitoring data. In another study, based on the extreme learning machine (ELM), Huang et al. [65] proposed a solution for the data loss of their bridge health monitoring system.

(3) Feature Engineering

Feature engineering is the process of transforming raw data into feature data that can be used for machine learning applications. This process includes feature construction, extraction and selection, and is essential in improving the accuracy and efficiency of machine learning models. In the context of SHM, feature engineering involves extracting relevant characteristics from the original monitoring data that reflect the health of structures. The characteristics that are extracted will vary, depending on the specific applications and the types of structures being monitored. Features extracted from monitoring data in SHM systems can vary, and commonly include frequency domain features, time–frequency domain features, statistical features, and mechanical features [66].

Yuan et al. [67] employed a combination of time domain, frequency domain, and statistical characteristics derived from an autoregressive model to identify anomalies in acceleration data. They used kernel principal component analysis (KPCA) to extract features from the data of the deceleration sensor located on the steel plate of the primary beam of Jiangyin Bridge. After extracting the features using KPCA, they investigated the anomaly identification and classification of sensor data sets. Yuan Shenfang et al. implemented the identification and classification of abnormal data by considering feature separability. This approach ensured that the extracted features were informative enough to accurately detect and categorize any anomalies present in the data. In a separate study, Ashish et al. [68] developed a system that integrates machine learning tools into mobile devices to monitor and identify bridge vibration signals in real time. Their approach extracted seismic, traffic, and other characteristic signals that are necessary for structural health monitoring and analysis. Researchers have developed a one-dimensional machine learning network as the central component of a SHM system's architecture that can run on small mobile devices such as smartphones. This approach has the advantages of small memory requirement and high operating efficiency, which are critical for resource-limited mobile devices, Compared to classical neural networks such as AlexNet and VGG16, the developed one-dimensional machine learning network exhibits the same accuracy in feature signal recognition, making it an efficient and reliable option for SHM systems. This system architecture that employs mobile devices as sensors and processors is distinguished by its low cost, offline availability, and low latency. The monitoring and recognition system for bridge vibration signals developed by combining the benefits of machine learning and mobile devices is of great significance in promoting high efficiency and portability of bridge health monitoring.

2. Structural Performance Evaluation

Evaluation and prediction of structural performance are conducted primarily from the perspectives of material performance prediction, structural deformation prediction, structural damage identification, structural state evaluation, and structural reliability analysis. However, the conventional methods for addressing these issues often prove to be time-consuming and inaccurate. Application of machine learning algorithms to the evaluation of structural performance can significantly enhance the efficacy and precision of the evaluation process. Moreover, as the dataset size increases, more samples can be used to train and improve the existing algorithms, leading to better model performance.

(1) Material Performance Prediction

In civil engineering, concrete is the most commonly used construction material. Various concrete material properties, including but not limited to elastic modulus and compressive strength, can be accurately predicted using machine learning algorithms. SVM is a popular algorithm widely employed to predict the performance of materials, including concrete.

Gong et al. [69] developed a prediction model to improve the accuracy of concrete compressive strength prediction. The model utilizes a SVM algorithm with three optimization methods, including the SVM model based on grid optimization (GRID-SVM), the SVM model based on particle swarm optimization (PSO-SVM), and the SVM model based on a genetic optimization algorithm (GA-SVM). The resultant model achieved good results, indicating the potential for adoption of such techniques in the design and evaluation of structures. The researchers first determined the optimal values for compromise parameter C and kernel width parameter E for a SVM model using a RBF kernel. The optimization process followed a specific sequence including (1) determining the number of training and test sets, and the ratio between them; (2) selecting prediction parameters; (3) standardizing the experimental data; and (4) determining the appropriate kernel function and kernel function parameters. After optimizing the SVM model and selecting the most appropriate RBF kernel function, three optimization methods, including grid optimization, particle swarm optimization, and genetic optimization algorithms, are employed to search for the optimal parameter values of C and E. According to the results presented by Gong et al., the three optimization models (GRID-SVM, PSO-SVM and GA-SVM) are capable of selecting appropriate SVM model parameters, which enhanced the predictability of concrete compressive strength. The correlation coefficients of the three optimization models were high, with the GRID-SVM model having the highest correlation coefficient of 0.9948, followed by the PSO-SVM model with a correlation coefficient of 0.9946, and finally the GA-SVM model with a correlation coefficient of 0.9930. Moreover, the mean square error of the GRID-SVM model is only 0.001, indicating excellent optimization effectiveness. These results demonstrated the viability of using machine learning algorithms to predict material properties.

Yan et al. [70] conducted a study on the prediction of the elastic modulus of normal-strength concrete and high-strength concrete using SVM models. The study found that SVM models have superior generalization ability and performance in predicting the elastic modulus of concrete, compared to the ANN models. Moreover, SVM was found to be highly effective in predicting both the compressive strength and the elastic modulus for all levels of concrete. In another study by Xin et al. [71], an iterative decision tree was used to predict the compressive strength of concrete. The researchers utilized test results to train the iterative decision tree. The high average accuracy of the predictions achieved in their study, close to 99%, underscored the potential of machine learning techniques to advance the field of civil engineering. In the study by Xiao et al. [72], the researchers developed an ANN model using a BP algorithm to evaluate the performance of concrete after freezing and thawing. The predicted relative dynamic modulus of elasticity is in excellent agreement with the experimental data, demonstrating notable advantages. The employment of ANN method to predict the relative dynamic modulus of elasticity is an extremely complex property influenced by several factors. One of the significant advantages of the ANN method is that it does not require a clear mathematical and physical model to provide relatively accurate results. Additionally, the ANN model can continue to learn and improve its accuracy as the number of samples grows. This self-learning feature is a distinct advantage of ANN over other conventional techniques, which often requires modifications to their structures or parameters as additional data becomes available. The ability of ANN to learn and adapt to new data without substantial reprogramming makes it ideal for analyzing large datasets and predicting properties of complex systems.

In the study by Han et al. [73], the researchers utilized SVM algorithms to predict the carbonation depth of organic film-forming concrete. One of the significant advantages of the SVM algorithm for this application is its ability to solve high-dimensional pattern recognition problems, making it suitable for analyzing complex data in the entire life cycle of materials. The test value of the prediction model correlated well with the curve of the predicted value, indicating limited model error. Upon further investigation, it was found that the predicted value of the BP neural network prediction model also showed good fitting with the curve of the test value. However, it was observed to have lower accuracy and reliability compared to the results obtained by the SVM algorithm, which illustrated the predictability of the SVM model. By developing an intelligent prediction model for the carbonation depth of coated concrete, it is possible to precisely and effectively predict the carbonation depth of concrete even when the protective coating failed. This model can help guide the application of protective coating to ensure the concrete is adequately protected. In the study by Liu et al. [74], the researchers merged the artificial bee colony (ABC) algorithm and the BP neural network to create a hybrid ABC-BP neural network. The aim was to improve the prediction accuracy of the neural network model in predicting the properties of concrete. In the study by Shen et al. [75], two algorithms, decision tree algorithm and logistic regression algorithm, were utilized to conduct non-destructive detection of multi-sensor steel bar corrosion. The study found that the accuracy of the logistic regression algorithm was superior to that of the decision tree algorithm.

(2) Structural Deformation Prediction

Machine learning algorithms are increasing being used in civil engineering applications to predict various types of structural deformation, such as slope deformation, tunnel surrounding rock deformation, and dam deformation. In the study conducted by Jia et al. [76]. the researchers employed chaotic time series to create training samples and applied the regression capabilities of SVM to accurately predict slope deformation. In the study by Duan et al. [77], the researchers utilized monitoring data as learning samples to predict exactly the settlement and deformation convergence of tunnel-encircling rock. Su et al. [78] developed a dam deformation prediction model by combining SVM with various techniques, including phase space reconstruction, wavelet analysis, and particle swarm optimization. In the study referenced, after selecting the observed deformation, the researchers utilized a combination of phase space reconstruction and Morlet wavelet basis function to prepare the input vector and kernel function for SVM, respectively, after which, the researcher utilized PSO algorithm to optimize the parameters of this SVM-based model. Based on the validation instance, it was concluded that the modeling efficiency and prediction precision have been enhanced, yielding positive results.

BP neural network algorithm has demonstrated superior adaptability in predicting structural deformation. Because many structural deformation problems involve multiple variables, nonlinearity and complexity, it is often challenging to develop accurate mathematical models to represent such systems. The neural network prediction model is built upon a robust and nonlinear system that possesses strong nonlinear mapping capabilities. Xue et al. [79] employed the BP neural network method to analyze and predict the deformation of the retaining structure of an urban deep foundation pit with high accuracy. In addition to the high accuracy achieved in predicting the deformation of the retaining structure, they also noted that the neural network model was more effective in describing the mechanism-relationship between retaining structure displacement and its influencing factors than the information model. In order to improve the performance of BP neural network, several scholars have focused on optimizing the selection of its parameters to avoid issues such as premature convergence and local extremum.

Zuan et al. [80] used a BP neural network to predict landslide deformation and found that the selection of network structure parameters had a significant impact on the prediction results. However, the standard BP algorithm suffers from some limitations such as slow convergence, sensitivity to initial values, and local optima. In response, several improved algorithms, MOBP, CGBP, QNBP and LMBP, have been proposed. These algorithms aimed to overcome the flaws of the standard BP algorithm to improve neural network's performance. Finally, the RBF algorithm was implemented, and the prediction results showed good agreement with the land-slide curve. Wei et al. [81] optimized the BP neural network using a genetic algorithm and applied it to the prediction of tunnel rock deformation. The results showed very little deviation between the predicted value and the actual experimental deformation value. Wang et al. [82] utilized the MPGA-BP neural network method to predict dam deformation, while Xu [83] applied the RBF neural network method to the same task. The simulation results using neural network methods shown precision and dependability, highlighting their potentials as valuable tools in engineering applications.

(3) Structural Damage Identification

Use of machine learning algorithms for structural damage identification has been studied extensively since the 1990s. Kudva et al. [84] conducted one of the earliest studies that used a neural network with two hidden layers to identify the damage to a stiffened aluminum plate. Their results showed that the method was able to accurately determine the location and extent of the damage.

Elkordy et al. [85] demonstrated the ability of BP neural network to effectively identify the damage in a 5-story steel-framed structure. Their results showed that the neural network maintained its noise resistance even when the training data was not entirely accurate. This highlighted the potential of BP neural network as a tool for accurately identifying structural damage in real-world applications. Figure 6.24 illustrates the process for identifying the damage to a bridge segment, a topic that has undergone extensive research in recent years. Hakim and Razak [86] employed a BP neural network and experimental modal data extracted by natural frequency to identify damage to a primary beam of a steel bridge. Sun et al. [87] used a suspension bridge model test to analyze model error, and established a high-precision finite element model. They then simulated various damage conditions to generate training samples and used a BP neural network to identify the location of the damage. The BP neural network was able to achieve an identification precision of 86% in detecting the location of the damage.

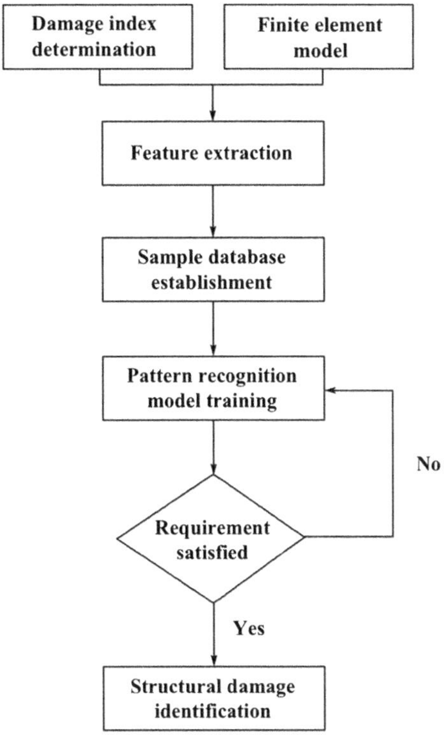

Fig. 6.24 Identification process of bridge damage mode

Gonzalez-Perez et al. [88] established a neural network based on modification of the bending stiffness of a vehicle bridge component, using the difference in modal strain energy as the input data and the stiffness as the output data. Using the neural network, the bending damage of the primary beam of the vehicle bridge was identified, and the neural network's generalization capability in the noise environment was validated. After assuming the structural damage area, Sun et al. [89] selected five damage scenarios for each of the ten deck structure components of the Kapshuimen cable-stayed bridge for numerical simulation. A neural network was then trained through obtaining the modal parameters under various working conditions and feeding them into the neural network as input variables, The obtained data demonstrated that the neural network can achieve improved identification results under the majority of working conditions.

Liu et al. [90] proposed a new wavelet neural network. The approximation effect of the neural network was enhanced based on gray-correlation analysis and local learning. It is proven to be superior to the conventional neural network in identifying damage to simply supported beam bridges. In another study, Liu et al. [91] employed the modal curvature to detect the location of the damage and an RBF neural network to determine the severity of the damage. Using a five-span, simply supported beam bridge as an example, this method's applicability was examined. In order to conduct probabilistic damage detection based on vibration measurement,

Yin et al. [92] selected the number of hidden layer neurons and the type of hidden layer transfer function concurrently, embedded multi-objective variables into the ANN using Bayesian reasoning for optimal design, and obtained a network model with an appropriate level of complexity. A numerical example of a steel truss bridge model was used to verify the efficiency of this method. Shu et al. [14] implemented a damage detection algorithm based on ANN with the statistical characteristics of structural dynamic response as input data and applied it to a simplified railway bridge model. Hasni et al. [93] proposed an AI-based method for detecting a steel bridge's cracking due to deformation fatigue.

Arangio et al. [94] evaluated the Bayesian neural network damage detection method using a data set consisting of bridge vibration data before and after damage. Tianjin Yonghe Bridge was taken as an example, which has been equipped with a continuous monitoring system, including 14 uniaxial accelerometers permanently installed on the bridge deck and one biaxial accelerometer fixed on the tower top to monitor its horizontal vibration. The anemometer installed at the top of the tower measured the wind speeds in three directions, while the temperature sensor installed at the girder's midspan measured the ambient temperature. For damage detection and analysis, the Intelligent Structure Research Center provided sensor data measured on the bridges on two different days. Then, the author developed a Bayesian neural network model, by utilizing the Bayesian model classification selection and choosing the optimal model structure. Various models were trained concurrently to approximate the time history of the structural response under healthy conditions. Then, the trained model was tested with the new inputs of the healthy and the damaged structures. Analysis of the prediction error increment led to the conclusion that performance of the damaged bridge was abnormal. Lastly, the results were compared with the conventional vibration-based structural identification technology, EFDD, and the conclusions were validated. With the same data set as the Bayesian neural network, the modal features of the structure were obtained, and the natural frequencies and mode shapes with and without the damage were compared, indicating that the frequency sensitive to the potential abnormalities or the damage symptoms was drastically reduced.

(4) Structural State Assessment

Structural state assessment involves using various testing methods, permitted by the structure, to obtain the internal information that can reflect the current working state of the structure. A state evaluation theory is then employed to assess the overall construction, operation, and other working states of the structure based on this information. The conventional methods for state evaluation of civil engineering structures often rely on modal analysis and analytic hierarchy, which can be time-consuming and require special expertise. However, with the advancement of machine learning and AI, these methods are becoming closer to AI-based approaches.

Chen et al. [95] proposed an ANN to estimate the state of reinforced concrete beam bridges based on grid and domain knowledge. The method incorporates the subjective judgment of a great number of bridge experts to establish complicated nonlinear relationships, leading to a more systematic determination of the condition of the bridge. Gehl et al. [96] proposed a multi-hazard vulnerability function derivation method based on the system reliability method and the Bayesian network. To identify specific failure mechanisms and damage states at the component level, the bridge system was broken down into its constituent parts. By breaking down the bridge system into its components, this method can provide a more thorough assessment of the vulnerability of each individual component and its contribution to overall system reliability. The incorporation of Bayesian network further improved the accuracy and applicability of the method. Setunge et al. [97] proposed a fault tree-based method for the comprehensive evaluation of system fault risk. This method can determine the likelihood of damage mechanisms in bridge components and the relative severity of the risk, as well as rank the overall service performance failure risk of different components in a bridge or a group of bridges. By evaluating the risk of each component and its contribution to overall system reliability, this method offers a more thorough assessment of bridge safety. This can help bridge engineers and managers to prioritize maintenance efforts and allocate resources more effectively, ultimately improving the overall safety and reliability of the bridge. Luo [98] introduced the Bayesian network method for effective calculation of probability in the comprehensive evaluation of the technical status of the double curved arch bridge. It was proven that the Bayesian network method can incorporate both quantitative and qualitative data to accurately assess the technical status of the bridge.

(5) Structural Reliability Analysis

Reliability indicators based on probability are essential in determining the dependability of a structure, as they provide a more accurate and comprehensive assessment of its performance and safety. Instead of a simple "yes" or "no" evaluation, probability measure can be used to assess the reliability of an engineering structure. This measure considers the probability of the engineering structure completing the predetermined function within the specified time and conditions, taking into account the potential impact of various environmental factors, loading conditions, and other factors that may affect the structure's performance. Currently, the matrix method, Monte Carlo method, and response surface method are the majority of reliability analysis techniques used.

Rocco and Moreno [99] were the pioneers in combining SVM with reliability analysis of the evaluation network system in 2002. Later, Hurtado and Alvarez [100] proposed a method that converted reliability analysis problem into pattern recognition problem, and used SVM in conjunction with finite element calculation for reliability analysis. Yu et al. [101] utilized the SVM regression method to approximate the structural limit state function and applied it to the reliability analysis of a cable-stayed bridge under the main span deflection failure mode. They compared the linear analysis and the nonlinear analysis results to evaluate the accuracy of the SVM regression method. In his doctoral dissertation, Bai [102] applied the SVM classification algorithm to reconstruct the system's limit state surface and conducted a reliability analysis of the cable-stayed bridge system.

Su et al. [103] analyzed the reliability of a large bridge using a BP neural network to display the implicit function in conjunction with the matrix method. However, due to the neural network's use of minimum variance criterion, it led to overfitting of the data during generalization, which adversely affected the reliability calculation results. Zhang et al. [104] utilized a neural network optimized by a genetic algorithm to conduct the reliability analysis of cable-stayed bridges. The practice significantly enhanced the calculation and analysis efficiencies. Zhu et al. [105] proposed a reliability evaluation method that combined RBF and MC methods for the reliability analysis of cable-stayed bridges. To apply this method, they established a cable-stayed bridge model for reliability analysis using ANSYS. Liu et al. [106] presented an improved method for estimating the deflection of simply supported beam bridges, using the modal flexibility analysis. The dynamic Bayesian network was then applied to evaluate the reliability of the deflection. By combining the monitoring results of various factors that may affect structural performance, such as modal frequency, mode shape, ambient temperature, and humidity, the dynamic Bayesian network provided a more comprehensive and accurate evaluation of the bridge's reliability and safety.

3. Disaster Early Warning and Recovery

The previous section provided an overview of the application of machine learning algorithms for bridge monitoring, data processing and performance evaluation. With the continued advancement of AI in recent years, machine learning has made significant strides in the study of bridge disasters caused by various factors such as strong winds, tsunamis, random waves, ship collisions and earthquakes. In the following section, we will describe in detail how these machine learning approaches are being utilized to enhance our understanding of these complex phenomena.

(1) Strong Wind

Recent years have witnessed the successful application of ANN method to study wind resistance of bridges and structures [107]. Wu et al. [108–110] employed ANN to predict the aerodynamic time history of bridges. To capture the nonlinear effects of aerodynamic forces, the researchers incorporated a quadratic term in the neurons. Jin et al. [111] utilized a convolutional neural network (CNN) to establish the relationship model between the cylinder pressure and the velocity field in fluid–structure interaction analysis. By using the CNN approach, the researchers were able to predict the structure wake velocity field by measuring only the pressure distribution on the bluff body. Lute et al. [112] utilized SVM to identify flutter derivatives and flutter critical wind speed of cable-stayed bridges. Similarly, Jung et al. [113] employed neural networks to identify flutter derivatives of various section forms. In China, Chen et al. [114] used ANN methods to identify flutter derivatives of rectangular sections in their study; Li et al. [115] used ANN system to identify the three-dimensional static component force coefficient and flutter critical wind speed of bridge sections.

The recognition process in machine learning-based bridge performance analysis is heavily reliant on experimental data. Issues such as insufficient sample size and inaccurate data can lead to a suboptimal recognition effect. Chen and Ge [116] integrated the database of existing long-span bridge wind tunnel test data to increase training data samples for neural networks and make them more accessible to users. The use of known data and ANN methods, with an improved error back-propagation algorithm, can be utilized to fit and identify three-dimensional static component force coefficients and eight flutter derivatives. This approach enables the prediction of parameters of similar sections based on existing aerodynamic parameter data samples. By separating the identification of aerodynamic parameters of specific sections of bridge structures from time-consuming tests and complex numerical simulation, this method can provide a more efficient and cost-effective approach to bridge design and maintenance.

One of the most notable characteristics of ANNs is their strong learning ability, with the learning content stored in the network's connection weight. Learning rules refer to the algorithm used to modify the weights of the network. In recent years, BP algorithm has been widely utilized in various disciplines. The entire neural network model's structure can be designed based on the inputs and outputs of the neural network, such as the flat box girder shown in Fig. 6.25. The number of cells in the hidden layer of an ANN is typically determined via experimentally adjusting the network structure and analyzing the training outcomes. However, recent research has shown that impact of the number of hidden layer units on the training outcomes is highly intricate and varies based on various factors such as data sample size and complexity. Hence, it is now recognized that different sample sizes require specialized numbers of hidden layer units.

Insufficient hidden units can result in an inaccurate representation of the sample rule, leading to a lack of accuracy and precision in the network's output. On the other hand, an excessive number of hidden units can lead to excessive noise, diminishing the network's inductive capability and making the internal law more difficult to extract. Moreover, it can result in the network simulating the input–output relationship with an overly complex nonlinearity, elongating the training time significantly. In the context of aerodynamic parameter analysis of bridge structures using neural networks, it is necessary to allow the network to discover the optimal parameters through the training process. The relationship between the random assignment times of the network and its training times is complex, and maximizing randomness without sacrificing sufficient training time is essential, as random assignment directly affects the training outcomes.

Vortex-induced vibration (VIV) is a common phenomenon that can occur on long-span bridges. Despite extensive research and development of semi-empirical models based on wind tunnel experiments, precisely simulating the VIV on full-scale structures remains a significant challenge. This is due to the complexity of the wind field, Reynolds number effect, 3D aeroelastic effect, uncertainty in dynamic structure models, among other factors. In a recent study by Li et al. [117], a data-driven machine learning method was proposed to predict VIV responses in a practical long-span suspension bridge. They developed a VIV response prediction model based on the VIV database collected over the last six years. The authors employed the decision tree algorithm to train a VIV pattern classification model, which mapped the relationship between the wind field characteristics and the VIV pattern. The model proposed by Li et al. [117] was capable of automatically identifying the VIV pattern and predicting

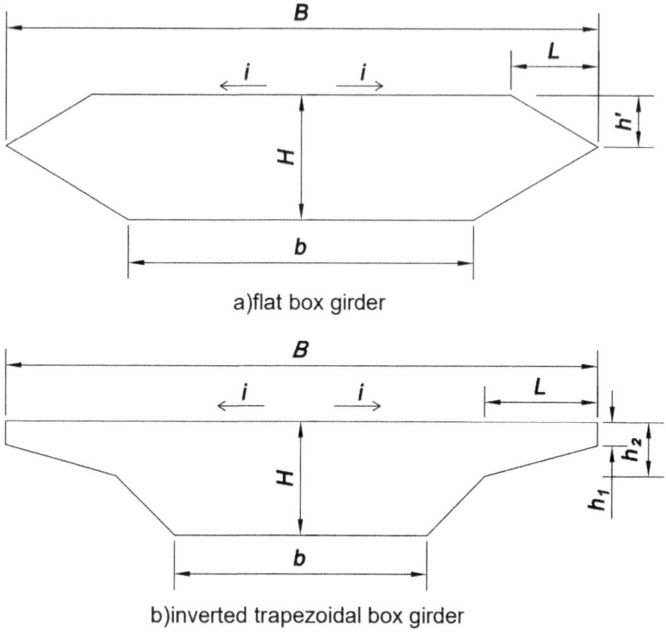

Fig. 6.25 Geometric characteristic parameters of bridge section

VIV responses in a long-span suspension bridge. To simulate the VIV response of the primary beam in the time domain, the support vector regression (SVR) algorithm was employed for various VIV modes. Wind velocity, wind direction and non-uniformity were identified as crucial parameters in predicting the entire VIV response process, including the development stage, stability stage and decline stage, as well as the VIV response amplitude.

(2) Tsunami

Tsunamis pose a significant hazards to coastal bridges (Fig. 6.26a). To address this issue, Mulia et al. [118] conducted a study on the prediction of tsunami waveforms in coastal regions using an extreme learning machine (ELM) algorithm. The ELM algorithm was utilized to rapidly predict the tsunami waveform using the tsunami signal recorded at specific locations. Its excellent training speed made it appropriate for near-field tsunami warning, as it can operate in real-time. In addition to its excellent training speed, the ELM algorithm can also function as a general function approximator, allowing it to quantify the nonlinearity of tsunami waveforms. This sets it apart from the standard inversion analysis methods used in many existing studies, which typically make the assumption of linearity. The ELM algorithm has been applied to the prediction of the 2011 Northeast tsunami early warning, with results indicating its higher precision compared to the standard methods without significantly increasing calculation time.

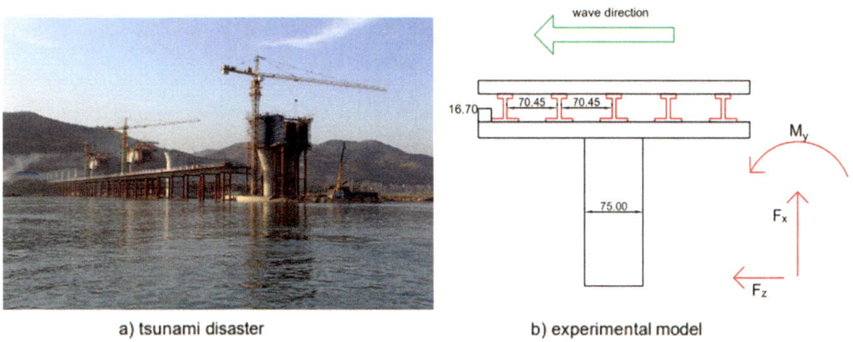

Fig. 6.26 Machine learning application in tsunami case (unit: cm)

Mazinani et al. [119] developed an ELM-based test method to estimate the destructive force of a tsunami on coastal bridges. The study focused on understanding and estimating the horizontal force, vertical displacement and overturning moment exerted by a tsunami on coastal bridge structures. The parameters of the bridge model consisted of three water depths, ten wave heights and four beam types, resulted in a total of 120 cases. The test utilized a 1:40 scale model of a concrete bridge and was conducted in a 24 m × 1.5 m × 2 m wave trough (Fig. 6.26b). Two six-axis load cells and four pressure sensors were installed on the bottom plate of the scaled-down bridge model to measure the force. In numerical simulation, the ELM-based prediction results were compared to those of genetic programming (GP) and ANN models. The experimental results demonstrated that the ELM model outperformed GP and ANN in terms of prediction accuracy and versatility. The ELM-based test method developed has potential to serve as a basis for further research and development of new strategies to accurately predict tsunami forces on coastal bridges. In conclusion, the experimental results of the study shown that ELM model can achieve favorable generalization performance in predicting the destructive force of tsunamis on coastal bridges, and do so thousands of times faster than the conventional learning algorithms in most cases. Based on these findings, ELM can be developed as an alternative method for predicting the destructive force of tsunamis on coastal bridges.

(3) Random Wind Waves

Long-span bridge response evaluation is often impeded by the uncertainty of multiple random factors. To address this issue, Fang et al. [120] conducted a study on the impact of random wind waves on two-span cable-stayed bridges, utilizing the Bayesian regularized back-propagation neural network (BRBPNN). Compared to the conventional BP neural network model (BPNN), BRBPNN can automatically optimize target parameters and adjust the weight of key input variables during model training, resulting in superior performance.

In the study by Fang et al. [120], a framework for dynamic analysis was proposed by integrating the wind-wave-bridge system with BRBPNN. The random response of

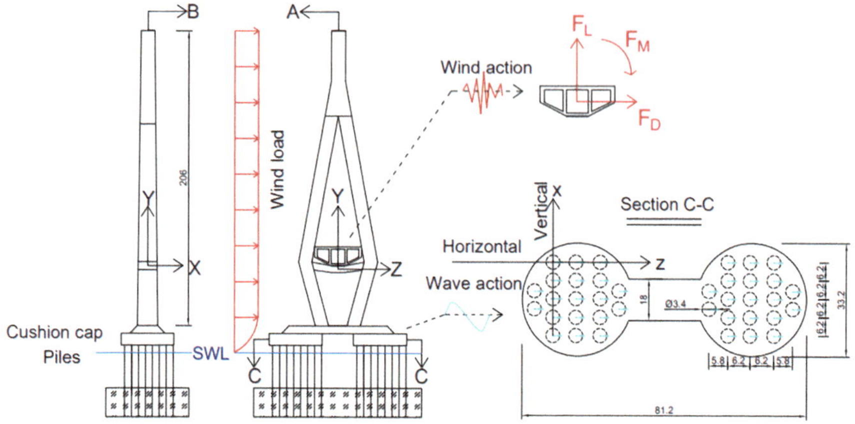

Fig. 6.27 Random wind wave actions (unit: m)

sea-crossing cable-stayed bridges to extreme wind-wave parameters was investigated using this framework. To consider both the steady wave wind field and the multi-directional random wave field, as shown in Fig. 6.27, a wind-wave bridge system comprising wind bridge and wave bridge interactions was constructed. The dynamic response of the bridge was then calculated using the Newmark method.

To reduce calculational costs, Fang et al. [120] introduced and evaluated the accuracy and efficiency of the BRBPNN model. The study involved performing a small-scale finite element analysis to obtain response statistics, which were then used as training samples for the neural networks. To investigate the wind and wave mechanisms, a power spectrum analysis of deterministic results was conducted. Ultimately, the BRBPNN model was utilized to perform an uncertainty analysis, which allowed for the correlation between the bridge's response and individual wind wave parameters to be determined. Results of the study demonstrated that the proposed framework was capable of capturing the nonlinear bridge responses caused by nonlinear wind and wave loads. However, there were significant differences compared to responses under wind alone. The vibration characteristics of bridges under small wind and wave loads were significantly impacted when exposed to wind waves. Under substantial wind wave loads, contribution of the bridge's vibration characteristics became more apparent. The uncertainty analysis conducted using the BRBPNN model revealed that the four wind wave parameters had a major impact on the beam, tower and underwater structure of the bridge.

(4) Ship Collision

Collision between a bridge and a ship is a dynamic, highly nonlinear and instantaneous process. Conventionally, calculation of ship-bridge collision force involves the equivalent static load specified in the design code or the time-consuming finite element method (FEM) that requires supercomputing resources. References [121, 122] proposed an alternative method that combined FEM and ANN. As shown in

Fig. 6.28, the radial basis function artificial neural network (RBFNN) was utilized to calculate the impact force following a ship-bridge collision. The neural networks were trained using learning samples that incorporated the ship's speed and mass as input vectors and the collision force as output vectors. Varying network parameters were used to train the neural networks, with learning samples derived from the results of finite element simulations. The error analysis of training and testing samples demonstrated that the proposed RBFNN had sufficient accuracy, and generated an input–output relationship that was consistent with typical empirical formulas. To improve calculation efficiency, the study utilized MATLAB software to develop a specialized toolbox.

Current research on bridge anti-collision technique focuses primarily on the bridge itself by using energy and force calculations to study various anti-collision devices

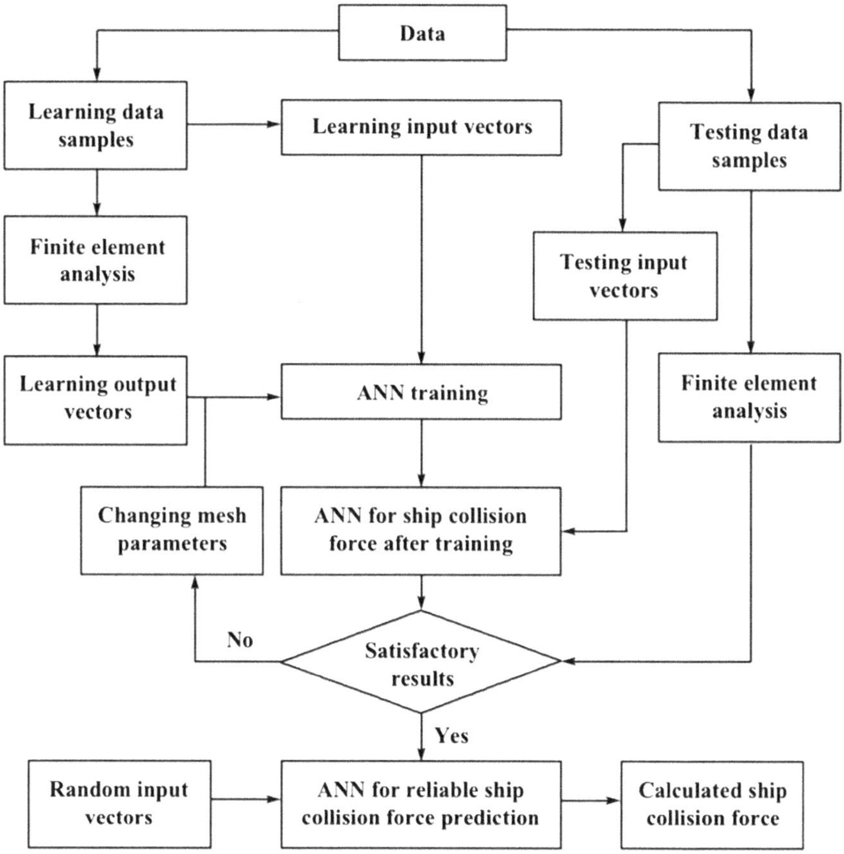

Fig. 6.28 Flow chart of ship collision force calculation using RBFNN

and strengthen the bridge's anti-collision capacity [123]. This approach can effectively reduce the degree of bridge damage in a ship collision, but it cannot prevent accidents entirely, as this bridge still has the potential to collapse.

Given the growing number of ship collision accidents in China, research on active bridge anti-collision technique is becoming increasingly vital. One approach involves installing a camera on the bridge, which captures the video image and detects the moving ship. The camera creates a virtual channel in the video image, and identifies whether the ship is yawing in real time. The main points are: (1) development of a computer-based detection and recognition algorithm for moving ships in complex inland waterways. Experimental tests are conducted to evaluate its performance; and (2) determination of the coordinates of the end points of the virtual channel line, which is created using template matching and the triangle similarity principle. For complex river terrain, the study constructs a linear virtual channel line between endpoints utilizing the OpenCV line drawing function. The distance between the ship and the virtual channel line is then computed in real time using the distance formula for points to a line. If the distance computed in real-time between the ship and the virtual channel line is less than the specified threshold, it indicates that the ship is yawing and an imminent ship collision is possible. Bezier curve is used to create a more accurate and curved virtual channel line between the endpoints. RGB value is used to extract the coordinates of pixel points on the curve, and their distance from the ship is determined in real-time. The minimum distance serves as a standard to determine whether the ship is yawing.

(5) Post-Earthquake Bridge Recovery

Kameshwar et al. [124] developed a bridge seismic recovery model based on the decision tree algorithm and evaluated its impact on the regional road network. The study utilized a data-driven approach to extract qualitative data regarding bridge damage, function and recovery time from reconnaissance reports of previous earthquakes. To augment the dataset, maintenance and repair observation results and expert opinions based on engineering judgment were also included. The level of bridge component damage served as the principal input of the decision tree model. It can be estimated using the component vulnerability model provided by the existing literature. Determining the actual traffic flow limit and its duration was a crucial component of the developed bridge seismic recovery model. It directly contributed to road network analysis, which was the primary function of the model. The developed model can provide recommendations for potential traffic restrictions resulting from bridge damage caused by earthquakes and other extreme events. Furthermore, as more information was obtained from surveys or empirical sources, the data-driven method can easily update the proposed model.

The study utilized a decision tree to determine the closure of a bridge based on the level of damage to bridge components. According to the decision tree shown in Fig. 6.29, there was an extremely high probability that a bridge will be closed if a bearing was severely or completely damaged. This was because a completely damaged bearing can cause the beam to shift or fall, which renders traffic unsafe or impractical. If the bearing damage was minimal, the possibility of closure was

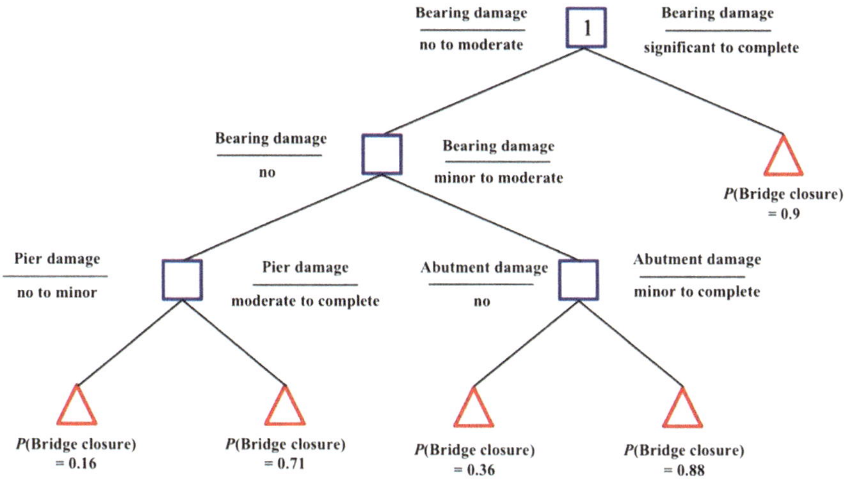

Fig. 6.29 Decision tree model for bridge closure decision

contingent on the degree of damage to the abutment and pier. Minor to moderate bearing damage combined with any abutment damage may result in the closure of the bridge, as excessive settlement of the beam or rotation of the abutment can destroy the abutment and render the bridge impassable. Even if the bearing was undamaged, the bridge could be shut down if the pier had suffered moderate to severe damage. However, the possibility of complete closure due to pier damage was lower than that of bearing or abutment damage, as partial bridge closure (i.e., lane/speed/load limit) was a possibility in this situation. Moreover, the bridge will not be shut down if the pier was only slightly damaged and the bearing was undamaged.

The proposed decision tree model was applied to both single bridges and regional road networks. To evaluate the impact of traffic restrictions on the traffic flow, the study assessed the impact on five fictitious bridges. This was necessary as the same restrictions may have varying effects on the bridge's use functions depending on traffic demand, number of lanes and speed limit. The results illustrated the limitations of existing bridge recovery models in accurately assessing the impact of traffic restrictions on critical infrastructures. The decision tree model was further applied to the regional road network in Memphis, Tennessee (Fig. 6.30). The evaluation considered the function and recovery of the post-earthquake bridge, taking into account the uncertainty caused by factors such as the bridge's damage state, function after the earthquake and the recovery time. A comparison between the road network level recovery curve proposed by the decision trees and by HAZUS, a United States-developed software for earthquake disaster loss analysis, showed that the HAZUS-based recovery curve lacked the gradual characteristics of road network recovery. Moreover, using the HAZUS-based recovery model can, in the majority of cases, lead to a severe overestimation of the total traffic time in the road network by several orders of magnitude, as compared to the decision tree-based model.

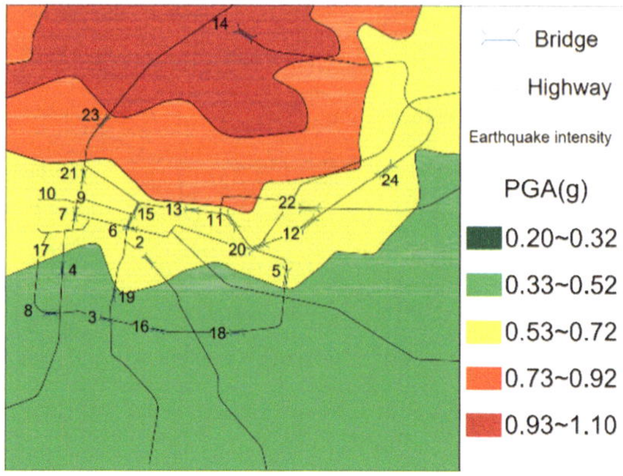

Fig. 6.30 Regional road network with fictitious earthquake and bridge distribution

The network-level analysis, facilitated by the use of functional and recovery models, can be employed to assess the indirect loss or sustainability of regional road networks. Future work may also focus on gathering additional data on bridge damage and recovery to account for multiple disasters, bridge types and bridge sizes in the decision tree model. This can result in developing models with low uncertainty.

4. Other Applications

(1) Bridge Design With Supervised Learning Algorithms

The choice of structural system is a critical aspect of bridge design. However, budget and time constraints often limit the design and selection scope. Engineers typically select the structural system of bridges based on their empirical evaluation of the project, leading to optimization and standardization of designs, and higher maintenance costs. Applicability of supervised learning algorithms in the preliminary design of bridges was investigated to address this issue [125]. The structural system was designed by analyzing over 600,000 bridges from the national bridge database. This approach ultimately improved the possibility of design optimization and standardization, leading to reduced maintenance costs. To determine the key attributes that influence the selection of bridge structure types, three feature selection methods were used. In addition to standard attributes such as bridge span and load rating, potentially useful attributes such as seismic intensity and historical data on material costs (reinforcement and concrete) were appended from the United States Geological Survey (USGS) database and the Engineering News Record.

SVM, a powerful classification algorithm in supervised learning, was used as a benchmark to evaluate the performance of decision trees (Fig. 6.31) and Bayesian networks (Fig. 6.32). The experimental results demonstrated that in all states, performance of the decision trees and Bayesian networks was comparable to or exceeded that of SVM. The single model that can predict all types of bridges in the United States has a significant disparity in prediction accuracy, rendering its recommendations on design type unreliable. In contrast, the performance and prediction accuracy of the model developed with data from a single state were evidently superior. To conduct

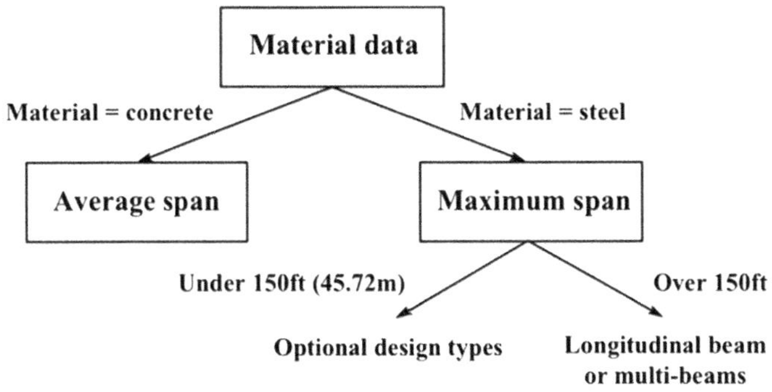

Fig. 6.31 Application of decision tree algorithms in bridge design

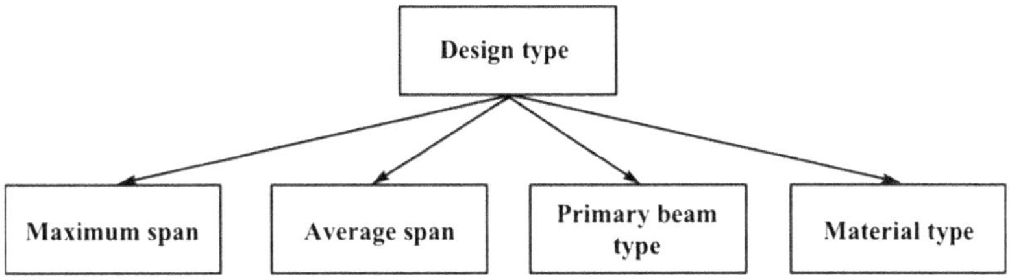

Fig. 6.32 Application of Bayesian networks in bridge design

the final type selection design of bridges based on the prediction results of supervised learning models, techniques such as advanced computing technology (e.g., topology optimization) and optimization technology (e.g., evolutionary computing) need to be further explored.

(2) Seismic Design and Evaluation of Bridges

Wang et al. [126] recently incorporated machine learning into seismic design and evaluation of beam bridges, outlining the implementation of a general idea (Fig. 6.33). Missions involving machine learning for bridge seismic resistance composed of two primary aspects: (1) collection of large quantity of bridge design data and establishing their data set; and (2) data mining, which involves processing of raw data, debugging or developing a reasonable machine learning algorithm model.

Introduction of machine learning is not intended to replace the conventional statistical analysis technology but rather to augment it. In probabilistic seismic design and evaluation methods based on bridge performance, the seismic vulnerability analysis is primarily conducted in the form of statistical regression probability using the nonlinear time history method, which is an effective tool for quantifying the seismic performance of bridges. However, there are two obstacles to analyzing the vulnerability of Chinese bridges. First, the probabilistic seismic demand model method makes it difficult to assume the seismic demand distribution of structures. Second,

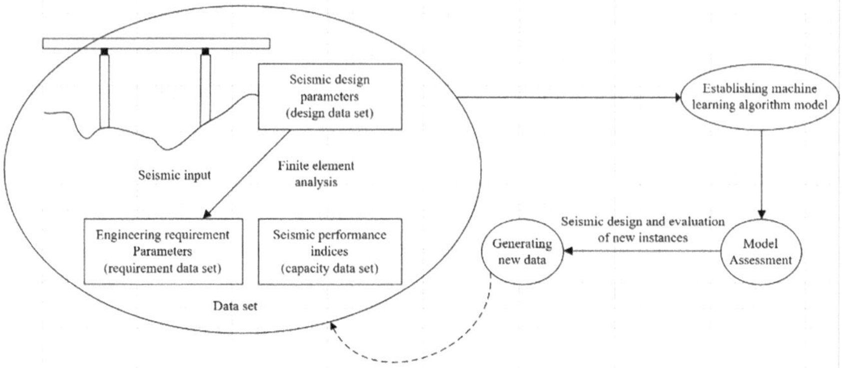

Fig. 6.33 Analysis of machine learning in seismic design and evaluation of beam bridges

accurately describing the nonlinear correlation between seismic demands of components poses a challenge when using the first-order limit method or Monte Carlo sampling method, both of which have been widely used to analyze the formation process from bridge components to system vulnerability in recent years. Furthermore, application of vulnerability technology to bridge seismic resistance based on flawless mathematical theory and stringent assumptions has reached maturity. Decades of bridge seismic design have yielded relevant experimental research and earthquake damage experience. Multiple studies on a wide range of medium and small span highway bridges have demonstrated high repeatability, statistical significance, and learnability. Given the fact that engineering requirements, such as performance-based seismic design and evaluation, can be rapidly and effectively encountered with data science, machine learning may be capable of producing results that cannot be attained with the conventional analysis methods.

6.3.2 Application Challenges

Due to the increasing amount of data, application of machine learning algorithms in the traditional industries has gradually become a hot topic of research, particularly in civil engineering. However, machine learning still faces many challenges in practical applications, primarily on two fronts: data and algorithm.

1. Data

(1) Insufficient Complete Labeled Data

Data is the driving force behind machine learning algorithms, and insufficient data is often a challenge. This issue frequently manifests as a lack of complete and accessible labeled data for bridge structure health monitoring. Throughout the majority of the monitoring period, bridge structures are in normal condition with no signs of damage or malfunction. While there is vast amount of monitoring data available, creating a complete case to build the corresponding annotation library can be challenging, making it difficult to apply the most effective supervised algorithm in machine learning. In this case, it is valuable to explore how to apply data mining to massive, unlabeled monitoring data.

In addition, the paper inspection report of the bridge contains a large amount of historical data, but data reading presents a challenge. When data is read manually, labor cost is the most immediate issue. Therefore, automatic reading of detection reports based on natural language processing technology is an area worth investigating.

(2) Unbalanced Data

In research, most algorithms are based on a fundamental assumption that the distribution of corresponding observation data results is balanced. However, in real-world applications, data are often distributed unevenly, leading to the "long-tail phenomenon." In the bridge structure health monitoring system, for instance, the proportion of valuable abnormal data among the collected data is extremely low, and the vast majority of the data are normal.

To address this issue, techniques such as sampling, new data generation and weighting, can be used to assign both positive and negative samples equal weight in the training process. The sampling method can transform an unbalanced data set into a balanced data set by sampling the training set. The new data generation process involves generating additional samples from existing samples. The weighting method solves data imbalances by designing different model misjudgment costs for different categories.

While there are numerous approaches to resolving the issue of data imbalance, this book only discusses the most prevalent ones. Each method has its own advantages and disadvantages, and the bridge health monitoring process must make judicious decisions and selections based on real-world situations.

(3) Low Data Quality

If the training set contains an excessive number of errors, outliers and noises, it will be increasingly challenging for the system to detect the rules underlying the data, and its performance will deteriorate as a result. Therefore, it is crucial to clean up the training data to ensure that the system is trained on accurate and high-quality data.

When compared to other typical industrial big data sets, monitoring data for bridge structures is unique and complex. This is due to the larger monitoring volume of bridge structures, which indicates a wide parameter space, a low density of measuring points, a lack of absolute data, and an increased vulnerability to random factors [69]. The primary targets of data cleansing are data errors, which can include default values, noise, outliers, and trend items. Data cleansing is the most time-consuming and labor-intensive link in the monitoring and analysis of bridge structure health. Therefore, to improve the efficiency of analyzing the big data of monitoring bridge health, it is crucial to enhance the level of automated data cleaning.

(4) Irrelevant Features

The learning task requires some attributes that may be essential, while other attributes may be useless. "Relevant features" generally refer to attributes that are beneficial for the current learning task, while "irrelevant features" relate to unrelated attributes. To enable the system to learn, it is necessary to include sufficient relevant features and limit irrelevant ones in the training data. In bridge structure health monitoring, the importance of complete monitoring data cannot be overstated. In addition to capturing response features that reflect the state of the bridge structure, such as strain, displacement and acceleration, the monitoring and recording of environmental conditions, such as wind speed and temperature, are also critical. Feature engineering is the process of selecting appropriate training features from multi-dimensional data to enable more effective completion of machine learning tasks. This process typically involves three steps: feature selection, feature extraction and feature construction, and it is vital to machine learning.

Feature selection involves selecting a subset from a given set of features. In the practical machine learning applications, if crucial features can be extracted from multi-dimensional data, the dimension disaster problem can be significantly mitigated. However, during the feature selection process, it is crucial to ensure that no critical features are omitted, as this can lead to missing data during the subsequent learning process and hinder performance. It is important to note that the interpretation of "irrelevant features" in feature selection is relative and refers only to those features that are irrelevant to the current learning task. Feature extraction involves transforming the original data into a set of features with defined physical or statistical significance. This process many include modifying the feature values to reduce their quantity or converting continuous features into discrete ones.

Feature construction is the process of manually creating new features from the original data. It involves analyzing data samples, evaluating the possible form and data structure of the problem, and extracting physically significant features from the original data. Assuming the original data is in table format, new features can be created through combining or mixing of attributes, or through decomposing or segmenting of the original features. The process of feature construction can be time-consuming, but it can significantly enhance the accuracy and efficiency of machine learning models.

2. Algorithm

As previously stated, the primary objective of machine learning is generalization ability. In practical applications, underfitting and overfitting are common issues. Underfitting typically occurs when the model's learning ability is limited, while the data's complexity is high, whereas overfitting usually happens in the opposite scenario. Essentially, both underfitting and overfitting are problems that arise when the model's capacity and training set do not match the complexity of the problem at hand. To achieve ideal generalization error, the model's capacity and the size of the training set must be proportional to the problem's complexity. To improve the efficiency of machine learning applications, it is necessary to select and design appropriate algorithms based on the specific problem setting.

In general, the selection of an appropriate algorithm can be challenging, but a particular procedure can be utilized to narrow down the choices. There is typically no single optimal algorithm that can produce the best results for all problems. In actual problem-solving, it may be necessary to test different algorithms repeatedly. In bridge engineering, the following two fundamental issues must be considered when applying machine learning algorithms to solve practical problems.

(1) Purpose of Using Machine Learning Algorithms

Before selecting an algorithm, it is crucial to precisely define the objective of the task. For discrete target variable, a classification algorithm should be chosen, such as in the diagnosis of sensor fault types, or the identification of bridge distress types. For continuous target variables, a regression algorithm should be used, such as when estimating the wind-induced dynamic internal force response of suspension bridge suspenders under different wind deflection angles. Unsupervised learning is suitable for tasks that do not require the prediction of a target value, such as extracting essential data from voluminous monitoring data. In contrast, reinforcement learning problems aim to optimize the objective function through interaction with the environment, such as in the optimization problem of bridge management's decision-making.

To select and design the most suitable algorithms, it is necessary to analyze the task requirements and scenarios. The "quality" of the model is relative, and depends not only on the choice of algorithm and data, but also on the specific task requirements. For example, in disaster early warning tasks, it is better to err on the side of caution and issue early warnings even if they are false positives. Missing any potential danger due to false negatives can have catastrophic consequences in terms of losses of life and property. Thus, in such critical scenarios, the model recall ratio becomes more crucial than the precision ratio. On the other hand, the model precision is more important when identifying the degree of bridge damage. Therefore, the first step in resolving the issue involves analyzing the objectives and requirements of the task in conjunction with the actual scenario.

(2) Data Collected and Its Features

The second step in selecting an appropriate algorithm is to gain a thorough understanding of the data and its characteristics. This understanding plays a crucial role in maximizing the efficiency of the chosen algorithm for a given problem. Some algorithms can handle small sample sets, whereas others require a substantial number of samples for effective training. For the study and resolution of the small sample fast learning problem, there is a specialized branch of machine learning, "new shot learning". The conventional machine learning algorithms may have fewer advantages than the depth learning methods for handling large datasets. For the recognition and analysis of visible distresses based on images, a deep convolutional neural network model can achieve more accurate classification results by combining the feature extraction and the classification functions. In time-series analyses, such as identifying overloaded vehicle load patterns, a recurrent neural network can effectively model the time-series data by considering the dependence on the time scale. It should be noted that deep learning requires a significant amount of data to achieve high accuracy. The conventional machine learning methods may be more appropriate than deep learning when the amount of training data is relatively small. In the cases where data annotation is challenging, weakly supervised learning techniques can reduce the algorithm's dependence on labeled data. After gaining comprehensive understanding of the data's characteristics, it may be necessary to combine multiple algorithms to achieve the desired accuracy and efficiency.

6.4 Summary

In general, AI creates new opportunities for the advancement of civil engineering. As an effective means of achieving AI, machine learning offers hope and opportunity for resolving bridge health monitoring issues. Nevertheless, due to the uniqueness of the field, the development of AI continues to face formidable obstacles. To better apply AI theory and technology to civil engineering and promote the development of intelligent civil engineering, it is necessary to conduct systematic scientific research and technology development in combination with the characteristics of civil engineering disciplines, and to establish the AI theoretical system for civil engineering.

References

1. Li KF, Wang YG (2018) What exactly is artificial intelligence. Science Grand View Park 2:48–49 (in Chinese)
2. Haugeland J (1985) Artificial intelligence: the very idea. Philos Rev 7:3–11
3. Kurzweil R, Richter R, Kurzweil R et al (1990) The age of intelligent machines. MIT Press, Cambridge
4. Zf CHEN (2016) Strong artificial intelligence and super intelligence: technical rationality and criticism. Sci Manage 36(5):25–33 (in Chinese)
5. Gelernter H (1963) Realization of a geometry-theorem proving machine. Comput Thought 134–152
6. Winograd S, Cowan JD (1964) Reliable computation in the presence of noise. Math Comput 18(87):1469–1477
7. Zf ZHOU (2016) Machine learning. Tsinghua University Press, Beijing (in Chinese)
8. Yu K, Jia L, Chen YQ et al (2013) Yesterday, today, and tomorrow of deep learning. J Comput Res Dev 50(9):1799–1804 (in Chinese)
9. Lee K-F (1988) Automatic speech recognition: the development of the SPHINX system. Springer Science and Business Media
10. Cooper GF, Aliferis CF, Ambrosino R et al (1997) An evaluation of machine-learning methods for predicting pneumonia mortality. Artif Intell Med 9(2):107–138
11. Wq SUN (2008) Beam bridge damage identification based on support vector machine. Southwest Jiaotong University, Chengdu (in Chinese)
12. Ma LY, Yuan TS (2018) Welding joint quality detection based on extreme learning machine. Comput Eng Appl 12:214–218,239 (in Chinese)

13. Li H (2012) Statistical learning methods. Tsinghua University Press, Beijing)
14. Shu J, Zhang Z, Gonzalez I et al (2013) The application of a damage detection method using Artificial Neural Network and train-induced vibrations on a simplified railway bridge model. Eng Struct 52:408–421
15. Fu CY, Shan DS, Li Q (2010) Static damage identification method based on support vector machine. China Railway Sci 31(5):47–53 (in Chinese)
16. Pan H, Azimi M, Yan F et al (2018) Time-frequency-based data-driven structural diagnosis and damage detection for cable-stayed bridges. J Bridg Eng 23(6):04018033
17. Huo LS, Li X, Yang YB et al (2016) Damage detection of structures for ambient loading based on cross correlation function amplitude and SVM. Shock Vib 1–12
18. Olshausen BA, Field DJ (1997) Sparse coding with an overcomplete basis set: a strategy employed by V1? Vision Res 37(23):3311–3325
19. Cury A, Cremona C, Dumoulin J (2012) Long-term monitoring of a PSC box girder bridge: operational modal analysis, data normalization and structural modification assessment. Mech Syst Signal Process 33:13–37
20. Liu JY, Jia XY (2017) A multi label classification algorithm using association rule mining. J Softw 28(11):2865–2878 (in Chinese)
21. Culurciello E, Bates J, Dundar A et al (2013) Clustering learning for robotic vision. Computer Science
22. Saxena DK, Deb K (2007) Non-linear dimensionality reduction procedures for certain large-dimensional multi-objective optimization problems: employing correntropy and a novel maximum variance unfolding. Evolut Multi-Criterion Optim 4403:772–787
23. Zhou YL, Maia NM, Sampaio RP et al (2017) Structural damage detection using transmissibility together with hierarchical clustering analysis and similarity measure. Struct Health Monit 16(6):711–731
24. NIE JG (2014) Research on finite element model correction algorithm for bridge structures based on virtual deformation method. Harbin Institute of Technology, Harbin (in Chinese)
25. Yang BY (2006) Application of association rule discovery in bridge management system. In: The 17th national bridge academic conference (in Chinese)
26. Hawkins DM (1980) Identification of outliers. Springer
27. An J, Cho S (2015) Variational autoencoder based anomaly detection using reconstruction probability. Spec Lect IE 2(1)
28. Malhotra P, Vig L, Shroff G et al (2015) Long short term memory networks for anomaly detection in time series. In: 23rd European symposium on artificial neural networks, computational intelligence and machine learning. ESANN 2015
29. Du M, Li F, Zheng G et al (2017) Deep log: anomaly detection and diagnosis from system logs through deep learning. In: ACM Sigsac conference on computer & communications security
30. Gibney E (2017) Self-taught AI is best yet at strategy game Go. Nature 10(1):68–74
31. Settles B (2011) From theories to queries: active learning in practice. Active Learn Challenge Challenges Mach Learn 6(1)
32. Zhu XJ (2005) University of Wisconsin-Madison Department of Computer Sciences
33. Tu EM, Yang J (2018) Overview of semi supervised learning theory and its research progress. J Shanghai Jiaotong Univ (Chin Ed) 52(10):1280–1291 (in Chinese)
34. Castelli V, Cover TM (1996) The relative value of labeled and unlabeled samples in pattern recognition with an unknown mixing parameter. IEEE Trans Inform Theor 42(6):2102–2117
35. Jiang YJ, Liu YP, Zhou LC et al (2019) Intelligent damage identification method based on bridge response statistical information. J Hefei Univ Technol 42(12):1667–1671 (in Chinese)
36. Vapnik V, Chervonenkis A (1964) A note on class of perceptron. Autom Rem Control 1964:24
37. An HP, Wu XG (2019) Research on multi position damage identification of beam bridges based on support vector machine. J Railway Sci Eng 5:16 (in Chinese)
38. Kourehli S (2017) Structural damage diagnosis using incomplete static responses and LS-SVM. Inverse Probl Sci Eng 25(3):418–433
39. Pearl J (1990) Probabilistic reasoning in intelligent systems: networks of plausible inference. Artif Intell 48(8):117–124

40. Arangio S, Beck J (2012) Bayesian neural networks for bridge integrity assessment. Struct Control Health Monit 19(1):3–21
41. Zhang ML, Zhou ZH (2007) ML-KNN: A lazy learning approach to multi-label learning. Pattern Recogn 40(7):2038–2048
42. Li XJ (2009) Research on KNN classification algorithm. Technol Inform 2009(31):81383 (in Chinese)
43. Sarmadi H, Karamodin A (2020) A novel anomaly detection method based on adaptive Mahalanobis-squared distance and one-class kNN rule for structural health monitoring under environmental effects. Mechan Syst Signal Process 140
44. Kohonen T. An introduction to neural computing. Neural Netw 1(1):3–16
45. McCulloch WS, Pitts W (1943) A logical calculus of the ideas immanent in nervous activity. Bull Math Biol 52(1–2):99–115
46. Broomhead DS, Lowe D (1988) Multivariable functional interpolation and adaptive networks. Complex Syst 2(3):321–355
47. Vincent P, Larochelle H, Bengio Y et al (2008) Extracting and composing robust features with denoising autoencoders. Mach Learn
48. Hosseini-Asl E, Zurada JM, Nasraoui O (2015) Deep learning of part-based representation of data using sparse autoencoders with nonnegativity constraints. IEEE Trans Neural Netw Learn Syst 27(12):2486–2498
49. Kingma DP, Welling M (2014) Auto-encoding variational bayes. CLR. ArXiv.org
50. Hinton GE, Osindero S, Teh YW (2006) A fast learning algorithm for deep belief nets. Neural Comput 18(7):1527–1554
51. Lecun Y, Bengio Y (1995) Convolutional networks for images, speech, and time-series. In: Handbook of brain theory & neural networks
52. Scarselli F, Gori M, Tsoi AC et al (2009) The graph neural network model. IEEE Trans Neural Netw 20(1):61–80
53. Bai B, Liu YT, Ma CC et al (2020) Graph neural network. Chin Sci Math 50(3):367–384 (in Chinese)
54. Liu Y, Bi D, Li ZX (2009) Optimization of sensor placement for large-span cable-stayed bridges based on genetic algorithm. J Southeast Univ (Nat Sci Edn) 39(4):825–829 (in Chinese)
55. Zhao JH, Zhnag L (2015) Research on sensor optimization layout and structural damage identification using particle swarm optimization algorithm. J Xi'an Jiaotong Univ 49(1):79–85 (in Chinese)
56. Yi TH, Li HN, Zhang XDJSC et al (2015) Health monitoring sensor placement optimization for canton tower using immune monkey algorithm. Struct Control Health Monit 22(1):123–138
57. Yuan C (2018) Research on the optimal placement method of bridge health monitoring sensors based on deep learning. Chongqing Jiaotong University, Chongqing (in Chinese)
58. Jabbari A, Jedermann R, Lang WJWAOS et al (2007) Application of computational intelligence for sensor fault detection and isolation. Int J Comput Inform Syst Sci 33:265–270
59. Zavaljevski N, Gross KC (2000) Sensor fault detection in nuclear power plants using multi-variate state estimation technique and support vector machines. Office Sci Tech Inform Tech Rep
60. Liu Z, Mrad N (2012) Validation of strain gauges for structural health monitoring with Bayesian belief networks. IEEE Sens J 13(1):400–407
61. Luo J, Song XY (2010) Error compensation of laser displacement sensor based on BP neural network. In: 2010 optical conference of the Chinese optical society, Tianjin, 2010 (in Chinese)
62. Zeng ZZ, Zhu W, Wang YN (2007) A sensor error compensation method based on orthogonal basis neural network algorithm. J Sens Technol 20(3):536–539 (in Chinese)
63. Fonollosa J, Vergara A, Huerta RJS et al (2013) Algorithmic mitigation of sensor failure: is sensor replacement really necessary? Sens Actuat B Chem B183:211–221
64. Zheng XZ, Chen GL, Zou T (2020) Cleaning of abnormal data in dam monitoring considering time series correlation. Hydropower 46(4):111–114, 125 (in Chinese)

65. Huang YW, Wu DG, Li J (2011) Data recovery of Structural health monitoring based on extreme learning machine. Comput Eng 37(16):241–243 (in Chinese)
66. Sun LM, Shang ZQ, Xia Y (2019) Research status and prospect of bridge structural health monitoring under the background of big data. China J Highway Transp 32(11):1–20 (in Chinese)
67. Yuan SF, Liang D, Gao N et al (2013) Research on bridge data anomaly diagnosis based on structural health monitoring system. J Univ Electron Sci Technol 42(1):69–74 (in Chinese)
68. Shrestha A, Dang J (2020) Deep learning-based real-time auto classification of smartphone measured bridge vibration data. Sensors 20(9):710
69. Gong Z, Zhang YM, Hu YJ et al (2016) Cubic meter compressive strength prediction of concrete. J Wuhan Univ Technol (Mater Sci Edn) 31(3):590–593
70. Yan KZ, Shi CJ (2010) Prediction of elastic modulus of normal and high strength concrete by support vector machine. Constr Bulid Mater 24(8):1479–1485
71. Zhang X, Zhang DX, Xu LL (2018) Application of iterative decision tree in predicting concrete compressive strength. Sichuan Build Sci 44(5):85–87 (in Chinese)
72. Xiao QH, Fan J (2013) Prediction of concrete frost resistance based on artificial neural network. Concrete 1:30–32 (in Chinese)
73. Han JJ, Nan SW, Wang JW (2020) Prediction model for carbonation depth of organic film-forming coating concrete based on support vector machine. Constr Technol 49(2):94–98 (in Chinese)
74. Liu BY, Wang X, Wan QW (2019) Predicting the degree of steel corrosion based on ABC-BP neural network. Synth Mater Aging Appl 48(5):54–58 (in Chinese)
75. Shen JW, Peng JP, Guo JQ et al (2019) Machine learning of multi-sensor non-destructive testing data for steel corrosion. Non-Destr Test 41(11):59–64 (in Chinese)
76. Jia L, Li Y, Xie Y (2013) Slope deformation prediction based on chaotic-SVM. Appl Mech Mater 353–356:673–677
77. Duan HT, Liu N, Guo W et al (2019) Prediction of surrounding rock deformation in highway tunnel based on support vector machine. Shanxi Archit 45(22):126–127 (in Chinese)
78. Su HZ, Li X, Yang BB et al (2018) Wavelet support vector machine-based prediction model of dam deformation. Mech Syst Signal Process 110:412–427
79. Xue Y, Xu L, Cao ZZ (2011) Research on deformation forecasting of deep foundation pit supporting structure based on BP neural networks. Adv Mater Res 250–253:2116–2119
80. Zuan P, Huang Y (2018) Prediction of sliding slope displacement based on intelligent algorithm. Wirel Pers Commun 102(4):3141–3157
81. Wei J, Qi J, Wu Y et al (2013) Prediction of the deformation of the surrounding rock around tunnels by GA-BP network model. Appl Mech Mater 256–259:1157–1160
82. Wang XY, Yang K, Shen CS (2017) Study on MPGA-BP of gravity dam deformation prediction. Math Probl Eng 6:1–13
83. Gh XU (2014) Application of RBF neural network in dam deformation prediction. Appl Mech Mater 3547:261–264
84. Kudva JN, Munir N, Tan P (1992) Damage detection in smart structures using neural networks and finite-element analysis. Smart Mater Struct 1(2):108
85. Elkordy MF, Chang K-C, Lee GC (1993) Neural networks trained by analytically simulated damage states. J Clin Rheumatol 7(2):130–145
86. Hakim S, Razak HA (2011) Application of artificial neural network on vibration test data for damage identification in bridge girder. Int J Phys Sci 6(35):965–970
87. Sun ZG, Chen YF, Shan Y et al (2014) Research on structural damage identification of suspension bridge based on model test. Eng Mechan 31(6):132–137 (in Chinese)
88. Gonzalez-Perez C, Valdes-Gonzalez J (2011) Identification of structural damage in a vehicular bridge using artificial neural networks. Struct Health Honitor Int J 10(1):33–48
89. Sun ZG, Gao ZM, Ni YQ (2006) Identification of damaged components and damage levels based on neural networks. Eng Mechan 23(2):18–22 (in Chinese)
90. Liu RY, Yu FH, Liu J (2009) Damage identification of simply supported beam bridges based on wavelet neural network. J Jilin Univ (Eng Edn) 39(S2):413–416 (in Chinese)

91. Liu HB, Jiao YB (2013) Damage identification of simply supported bridge based on RBF neural network. In: International conference on graphic and image processing, 87681B-87681B-7
92. Yin T, Zhu HP (2018) Probabilistic damage detection of a steel truss bridge model by optimally designed Bayesian neural network. Sensors 18(10)
93. Hasni H, Alavi AH, Jiao PC et al (2017) Detection of fatigue cracking in steel bridge girders: a support vector machine approach. Arch Civil Mechan Eng 17(3):609–622
94. Arangio S, Bontempi F (2015) Structural health monitoring of a cable-stayed bridge with Bayesian neural networks. Struct Infrastruct Eng 11(4):575–587
95. Chen M (2009) ANN approach for existing bridge evaluation based on grid and domain knowledge. Los Alamitos: IEEE Computer Soc
96. Gehl P, D'Ayala D (2016) Development of Bayesian networks for the multi-hazard fragility assessment of bridge systems. Struct Saf 60:37–46
97. Setunge S, Zhu WQ, Gravina R et al (2016) Fault-tree-based integrated approach of assessing the risk of failure of deteriorated reinforced-concrete bridges. J Perform Construct Facilit 30(3):12
98. XI LUO (2019) Comprehensive evaluation of technical status and load test of reinforced concrete double curvature arch bridge based on Bayesian network. Highway Transp Technol (Appl Technol Edn) 15(10):177–180 (in Chinese)
99. Rocco CM, Moreno JA (2002) Fast Monte Carlo reliability evaluation using support vector machine. Reliab Eng Syst Saf 76(3):237–243
100. Hurtado JE, Alvarez DA (2003) Classification approach for reliability analysis with stochastic finite-element modeling. J Struct Eng-Asce 129(8):1141–1149
101. Yu XL, Yan QS, Li W et al (2010) Reliability analysis of large-span cable-stayed bridges based on support vector machine. Highway 10:82–86 (in Chinese)
102. Bai B (2015) Research on system reliability of construction and operation process of long span steel cable-stayed bridge. Southwest Jiaotong University, Chengdu (in Chinese)
103. Su H, Cheng J (2009) Application of neural network technology in reliability analysis of prestressed concrete bridges. Struct Eng 25(2):71–76 (in Chinese)
104. Zhang JR, Liu Y (1999) Application of genetic algorithm in reliability analysis of cable pylon of cable-stayed bridges. In: 1999 bridge symposium of the bridge and structural engineering. Society of the China Highway Society, Xiamen (in Chinese)
105. Zhu JS, Xiao RC, He LZ (2007) Research on intelligent reliability evaluation method for long span cable-stayed bridges. Chin Civil Eng J 40(5):41–48 (in Chinese)
106. Liu HB, He X, Jiao YB et al (2019) Reliability assessment of deflection limit state of a simply supported bridge using vibration data and dynamic Bayesian network inference. Sensors 19(4):29
107. Hu G, Kwok KCS (2020) Predicting wind pressures around circular cylinders using machine learning techniques. J Wind Eng Ind Aerodyn
108. Wu T, Kareem A (2011) Modeling hysteretic nonlinear behavior of bridge aerodynamics via cellular automata nested neural network. J Wind Eng Ind Aerodyn 99(4):378–388
109. Snaiki R, Wu T (2019) Knowledge-enhanced deep learning for simulation of tropical cyclone boundary-layer winds. J Wind Eng Ind Aerodyn
110. Li T, Wu T, Liu Z (2020) Nonlinear unsteady bridge aerodynamics: reduced-order modeling based on deep LSTM networks. J Wind Eng Ind Aerodyn
111. Jin XW, Cheng P, Chen WL et al (2018) Prediction model of velocity field around circular cylinder over various Reynolds numbers by fusion convolutional neural networks based on pressure on the cylinder. Phys Fluids 30(4)
112. Lute V, Upadhyay A, Singh KK (2009) Support vector machine based aerodynamic analysis of cable stayed bridges. Adv Eng Softw 40(9): 830–835
113. Jung SM, Ghaboussi J, Kwon SD (2004) Estimation of aeroelastic parameters of bridge decks using neural networks. J Eng Mech 130(11):1356–1364
114. Chen CH, Wu JC, Chen JH (2008) Prediction of flutter derivatives by artificial neural networks. J Wind Eng Ind Aerodyn 96(10–11):1925–1937

115. Li Q, Yang XW (2000) Artificial neural network method for analyzing the aerodynamic characteristics of bridge main beam sections. Central South Highway Eng 3:56–59 (in Chinese)
116. Chen NY, Ge YJ (2019) Identification of aerodynamic parameters of typical bridge sections based on artificial neural networks. China Civil Eng J 52(8):91–97, 128 (in Chinese)
117. Li S, Laima SJ, Li H (2018) Data-driven modeling of vortex-induced vibration of a long-span suspension bridge using decision tree learning and support vector regression. J Wind Eng Industr Aerodyn 172:196–211
118. Mulia IE, Asano T, Nagayama A (2016) Real-time forecasting of near-field tsunami waveforms at coastal areas using a regularized extreme learning machine. Coast Eng 109:1–8
119. Mazinani I, Ismail Z, Shamshirband S et al (2016) Estimation of tsunami bore forces on a coastal bridge using an extreme learning machine. Entropy 2016 18(5)
120. Fang C, Tang H, Li Y et al (2019) Effects of random winds and waves on a long-span cross-sea bridge using Bayesian regularized back propagation neural network. Adv Struct Eng 23(4):733–748
121. Fan W, Yuan WC, Fan QW (2008) Calculation method of ship collision force on bridge using artificial neural network. J Zhejiang Univ-Sci A 9(5):614–623
122. Wj WENG (2011) Research on ship collision prevention methods and applications for large bridges. J Ningbo Univ (Sci Eng Edn) 24(4):106–111 (in Chinese)
123. Zhao YT, Cao FQ (2015) Research on ship collision prevention control technology for bridges. China Water Transp (Second Half of the Month) 15(9):237–239 (in Chinese)
124. Kameshwar S, Misra S, Padgett JE (2019) Decision tree based bridge restoration models for extreme event performance assessment of regional road networks. Struct Infrastruct Eng 16(3):431–451
125. Jootoo A, Lattanzi D (2017) Bridge type classification: Supervised learning on a modified NBI data set. J Comput Civ Eng 31(6):1–11
126. Wang KH, Lu GY, Zhang PP (2019) Research on seismic design evaluation method for small and medium span highway beam bridges based on machine learning. Highway Transp Technol 36(2):74–8 (in Chinese)

Chapter 7
Computer Vision for Bridge Inspection and Monitoring

Abstract Bridge inspection mostly rely on visual assessment in practice. Therefore, digital image-based structural inspection has been proposed as early as in 1980s and become the most fruitful computer-science aided field in Civil Engineering. The approach, termed either computer-vision or machine-vision method, starts with the use of digital cameras or other imaging devices, then an image analysis or understanding algorithm is adopted to detect, classify, and even localize structural surface damage. Moreover, by collecting continuous imagery data in the spatial or temporal domain, vision methods can be adapted to measure structural motions in terms of local deformation or structural vibration. This chapter introduces the computer vision based inspection and monitoring technologies for assessing bridge structures. Specifically, this chapter introduces basic concepts and history of computer vision, conventional methods, deep learning methods, and applications for bridge damage inspection and structural health monitoring with real examples.

7.1 Overview of Computer Vision

7.1.1 Basic Concepts

Humans and animals can receive the electromagnetic wave (visible light) stimulations within a specific wavelength range in the external environment via the visual system's sensory organs (eyes) and obtain the vision after encoding, processing and analyzing the received signal by the relevant portions of the central system. Research shows that humans obtain more than 80% of their information visually, making vision the most important human sensing. Computer vision is intended to simulate the biological visual function using computers and related technologies. Its primary objective is to obtain 3D information of a scene from its collected images or videos.

As a commonly used phrase in academic and engineering circles, it is difficult to define computer vision precisely. Computer vision is, in a general sense, a subject that "endows computer with visual capacity of the biological vision system". However, natural biological vision cannot be precisely described, and the visual perception

G. Wu et al., *Intelligent Bridge Maintenance and Management*, Springer Tracts in Civil Engineering, https://doi.org/10.1007/978-981-97-3827-4_7

abilities of various organisms also vary from case to case. Consequently, this definition is not complete and cannot fully summarize the research and the progress within computer vision. According to Wikipedia, "computer vision is the theory of automatically extracting, interpreting and comprehending meaningful information from a single image or an image sequence". It involves development of the theoretical and the algorithmic underpinning for the autonomous visual perception. Computer vision is the process of arranging, identifying, and interpreting visual data in order to express and comprehend the surrounding environment. We can comprehend computer vision from both the scientific and technological perspectives. Computer vision, as a scientific subject, refers to the extraction of information from the image data. In this context, the image data include typical photographs, video sequences, images synthesized by several cameras, and multidimensional data from medical scanners. Computer vision is also a technical discipline that tries to automate a certain process by applying appropriate theories to computer vision systems.

7.1.2 Related Methods and Domain

Computer Vision in principle is a branch of Artificial Intelligence. It is special in that its basis for learning or inferring is imagery data or structured data points in terms of spatial and temporal coordinates. As such, Computer Vision is closely related to Digital Image Processing or Image Analysis methods. We elaborate their relations as follows.

Image processing typically refers to analyzing and processing images on a computer to produce the needed results by transforming images and extracting graphic characteristics from an image. It may also involve image compression and reconstruction. Typical image processing techniques can be categorized as follows:

Image transformation. Processing images in the spatial domain will directly utilize extensive computing resources because of the large image arrays. Therefore, various image transformation techniques, such as Fourier transform, Walsh transform, discrete cosine transform and other indirect processing technologies, are frequently used to transform the images in the spatial domain to those in the transformed domain, which not only reduces computation but also generates more effective processing results (for example, Fourier transform can put digital filtering into practice in the frequency domain).

Image compression and coding. Image compression and coding technologies can reduce the amount of data (bits) needed for defining images, hence reducing the time required for image transmission and processing as well as the image storage space. Compression can be undertaken either with no distortion or with admissible distortion.

Image enhancement and restoration. Image enhancement and restoration aims to increase the image clarity by removing the noise and enhancing the image definition. Image enhancement does not pay attention to the cause of image degradation but focuses on the elements of the image that are of interest.

Different from computer vision technologies, image processing technologies generally cannot identify the image's meaning, and image processing is performed one image after another. In many instances, people incorrectly equate "image processing" with "computer vision". Computer vision may employ image processing technologies for image preprocessing, although image processing is not the main focus of computer vision. In addition, it is not incorrect if one refers to many advanced image understanding methods as image processing methods. Indeed, if image understanding methods are added in the middle between image processing methods and computer vision methods, the boundaries are not a clear cut.

Computer Vision belongs to the domain of AI. Therefore, many vision models are traced back to general machine learning and AI models. Convolution neural network, one of the most widely used models for image recognition, is essentially a deep learning artificial intelligence algorithm. Statistically, nearly half of the research in the field of deep learning is related to computer vision. However, there are still significant differences between computer vision and artificial intelligence. In contrast to computer vision, research in the field of artificial intelligence focuses on reasoning and decision-making rather than analysis and processing of images. Currently, computer vision technology still focuses on image information expression and target recognition. Computer vision is a subject related to image information organization, object and scene recognition, and interpretation of events using images (videos) as input data, with the ultimate objective of expressing and comprehending the surrounding world. From current research status, computer vision mainly focuses on the organization and recognition of image information, and the interpretation of events is rarely involved, or at least at a very preliminary stage.

An illustration of the relations among image processing, computer vision, and AI is depicted in Fig. 7.1. It should be noted that with the development and progress of deep learning technology, in some frontier fields, the differences between computer vision, image processing and artificial intelligence are not so clear. Taking the widely popular convolutional neural network as an example, it can directly extract and process the features of the original image, realizing direct transformation from image to knowledge. However, convolutional neural network still belongs to the category of artificial intelligence. With further advancement of computer science, we believe there will be more and more intersections among the three of them.

7.1.3 Historical Review of Computer Vision

This section introduces the history of computer vision in terms of both algorithmic and application oriented development, in an effort to provide the readers a general understanding of computer vision as a technology.

It is generally agreed that computer vision originated in the middle and late 1960s. During that period, MIT launched the Summer Vision Project, which assigned students the task of designing a program to enable computers to understand what they saw, and the task had to be completed within their summer vacation. The endeavor

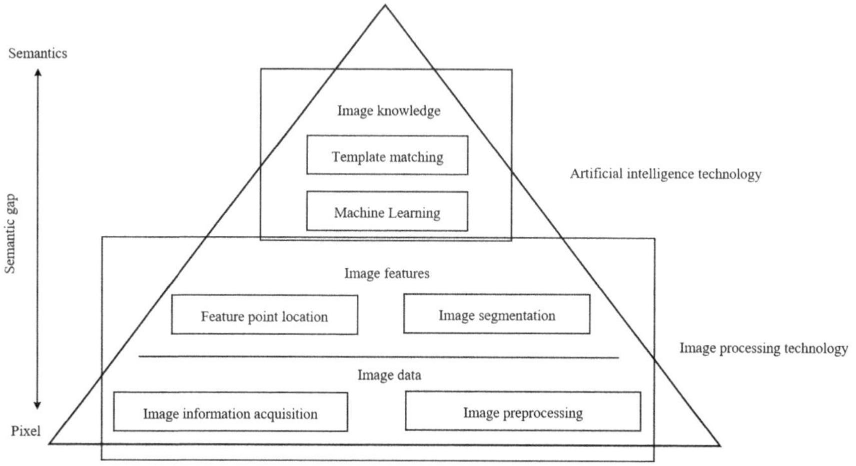

Fig. 7.1 Relationship between computer vision, image processing and artificial intelligence [1]

ultimately failed, but this summer project is considered to be the commencement of computer vision as an independent subject.

Researchers at early stages generally supposed that solving visual issues was analogous to processing auditory information and simpler than addressing more advanced intelligent problems (such as planning learning). People have gradually clarified the distinction between computer vision technology and conventional digital image processing technology during the continuous development of computer vision technology: computer vision technology must objectively characterize the three-dimensional scenes and objects captured in the images as the foundation for a more comprehensive understanding of image content. In order to fulfill this purpose, researchers at early stages used boundary lines or topologies of two-dimensional line segments to describe the three-dimensional objects. Their researches focused on describing image geometric characteristics. This approach is considered to be the preferred extraction method for many underlying image features because this employed type of image features is more intuitive. In conjunction with the well-developed border detection technology at that time, the detection and description of image geometric features also made significant strides forward. Nowadays, the geometric features of images still play an important role as ones that can be swiftly and effectively identified in the popular target recognition field.

In the 1980s, the development of computer vision technology was primarily focused on the mathematical theoretical models. Meanwhile, the quantitative analysis methods of image scenes were gradually diversified, and numerous mathematical models of contemporary popular computer vision algorithms can be traced back to researches in that period. For instance, the image pyramid technology has gained popularity in image fusion and target query. Wavelet technology, which originated in the late 1980s, has also consistently evolved and proved its benefits in image processing. During this period, researchers also found that numerous image processing technologies, including binocular stereo vision, boundary detection and modeling technology, can be characterized uniformly as a class of problems under the concept of variable optimization. At the same time, discrete Markov random field, Kalman filter, and other technologies have been developed and introduced into the field of computer vision. These innovative technologies have significantly altered the field of computer vision.

In the 1990s, computer vision technology achieved significant advancements in application with the emergence and maturing of several techniques meeting actual requirements, such as fingerprint recognition, optical character recognition and face recognition. In the late 1990s, machine learning technology based on statistic models emerged and was initially applied to facial recognition and linear dynamic system-based curve tracking. These applications indicated that statistical theory is widely utilized in computer vision.

At the turn of the twenty-first century, numerous theories and applications of pattern recognition technologies based on learning theory, template matching, etc. have emerged and evolved via absorbing various practical technologies developed in the 1990s. In recent years, the fusion of various image features, semantic information and pattern recognition methods has become the mainstream of computer vision development, which can analyze and describe complex scene in the image or the object information, more comprehensively. Nonetheless, computer vision methods prior to 2010 are not "end-to-end", which means that typical vision methods need a combination of feature extraction, selection, learning model training and testing, which are hand-picked or optimized by their own.

The milestone comes around 2012, which marked the significant shift in the field of computer vision during the ImageNet Large Scale Visual Recognition Challenge (ILSVRC). The representative work that signifies this transformation is "AlexNet" by Alex Krizhevsky, Ilya Sutskever, and Geoffrey Hinton. AlexNet outperformed the traditional computer vision methods by achieving a top-5 error rate of 15.3%, a significant improvement over the 26.2% achieved by the second-place model. The signature of this AlexNet model is a deep convolutional neural network (CNN), which firstly provides an end-to-end network model that combines image feature extraction and object classification. Since then, a surge of deep learning based computer vision methods and models have been developed. This chapter introduces the general computer vision technology with emphasis on this recent development.

7.1.4 Related Opensource Software

The aim of open source is for users to be able to more effectively use, copy, distribute, study and improve the software. In the realm of computer vision algorithms, the emergence of open-source software can be traced back to 1960s. The development of open-source software enables the rapid application of classical algorithms, dramatically speeding the deployment of them. Consequently, this section will introduce a number of open-source software platforms that are widely utilized in the field of computer vision. By studying and utilizing these platforms, computer vision-related applications can be developed quickly.

OpenCV

OpenCV was initially developed by Intel in 1999 and currently funded by Willow Garage. OpenCV is a cross-platform computer vision library licensed under BSD (Berkeley Software Distribution) and is compatible with Linux, Windows and Mac OS. Composed of a series of C functions, a handful of C++ languages and language interfaces such as Python, Ruby, MATLAB, it is lightweight and efficient, and realizes many common algorithms in image processing and computer vision. As a result, it has been extensively utilized in numerous fields related to digital image technology, such as human–computer interaction, target recognition, image segmentation, face recognition, motion recognition, motion tracking, etc.

Tesseract

Tesseract was provided for the open-source community by HP in 2005; later maintained, continuously enhanced, and optimized by Google. Tesseract is an open-source OCR (Optical Character Recognition) program built in C++ and based on the Apache License 2.0. Tesseract is user-friendly due to its support of C++, Python and Java interfaces. Due to its complete functions and trainable characteristics, it has attracted wide attention after opening its source, but in recent years has been gradually surpassed by other OCR based on deep learning.

TensorFlow Object Detection API

In 2017, based on the Apache License 2.0 protocol and using Python, Google built TensorFlow Object Detection API, which is an open-source target detection algorithm library. It includes SSD, Faster RCNN, R-FCN, Mask RCNN and other algorithms. It can be used to construct, train and deploy target detection algorithm models with ease.

Detectron

Detectron is an open source target detection and segmentation algorithm library based on Apache License 2.0 protocol, written by Facebook AI Research on Caffe2 in 2018, using C++ Python. It includes some of the most advanced algorithms for target recognition and segmentation, such as Fast R-CNN, Faster R-CNN, R-FCN, RPN, Mask R-CNN, RetinaNet, etc. It provides experimental results on multiple classical datasets with different backbone networks, different learning rates and batch sizes. Researchers can quickly test and evaluate new algorithms using its pre-training models.

InsightFace

InsightFace is a face recognition algorithm library based on MIT protocol, written by DeepInsight on MXnet in 2018, using Python. It contains classic loss functions used for face recognition, such as Softmax, Sphereface, CosineFace, ArcFace, Combined

Margin and TripletLoss, and demonstrated satisfying performance on public face recognition datasets.

7.1.5 General Applications

In recent years, computer vision combined with deep learning technology has gradually matured and been applied in all walks of life. We will summarize a number of currently mature application scenarios for computer vision technologies, including face recognition, video analysis, automatic driving, industrial vision detection, medical imaging diagnosis, character recognition, etc. We will focus on the application of computer vision technology in the field of civil engineering in the following chapters.

Face recognition: Face recognition is currently one of the most prevalent application scenarios within AI-based vision and image field. It is a biometric recognition technology based on human facial feature information. At present, face recognition has been widely used in many fields, for example, electronic payments employing face recognition.

Video analysis: The integration of image processing with artificial intelligence allows for more efficient video analysis and processing. By recognizing the targets, motion tracks, etc. in videos, additional analysis of behaviors and abnormal occurrences may be realized to provide cost-effective real-time monitoring.

Automatic driving: The automobile industry is a major implementor of artificial intelligence technologies. Increasingly more and more driving assistance functions are applying artificial intelligence technology, the majority of which are based on computer vision and image processing technologies.

Industrial vision inspection: Machine vision can quickly obtain and automatically process a large amount of information. In the automatic production process, machine vision system is widely used in the fields of working condition monitoring, final inspection, quality control, etc. It can be used in some dangerous working environments or scenarios where human vision cannot meet the requirements.

Medical image diagnosis: More than ninety percent of medical data is derived from medical images. The medical image field has massive data that can facilitate deep learning. Medical image diagnosis can aid the physicians and enhance their diagnostic efficacy.

Character recognition: Computer character recognition, commonly known as optical character recognition, employs optical technology and computer technology to read or write characters on paper and convert them into a format that can be accepted by computers and understood by people. This is a crucial technique for quick text typing.

7.2 Review of Computer Vision Methods

The ultimate objective of computer vision is to utilize computers to simulate human vision function, which includes comprehending and responding to visual input. There is no distinct border among basic image processing, image analysis and what is called computer vision. We can divide the entity of computer vision into three typical computing processes, namely, low-level processing, intermediate processing and advanced processing. Low-level processing involves fundamental processes, such as image preprocessing to remove the noise, contrast enhancement and image sharpening. Low-level processing features the fact that both the input and output are images. Intermediate processing is intended to perform numerous tasks, including classification, recognition and segmentation of objects (i.e., dividing an image into different regions or targets), so as to reduce the description of these targets and make them more suitable for computer processing. Intermediate image processing has the characteristic that its output is the features (such as the edges, contours and identification of objects) extracted from these images while the input is images. Advanced processing involves "understanding" the whole identified object, including understanding its appearance features and geometric features, and performing visual related cognitive functions at the far end of the continuum.

From the perspective of object model, computer vision can be divided into two major sub-categories: statistics-based and geometry-based computer vision. The former is typical of target recognition, while the latter has multi-view 3D reconstruction. It is worth noticing that both are generally based on low-level operations. There are large differences between different sub-categories in accordance with intermediate processing tasks, or it depends on the feature extraction method. Within the scope of advanced processing and comprehension, the two major computer vision sub-categories tend to be integrated (such as machine vision applications in automatic driving). In particular, the difference between geometry and statistics becomes a methodological matter of science involving philosophical speculation in some cases.

This section focuses on some important low-level processing, intermediate processing and advanced processing, as shown in Fig. 7.2. It is recommended that the readers learn with references to the digital image processing courses. In addition, advanced processing approaches involving pattern and geometry fusion are prevalent in the recent literature and studies.

7.2.1 *Basic Image Transformation Methods*

(1) Spatial Domain

Spatial domain refers to the image itself, and spatial domain-based image processing approaches are directly based on pixel operations within the image. Spatial domain processing can be divided into two main categories: grayscale transformation and spatial filtering. Grayscale transformation operates on a single pixel of the image,

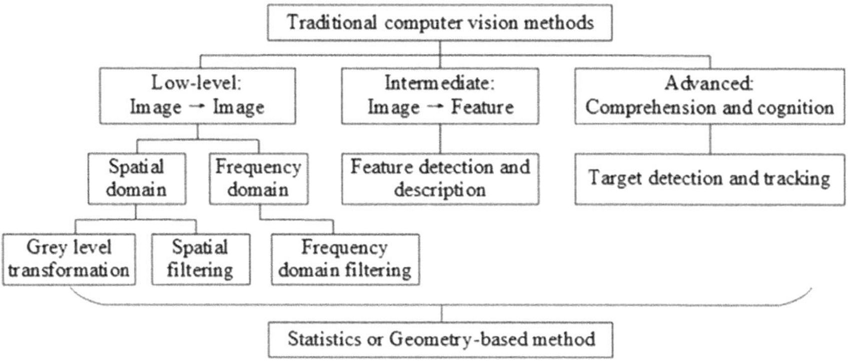

Fig. 7.2 Conventional computer vision methods

mainly for the purpose of contrast and threshold processing. Spatial filtering involves operations to improve performance, such as sharpening the image through processing the neighborhood of each pixel in the image.

Grayscale Transformation

Grayscale transformation is the simplest image processing technology, and can be regarded as a spatial filtering method with a neighborhood size of 1×1. Here are the fundamental grayscale transformation functions:

Image reversal. By inverting the grayscale level of an image, the equivalent background color of the picture can be obtained.

Log transformation. The low grayscale value with narrow range in the input is mapped to the grayscale value with wide range in the output. On the contrary, the high input grayscale value is also processed in this way. The antilog transformation has the opposite effect.

Power law (gamma) transformation. Its basic mathematical expression is $s = cr^{\gamma}$. Similar to the logarithmic case, the power law curve of partial γ values maps a narrow range of dark low input values to a wide range of output values, and conversely, the same situation holds true for high grayscale input values. The difference is that the power law (gamma) transformation will simply yield a family of possible transformation curves as the value of γ changes.

Piecewise transformation function. The piecewise function can be arbitrary and complex, and the actual implementation of some important transformations can only be explicitly expressed by the piecewise function, such as contrast stretching, grayscale level layering, bit plane layering, etc.

Spatial Filtering

Spatial filtering is one of the main tools widely used in image processing. The term filtering, following the conventional frequency domain processing technique, refers to accepting (passing) or rejecting certain frequency components. For example, a filter that let low frequency components to pass through is called a low-pass filter, and the effect of low-pass filter is to blur (or smoothen) the image. We can use spatial filters (also known as spatial masks, cores, templates and windows) to directly act on the image itself to achieve similar smoothing.

Spatial filtering consists of a predefined operations performed on image pixels surrounded by the neighborhood. For a new pixel with its coordinates equal to those of the neighborhood center to be generated by filtering, the value of the pixel should be the result of filtering. The center of the filter accesses each pixel in the input image to generate a processed (filtered) image. If the operation performed on the image pixel is linear, the filter is called a linear spatial filter, otherwise it is called a nonlinear spatial filter.

(2) Frequency Domain

Frequency domain filtering (wave number) refers to the spatial frequency of gray value of image pixel changing with position. The distribution characteristics of information within the image are represented by the spectrum. The spatial changes of image pixel values can be decomposed into linear superposition of simple harmonic functions with different amplitudes, spatial frequencies and phases. The composition and distribution of various spatial frequency components in the image are called spatial spectrum. This kind of decomposition, processing and analysis of spatial frequency features of images is called frequency domain processing. The most typical frequency domain processing is Fourier transform which can transform an image from the spatial domain to the frequency domain containing only different frequency information. The grayscale mutation part, the area with complex image structure, image details, interference noise and other information on the original image are concentrated in the high-frequency area, while the information on the area with gentle gray change on the original image is concentrated in the low-frequency area.

Basic application steps of frequency domain filtering are as follows: input → image preprocessing → image Fourier transform → filter → inverse Fourier transform → image cropping → output.

Low-pass filter can achieve the purpose of blurring the image (frequency domain smoothening) through high-frequency attenuation. Typical low-pass filters include the ideal low-pass filter, Butterworth low-pass filter and Gaussian low-pass filter.

High-pass filter can achieve the purpose of image sharpening (edge fusion is related to high frequency) by attenuating low frequency. Typical high-pass filters include the ideal high-pass filter, Butterworth-high pass filter and Gaussian-high pass filter.

In many applications, selective filtering is most-often required, and its interest is to deal with the specified frequency band or small area of the frequency rectangle. The first type of filter is called band stop filter or band-pass filter, and the second type of filter is called notch filter.

7.2.2 Low-Level Vision Methods

Conventional computer vision technology is a collection of extensive algorithms, among which an important link is to carry out subsequent target recognition according to the attributes of objects in the image. On the one hand, the attribute can be the source image itself (such as the source image used as the matching template according to the template matching method). On the other hand, attributes can also be features extracted from the source image. Generally speaking, features are meaningful image regions, which have unique features and are easy to identify. The main purpose of feature extraction is to extract some valuable information from the image. In the process of feature selection, it is necessary to fully consider the relevance of the target so as to extract features that can better describe the target category, and thus affect the accuracy of target detection. The accuracy of feature extraction largely depends on the design and flexibility of the algorithm utilized. According to the type and quality of the input image, different algorithms have different application levels. The image features used can be roughly divided into three categories: edges, corners and spots.

Edge Detection

Edge detection is an extremely important method to analyze images in image processing and computer vision. Its purpose is to find a set of pixels with obvious brightness changes in the image, which often shows the outline of the image. If the edge in the image can be accurately measured and located, it means that the actual object can be located and measured, including the area, diameter and shape of the object. In the actual image acquisition, when one or more of the following four conditions are met, the image will appear as an edge: ① Discontinuities in depth (objects in planes of different heights); ② Discontinuities in surface direction (such as two different faces of a cube); ③ Different object materials; and ④ Different lighting in the scene.

Figure 7.3 shows the grayscale value display effect of the seven pixels in the horizontal direction of the image. It is apparent that there is an edge between the fourth and fifth pixels because there is a strong grayscale jump between them. In actual edge detection, the edges are far more complicated and less obvious, so we need to use the corresponding thresholds to distinguish them.

Basic edge detection methods include:

Roberts edge detection operator. Roberts operator is the simplest operator which can measure the two-dimensional spatial gradient of an image simply and quickly, and highlight the high spatial frequency region corresponding to the edge. In general, the

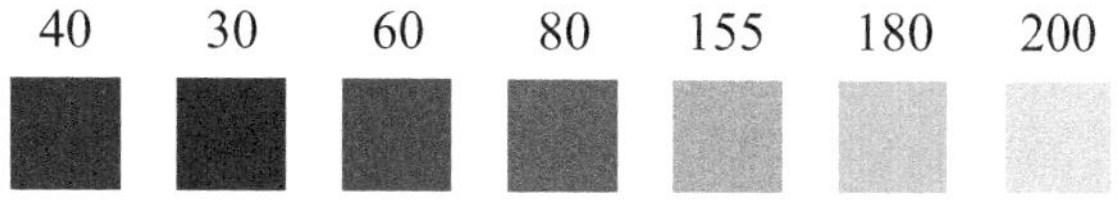

Fig. 7.3 Display effect of gray values of 7 pixels in the horizontal direction of the image

input and the output operators are both grayscale images, and the pixel value at each point in the output represents the estimated absolute value of the spatial gradient of the input image at that point. Roberts operator has high positioning accuracy and good effect in locating horizontal and vertical directions, but it is sensitive to noise.

Prewitt edge detection operator. The principle of Prewitt operator is to use two direction templates to perform neighborhood convolution with the image in image space. One of the two direction templates detects horizontal edges and the other detects vertical ones. Compared with Roberts operator, Prewitt operator can suppress noise. Its principle of suppressing noise is to average the pixels, so it can process images with noise better. However, averaging the pixels is equivalent to low-pass filtering of the image, so Prewitt operator is not as effective as Roberts operator in edge positioning.

Sobel edge detection operator. Sobel operator performs two-dimensional spatial gradient measurement on the image, so it emphasizes the high spatial frequency region corresponding to the edge. Usually, it is used to find the approximate absolute gradient value of each point in the input grayscale image.

Canny edge detection. Canny edge detection is a multi-stage optimization method. The algorithm can be divided into the following five steps: a. Using Gaussian filter to smoothen the image and filter out noise; b. Calculating the gradient strength and the direction of each pixel in the image; c. Applying non-maximum suppression to eliminate the spurious effect caused by edge detection; d. Applying double threshold detection to determine the real and the potential edges; and e. Suppressing isolated weak edges to complete the edge detection.

Corner Detection

A corner is usually defined as the intersection of two edges. In a broad sense, corners refer to image points with specific features, which have specific coordinates in the image and some mathematical features (such as local maximum or minimum grayscale level). If a small change of a point in any direction will cause a large change in the grayscale level, we can regard the point as a corner.

At present, corner detection algorithms can be divided into three types: corner detection based on grayscale image, corner detection based on binary image and corner detection based on contour. Corner detection algorithms are popular detection algorithms and a classic conventional detection method—Harris corner detection method will be introduced here.

Harris feature angle was first proposed by Chris Harris and Mike Stephens in A Combined Corner and Edge Detector [3]. As shown in Fig. 7.4, the basic idea of Harris corner detection is to use a fixed window to slide the image in any direction. Via comparing the variation of pixel grayscale level in the window before and after sliding, if the sliding in any direction causes a large change of grayscale level, it is considered that corner exists in the window.

Harris corner detection method has the following characteristics:

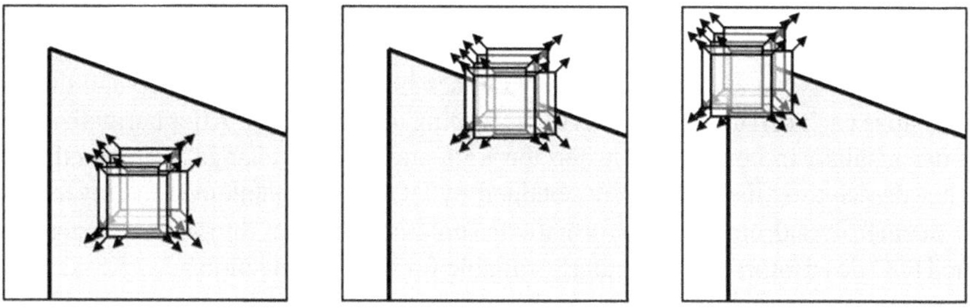

Fig. 7.4 Schematic diagram of Harris corner detection

The parameters are adjustable. By setting its internal parameter k, the sensitivity of corner detection can be adjusted to increase or decrease the number of corners to be detected.

It is insensitive to changes in brightness and contrast.

It has rotation invariance.

It has no scale invariance.

In subsequent research, scholars also improved the Harris corner detection method, developed multi-scale Harris corner detection method with scale invariance in combination with Gaussian scale space representation. They also developed Harris Affine checkpoint detection method with affine invariance, which can deal with obvious affine transformations, including large-scale changes and obvious visual angle changes. Readers can refer to related articles and monographs for more information.

(3) Spot Detection

Spot usually refers to the area with color and grayscale difference from those of the surrounding area. As the spot covers a region, it has better stability and stronger anti-noise ability than the usual corners, so it plays a very important role in image registration. The classical spot detection methods include HOG feature detection method, SIFT feature detection method, etc. The principles and effects of these methods are briefly introduced below.

Histogram of Orientated Gradient (HOG)

HOG feature detection method constructs features by calculating and counting histogram of gradient direction of the image's local area. This method was first published in the international conference on CVPR (Computer Vision and Pattern Recognition) by Navneet Dalal and Bill Triggs in 2005 for pedestrian detection in still images or videos [4]. The core idea of HOG is that for the target in an image, the shape of the local target can be well described by its local gradient direction and gradient strength distribution characteristics. Its basic application process is: (1) divide the whole image into small connected areas (called cell units); (2) collect the gradient or edge direction histograms of each pixel in the cell unit; and (3) combine these histograms to form a feature descriptor. In order to improve their accuracy, these local histograms can be contrast-normalized in a large area (block) of the image. The method used is to calculate the density of each histogram in this block, and then to normalize each cell unit in the interval according to this density. After normalization, better results can be obtained when the light and shadow change. Compared with other descriptors, the descriptors obtained by HOG well maintained the invariance of geometric and optical transformations (unless the object direction changes), so the HOG descriptors are particularly suitable for human detection.

Scale Invariant Feature Transform (SIFT)

The SIFT method was published in ICCV (International Conference on Computer Vision) by David Lowe in 1999 [5]. It attracted much attention because the algorithm is invariant to rotation, scaling, brightness changes, and also stable to some extent against the view angle changes, affine transformation and noise.

The idea of SIFT feature detection method is not complicated. It mainly includes four steps.

Scale space extremum detection. Gaussian difference function is used to compute and search the image positions on all scales, and to identify the potential points of interest with the same scale and direction.

Key point positioning. The location and scale of each candidate position are determined by a fine fitted model, and the selection of key points depends on their stability.

Direction matching. Based on the gradient direction of the local image, each key point position is assigned to one or more directions. All subsequent operations on image data are transformations relative to the direction, scale and position of the key points, thus providing invariance for these transformations.

Key point description. Local image gradients are calculated at the selected scale in the area around each key point. These gradients are transformed into a representation that allows for relatively large local shape deformations and lighting changes.

The feature descriptors used by SIFT and HOG share many similarities. As shown in Fig. 7.5, they take the detected key points as the center, select a 16×16 neighborhood, and divide this neighborhood into 4×4 sub-regions, then divide the gradient direction into 8 intervals, and then obtains a feature vector of $4 \times 4 \times 8 = 128$ dimension in each sub-region. The magnitude of the vector element is the interval weight of each gradient direction. After obtaining the feature vector, the neighborhood of it should be normalized, where the normalization direction is the main direction for calculating the key point of the neighborhood, and the neighborhood is rotated to a specific direction according to the main direction, which provides the feature with the rotation invariance. Then the neighborhood is specifically scaled according to the size of each pixel in the neighborhood to make the feature descriptor have scale invariance.

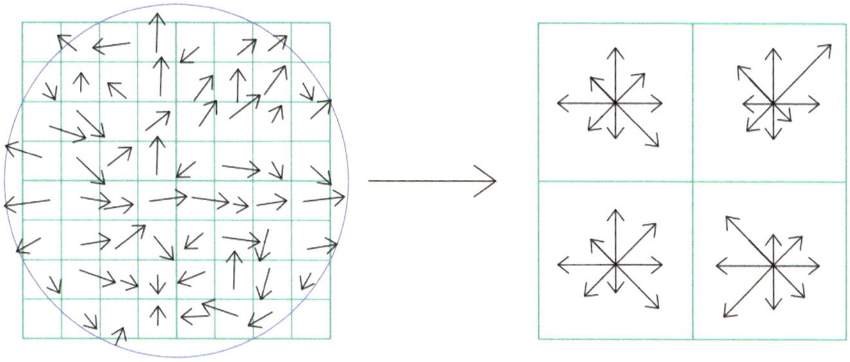

Fig. 7.5 SIFT local feature description [5]

(1) Target Recognition and Tracking

Conventional computer vision technology includes the recognition of various regions in an image, which are called targets or patterns. The current main target recognition method is intended to use quantitative features to describe various patterns, such as length, area, texture, etc. Later, the target is tracked by matching it in continuous images or directly using the relationship between video frames. The common target recognition and tracking methods in the field of civil engineering are introduced below.

Template Matching Technique

Template matching technique is one of the most effective image processing techniques for target tracking. Template matching is a computation-intensive process whose purpose is to locate the template in the image.

The key content of template-based matching technology is matching, and the classical method is to use enumeration to match image strength directly. To identify matching areas, the template image is compared to the source image by moving the template one pixel at a time (from left to right and from top to bottom). A metric is calculated at each position to represent the similarity of the specific area between the template image and the source image; then, the similarity of each position is stored in turn and the position of the template in the source image can be determined by searching the peak position of the similarity distribution. The existing correlation standards or similarity indicators/metrics using visual sensors are generally divided into two categories: ① Cross Correlation (CC) criteria, including CC, Normalized Cross Correlation (NCC) and Zero Normalized Cross Correlation (ZNCC); and ② Sum of Squared Differences (SSD) related criteria, including SSD, Normalized Sum of Squared Differences (NSSD) and Zero Normalized Sum of Squared Differences (ZNSSD).

Research shows that (1) ZNCC and ZNSSD correlation criteria have the strongest robustness against noise, and are insensitive to illumination offset and linear scale;

(2) NCC and NSSD related criteria are insensitive to the linear scale of illumination, but sensitive to the offset of illumination; and (3) CC and SSD related criteria are sensitive to all lighting fluctuations.

(2) Digital Image Correlation

Digital image correlation (DIC) is a non-contact, high-precision optical measurement technique based on photoelectric imaging and computer automatic or semi-automatic processing. It extracts the deformation (displacement and strain) information from the specimen by comparing the changes in information of the images acquired before and after deformation of the specimen surfaced. Digital image correlation is also called Digital Speckle Correlation Method (DSCM) or Digital Image/Speckle Correlation (DISC), and Electronic Speckle Photography (ESP). It was first proposed by Yamagichi [6], Peters and Ranson [7] in the early 1980s.

The core method with DIC technique is the correlation matching operation. Correlation matching can be performed in both image spatial domain and frequency domain, which are called spatial correlation operation and frequency domain correlation operation, respectively. Frequency domain method has low measurement accuracy because it does not allow deformation of the image subregion. Spatial correlation calculation can achieve higher deformation measurement accuracy compared with the frequency domain correlation calculation. Therefore, the vast majority of relevant literature adopts the methods based on spatial domain [8]. The general process of correlation matching operation is: define correlation function → optimize correlation function → obtain sub-pixel displacement.

The two-dimensional digital image correlation method can only measure the in-plane displacement of a planar object, and have certain requirements for the measured object, deformation state of the measured object and layout of the optical path. Its measurement accuracy is affected not only by the hardware (camera system), but also by the correlation function, correlation function optimization algorithm, sub-pixel gray interpolation method, speckle image, system noise, correlation calculation window size and other factors in the algorithm implementation process. In 1993, Luo, Sutton et al. [9] used two cameras to image the surface of the measured object from different angles. First, they calibrated the binocular stereo vision model to obtain the intrinsic and external parameters of the two cameras, and then directly used the two-dimensional digital image correlation matching algorithm to obtain the parallax of the corresponding points between the left and the right images. Using the parallax data of each point and the pre-obtained calibration parameters, the three-dimensional topography of the object surface can be recovered. The three-dimensional displacement distribution of the whole field can be obtained by comparing the changes of three-dimensional morphology of all points in the measurement area before and after the loading. The 3D digital image correlation method can be used to measure the 3D topography and 3D deformation of both plane and curved surfaces. Although its application is relatively complex compared with the 2D digital image correlation method, the 3D digital image correlation method has a broader application prospect in civil engineering.

In addition to the 3D digital image correlation methods based on the binocular stereo vision principle, deformation of the cube unit can be tracked using high spatial resolution X-ray computed tomography (X-CT), and the synchronous radiation CT has been used to study the deformation field inside the structure [10, 11]. These related methods, called digital volume, provide a new means to observe the internal deformation field in materials.

(3) Moving Object Tracking

Moving object tracking refers to finding targeted moving objects (including their positions, velocities, accelerations and other motion parameters) in each image of an image sequence. Moving object detection is the basis of target recognition and tracking. Moving object detection methods include the following main categories:

Inter-frame difference method. The adjacent frame difference method is a moving object detection method that obtains the position and shape of moving objects by calculating the difference between two adjacent frames. It is of strong adaptability to the environment, especially to the change of light. However, because the textures, gray levels and other information of the pixels on the moving object are similar to one another, the inter-frame difference method cannot detect the complete object well but only obtain part of the information of the moving object. It is also not sensitive to slow-moving object, causing certain limitations on this method. The inter-frame difference method is usually not used alone in object detection, but often combined with other detection methods in used.

Optical flow method. Optical flow is the description of motion information of image brightness. Optical flow technique is based on the premise that the adjacent points in the image shares similar brightness, and that means the brightness in the image changes smoothly. The optical flow method maps the rate information of the luminance points on the image surface in three-dimensional space to an approximate two-dimensional motion field for calculation. In object detection process, motion detection based on optical flow method uses the flux vector's characteristics of moving objects changing with time to detect moving regions in image sequences. Due to the use of optical flow information, we can obtain complete motion information, detect different moving objects and other prospects from the background, and even detect a part of moving objects. Therefore, the detection of independent moving objects in the process of single camera movement can be realized. However, the detection algorithm is computationally expensive, complex and time-consuming for the optical flow method.

Background subtraction method. The background subtraction method is a technique that uses the difference between the current image and the background image to detect the moving area. Generally, it can provide the most complete feature data, but it is particularly sensitive to changes in dynamic scenes (such as weather, light, background disturbance, etc.). The shadow of moving objects will also affect the accuracy of the detection results and the tracking accuracy. The basic idea is to first obtain a background model, and then subtract the current frame from the background model. If the difference between the pixels is greater than a certain threshold, it is regarded that the pixel belongs to a moving object, otherwise it belongs to the background image.

The tracking of moving objects is to locate the object in the image sequence through effective expression of the object. According to the expression and similarity measurement of moving objects, moving object tracking algorithms include the following main categories:

① Tracking based on active contour. Active contour model Snake is a deformable curve defined in the image domain. By minimizing its energy function, the dynamic contour gradually adjusts its shape to keep consistent with the object contour. This deformable curve is also called Snake curve. Snake technique can deal with arbitrary deformation of objects with arbitrary shape. First, the segmented object boundary is used as the initial template for tracking. Then, the objective function representing the real boundary of the object is determined. Finally, the initial contour gradually moves towards the real boundary of the object by reducing the value of the objective function. This method has the advantages that it not only considers the grayscale information from the image, but also considers the geometric information of the overall contour, which enhances the reliability of tracking. The tracking process is actually an optimization process, which requires a large amount of computation. However, for fast moving objects or large deformation cases, the tracking effect is not satisfactory.

② Feature based tracking. The tracking method based on feature matching does not consider the overall characteristics of the moving object, but only tracks the object through some distinctive features of the object image. Assuming that each moving object can be represented by a unique feature set, catching the corresponding feature set is considered as tracking the moving object. In addition to using a single feature to achieve tracking, multiple feature information can also be fused together as tracking features. The advantage of this method is that it is insensitive to the scale, deformation, brightness and other changes with the moving object. Even if a part of the object is occluded, as long as other features can be seen, the tracking task can be completed.

③ Region-based tracking. Region-based tracking algorithm is basically: (a) The template containing the object can be obtained by image segmentation or determined manually in advance. The template is usually a rectangle slightly larger than the object, but an irregular shape is also possible; and (b) In the sequence images, the correlation algorithm is used to track the object. The advantage of this method is that when the object is not occluded, the tracking is very stable and its accuracy is very high. But its disadvantage is that it is time-consuming, especially when the search area is large. Besides, the algorithm requires that the object deformation is not large, and there should not be too much occlusion, otherwise decline of the correlation accuracy will cause loss of the object.

7.2.3 Deep Learning Methods

Deep learning is a branch or sub field of machine learning. The term deep learning (DL) was first introduced into machine learning (ML) in 1986, and was later used in artificial neural networks in 2000. Deep neural network is based on multi-level nonlinear information processing, and learn data characteristics on multiple abstract levels accordingly. In 2006, the concept of deep learning was put forward in the field of artificial intelligence. Two solutions, namely "unsupervised pre training initialized weights" and "supervised training fine-tuning", were used to solve the gradient problem in the deep network and initiated the application of deep learning in academia and industry [12]. In the field of computer vision, large-scale datasets combined with the powerful expression ability of deep convolution neural networks (CNNs) can constitute ultra-accurate robust models, which brings revolutionary progress to computer vision technology and is also a hot spot in current AI research.

(1) Convolutional Neural Network

Convolutional neural network is a kind of deep learning technique which has been widely used in many fields and achieved excellent results. Convolution neural network uses convolution operation in one or more layers to replace the connections between layers of conventional neural network. In nature, convolution is a special linear operation. In the field of image processing, convolution operation refers to two-dimensional convolution operation. Figure 7.6 shows an example of convolution operation on two-dimensional tensor (the kernel is not flipped). Convolution operation improves the properties of neural network from three aspects: sparse interaction, parameter sharing and equivariant representation, which endow convolution neural network with excellent performance. These useful concepts are introduced below.

The relationship between the input and the output on layers of the conventional neural network depends on the matrix multiplication, that is, using the product of the weight matrix and the input vector of a layer to describe the relationship between

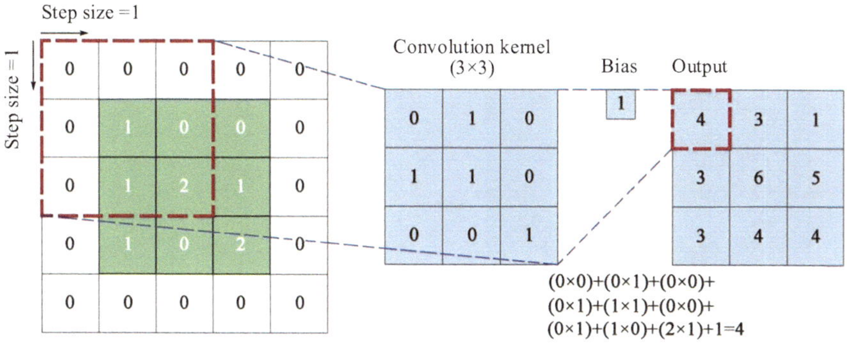

Fig. 7.6 Example of two-dimensional tensor convolution operation (kernel is not flipped)

the input and the output of this layer, which indicates that each output unit directly interacts with all input units. Convolution neural network has the property of sparse interaction or sparse connection, which is realized by making the convolution kernel smaller than the input matrix. Parameter sharing is another feature of the convolutional neural network [13] specifically referring to the computation of some functions using the same parameters in the operation process of the convolutional neural network model. Each element of the weight matrix of a layer in the conventional neural network is used only once when computing the output of the layer. Each element of the weight matrix only represents the connection between a certain neuron of a layer and its related input in the network. Parameter sharing can be regarded as adding "constraints" to the weights of neurons. The weights applied to one input unit are also applied to the connection between other input units and weights. Each convolution kernel is applied to every position of the input tensor during the process of convolutional neural network output. For a certain layer of the convolutional neural network, the property of parameter sharing means that it is not necessary to obtain the corresponding weights of each input and all neurons of a layer in the process of network training iteration, but only a relatively small number of weights, which makes the storage of network weights more efficient. It also reduces the memory requirement of the network model during the running process.

There are many kinds of pooling functions, such as maximum pooling function, average pooling function, etc. The maximum pooling function replaces a rectangular area with the maximum value in the rectangular area, which is common in image recognition tasks. The average pooling function is another common pooling function, which replaces the value of a rectangular area with its average value. Compared with the maximum pooling function, the average pooling function can retain more network output characteristics. Figure 7.7 shows the pooling operation diagram.

Pooling functions have two important properties for convolutional neural networks: ① Pooling functions can reduce the complexity of input. For example, in two-dimensional input, the kernel size is 2 × 2. The pooling function with a step

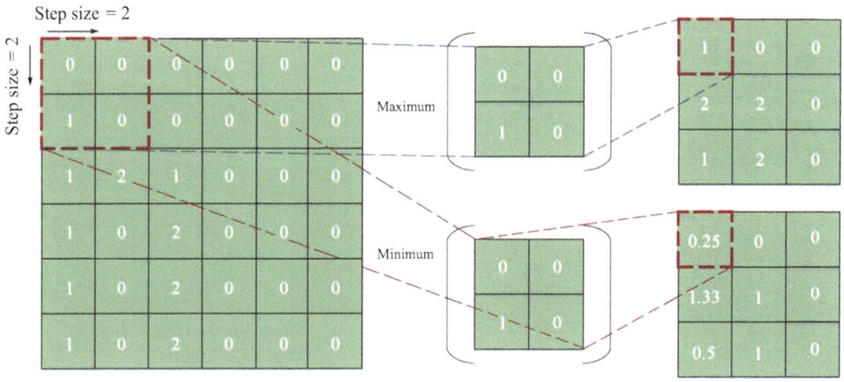

Fig. 7.7 Diagram of pooling operation

size of 2 changes the input to 1/4 of the original. This can improve the efficiency of the network and reduce the running time of the network. ② Pooling function increases the robustness of input. Specifically, when a small disturbance occurs to the input, such as a small translation of the input, most results of the pooled layer input will not change.

Deep neural network contains many nonlinear hidden layers which enables it to fit the complex functional relationship between input and output. However, in the case of limited training data, redundant fitted function relations are formed because of data noise, which is called the overfitting of neural network. Large-scale neural networks are not only prone to overfitting phenomenon, but also not suitable for using neural network integration method because the training and the use of neural networks require a large amount of computation. Dropout is an effective method to solve this problems [14]. The core idea of this method is that in the process of training neural networks, each iteration randomly discards some units, which can weaken the joint adaptability between neurons. Figure 7.8 is a schematic diagram of the Dropout method. Figure 7.8a is a neuron with two hidden layers. Figure 7.8b is a "light" neural network obtained by using the Dropout method on the neural network presented in Fig. 7.8a.

(2) Application Scenario of Convolutional Neural Network

The main applications of computer vision technology based on deep learning include classification, detection, segmentation, enhancement and restoration, action recognition, etc. Among them, the main idea of classification is to extract features from images through CNN, and then classify them through Sigmoid or Softmax classification layer. After removing the classification layer, the classification network can become a feature extraction network, which can be expanded to other tasks. If feature extraction network is added to coordinate regression branch, combined with sliding window method or candidate box generation network, it can be expanded to detection network. If classification layer is replaced by convolution layer, then it can carry out

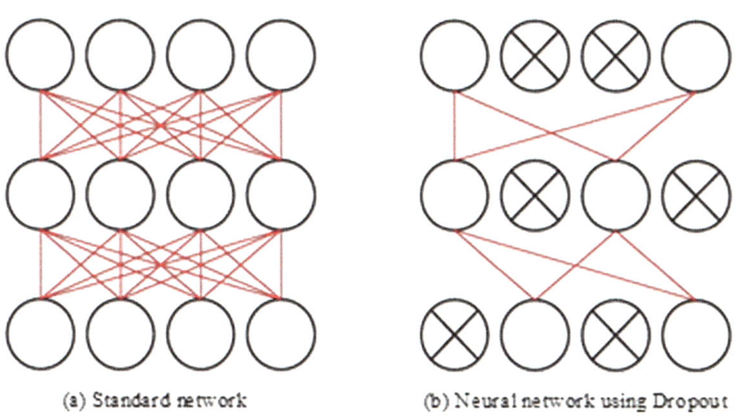

Fig. 7.8 Diagram of dropout method

pixel level classification which is known as the semantic segmentation network. In the field of civil engineering, image classification, object detection and segmentation are the most popular application scenarios of computer vision technology.

(1) Image Classification

Image classification is the task of assigning a label to an image from a given classification set. The main idea of image classification based on depth learning is to use convolutional neural network to extract and classify features from a large number of data sets with known labels. In 1998, LeNet convolution neural network first used multi-layer artificial convolution structure to recognize handwritten digits [14]. LeNet used convolution layer to extract features, with the pooling layer for down sampling, and the full connection layer and the active layer for classification. In 2012, AlexNet [15] used 8-layer CNN network structure to classify images and won the championship in the ImageNet image classification competition. The model adopts effective classification methods such as data expansion, local response normalization, ReLU function, pooling and Dropout to avoid overfitting, which has a great impact on all future in-depth learning models.

In 2014, GoogLeNet [16] series network Inception v1, Inception v2 and Inception v3 were successively proposed. First, Inception v1 (Fig. 7.9) adopts the "separation-fusion" strategy to reduce the dimension of input features by 1 × 1 convolution, and then pass 1 × 1 convolution, 3 × 3 convolution, 5 × 5 convolution and 3 × 3 pooling operation, and finally conduct image mosaicing along the channel. This design increases the width and depth of the network, reduces the parameters and computation. Inception v2 divides 5 × 5 convolution into two 3 × 3 convolution, and uses Batch Normalization (BN) to speed up network convergence and to some extent, reduces the risk of network overfitting. Inception v3 adopts asymmetric convolution to decompose N × N convolution into 1 × N convolution and N × 1 convolution, and adds feature channels in front of the pooling layer. Each layer of Inception v3 uses a different structure, so its costs are high. In 2014, VGG further proved that the depth of network structure can improve the accuracy of image classification.

In 2015, ResNet [17] adopted a deep residual structure to solve the problem of gradient disappearance in the deep network by learning the "residual", and achieved a high classification accuracy. Before ResNet, the accuracy was usually improved by stacking the depth of the network; but simply increasing the number of network layers would lead to gradient disappearance and gradient explosion. The main idea of ResNet is to add a direct connection channel in the network, allowing a certain proportion of the output of the previous network layer to be retained, and the original input information to be directly transmitted to the subsequent layers. The first 1 × 1 convolution of ResNet's basic Bottleneck module reduces its 256-dimension channels into 64-dimension ones, then, it does a 3 × 3 convolution operation, and finally, recovers through 1 × 1 convolution. In general, it greatly reduces the number of parameters compared to other networks.

Since then, excellent classification networks such as Wide ResNet [18] and ResNeXt [19] have been proposed one after another. The design idea of advantageous modules has greatly promoted the development of image classification techniques.

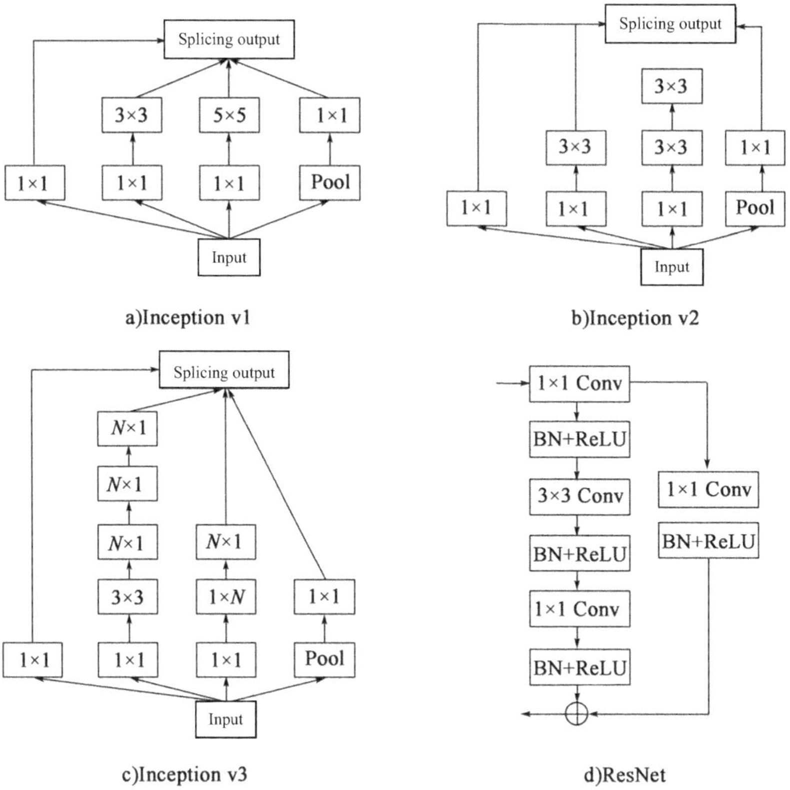

Fig. 7.9 Diagram of inception network structure [16]

Wide-ResNet improves on the basis of ResNet by increasing the number of convolution kernels in each residual block, replacing the previous 1 × 1 convolution → 3 × 3 convolution → 1 × 1 convolution with two 3 × 3 convolutions. ResNeXt learnt from the idea of group convolution. Each module adopts the operation of "partition–transform–integration", and divides the input features into N parts along the channel (N = 32/64). Then, each branch independently carries out the operation of 1 × 1 convolution → 3 × 3 convolution → 1 × 1 convolution, and finally adds the results of each branch.

(2) Object Detection

The task of object detection is to find out the objects of interest in the image or video, and identify their positions and sizes at the same time. Object detection is one of the core issues in the field of machine vision. Since the introduction of deep learning, the development of object detection has mainly focused on two topics: Two Stage algorithm (such as R-CNN [20] series) and One Stage algorithm (such as SSD). The main difference between the two algorithms is that the Two Stage algorithm (represented by Faster R-CNN) needs to first generate Proposal (a candidate box

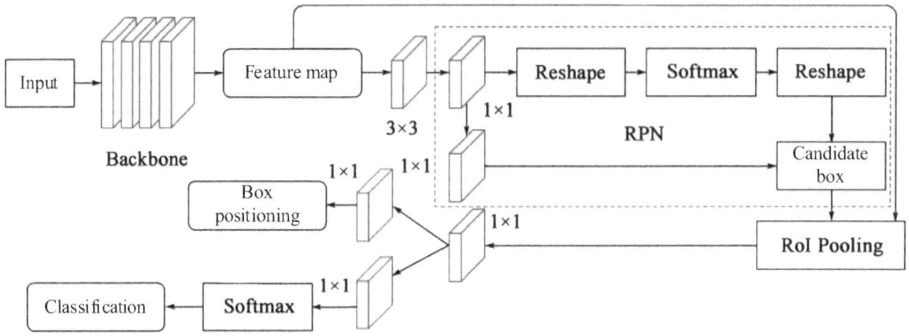

Fig. 7.10 Diagram of faster R-CNN network structure

that may contain the object to be detected), and then perform fine-grained object detection, while the One Stage algorithm (represented by SSD) will directly extract features in the network to predict the classification and location of the objects.

Faster R-CNN

The whole Faster R-CNN can be divided into three parts: feature extraction network (Backbone), Region Proposal Networks (RPN), and RoI Pooling and classification network, as shown in Fig. 7.10.

Feature extraction network: Faster R-CNN uses convolutional neural networks such as VGG and ResNet to extract the features of the whole picture. The basic structure of convolutional neural network generally includes convolution layer, pooling layer and activation layer. Taking VGG16 as an example, via removing the fully connected layer and only retaining the convolution layer, there will be 13 convolution layers,13 activation layers and 4 pooling layers. A matrix with the original size of M × N will definitely have a size of M/16 × N/16 after it is processed by neural network, so that the feature map extracted by the convolution network can have one-to-one correspondence with the coordinates of the original map.

Region Proposal Networks (RPN): RPN network is actually divided into two branches. The upper branch obtains positive samples (object) and negative samples (background) through Softmax classification Anchors. The lower branch uses 1 × 1 convolutional regression corresponding to the coordinate offset of Anchors. The final Proposal layer is responsible for synthesizing positive samples and corresponding coordinate offset to obtain the candidate object boxes. It is also responsible for eliminating those candidate object boxes that are too small or beyond the boundary and using non- maximum suppression for filtering.

RoI Pooling and classification network: RPN will generate candidate object areas with inconsistent sizes in the feature map, while the classification network contains a full connection layer, the input of which is fixed, and the input size of the classification network in Faster R-CNN is fixed at 7 × 7. Therefore, for the input feature map of any size, a 7 × 7 grid will be used to cover the original area, and the Max Pooling value of the current grid coverage area will be obtained for each grid of 7 × 7, so that the candidate area of any size will be pooled into a size of 7 × 7. The feature map of the candidate area after RoI Pooling operation will be sent to the classification branch respectively for classification and sent to the regression branch for coordinate prediction.

Loss function: The output of the two stages of Faster R-CNN, RPN and subsequent classification networks are both coordinate regression values and classification values, so the loss functions of the two networks can be expressed by the following formula:

$$L(p_i,t_i) = \frac{1}{N_{\text{cls}}}\sum_i L_{\text{cls}}(p_i,p_i^*) + \lambda\frac{1}{N_{\text{reg}}}\sum_i p_i^* L_{\text{reg}}(t_i,t_i^*) \tag{7.1}$$

where, L_{cls} is the classification loss function, and different loss functions can be used for binary classification and multi classification; λ is a factor used to balance the proportion of regression and classification losses; L_{reg} is a regression loss function and only the sample box is positive, p_i* is the control in front to generate the regression loss of the detection box, and L_{reg} regression loss function uses Smooth L_1 loss function.

Training process: The convolution layer of feature extraction is collectively called Backbone, and the training process is divided into the following steps. (1) Backbone uses the initialization weight of the ImageNet pre-trained model, and uses 256 sampled positive and negative Anchors boxes to start training the RPN network. Backbone weights also participate in fine-tuning. (2) Use the RPN trained at first stage to generate a positive case prediction box for Fast R-CNN classification network training. At this time, the Backbone weight value is also initialized using the ImageNet pre trained model. In the second step, RPN and Fast R-CNN use two Backbones without sharing. (3) Using the corresponding Backbone weight of the Fast R-CNN network trained in the second step to initialize the Backbone before the RPN network. The RPN part is initialized using the training result of the first step. The third step only fine-tunes the weights in the RPN. In the third step, Fast R-CNN and RPN began to share Backbone. (4) The weights of Backbone and RPN do not change any more. The Fast R-CNN is initialized by partial results of the Fast R-CNN trained in the second step. Fine-tuning training is performed again for real end-to-end training.

SSD

Single Shot MultiBox Detector (SSD) [22] uses a CNN network for detection but uses a multi-scale feature map, whose network architecture is as shown in Fig. 7.11. Using VGG16 as feature extraction network, SSD replaces the full connection layers FC6 and FC7 of VGG16 with convolution layers Conv6 and Conv7, and changes the pooling layer Pool5 from the original stride = 2 to stride = 1, which expands the convolution's field of vision exponentially without increasing the parameter and model complexity. Then, remove the Dropout layer and the FC8 layer, and add a series of new convolution layers.

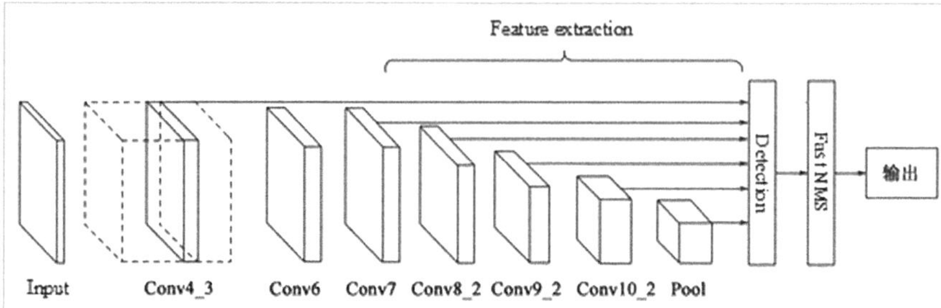

Fig. 7.11 Framework of SSD network structure

The core design philosophy of SSD is summarized in the following three points:

(a) Multiscale feature map is used for detection. SSD performs object classification and location regression on the feature maps with different sizes of Conv4_3/Conv7/Conv8_2/Conv9_2 respectively. The feature maps with small receptive field of view are used to detect small objects while the feature maps with large receptive field of view are used to detect larger objects. The accuracy is so improved.

(b) Convolution is used for detection. SSD directly uses convolution to extract detection results from different feature maps. For feature maps with shape of $M \times M \times P$, a relatively small $3 \times 3 \times P$ convolution kernel can be used to obtain the detection value.

(c) Set prior boxes. SSD outputs a set of independent detection values for each prior box of each Anchor unit, corresponding to a bounding box, which can be divided into two main parts. The first part is the confidence or score of each category. SSD takes background as a special category. If there are c categories of detection objects, SSD actually needs to predict $c + 1$ confidence values, among which the first confidence refers to the score without object or belonging to the background. The second part is the location information of the bounding box, including four values (cx,cy,w,h), representing the central coordinates, width and height of the bounding box, respectively.

The loss function in SSD is defined as the weighted sum of location error L_{loc} and confidence error L_{conf}:

$$L(x,c,l,g) = \frac{1}{N}[L_{\text{conf}}(x,c) + \alpha L_{\text{loc}}(x,l,g)] \tag{7.2}$$

where, N is the number of positive samples of prior box; c is the predictive value of category confidence; l is the location prediction value of the bounding box corresponding to the prior box; and g is the location parameter of the Ground Truth. The weight coefficient is set to 1 through cross validation.

(3) Segmentation

The segmentation task is to find a segmentation map from RGB color images or grayscale images, including the category annotation of each pixel. At present, the most popular image segmentation model is the encoder—decoder structure. In the encoder, the spatial resolution of input is reduced by down sampling, thus generating a feature map with low resolution. The decoder then up-samples these feature descriptions and restores them to the full-resolution segmentation image. In 2014, for first time in the field of semantic segmentation, Fully Convolutional Networks (FCN) [23] replaced the last full connection layer of the classification network with corresponding convolution operations, and realized pixel level classification of images, that is, semantic segmentation. FCN is the pioneering work of semantic segmentation. Inspired by FCN and Faster R-CNN, in 2017, Mask R-CNN [24] expanded on the basis of Faster R-CNN, added new semantic segmentation FCN branches, proposed to simultaneously detect and segment object boxes, and realized instance segmentation. Mask R-CNN has also become a classic network for instance segmentation.

FCN

As shown in Fig. 7.12, the overall FCN network [23] structure is divided into two parts, namely, the full convolution part and the deconvolution part.

Full convolution feature extraction: The full-connection layer in CNN network is converted into convolution layer to extract features. Take AlexNet as an example, assuming that the input of the network is 512 × 512 × 3 images, the original AlexNet uses two full-connection layers with the size of 4096, and the last full-connection layer with 1000 neurons is used to calculate the classification score. FCN replaces the full connection layer with a 1 × 1 convolution layer, which is equivalent to applying a full connection layer to each pixel to predict the category results from 4096-dimensional features.

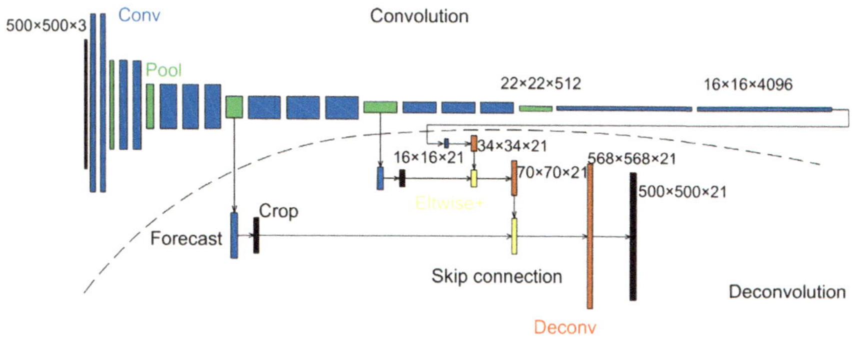

Fig. 7.12 Diagram of FCN network structure [25]

Deconvolution up-sampling: After multiple convolution and pooling, the resulting image becomes smaller continuously with lower and lower resolution. When the image reaches $H/32 \times W/32$, which is the smallest layer, the generated image is the most important high-dimensional feature image. Here, FCN uses up-sampling (deconvolution) to restore the image to the original resolution.

Skip connection structure: After the fifth pooling layer of the convolutional network, the resolution of the image becomes 1/32 of the original. If deconvolution is used to directly up-sample the size of the original image, the results obtained will not be accurate enough, and some details cannot be recovered. Therefore, amplify the output of the fourth layer and the third layer with 16 times and 8 times deconvolution, respectively, and add the obtained results together to make the final results more accurate. In this way, both local and global information are taken into account.

The training process of FCN is divided into four stages: ① Initialize the classical feature extraction network (such as VGG, etc.), and replace the last two full connection layers with the convolution layer; ② The segmentation mini-graph ($16 \times 16 \times 21$) is predicted from the feature mini-graph ($16 \times 16 \times 4096$) and then directly up-sampled into the large image. The deconvolution step size is 32, and the network obtained by this step is called FCN-32s; ③ The up-sampling is completed in two times. Before the second up-sampling, the prediction results of the fourth pooling layer are integrated, and the accuracy is improved by using a skip structure. The step size of the second deconvolution is 16. The network obtained in this step is called FCN-16s; and ④ The up-sampling is completed in three times, and the prediction results of the third pooling layer are further integrated. The step size of the third deconvolution is 8, and the network obtained in this step is denoted as FCN-8s.

Inspired by FCN, various semantic segmentation networks have emerged one after another. In 2015, the U-Net convolutional neural network [26] improved the above full convolution structure by expanding the capacity of the decoder module. On the basis of FCN, it adopted a symmetrical coding and decoding structure design. In the up-sampling stage, the output of the previous layer and the next layer were connected to the convolution layer after channel dimension splicing to achieve feature fusion of low and high dimensions, which achieved high accuracy in the field of medical image segmentation. In 2015, DeepLab series networks [27] were proposed successively. DeepLab v1 changed the Stride step size of Pool4 and Pool5 layers of VGG network from 2 to 1, thus making the sub-sampling multiple change from 32 to 8 and retaining the details. DeepLab v2 innovatively put forward the Atrous Convolution and the Atrous Spatial Pyramid Pooling (ASPP) module to expand the perception vision and capture the image context content information. In 2016, RefineNet [28] refined the low resolution semantic features in a recursive way to generate a semantic feature map with high resolution. All modules use the long range residual connection of direct mapping to combine the information lost in the down sampling process, so as to improve the high-resolution prediction results. In 2017, Large Kernel Matters convolutional network [29] introduced the residual module of boundary refinement, thus solving the "local" and "classification" problems and further improving the positioning results near the object boundary.

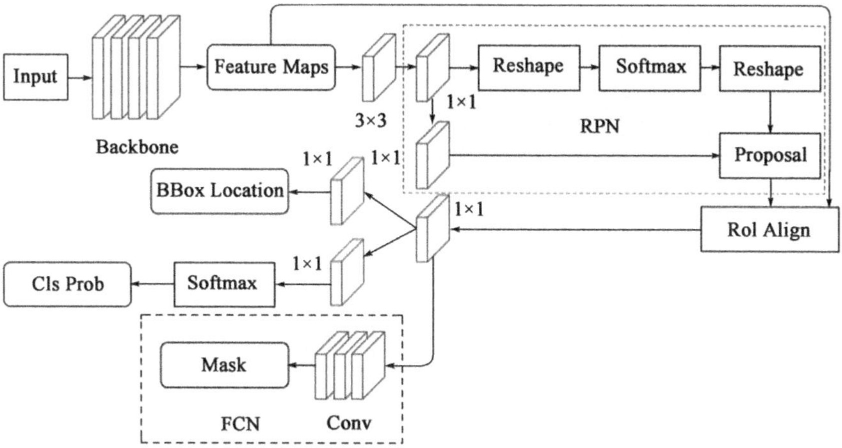

Fig. 7.13 Diagram of mask R-CNN network structure [30]

Mask R-CNN

Adding semantic segmentation branches, Mask R-CNN [30] expanded from Faster R-CNN [21], and can therefore detect and segment the object box at the same time. The overall structure is shown in Fig. 7.13. Mask R-CNN uses RoI Align Pooling to replace RoI Pooling in Faster R-CNN to improve the accuracy. FCN branches are added to perform semantic segmentation on each candidate box with classification and coordinate regression performed at the same time.

RoI Align pooling: Compared with RoI Pooling, RoI Align uses the bilinear interpolation method to obtain the image values on the pixel points whose coordinates are floating points, thus transforming the entire feature aggregation process into a continuous operation: ① Traverse each candidate region, leaving the floating-point boundary unquantized; ② Divide the candidate region into 7×7 units, and the boundary of each unit is not quantified; ③ Calculate the positions of four fixed coordinates in each unit. Using bilinear interpolation method to calculate the values of these four positions, and then carry out the maximum pooling operation.

FCN branch: The RoI feature map generated by RPN network is $7 \times 7 \times 1024$. Due to the previous multiple convolution and pooling, the corresponding resolution is reduced. The FCN branch first uses deconvolution to improve the resolution and reduces the number of channels to $14 \times 14 \times 256$ at the same time, and finally outputs a $14 \times 14 \times N$ Mask template. N is the number of categories.

Due to the addition of FCN branches, the loss function adds segmentation loss on the basis of Faster R-CNN, which can be formulated as:

$$L = L_{\text{cls}} + L_{\text{box}} + L_{\text{mask}} \tag{7.3}$$

where the definitions of L_{cls} and L_{box} are the same as those of Faster R-CNN. For L_{mask}, assuming a total of K categories, the output dimension of the Mask segmentation branch is $K \times M \times M$. K binary masks will be output for each point in the $M \times M$ feature map (with Sigmoid output for each category). Mask R-CNN is a very flexible framework, which can add different branches for different tasks, and can complete multiple tasks such as object classification, object detection, semantic segmentation, instance segmentation, human posture recognition, etc.

7.3 Applications for Bridge Maintenance

7.3.1 Overview

Different from the conventional sensors such as contact sensor, GPS, radar, etc., the data acquired by visual sensor corresponds to the content observed by the maintenance personnel. In addition, the information contained in a single collection is richer and more thorough from the visual sensors. Consequently, the implementation of visual sensors can significantly minimize the amount of human resources required for bridge operation and maintenance, and increase the effectiveness of bridge operation and maintenance. In bridge engineering, computer vision is utilized primarily for the following aspects:

Damage identification: Bridge damage identification based on computer vision has received the attention of relevant scholars for a long time. It has been proven that it is possible to identify the surface defects of bridge structures using images. In order to achieve better recognition effect, researchers have enhanced the damage recognition method based on computer vision from the aspects of acquisition equipment and processing algorithms. Visual sensors installed on various platforms, including robots and unmanned aerial vehicles, are used for image acquisition, and computer vision technologies and artificial intelligence algorithms are combined to automatically identify cracks or defects.

Displacement measurement and vibration monitoring: Vision-based displacement and vibration monitoring techniques have been utilized in civil engineering for over a decade. Typical visual vibration measurement methods include optical interference-based vibration measurement method, target tracking approach and digital image correlation method.

Modal identification: Modal parameter identification method based on computer vision has become a research hot topic in recent years. The structural modal parameters can be directly extracted according to photogrammetry without the pretreatment of structural surfaces. The development of this technology provides a new way for structural dynamic monitoring.

Load identification: Load is an important input for monitoring bridge operation and maintenance. The vehicle loads on bridges can be measured by identifying vehicle targets in order to assess the bridge conditions.

7.3.2 Damage Detection

(1) Cracks of Concrete Bridges

Throughout the concrete bridge operation, the structural surfaces will be subjected to varying degrees of cracks, fall-off, flaws and other damages due to the environmental temperature variations, long-term load stress imbalances, corrosion and concrete shrinkage. Among them, crack is a common damage affecting the bridge operation, maintenance and management that significantly reduces the bridge's bearing capacity and longevity. Components started to be damaged when the crack evolved and grew. Therefore, regular monitoring detection for bridge cracks is crucial for bridge maintenance and management. With the development of digital image processing techniques and automatic image acquisition equipments, various effective non-contact bridge crack detection techniques have emerged. Deep learning in combination with UAV has gradually become the preferred method for bridge crack detection, especially for medium- and long-span bridges.

(2) Crack Detection

With the development of computer graphics technology, applications of deep learning in image classification, detection and semantic segmentation have been developed rapidly in the field of crack detection. The employed networks belong to the common ones. Deep learning-based crack detection methods can be classified into pixel-level and regional-level categories: ① crack location; and ② crack segmentation. As for crack location, the early method is to establish the CNN model to classify and identify the crack in the dataset. Then, the sliding window method is adopted to divide the crack images into regular small grid, and then the results are scanned to obtain the rough location of the crack region. Cha et al. used the convolutional neural networks for the first time to detect concrete crack damage, as shown in Fig. 7.14. The five-layer CNN model is established to roughly identify and locate the cracks using sliding window inspection [31]. Li et al. [32] created a multi-scale crack classification model named Deep Bridge Crack Classify (DBCC) to detect bridge cracks, using an image pyramid and a Region of Interest (ROI) to improve the performance of the sliding window algorithm. With the development of object detection technology, the object detection and area location are combined into one process, greatly improving detection efficiency. Wu et al. [33] proposed a Faster R-CNN network using a Region Proposal Network (RPN) instead of the sliding window method, to create the candidate regions for crack identification in tunnel images. This method can partly solve the problem of low quality tunnel images and realize the rapid detection, accurate location and labeling of tunnel image cracks. As for crack

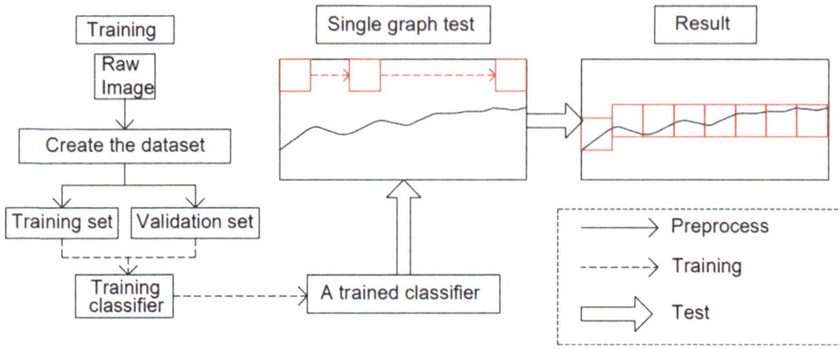

Fig. 7.14 Flowchart of crack detection technique using CNN and slide windows

segmentation, the main process is to replace the fully connected layer of classification network with the convolution layer and perform deconvolution up-sampling on the low-resolution feature map to generate a probability map with the same resolution as the original image to achieve classification at the pixel level. Wang et al. [34] introduced the fully convolutional neural network into image crack detection to establish the crack FCN model, which can solve the loss of local information and improve the ability of refining the distinction of the FCN model. The proposed approaches include increasing the resolution and depth of the FCN network as well as adding a higher-scale deconvolution layer after the network to expand fine local details. Ji et al. [35] adopted U-Net convolutional network to realize pixel level and small sample crack inspection. This method used multi-layer convolution to extract crack features automatically. It utilized the superposition of shallow and deep networks to realize the fusion of local and abstract features, which can well preserve the detailed features of cracks.

(3) Crack Measurement

The crack image, i.e., a geometric figure in nature, has a series of geometric features. To take into ful account the various parameters of the cracks, the calculated geometric properties of the cracks, such as crack length, width, and area, are discussed in this section. Moreover, the crack width refers to the crack width of any position, the average crack width, and the maximum crack width.

The common calculating method for crack length is the method based on skeletonization operation. Its main operation contains two steps: crack skeletonization operation and pixel statistics. The skeletonization operation transforms the crack into a thin line with a single-pixel width. Within this context, the crack length is determined by the number of pixels along the crack skeleton line in the statistical diagram.

The width of the crack includes the average width, the maximum width and the width at any point. The most popular methods for calculating crack width are the horizontal method, the minimum distance method and the triangle approximation approach. Firstly, using specific techniques, the horizontal method changes the crack direction to vertical direction. Then, a horizontal line is adopted to intersect the crack from the top to the bottom to measure the crack width at each intersection [36]. The minimum distance method calculates the crack width by finding the minimum distance, and the crack width at any point is calculated via obtaining the fracture skeleton line. At a certain point on the skeleton line, the nearest point with a gray value of 0 is found on both sides, and the sum of the minimum distance on both sides is considered as the crack width. The triangle approximation method is an innovative fracture width measurement method, using horizontal and vertical lines passing through a point to intersect the crack. Then, the horizontal and vertical widths are calculated to form a triangle. Furthermore, the triangle's hypotenuse length is the crack width, calculated by the Pythagorean theorem.

For the tiny crack in an image, the conventional crack width measurement method relying on the number of pixels is difficult to meet the accuracy requirements. Professor Zhang's team from Southeast University [37] developed a crack width measurement technique based on Zernike orthogonal moment. When Zernike orthogonal moment is used to calculate the line width, it is assumed that the two edges of the line width of the crack are ideal steps. Hence, the line width calculation model based on the Zernike orthogonal moment can be illustrated in Fig. 7.15, where g is the background grayscale value and h is the grayscale step value. Thus, $g + h$ is the line width gray value and $l_1 + l_2$ is the line width value. As a result, Fig. 7.15b is the result of rotating the line parallel to the Y-axis. The width of the left part of the Y-axis is l_1, and the width of the right part is l_2. The orthogonal moment is obtained by revealing l_1 and l_2, respectively, then by adding them. Figure 7.15b is obtained by rotating Fig. 7.15a by the degree ψ in a clockwise direction. In order to solve l_1 and l_2 respectively, the left and right halves of the Y-axis in Fig. 7.15b are mirrored with the Y-axis as the axis of symmetry. The symmetrical line widths obtained from mirroring Fig. 7.15b are shown in Fig. 7.15c, d, and the same ideas apply to their solution. Since the image is made up of discrete pixels, the column with the highest cumulative gray value of the line width after rotation (the brightest column) is picked as the central column. The separate mirror operation resulted in two symmetrical images. The rotated image's second- and fourth-order Zernike orthogonal moment $f'(x, y)$ are used to determine the line width $2l$. Then, l_1 and l_2 are calculated, respectively. The crack width at the site is then equal to $l_1 + l_2$.

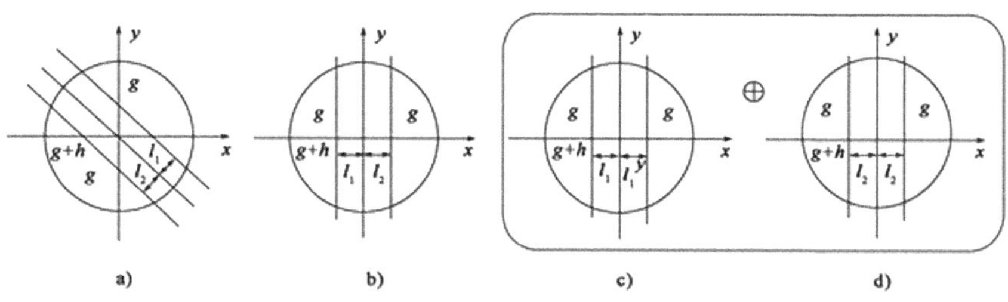

Fig. 7.15 Zernike-moment of thin-crack width [37]

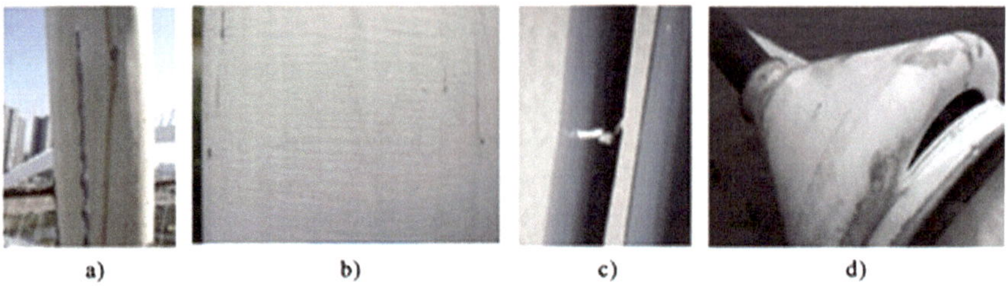

Fig. 7.16 Typical apparent damages of cable outer protective sheath

(4) Cable Damages

As the main bearing component of cable-stayed bridges, performance of the cable is crucial to the safety and longevity of the bridge. The cable comprises three parts: anchor, cable body and protection device. The types of protection devices contain sheaths, waterproof covers and others. The main damage types of the sheath are cracks, holes, pits, chaps and so on, as shown in Fig. 7.16a–c. Main damage to the waterproof covers is the fracture at the connection between the cable body and the lower anchor head, which is the primary manifestation of waterproof cover cracking.

The existing detection methods for cable jacket apparent damage include manual visual detection, bridge inspection vehicle detection, UAV detection and cable inspection robot detection. Manual visual detection refers to the inspectors on the bridge deck using high-power telescope or camera to observe the cable surface from a distance. This method does not need to rely on large mechanical equipments, and the implementation is much more convenient compared with other methods. However, the cable jacket surface is distributed with joints and wires, and the subjective error of manual visual judgment is therefore large. Bridge detection vehicle inspection refers to the method that relies on the vehicle with lifting bracket that can carry the bridge inspectors to move along the cable direction to inspect the cable. This method can check the apparent distresses of the cable jacket clearly and accurately, and can also facilitate the inspector to clean and maintain the cable jacket. However, the cost of bridge detection vehicle is high, so this method is generally used for the detection of medium and large inclined bridge cables. UAV detection means that the inspectors use UAV to photograph the cable along the direction of the cable. Existing commercial UAVs (such as DJI) can realize the path planning cruise, and are lightweight, flexible and highly maneuverable. However, the UAV generally has to keep a safe distance from the target (greater than 2 m). It cannot photograph close to the cable, resulting in low resolution of the picture shot. It cannot photograph around the loop of the cable either. Therefore, it is only suitable for quick cable detection of large cable-stayed bridges. Cable inspection robot detection refers to the development of a special robot which can climb along the cable direction at a certain speed and simultaneously photograph, transmit and recognize the surface defects of the cable jacket with multiple cameras. This method can comprehensively capture the cable jacket's apparent information and can fulfill other tasks, such as embedding a magnetic flux

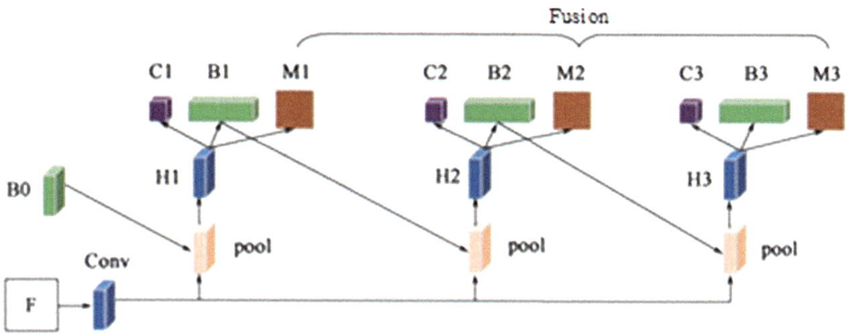

Fig. 7.17 CMR network structure [38]

sensor to probe the internal condition of the cable. It is currently the most suitable approach for addressing cable detection.

The authors [38] proposed a new network CMR (Cascade Mask RCNN) in the Mask. Based on RCNN, deformable convolution is introduced to increase the sensory field of view. Three cascades of RCNN heads are set and gradually increased threshold values are adopted. The scoring mechanism of the Mask part is thereby improved, that is, the IoU index is used for scoring rather than directly consistent with the score of the coordinate box, to improve the accuracy of positioning and identification of cable distress. For the extracted distress area, the physical size was measured by combining the skeleton extraction algorithm and camera parameters. The CMR network structure is displayed in Fig. 7.17.

The authors' team [38] obtained 7,040 original photographs of cable covers with evident damages using a cable detection robot. These images were all 1408 pixels × 1152 pixels. In order to retain more defect information, the captured image was reduced into a smaller image with resolution of 512 pixels × 512 pixels, which contained minor cracks or scratches, and were divided into training dataset, verification dataset and test dataset, according to the ratio of 8:1:1.

After obtaining the mask information of the distress, it is necessary to estimate the pixel width and length of the distress to facilitate the evaluation of cable maintenance. In practice, the real length and width of the affected area can be roughly estimated by combining the focal length of the camera and the distance between the camera and the outer surface of the cable. Estimating the physical size of the distress area is divided into two steps: apparent distress skeleton extraction and pixel feature extraction. Pixel feature extraction includes length and width extractions, as shown in Fig. 7.18. Skeleton extraction is the main step for estimating apparent damaged pixels' features, and there are many existing methods available for use. Here, a fast centerline extraction method based on template matching is used. The basic idea is to determine the removal of 8 fields of each pixel. The center line is essentially a thin line with a single pixel width. In this setting, the number of pixels in the center line is the pixel length of the apparent distress area (Fig. 7.19).

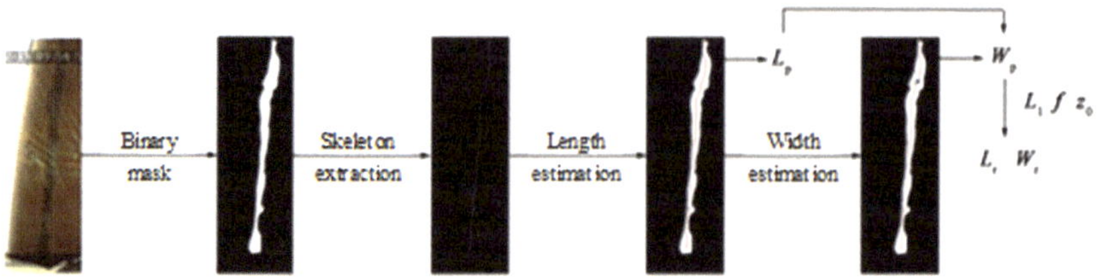

Fig. 7.18 Flow chart for physical feature extraction of apparent damages [38]

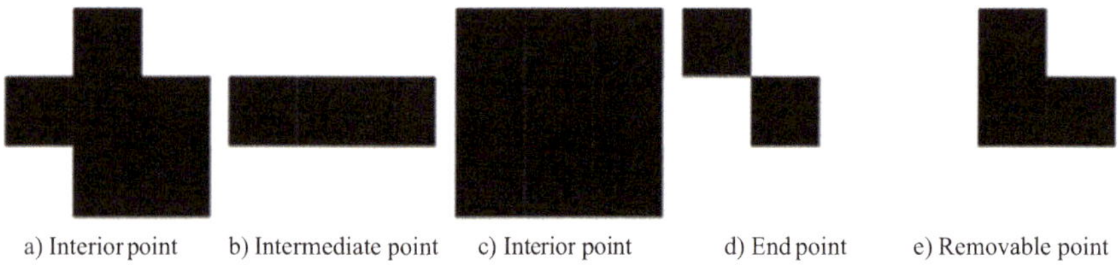

Fig. 7.19 Distribution map of 8 field points of one pixel

The eight direction traversal method is used to compute the pixel width. First, determine the starting seed point of measurement according to the center line, and search along eight directions starting from this seed point. If it is an apparent distress area point, the pixel width in this direction will be increased by one until it reaches the edge of the pixel, and continues searching for the next direction. The angles in the eight directions are 0°, 45°, 90°, 135°, 180°, 225°, 270° and 315°. Then, these pixel widths are combined along the bisector directions of the four quadrants, and the minimum value is taken as the final pixel width. Based on the focal length of the camera and the distance from the camera to the outer surface of the stay cable, the true length and width of the apparent distress area can be calculated by the following formula:

$$L_r = \frac{z_0 \times L_1}{f},\ W_r = \frac{z_0 \times W_1}{f} \tag{7.4}$$

where L_1 is pixel length; W_1 is pixel width; z_0 denotes the distance from the camera to the surface of the inclined cable; and f represents the focal distance of the camera.

Figure 7.18 depicts the entire procedure. The physical feature extraction method described above can be used to process the CMR model's segmentation results, as shown in Table 7.1. This method can precisely determine the center line and the maximum width of the damaged region.

(5) Corrosion of Steel Bridges

The authors' team [39] trained the convolutional neural network model (GoogLeNet [16]) using photos of the bridge damages obtained from the recent bridge inspection

Table 7.1 Some physical feature extraction results for distress areas [38]

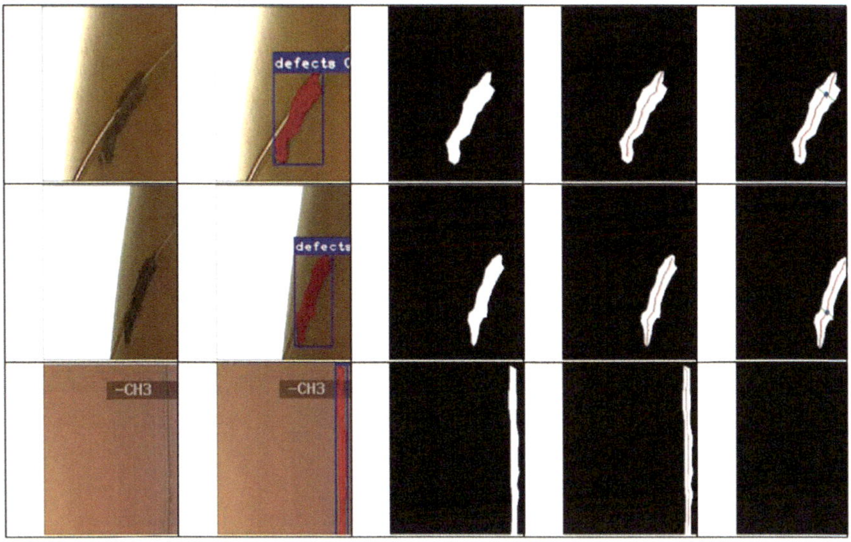

reports in Japan, evaluated the accuracy of the damage inspection machine and tested the image of the steel bridges taken by UAVs. During the database preparation stage, damage images were first taken from the inspection report and manually labeled into several damage categories. Following this, image processing was used to supplement the data. Schematic diagram for constructing the dataset is displayed in Fig. 7.20.

The trained model is used to identify the damages in the UAV's image of a steel bridge. Figure 7.21 displays schematic diagram of the identification results. As apparent from the figure, many damages were successfully identified, but there were also many misidentifications typically brought about by overfitting of the convolutional neural network model due to the limited training dataset.

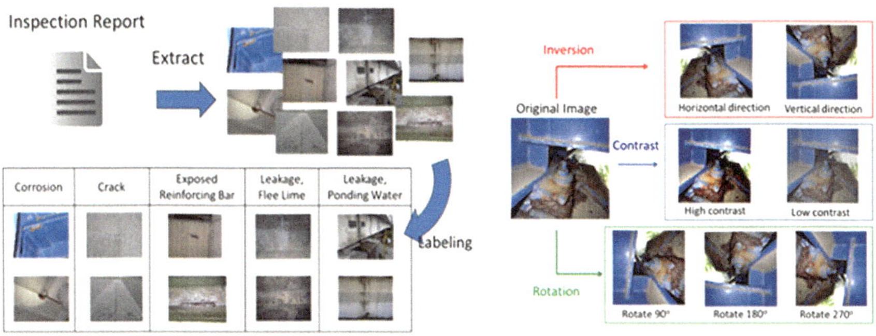

Fig. 7.20 Schematic diagram for constructing the dataset [39]

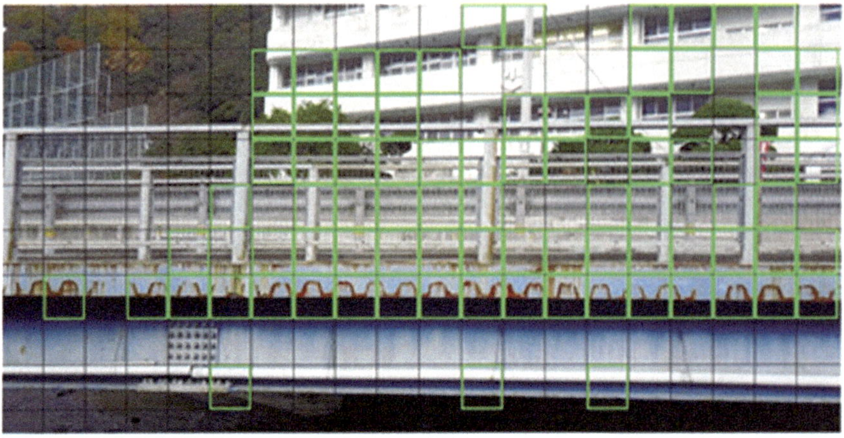

Fig. 7.21 Corrosion identification of steel bridge based on image classification [39]

In the subsequent study [40], the authors made manual annotations on corrosion images of steel bridges to form datasets. They proposed an image segmentation method based on "clip-collage" to improve the accuracy of steel bridge damage detection. Figure 7.22 shows the schematic diagram of network training. The steel bridge corrosion picture and its corresponding label are trained in the FCN network with the same clipping method. In the prediction stage, the clipped picture is predicted, and then the predicted image is spliced with the opposite clipping pattern. Figure 7.23 shows the identification result of a steel bridge.

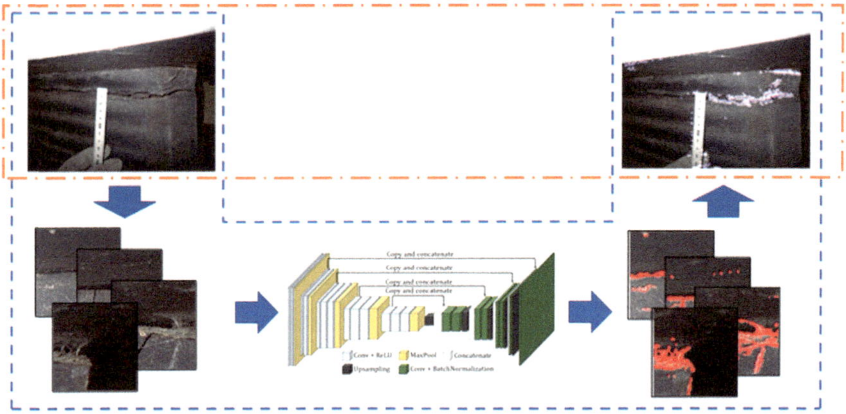

Fig. 7.22 Schematic diagram for image segmentation technique based on "clipping and junction" [40]

Fig. 7.23 Corrosion identification of steel bridge by image segmentation technique based on "clipping and junction" [48]

(6) Bearing Damages

With the continuous expansion of the road network in our country, the number of bridges is increasing year by year, and the bridge maintenance workload is huge. The bridge bearing is an important component connecting the upper and the lower structures of the bridge. After the bearing damage occurs, it will affect the bridge structure's mechanical state and the traffic safety if it is not found and treated in time. Bridge bearing inspection is an indispensable link in bridge maintenance, and the primary way of current inspection of bridge bearings is manual detection. This method is not only time-consuming and laborious, it will also affect the traffic.

The bridge bearing can transfer loads and provide moderate constraints. Its role is to effectively transfer various loads of the bridge superstructure to the bridge substructure, and to make the bridge superstructure rotate freely for the necessary free deformation; Bridge rubber bearing should have the ability of vertical bearing, rotational deformation, shear deformation, or sliding under the action of bridge load. Figure 7.24 depicts the functions of the bridge bearing.

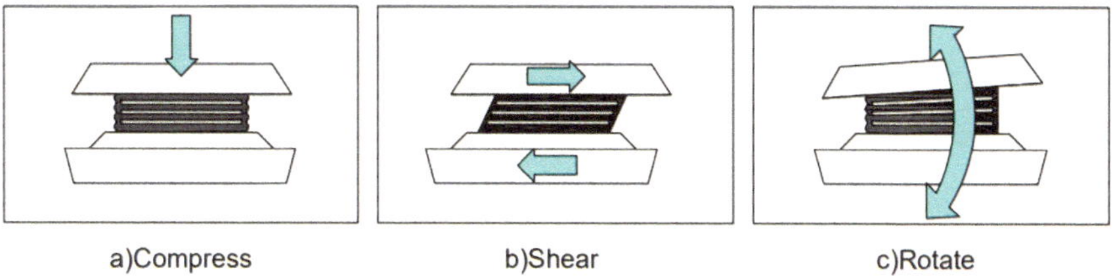

Fig. 7.24 Schematic diagram for functions of bridge support [41]

Based on the image processing technology of artificial intelligence deep learning, the authors [42] conducted a study on the automatic recognition of bridge-bearing damages. According to the characteristics of images of bridge bearing, the training database is amplified utilizing image processing on the prerequisite that the damage characteristics of bridge bearing do not change. Image processing methods mainly include horizontal inversion, vertical inversion, counter-clockwise rotation, clockwise rotation and use of a principal component analysis algorithm to adjust the color channel of the image, which provides the basis for training of a convolutional neural network. Table 7.2 shows the dataset of bearing damages.

The convolutional neural network structure uses the classical VGG16 network, and its output layer is adjusted to adapt to the classification of bearing distresses. First layer of the network is the input layer. VGG16 network requires the resolution of the input image to be 224 pixels × 224 pixels. The image is encoded into a vector through convolution, pooling and other operations, and this vector is input into the full connection layer to obtain the required classification results. In the classic VGG16 network, a BN layer is added after the full connection layer, except for the output layer, making the network training easier to converge. The network training uses transfer learning to improve the recognition accuracy of the model. The network pre-training obtains the weight of the convolutional layer of the network in the ImageNet dataset. Figure 7.25 shows the curve relationship between the prediction accuracy of the network and the number of network iterations when the network is re-trained to all fully connected layers.

The authors [43] also carried out relevant research on bearing crack segmentation technology. The dataset was marked by artificial pixel level, and the images in the dataset were clipped into small pictures to train the full convolutional neural network (FCN). The trained FCN model can mark the bearing cracks, and the convolutional neural network (CNN) is used to pre-classify the bearing and background to improve the identification accuracy of the FCN model.

CNN model and FCN model are trained, respectively. Figure 7.26 shows the network training diagram of CNN and FCN. In the prediction stage, the convolution neural network classification model first distinguishes the bearing from the background, and then the FCN model marks the bearing cracks. Figure 7.27 is a

Table 7.2 Support distress dataset

Classification	Number of images in the original training dataset	Number of images in the expanded training dataset	Number of images in the verification dataset
Normal support	310	1860	78
Shear deformation	308	1848	77
Circumferential cracking	309	1854	77
Total	927	5562	232

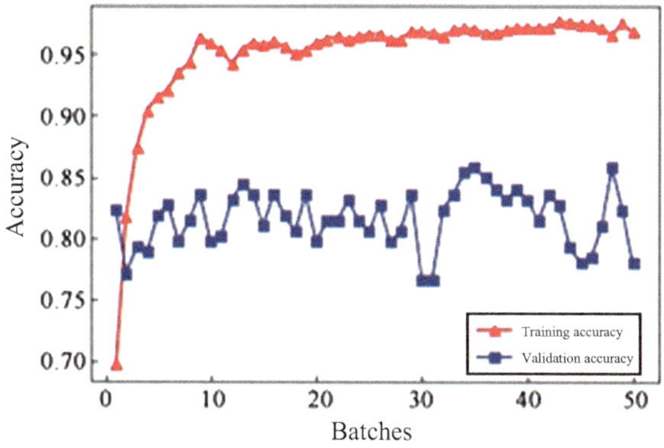

Fig. 7.25 Curve correlation between prediction precision of the network and the number of network iterations

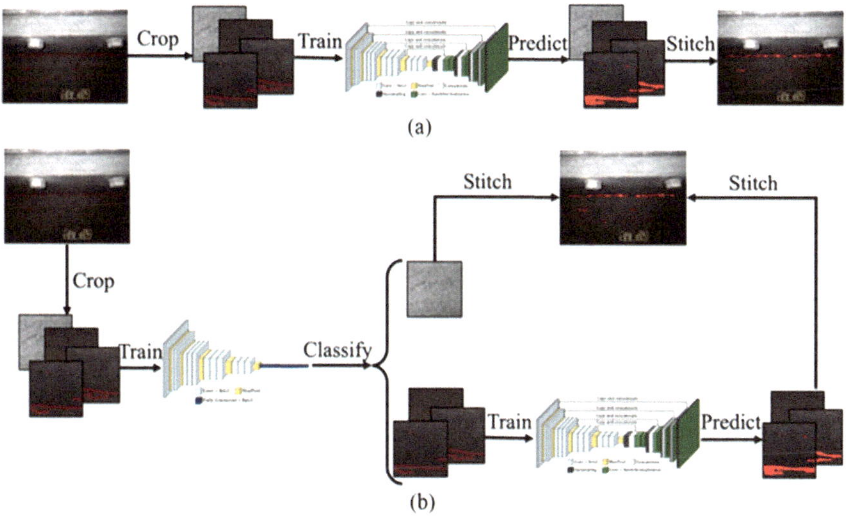

Fig. 7.26 Schematic diagram for training of CNN and FCN [43]

schematic diagram of the prediction results. It can be seen from the figure that CNN model can improve the prediction accuracy of the network. However, accuracy of the CNN model itself will also have an effect on the model's final prediction results.

(7) Apparent Damage to Underwater Concrete Pier Columns

In order to solve the bottleneck problems of underwater structure detection, such as weak detection ability, low performance and poor detection effect, Professor Wu Gang of Southeast University, China, designed and developed a new concrete pier column apparent damage detection sensor, Contact-Visual Sensing Device (CVSD),

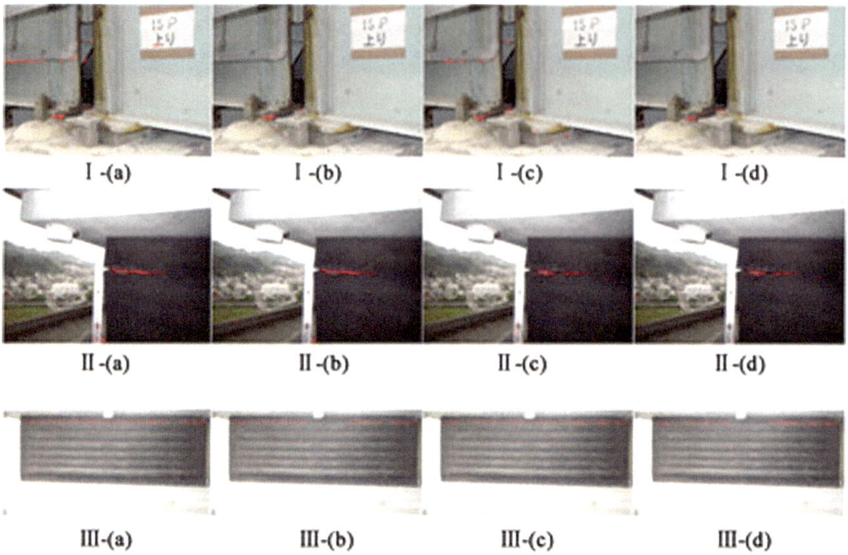

Fig. 7.27 The schematic diagram for the predicted results. **a** FCN/40 epoch; **b** CNN + FCN/140 epoch; **c** FCN/150 epoch; **d** CNN + FCN/150

which is associated to a working machinery for measurement in extreme underwater environments. Details of this device are described in Chap. 4. A large number of test results obtained in the real underwater environment were simulated, and showed that the equipment can realize rapid identification and quantitative analysis of concrete surface distresses (accurate to 0.1 mm cracks) in turbidity or similar environment with very low visibility (turbidity unit above 400NTU). It thus effectively solved the current inspection problems concerning underwater environment that conventional manual detection or other imaging equipment (Sonar, array camera, etc.) encountered, e.g., difficult imaging and low accuracy. It provides a new idea and technical means for underwater pier damage detection.

A reinforced concrete column was fabricated and tested in the laboratory to verify the feasibility of Contact-Visual Sensing Device (CVSD) and the prototype contact imaging system (Fig. 7.28). The column was loaded with low cyclic loads prior to underwater inspection. In the final failure state, many different types of damage were observed from the column surface, including cracks, defects, spalling, etc. It is worth noting that the damage simulated for this test was not exactly the actual pier damage. However, the damage type with respect to its apparent morphology is similar to that of actual underwater damages.

In order to achieve a refined description of the apparent damage of underwater concrete structures, Professor Wu Gang's team also proposed a deep learning (DL) based surface damage recognition algorithm for underwater structures. At the same time, in order to better adapt to the needs of underwater structural damage detection, the algorithm proposed several improved methods suitable for underwater distress

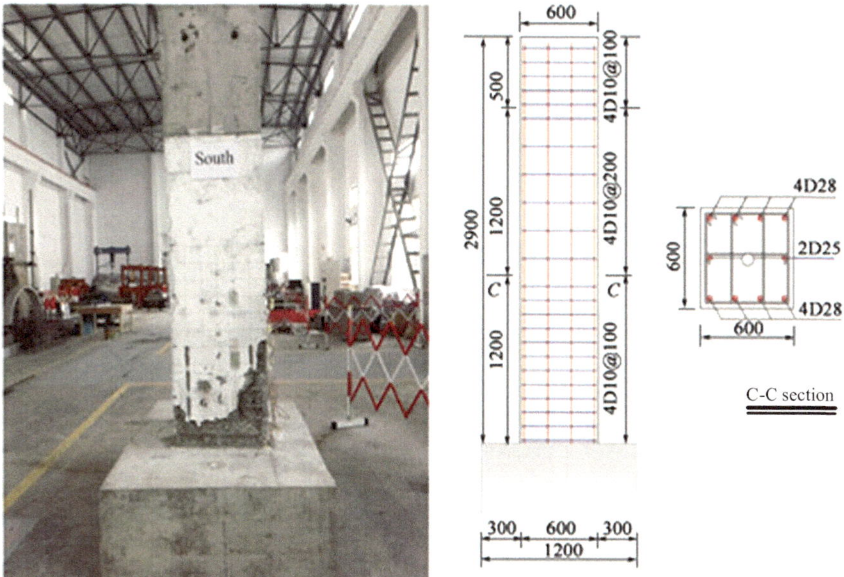

Reinforced concrete column

a) the appearance after loading

b) the reinforcement drawing of the original column

Fig. 7.28 Configuration of tested concrete column (unit: cm)

detection by modifying the fully connected convolutional neural network (FC-CNN or FCN) proposed in the literature [44, 45], such as redefining the loss function, etc. The algorithms related to the FCN model was mainly intended to achieve three objectives: image classification, object localization, and semantic segmentation. Image classification refers to the prediction of the entire input, that is, prediction of the objects in the image. Then, image classification can provide a sorted list of many objects in the image. Object localization provides object classes and additional information about the spatial location of these objects, such as centrosomes or bounding boxes. Semantic segmentation makes intensive predictive reasoning for each pixel of a class labeled with its closed object or region. According to the steps of image classification, object localization and semantic segmentation, the defect recognition algorithm can be developed from coarse-grained inference to fine-grained inference.

In the damage identification process, we built an FCN architecture that combines semantic information from the deep rough layer with appearance information from the shallow fine layer to produce accurate and detailed segmentation of underwater damage, namely cracks and defects. Figure 7.29 shows the workflow of the proposed FCN framework. The FCN network was trained using 20,000 manually labeled images (via manual semantic segmentation). Specifically, 5000 images were collected for each scene to increase the scene's complexity. In each imaging scene, in situ refraction and reflection due to changes in water flow and contact pressure

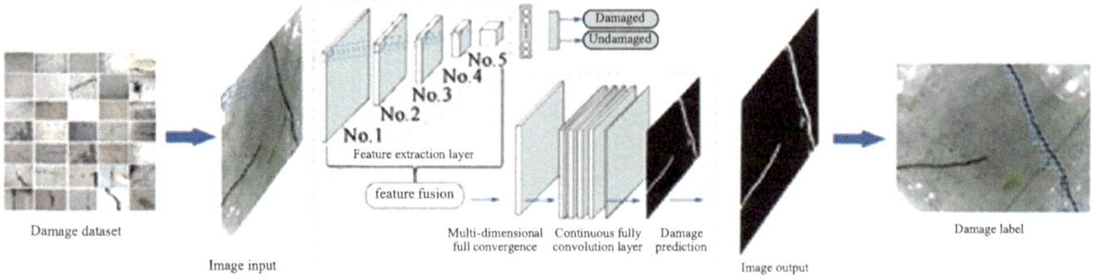

Fig. 7.29 Workflow of FCN framework

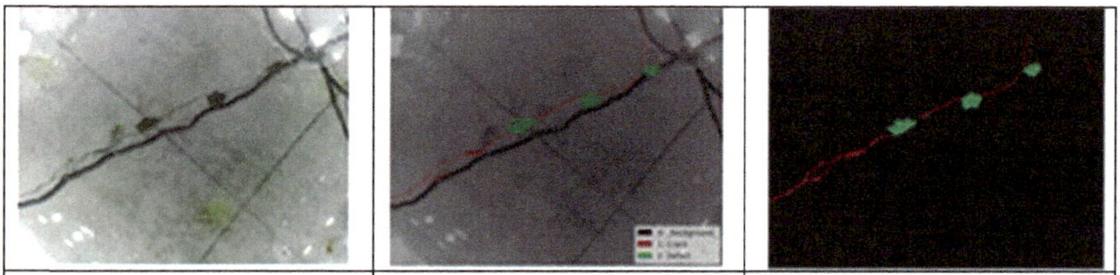

a) original image | b) calibration of the damage type | c) the result of identification

Fig. 7.30 Identification results of the "crack" damages

make lighting conditions and the appearance of water artifacts in each image different. Therefore, the generated DL model is only suitable for this image dataset. In addition, in the learning process (training), it not only learns the bottom 5 damage scenes but also the complex image noise on a desktop computer (Intel i5-8400@3.60 Hz CPU, 16 GB RAM, NVIDIA GTX-1060GPU).

Figures 7.30, 7.31 and 7.32 show the recognition results after semantic target detection and segmentation using the trained FCN model. It is apparent from the figures that the results identified by this algorithm are very close to the original damages with respect to the damage distribution and damage details. For both macroscopic damage detection (such as cavitation defects or necking) and microscopic damage detection (such as cracks), the FCN model can identify high-resolution damage images in extreme environments (such as underwater environments with low visibility and fast velocity). Besides, when the proposed FCN model was used as a quantitative damage detection tool, it was also able to highlight the advantages of the detectability and fine description of these tactile images.

7.3.3 Deformation Identification

(1) Displacement

DIC can achieve full field displacement measurement at sub-pixel precision through sub-pixel matching calculation. Early 2D-DIC can obtain the horizontal and the

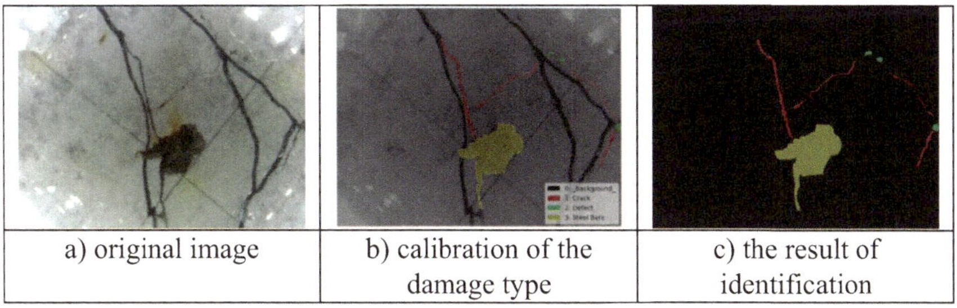

Fig. 7.31 Identification results of the "defect" damages

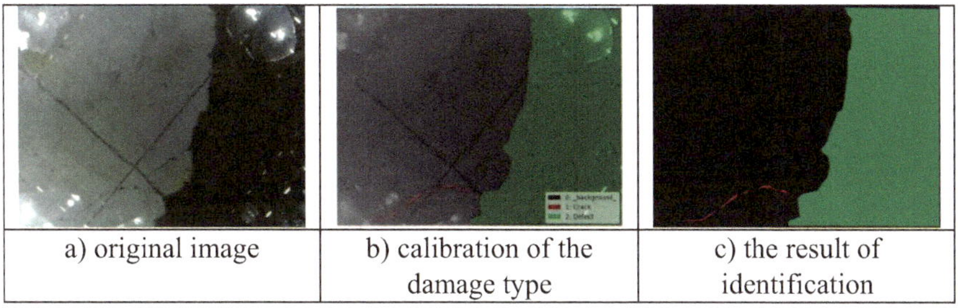

Fig. 7.32 Identification results of the "necking" damages

vertical displacements of any point in the object's plane measured in the image. With the emergence of 3D-DIC, DIC can obtain the displacement of the measured object in any direction in three-dimensional space. LaSalle Causeway lifting bridge, an over 100-year-old highway bridge, is located in Kingston, Ontario, Canada. Hoag et al. [46] used DIC technique to monitor the displacement of the bottom string rod and lifting of the end of the LaSalle Causeway lifting bridge before and after restoration, during the static load test, and under conventional traffic loads. In the static load test, four dump trucks loaded with gravel were parked on the middle-span of the bridge successively after being weighed, in a symmetrical position relative to the midpoint. The total weight of one vehicle was about 24,000 kg (235.5 kN). A single-lens reflex camera (Canon EOS Rebel t3i) equipped with a 180 mm focal length lens was mounted on the south side of the bridge, aimed at the mid-span, to capture the static images of the bridge. Ten images were taken at each loading stage, and they were combined into one image by averaging the pixel intensity to reduce errors due to camera shake.

Figure 7.33a shows the position of each truck at each load case. In condition 1: measurements were undertaken on the positions of all reference points with no external load. In condition 2, truck one placed its front wheel on the bridge to test whether the bridge was in contact with the support without loading. One truck was added for each subsequent working condition. Condition 7 was completely unloaded, and then the loading process of the bridge was repeated. At the end of the test, the

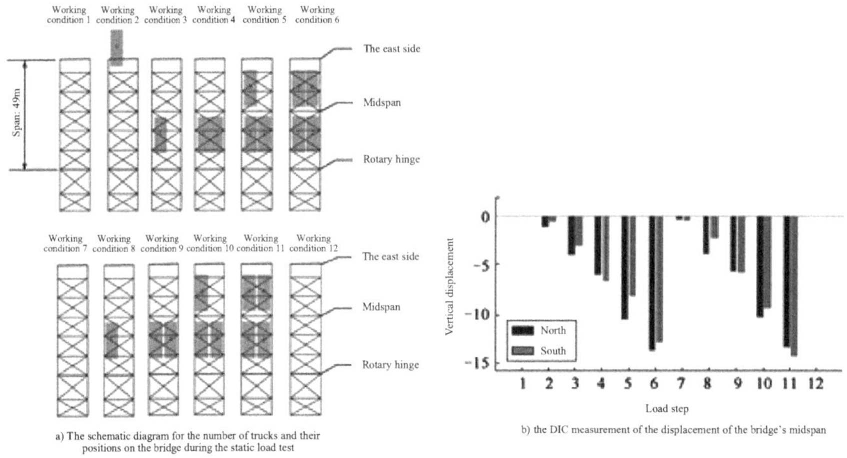

Fig. 7.33 Setting and measurement results of the conditions

bridge was completely unloaded again. In condition 12, the bridge was inspected for any permanent deformation. Figure 7.33b shows the vertical displacement measured in the mid-span of the bridge at each load position in the static test. The maximum vertical displacement of the bridge was found in load condition 5 (four trucks were concentrated in the middle span), which was 13.8 mm. When the trucks were evenly distributed on the bridge, such as in conditions 4 and 6, the displacement on both sides was similar. The maximum displacement difference between the north and south sides was only 0.9 mm for conditions 4 and 6. When the bridge's load was asymmetric, the vertical displacement on the side with large load was expected to be larger than on the other side. In condition 5, there were two trucks on the north side of the bridge and only one truck on the south side. The deflection on the north side of the bridge was 10.5 mm, while the deflection on the south side was only 8.2 mm, with a difference of more than 20%.

(2) Strain

Compared with the conventional strain gauges, the DIC method can obtain the continuous strain field on the specimen surface. This method first uses digital imaging technique (such as optical imaging, electronic imaging, tomography, etc.) to obtain digital images of objects in different states, and then uses computer technology, including intensity or region-based matching algorithm and numerical differential method, to realize the identification, matching, tracking and comparison of the target region (Fig. 7.34), and finally extracts the accurate full-field displacement and strain responses. DIC technique is currently been improved to adapt to different application scenarios.

Wattrisse et al. [47] used DIC to observe the strain fields of steel structures and analyzed the strain measurement accuracy, finding that the accuracy could meet the observation requirements of the strain concentration area. Based on the evolution

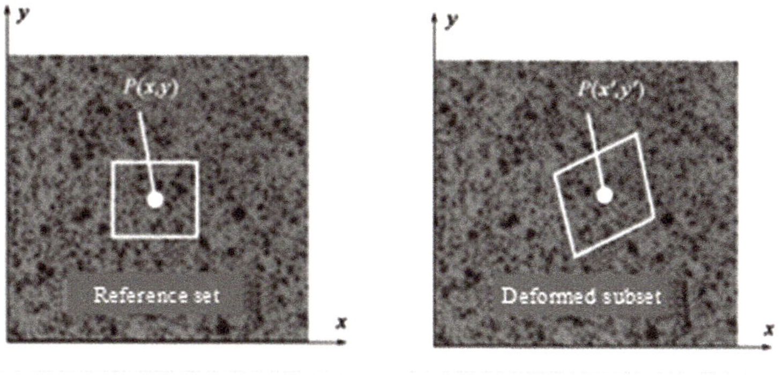

Fig. 7.34 Matching and calculation of the subsets by DIC before and after deformation

process of strain concentration obtained by DIC, they conducted an in-depth analysis on the neck shrinkage process of steel. Tung et al. [48] used DIC to measure the tube's deformation and verified the DIC's ability to measure strain in small areas. The test results showed that the DIC strain field could demonstrate the strain increase caused by local buckling. Pang [49] used 3D-DIC technique to measure the bonding properties of carbon fiber reinforced polymer (CFRP) -steel bonding interface. The test results included the failure mode, ultimate load, strain distribution and displacement distribution on the specimen surface. Figures 7.35a–h shows the strain evolution on the surface of the CFRP-steel specimens at different loading stages. In the bonding stage, the strain gradient near the loading end is large, as shown in Fig. 7.35a–d, and the deformation is mainly concentrated near the loading end. The distribution range of strain gradient also increased with the increase of loading, indicating that the stress transfer length gradually became longer. Subsequently, the range of high strain gradient moved down along the length of CFRP, that is, to the free end of the interface during the debonding stage, as shown in Fig. 7.35e–h. An obvious interface debonding process was observed.

Professor Xiaoyuan He's group at Southeast University, China, independently developed a complete set of DIC systems and conducted in-depth research on various issues existing with the technique. To increase the computational and measurement accuracy of DIC, the matching precision of sub-pixel pictures was further

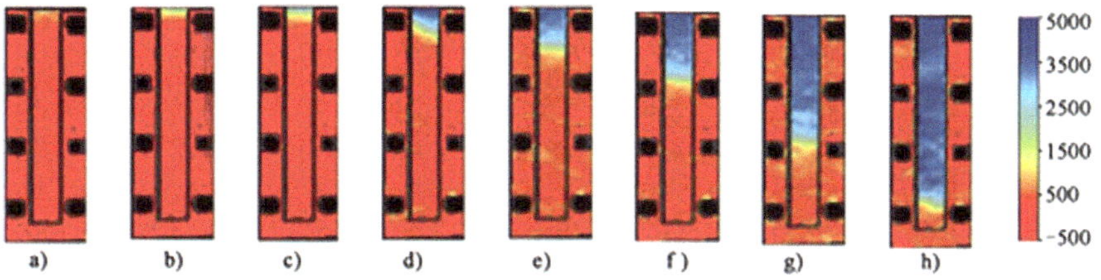

Fig. 7.35 Strain cloud of the bonding DIC observation test specimen [49]

studied based on the up-sampling method [50]. The accuracy of DIC strain measurement has been experimentally investigated, and the results indicated that telecentric lens imaging can help to improve the strain measurement accuracy [51]. A rigid body displacement elimination approach [52] was presented based on the rotation angle and force deformation features determined by the double-point displacement. Depending on the mechanical deformation characteristics of concrete, a method based on displacement field was proposed to provide the strain distribution. The actual test results showed that this method can better meet the requirements of engineering experiments.

(3) Fracture Process Zone

Based on the strain field information of the structural surface obtained by DIC, it can be found through three-dimensional visualization that there is a strong strain concentration area near the crack (Fig. 7.36). In concrete fracture mechanics, this area is called the fracture process zone (FPZ), which is the area where microcracks are developed and joined together to eventually form new macroscopic cracks. This area plays a very important role in the development of concrete cracks and structural deformation, and DIC technique can be used to identify the concrete cracks and the area effectively. By developing the processing program for the strain field and the displacement field obtained by DIC, the computer system can identify and track the processing area around the main fractures to obtain its evolution characteristics.

Fan et al. [54] conducted an in-depth study on the different evolution processes and characteristics of the fracture process zone of concrete members under various loads, such as monotony, reciprocation and fatigue, through the DIC method. They developed DIC crack identification, fracture process tracking and measurement, development process visualization, and other techniques. They analyzed the concrete structure's failure process by integrating the structure's mechanical performance and discovered a clear correlation between the development of fracture process zone and change in structural bearing capacity. The research on the formation stage and development characteristics of the fracture process zone will contribute more specific test data to the development of the concrete fracture theory. The basic operation process is shown in Fig. 7.37, with main steps including speckle pattern making, camera calibration, digital image acquisition and displacement strain field calculation.

7.3.4 *Structural Motion Monitoring*

The application of structural health monitoring (SHM) is mostly based on the measurement of dynamic responses of the structure. Vibration-based SHM is one of the most popular methods to evaluate the safety of civil infrastructures. The conventional vibration measurement of bridge structure requires the sensors to be in contact with the measured structure or component, which will add load to the light-weight structure and therefore cause measurement error. For long-term applications, wired sensors need cables for power supply, and wireless sensors require

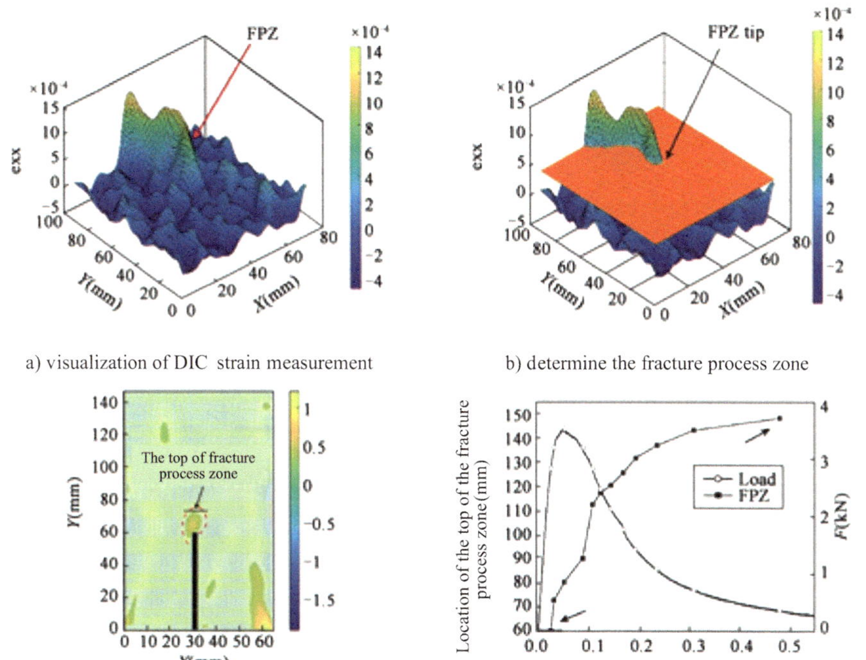

a) visualization of DIC strain measurement

b) determine the fracture process zone

c) schematic diagram of surface fracture process zone formation

d) failure process and fracture of structure

Fig. 7.36 DIC information of the strain field during fracture process [53]

regular battery replacement, which are both expensive and time-consuming. In addition, these arranged sensors are usually limited in number, discrete in position and provide unsatisfactorily low spatial resolution, which is difficult to meet the needs of modal-based damage location and model updating of large-scale structures. Non-contact measurement methods, such as laser vibration scanners, can provide high-resolution sensing capabilities with no additional load on the structure. However, these measuring devices are relatively expensive and require continuous measurements, which usually take a long time. With the improvement of camera hardware and the development of image processing technique, the camera-based non-contact measurement method has received wide concern for its advantages of low cost, flexible layout, simultaneous multi-point measurement of structures and high spatial resolution. It has already been successfully applied in structural vibration measurement and subsequent modal analysis, system identification and cable force evaluation.

(1) Bridge Structural Vibration

Generally speaking, a state of the structure corresponds to a dynamic characteristic, and any change in the physical characteristics will cause a change in the dynamic

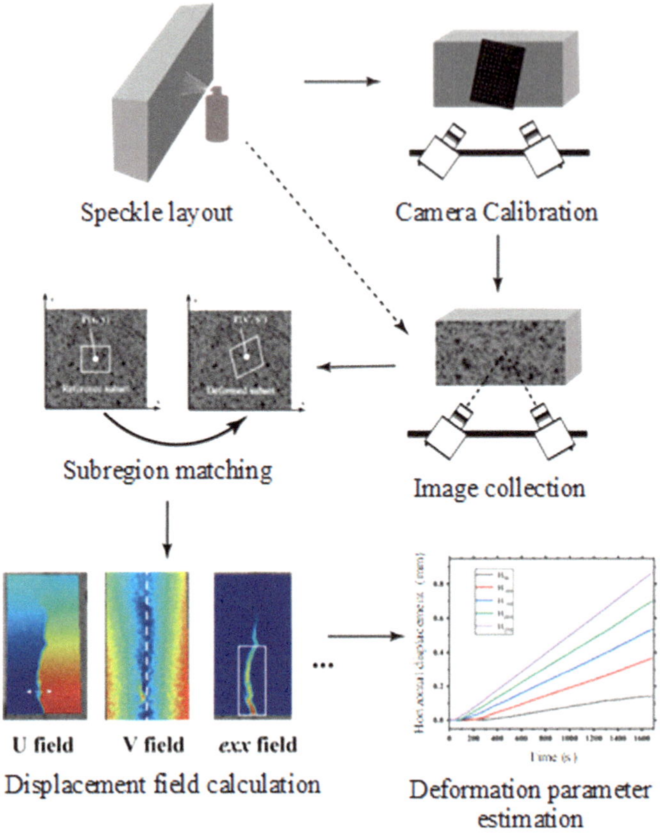

Fig. 7.37 DIC fundamental operating procedure

response of the system. In the past 20 years, vibration-based structure health monitoring system is intended to evaluate the structure's health status by analyzing the structure's vibration behaviour.

Mas et al. [55] proved that the vibration frequency of all vibration targets in the RoI sequence could be obtained simultaneously by analyzing high-speed video sequences and developed a synchronous measurement method of multi-point vibration frequency. Feng et al. [56] developed a single-camera structure multi-point synchronous measurement vision sensor by using sampling cross-correlation technique (UCC) and positioning code matching technique (OCM). It is proven that a camera can be used to accurately measure the dynamic displacement response of a series of points, and determine the natural frequency and mode shape simultaneously. The test results agreed with the data measured using the laser displacement sensor. Then, through field test of Manhattan Bridge [57], performance of the multi-point real-time displacement remote measurement system based on vision sensor was tested. Yoon et al. [58] proposed a non-marked displacement measurement method based on consumer cameras utilizing the optical flow estimation method. The indoor vibration experiment was carried out on a six-layer frame structural model, and the modal parameters identified were consistent with those identified by the accelerometer method. Yang et al. [59] demonstrated the potential of pure output modal identification at low frame rates by using aliasing-free or temporary aliasing video measurements through experiments on laboratory-scale structures. Han et al. [60] proposed a structural dynamic displacement measurement method combining machine vision and uncrewed aerial vehicles (UAV). Laser spots were projected on the structure's surface as a stationary reference by placing laser lights on the stationary ground. The movement of laser spots and target measuring points on the structure's surface was photographed by UAV at close range. The absolute displacement of the target measuring point was so obtained. The measurement results from long-span suspension bridges showed that the dynamic structural displacement obtained by this method was in good agreement with the reference values.

In 2013, Wadhwa et al. [61] proposed a phase-based video "MotionMagnification" technique, which was developed based on the optical flow method and complex manipulable pyramid. They could extract and magnify tiny motions of specific frequencies in videos. Since then, many experts and scholars in the field of civil engineering have used this technique to carry out relevant applications in housing construction, infrastructure and other aspects. Chen et al. [62] first used the motion amplification technique to visualize and quantify vibration patterns of simple structures captured by high-speed cameras and verified the accuracy of this method by comparing it with the data measured by a laser vibration meter and accelerometer. Yang et al. [63] used multi-scale pyramid decomposition and representation methods, unsupervised learning methods and blind source separation (BSS) to measure the resonance frequency, damping ratio and high-resolution modal shape from video measurement of the structure by extracting, modeling and processing the phase of spatiotemporal pixels in the entire field. They further adopted a three-story building structure model for experimental verification.

Structural damage detection methods based on vibration parameters were also gradually developed. Through combining knowledge of damage mechanics, structural dynamics and other structural knowledge, quantified structural vibration and structural health status evaluation approaches are well established. The vision-based detection method makes it possible to collect complete structural vibration data. The method can provide a better data basis for realizing more accurate damage detection and location, and model updates. Cha et al. [64] measured the cantilever beam's displacement using a phase-based optical flow algorithm and used an untracked Kalman filter to remove the noise in displacement measurement. Under the premise of known structural mass, the current structural stiffness, damping coefficient and other structural characteristics were identified to detect the structural damage. Feng et al.'s experiment [56] proved that the smooth mode shape extracted from the full-field displacement response measured by a single camera could detect and locate structural damage of a simple beam based on the mode curvature index. Dworakowski et al. [65] obtained deflection curves of small experimental beams through digital image correlation (DIC) and evaluated two beam damage detection algorithms based on deflection shapes. Yang et al. [66], based on earlier proposed method of effectively extracting full-field high-resolution dynamic parameters of structures from videos, further studied the feasibility of damage detection based on the video output of current structures without knowing the intact structure information.

Professor Zhang Jian's group in Southeast University, China, has developed a vision-based bridge deflection measurement method [67] which can track the natural features of structure without installing artificial targets on the structure. A simplified and fast Hessian detector and a preprocessing-based random sampling consistency (RANSAC) method were proposed for the small image scale change and small deformation degree of small and medium-sized bridges. The following is a brief description of the bridge deflection measurement method. Using this method, the specific flow of deflection measurement is as follows:

(1) Demarcate the scaling factors (mm/pixel) for two-dimensional optical measurement;
(2) Estimate the maximum deformation amplitude based on engineering experience and design value;
(3) Inspect the features using the fast Hessian detector, and describe the features using BRISK descriptor;
(4) Match the features using KNN;
(5) Remove the mismatches using improved RANSAC;
(6) Transform pixel distance to actual physical distance using the scaling factor s.

The field test focused on the vertical displacement measurement of the 1/4 main span of the Hainan Bridge (see Fig. 7.38a). The bridge is a stayed tie arch structure with a 100 m long and 50 m wide main span. In addition, conventional contact sensors, including an inclinometer and a liquid level gauge, were used for comparison tests. The camera was installed beside the inclinometer (Fig. 7.38b), and the lens was focused on the railings near the bridge 26 m away from the camera (Fig. 7.38c). An industrial camera (UI-3370CP-MGL) equipped with a 200 mm lens was used for this measurement.

The test results showed that the simplified Hessian detector can save 30% of detection time. The improved RANSAC was faster than the original RANSAC, and the preprocessing helped to improve the matching accuracy. Compared with the conventional measurement methods, relative error of the measurement results using the improved deflection measurement algorithm is less than 5% in most cases.

(2) Cable Vibration of Bridges

The cable is a crucial load-bearing component of a cable-stayed bridge, so it is necessary to detect and evaluate its working state during construction and operation. The conventional measurement devices and methods for the cable force mainly include resistance strain gauge, vibrating series crossflow meter, pressure ring, vibration frequency method, ultrasonic guided wave method, fiber Bragg grating sensing technique and magneto-elastic effect method [68]. Among them, the widely used vibration frequency method uses the relationship between the cable's vibration frequency of each order and its internal tension to determine the cable force. The vibration

a) the tested bridge

b) the setting of the optical system c) the scene of the camera

Fig. 7.38 The field test [67]

frequency of the cable can be obtained by the acceleration signal measured by the vibrator or the accelerometer installed at a certain position of the cable through the spectrum analysis. However, these methods usually require the installation of sensors on the cable and the arrangement of data transmission lines, which is time-consuming, laborious and costly, and also has to face many difficulties in maintenance and management. Using computer vision technology instead of accelerometers and other sensors, the natural frequency of the cable can be more effectively obtained, which becomes a new vibration measurement method to estimate the cable force.

(1) Research Status

Ji et al. [69] proposed an image-based vibration measurement method for cables. A commercial video camera was used to remotely collect the digital image sequence from the vibrating cable. The optical flow method was used to compute the change in the light intensity of any selected region of interest (RoI) in the cable image sequence. The optical loss indicated the direction of vibration of the cable and can be used to estimate the displacement of the cable. However, vibration of the cable in the plane, position of the camera and both ends of the cable are known, but the sag of the cable was not obvious for the study by Ji et al. [69]. Caetano et al.

[70] used the optical flow method to remotely measure the vibration of cable of the Guadiana cable-stayed bridge, and the measurement results were consistent with those using accelerometer. Chen [71] proposed a method to attach circular marks onto the cable as measurement targets due to the cumbersome calculation of the optical flow method. However, sometimes the installation of additional marks was difficult, so he proposed to take the midpoint of the cable's edge line in each image as the pseudo-target. Since the coordinates of the midpoint of a line were easier to calculate than those of the center of a circle, this more convenient non-target technique was even more computationally efficient than its analogue with the target. Kim et al. [72, 73] developed a vision-based monitoring system that used normalized cross-correlation template matching to remotely measure the dynamic characteristics of stay cables. The template matching approach allowed the object itself to be used as a template without a target, avoiding the installation of a human target. The disadvantage of this method was that the images can spread and deteriorate in the foggy, rainy nights and backlit conditions, making reliable data difficult to obtain.

At the same time, a correction algorithm based on the fixed object in the image was proposed to eliminate the camera motion caused by external vibration. Because of insufficient accuracy of the conventional vibration frequency method to estimate cable force, which is only based on the cable frequency, Chen et al. [74] proposed a cable force measurement method combining mode shape ratio and mode frequency. Multiple cameras were used to measure the ambient vibration synchronously at different cable positions to identify the mode shape ratio. Zhao et al. [75] innovated the smartphone as a video acquisition tool. In the RoI of the video image sequence captured by the smartphone camera, the cable boundary was selected as the tracking target to identify the dynamic characteristics of the cable according to its dynamic displacement response in the frequency domain. A moving average filter was used to eliminate the noise generated by the smartphone camera jitter during measurement, avoiding the use of tripod. Xu et al. [76] proposed a multi-point displacement measurement system based on commercial cameras and low-cost and non-contact vision. The system adopted a cable tracking method based on edge detection. The cable tracking consisted of two steps: edge detection and motion estimation. The purpose of edge detection was to determine the position of the cable in a small subset window, and the motion of the cable was estimated by the distance between the extracted two edges. Du et al. [77] adopted two measurement methods based on digital image technology, namely digital image processing (DIP) and digital image correlation (DIC), to identify the single-point and multi-point cable forces in the images captured by the camera. The minimum deviation of single point cable force measurement was 0.24% and the maximum deviation was 2.69%. The multi-point cable force measurement error was about 3.05%, for which the measurement accuracy was improved compared with Kim et al. [72]'s method and the feasibility of cable force measurement based on computer vision was demonstrated. Feng et al. [78] proposed a new non-contact vision sensor method for cable force estimation. The template matching algorithm was used to measure the cable force of a stadium in Florida. The results were consistent with the results obtained using accelerometers, and the maximum error was 5.6%.

(2) Case Studies

The following is a detailed introduction to the measurement system proposed by Feng et al. [78] and a cable force measurement case, so that readers can comprehend the process of cable force estimation using computer vision. The cable force measurement system is based on the subpixel orientation code matching method created by Feng et al., with the following procedure:

(1) Test setup. Place the camera on a tripod away from the cable and link it to a laptop with real-time picture processing software.
(2) Cable vibration measurement. The starting section of the cable surface is chosen as the tracking template. The measured vibrations are collected in real-time utilizing software based on the subpixel OCM algorithm by searching in the predetermined region of interest (RoI).
(3) Fourier transform and cable tension estimate.
(4) Given the cable's geometric shape and material parameters, the cable tension force is determined by the relationship between the cable's vibration frequency and its internal tension.

In order to verify the effectiveness of the cable force measurement system, cable force estimation of the Hard Rock Stadium under construction in Florida was performed. Figure 7.39 shows the vision sensor system. The camera used a CMOS sensor with a maximum resolution of 1280 × 1024 pixels and a maximum speed of 150 fps per second. Focal length of the lens was 16–160 mm. Since the base frequency of cable in most civilian infrastructure is below 10 Hz, a sampling rate of 50 fps was used.

In Fig. 7.39a, the cable segment approximately halfway across the length of the red box was identified as the motion tracking template. The results showed that the cable force measured by the vision sensor was the same as that measured by the

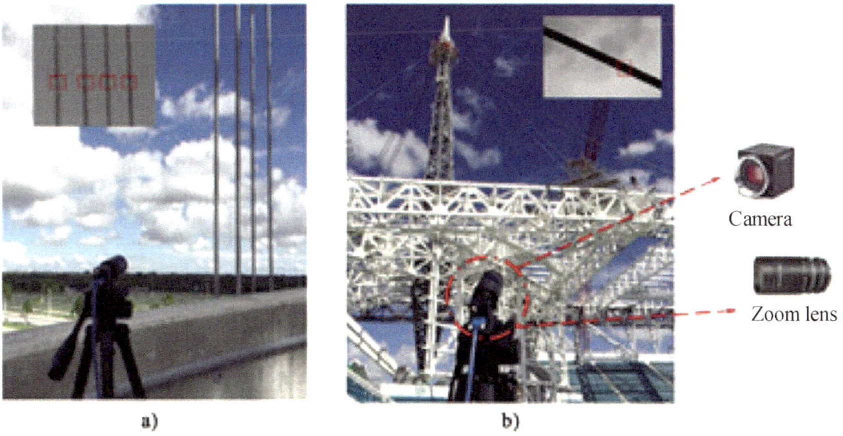

Fig. 7.39 System of vision sensors [78]

pressure ring, and the maximum error was 5.6%. It can be seen that this visual-based cable force measurement method is a convenient and low-cost method for long-term or periodic monitoring of cable structures.

It should be noted that although the vision-based sensor system has many advantages, several technical problems in field application still need to resolve in future research, such as thermal air fog, low light level, optical noise and measurement errors caused by ground/camera motion. With the development of high-performance cameras, computer vision technology will become more reliable and convenient in bridge health monitoring, especially cable stress monitoring.

7.4 Summary

This chapter starts with the basic concept and development of computer vision. It introduces the related open-source software, current main application scenarios and application in bridge operation and maintenance. According to the logical method of processing tasks, computer vision technology is divided into the conventional computer vision technology based on image feature transformation and the deep learning-based computer vision technology. In the conventional computational vision technology, the basic transformation of an image, feature detection and description, target recognition and tracking methods are introduced. Deep learning-based computer vision technology is introduced focusing on the convolutional neural network and its application scenarios. Furthermore, the application of computer vision technology in identifying and measuring apparent structural damage, static deformation and structural vibration is introduced.

In general, the development of computer vision technology provides a new, more economical and convenient solution for bridge inspection and monitoring. Under current development prospect of new infrastructures, it is imperative to develop disciplines by integrating interdisciplinary knowledges. However, the development of interdisciplines knowledge requires more comprehensive knowledge and more comprehensive perspective. In order to enable computer vision technology better serve the civil engineering industry and promote the development of intelligent and modern civil engineering, it is necessary to conduct more in-depth scientific research according to the requirements of construction, monitoring, testing, as well as the existing difficulties and pains of the current conventional technology.

References

1. Xu ZJ, Wang J, Liu Y et al (2012) Current situation and prospect of computer vision core technology. J Xi'an Univ Post Telecommun 17(6):1–8 (in Chinese)
2. Gonzales RC, Woods RE (2011) Digital image processing, 3rd edn. Electronic Industry Press, Beijing (in Chinese)

3. Chris H, Stephens M (1988) A combined corner and edge detector. In: Alvey vision conference, vol 15, no 50
4. Dalal N, Triggs B (2005) Histograms of oriented gradients for human detection. Comput Vis Pattern Recogn 2005:886–893
5. Lowe DG (2004) Distinctive Image Features from Scale-Invariant Keypoints. Int J Comput Vision 60(2):91–110
6. Yamaguchi I (1982) Simplified laser-speckle strain gauge. Opt Eng 21(3):213–236
7. Peters WH, Ranson WF (1982) Digital imaging techniques in experimental stress analysis. Opt Eng 21(3):427–431
8. Sutton M, McNeill S, Helm J et al (2000) Advances in two-dimensional and three-dimensional computer vision. Top Appl Phys 77:323–372
9. Luo P, Chao Y, Sutton M et al (1993) Accurate measurement of three-dimensional displacement in deformable bodies using computer vision. Exp Mech 33(2):123–132
10. Bay B, Smith T, Fyhrie D et al (1999) Digital volume correlation: three-dimensional strain mapping using X-ray tomography. Exp Mech 39(3):217–226
11. Smith T, Bay B, Rashid M (2002) Digital volume correlation including rotational degrees of freedom during minimization. Exp Mech 42(3):272–278
12. Hinton GE, Osindero S, Teh Y-W (2006) A fast learning algorithm for deep belief nets. Neural Comput 18(7):1527–1554
13. Goodfellow L et al (2017) Deep learning. Posts and Telecommunications Press, Beijing (in Chinese)
14. Lecun Y. LeNet-5, convolutional neural networks. http://yann.lecun.com
15. Krizhevsky A, Sutskever I, Hinton GE (2017) ImageNet classification with deep convolutional neural networks. Commun ACM 60(6):84–90
16. Szegedy C, Liu W, Jia Y et al (2015) Going deeper with convolutions. In: IEEE conference on computer vision and pattern recognition (CVPR)
17. He K, Zhang X, Ren S et al (2016) Deep residual learning for image recognition. In: IEEE conference on computer vision & pattern recognition
18. Zagoruyko S, Komodakis N (2016) Wide residual networks. arXiv preprint arXiv:1605.07146
19. Xie S, Girshick R, Dollár P et al (2017) Aggregated residual transformations for deep neural networks. In: 30th IEEE/CVF conference on computer vision and pattern recognition (CVPR)
20. Wang N, Li S, Gupta A et al (2015) Transferring rich feature hierarchies for robust visual tracking. Comput Sci
21. Ren SQ, He KM, Girshick R et al (2017) Faster R-CNN: towards real-time object detection with region proposal networks. IEEE Trans Pattern Anal Mach Intell 39(6):1137–1149
22. Liu W, Anguelov D, Erhan D et al (2016) SSD: single shot multibox detector . In: 14th European Conference on Computer Vision (ECCV)
23. Long J, Shelhamer E, Darrell T (2014) Fully convolutional networks for semantic segmentation. IEEE Trans Pattern Anal Mach Intell 39(4):640–651
24. Avril S, Bonnet M, Bretelle A-S et al (2008) Overview of identification methods of mechanical parameters based on full-field measurements. Exp Mechan 48(4):381–402
25. Image Segmentation (2018) Fully convolutional networks for semantic segmentation. https://blog.csdn.net/nclgsj1028/article/details/79497121?ops_request_misc=&request_id=&biz_id=102&utm_term=%E3%80%90%E5%9B%BE%E5%83%8F%E5%88%86%E5%89%B2%E3%80%91Fully%20Convolutional%20Netw&utm_medium=distribute.pc_search_result.none-task-blog-2~all~sobaiduweb~default-3-79497121.first_rank_v2_pc_rank_v29&spm=1018.2226.3001.4187
26. Ronneberger O, Fischer P, Brox T (2015) U-Net: convolutional networks for biomedical image segmentation. Springer International Publishing
27. Chen LC, Papandreou G, Kokkinos I et al (2018) DeepLab: semantic image segmentation with deep convolutional nets, atrous convolution, and fully connected CRFs. IEEE Trans Pattern Anal Mach Intell 40(4):834–848
28. Lin G, Milan A, Shen C et al (2017) RefineNet: multi-path refinement networks for high-resolution semantic segmentation. In: 30th IEEE/CVF Conference on Computer Vision and Pattern Recognition (CVPR)

29. Peng C, Zhang X, Yu G et al (2017) Large Kernel matters—improve semantic segmentation by global convolutional network . In: 30th IEEE/CVF conference on computer vision and pattern recognition (CVPR)
30. He K, Georgia G, Piotr D et al (2017) Mask R-CNN. In: 16th IEEE international conference on computer vision (ICCV)
31. Cha YJ, Choi W, BüYüKöZTüRK O (2017) Deep learning-based crack damage detection using convolutional neural networks. Comput-Aided Civil Infrastruct Eng 32(5):361–378
32. Li LF, Ma WF, Li L et al (2019) Research on bridge crack detection algorithm based on deep learning. Acta Automatica Sinica 45(09):1727–1742 (in Chinese)
33. Wu HH, Wang AH, Wang HD (2019) Tunnel image crack detection based on Fast R-CNN. J Taiyuan Univ Sci Technol 40(03):165–168 (in Chinese)
34. Wang S, Wu X, Zhang YH et al (2018) Full convolutional network image crack detection based on deep learning. J Comput-Aid Des Comput Graph 30(05):859–867 (in Chinese)
35. Ji J, Wu L, Chen Z et al (2018) Automated pixel-level surface crack detection using U-Net. In: International conference on multi-disciplinary trends in artificial intelligence, Hanoi, Vietnam, 18–20 Nov 2018
36. Yi L (2006) Measurement and analysis of concrete surface crack characteristics based on digital image processing technology. Southeast University, Nanjing (in Chinese)
37. Ni F, Zhang J, Chen Z (2019) Zernike-moment measurement of thin-crack width in images enabled by dual-scale deep learning. Comput Aided Civ Inf 34:367–384
38. Dong B (2020) Study on the identification method of cable sheath apparent diseases and cable force time-varying for cable-stayed bridges. Southeast University, Nanjing (in Chinese)
39. Dang J, Shrestha A, Chun P-J (2018) Bridge damage detection using unmanned Arial vehicle and deep convolutional neural network. In: Asia conference on earthquake engineering
40. Shi J, Zuo R, Dang J (2019) Bridge damage classification and detection using fully convolutional neural network based on images from UAVs. In: Experimental vibration analysis for civil engineering structures
41. Jiangsu Provincial Local Standards (2012) Technical standard for disease assessment of rubber bearings for highway bridges: DB 32/T 2172—2012. Jiangsu Provincial Bureau of Quality and Technical Supervision, Nanjing (in Chinese)
42. Cui MD (2018) Research on key technologies of automatic identification of bridge bearing diseases based on image processing. Southeast University, Nanjing (in Chinese)
43. Shi Y, Cui MD, Zuo RZ et al (2020) Bridge damage detection and segmentation by deep learning for bridge UAV inspection. In: World conference on earthquake engineering
44. Dolz J, Desrosiers C, Ben AI (2018) 3D fully convolutional networks for subcortical segmentation in MRI: a large-scale study. Neuroimage 170:456–470
45. Shelhamer E, Long J, Darrell T (2017) Fully convolutional networks for semantic segmentation. IEEE Trans Pattern Anal Mach Intell 39:640–651
46. Hoag A, Hoult N, Take A (2017) Assessment of a bascule lift bridge using digital image correlation. Proc Inst Civil Eng Bridge Eng 170(3):168–180
47. Wattrisse B, Chrysochoos A, Muracciole J-M et al (2001) Analysis of strain localization during tensile tests by digital image correlation. Exp Mech 41(1):29–39
48. Tung SH, Sui CH (2010) Application of digital-image-correlation techniques in analysing cracked cylindrical pipes 35(5):557–567
49. Pang YY, Wu G, Wang HT et al (2019) Experimental study on the bond behavior of the CFRP-steel interface under the freeze-thaw cycles. J Compos Mater 54(1):13–29
50. Wang YY, Heng W, Shao XX et al (2017) A method for improving the accuracy of digital image correlation based on upsampling processing. In: China mechanics conference-2017 (in Chinese)
51. Zhu FP, Lei D, He XY (2012) Research on the accuracy of full field strain measurement based on digital image correlation method. In: Abstracts of the 13th national academic conference on experimental mechanics (in Chinese)
52. Dai YQ, Sun ZY, Wu G et al (2010) Full-field deformation measurement of concrete based on digital image correlation method. J Southeast Univ (Nat Sci Edn) 40(4):829–834 (in Chinese)

53. Li ST, Fan XQ, Chen XD et al (2020) Development of fracture process zone in full-graded dam concrete under three-point bending by DIC and acoustic emission. Eng Fract Mech 230(3):106972
54. Fan XQ, Li ST, Chen XD et al (2020) Fracture behaviour analysis of the full-graded concrete based on digital image correlation and acoustic emission technique. Fatigue Fract Eng Mater Struct 43(6):1274–1289
55. Mas D, Ferrer B, Acevedo P et al (2016) Methods and algorithms for video-based multi-point frequency measuring and mapping. Measurement 85:164–174
56. Feng DM, Feng MQ (2016) Vision-based multipoint displacement measurement for structural health monitoring. Struct Control Health Monit 23(5):876–890
57. Feng DM, Feng MQ (2017) Experimental validation of cost-effective vision-based structural health monitoring. Mech Syst Signal Process 88:199–211
58. Yoon H, Elanwar H, Choi H et al (2016) Target-free approach for vision-based structural system identification using consumer-grade cameras. Struct Control Health Monit 23(12):1405–1416
59. Yang YC, Dorn C, Mancini T et al (2018) Spatiotemporal video-domain high-fidelity simulation and realistic visualization of full-field dynamic responses of structures by a combination of high-spatial-resolution modal model and video motion manipulations. Struct Control Health Monit 25(8):e2193
60. Han Y, Wu G, Feng D (2022) Vision-based displacement measurement using an unmanned aerial vehicle. Struct Control Health Monit e3025. https://doi.org/10.1002/stc.3025
61. Wadhwa N, Rubinstein M, Durand F et al (2013) Phase-based video motion processing. ACM Trans Graph 32(4):1–10
62. Chen JG, Wadhwa N, Cha Y-J et al (2015) Modal identification of simple structures with high-speed video using motion magnification. J Sound Vib 345:58–71
63. Yang YC, Dorn C, Mancini T et al (2017) Blind identification of full-field vibration modes from video measurements with phase-based video motion magnification. Mech Syst Signal Process 85:567–590
64. Cha YJ, Chen JG, BüYüKöZTüRK O (2017) Output-only computer vision based damage detection using phase-based optical flow and unscented Kalman filters. Eng Struct 132(1):300–313
65. Dworakowski Z, Kohut P, Gallina A et al (2016) Vision-based algorithms for damage detection and localization in structural health monitoring. Struct Control Health Monit 23(1):35–50
66. Yang YC, Dorn C, Mancini T et al (2018) Reference-free detection of minute, non-visible, damage using full-field, high-resolution mode shapes output-only identified from digital videos of structures. Struct Health Monit 17(3):514–531
67. Yu SS, Zhang J (2020) Fast bridge deflection monitoring through an improved feature tracing algorithm. Comput-Aided Civil Infrastruct Eng 35(3):292–302
68. Dong CZ (2016) Bridge health monitoring and condition evaluation based on machine vision. Zhejiang University, Hangzhou (in Chinese)
69. Ji YF, Chang CC (2008) Nontarget image-based technique for small cable vibration measurement. J Bridg Eng 13(1):34–42
70. Caetano E, Silva S, Bateira J (2011) A vision system for vibration monitoring of civil engineering structures. Exp Tech 35(4):74–82
71. Chen C-C, Tseng H-Z, Wu W-H et al (2012) Modal frequency identification of stay cables with ambient vibration measurements based on nontarget image processing techniques. Adv Struct Eng 15(6):929–942
72. Kim SW, Jeon BG, Kim NS et al (2013) Vision-based monitoring system for evaluating cable tensile forces on a cable-stayed bridge. Struct Health Monit 12(5–6):440–456
73. Kim SW, Kim NS (2013) Dynamic characteristics of suspension bridge hanger cables using digital image processing. NDT & E Int Independ Nondestr Test Evaluat 59:25–33
74. Chen CC, Wu WH, Tseng HZ et al (2015) Application of digital photogrammetry techniques in identifying the mode shape ratios of stay cables with multiple camcorders. Measurement 75:134–146

75. Zhao XF, Ri K, Wang NN (2017) Experimental verification for cable force estimation using handheld shooting of smartphones. J Sens 2017:1–13
76. Xu Y, Brownjohn JMW (2018) Review of machine-vision based methodologies for displacement measurement in civil structures. J Civil Struct Health Monit 8(1):91–110
77. Du WK, Lei D, Bai PX et al (2020) Dynamic measurement of stay-cable force using digital image techniques. Measurement 151:107211
78. Feng DM, Scarangello T, Feng MQ et al (2017) Cable tension force estimate using novel noncontact vision-based sensor. Measurement 99:44–52

Chapter 8
Data Fusion and Digital Modeling

Abstract The concept of data fusion originated in the 1970s and was first applied in the military. With the growing complexity of bridge structures, single-source data could not satisfy the demand of bridge health monitoring. Hence, it was essential to synthesize data from different sensors in order to evaluate the bridge condition more precisely. Therefore, data fusion plays a vital role in making decisions for bridge maintenance. The current bridge health monitoring data centers have collected numerous types of data, hence it is imperative to perform data fusion. This chapter will present a systematic overview of the data fusion algorithms, data fusion-based condition assessment methods, and data fusion platforms within the intelligent bridge operation and maintenance framework. It should be noted that, in a strict sense, data implies unprocessed unstructured numerical sets; while information implies processed data with significance. The general notion of data covers raw data (sensor data), processed data and validated data (that is, information). In this chapter, our data refers to the general notion of data unless otherwise stated.

8.1 Data Fusion

No single sensor can obtain all aspects of the target data, and data loss is possible during the transmission process challenging accuracy of the judgment based on the detection attributes of a single sensor. Hence, fusing multi-source data (such as detection and monitoring data) is essential in making scientific decisions in bridge maintenance. Data fusion-driven decision-making requires building a data fusion platform to make better use of the existing data. Generally speaking, a data fusion platform comprises an algorithm platform, a physical entity platform, and a physical and digital fusion platform.

G. Wu et al., *Intelligent Bridge Maintenance and Management*, Springer Tracts in Civil Engineering, https://doi.org/10.1007/978-981-97-3827-4_8

8.1.1 Development of Data Fusion

The rapid development of computer and digital technology in the 1970s has led to the emergence of data fusion as an important concept in military applications. It involves "fusion processing" of data from various sensors to obtain data more accurately than from a single sensor. Data fusion is a basic and ubiquitous function of humans and other biological systems, as they can instinctively integrate the signals (sights, sounds, smells and touch) from various sensory organs (eyes, ears, nose and limbs) with their prior knowledge to make judgments about the environments and the events. Humans can detect various physical phenomena in different spatial ranges due to the different measurement characteristics of their senses, and transform them into valuable interpretations of the environment through fusion processing of different features. Multi-source data fusion is a simulation of the human brain's functions to comprehensively process complex problems. However, in social and economic activities, responsible governments and enterprises also need to collect a large amount of data and gather different viewpoints in order to make decisions in line with the objective laws.

So far, researchers have proposed various functional models of information fusion according to the different input information abstractions or fusion outputs. These models include data level (or pixel level) fusion, decision level fusion and feature level fusion. Data level fusion can preserve more field data and provide information that other levels cannot provide. However, it has many drawbacks, such as the requirement of many sensors, large data communication capacity, high processing cost, long processing time, poor real-time performance and inferior anti-interference ability. Pixel-level image fusion is a typical example of data level fusion. In contrast, decision level fusion has low requirements for information transmission bandwidth, small communication capacity, strong anti-interference ability and low processing cost of the fusion center. But it suffers from high pre-treatment costs and large information loss. Feature level fusion is in between data level fusion and decision level fusion. These three models can be applied at different levels, for instance, they can be fused at both feature and decision levels when conducting distributed detection fusion.

In bridge maintenance, data from various sources (such as point clouds, seismic waves, images, and finite element analysis) should also be integrated in order to make scientific decisions. For example, in Chap. 5 of this book, the "Jiangsu Changda-Bridge Health Monitoring Data Center" established by Sujiaoke Group in 2010 gathered various data types such as bridge detection and monitoring data. Therefore, research on data fusion has become a practical demand for bridge maintenance and management. Figure 8.1 shows the application of data fusion to bridge maintenance knowledge decision-making.

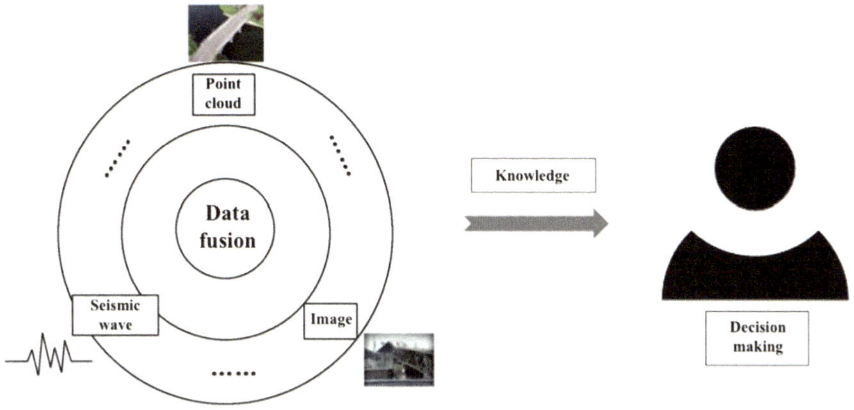

Fig. 8.1 Application of data fusion to bridge maintenance knowledge decision-making

8.1.2 Data Fusion Platform

Data fusion and information fusion are similar terms; but sometimes, data fusion refers to combining raw data without processing, while information fusion refers to combining pre-processed data. In this sense, information fusion is more abstract and high-level than data fusion. However, data fusion can also refer to the general concept of fusing different types of data, such as decision fusion, data integration, multi-sensor data fusion, sensor fusion and other related concepts that we termed data fusion in this book. The goal of using multi-source information is to make better decisions and to build a platform favorable for optimizing decision-making. In this context, the data fusion algorithm is the core; the BIM model is a physical platform that can sense and display the state of physical entities by integrating multi-source data; the digital twin platform can not only sense the state of physical entities through data fusion but also interact with the real world. Figure 8.2 shows how the data fusion algorithm, the BIM model and the digital twin platform are related.

(1) Platform established on the basic and the special algorithms. This platform focuses on fusing data from multiple sources and modes, and usually does not involve the representation and the dynamic update of the physical entities. It includes basic data fusion algorithms such as Bayes fusion, evidence theory, data-driven fusion, etc. At the same time, sensor management affects the efficiency of data fusion, so it should also be included as a part of the data fusion algorithm platform.
(2) Physical entity model platform. This platform focuses on synthesizing information about the physical entities, mainly the geometric, spatial and operational information. For example, there are platforms based on GIS and remote sensing, and platforms based on geometric models, entity module information and operating information, such as the BIM model. The integration of large sets of monitoring data from various sensors into the bridge BIM model can help bridge

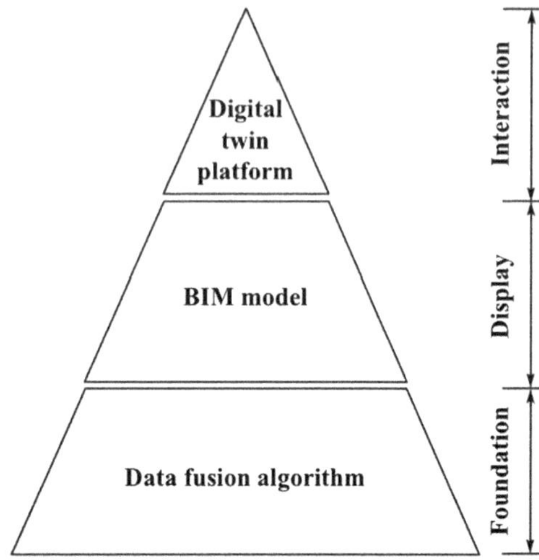

Fig. 8.2 Relationship between data fusion algorithm, BIM model and digital twin platform

managers better understand the health condition of the bridge. BIM plays a role in integrating and presenting the bridge information to the decision-makers.

(3) Physical and digital integration platform. This platform consists of the previous two platforms. It focuses on digitally constructing the physical system principle (the input and the output, or the load and the response) to create a digital replica of the physical system. Its typical representative is Digital Twin Model (DTM). DTM is the integration of 3D models, finite element analysis, detection and monitoring data, etc., which can sense and predict the state of the physical entities, regulate the behaviors of the physical objects, interact with the physical models in the real world, and update the state of digital models in real time. Developing digital twin systems for civil engineering is a huge challenge, but it has transformative significance.

8.2 Data Fusion Theory

8.2.1 Probability Fusion Theory

Probabilistic methods use probability distributions or probability density functions to mimic the uncertainty of data. These methods rely on Bayes estimation, which is based on Bayes theory and has many applications. To apply Bayes reasoning to multi-source data fusion, the system's possible decisions must be independent of each other. This allows the decisions to be treated as a sample space, and the Bayes reasoning to be used to make decisions.

Bayes theorem: Suppose that S is the sample space, B is an event of the sample space, and A_1, A_2,... is a partition of the sample space that satisfies the following conditions:

(1) $A_i \cap A_j = \emptyset (i \neq j)$
(2) $A_1 \cup A_2 \cup \ldots = S$;
(3) $P(A_i) > 0 (i = 1, 2, 3, \ldots)$.

The general form of Bayes probability formula is:

$$P(A_i|B) = \frac{P(B|A_i)P(A_i)}{\sum_j P(B|A_i)P(A_i)} \tag{8.1}$$

Each sensor converts its observations of the target into probability assignments (decision values). These independent decisions are integrated into a unified recognition framework and processed by Bayes' theorem. The system then makes decisions based on inferential rules, such as maximum posterior decisions. The basic idea of Bayes' reasoning can be summarized as the following procedures [1]:

(1) Obtain the probability density expression and the prior probability of each decision according to prior knowledge or statistics;
(2) Use Bayes' formula to find the posterior probability;
(3) Make judgments and decisions according to the posterior probability.

As shown in Fig. 8.3, n different sensors are utilized to monitor the same target, and the target has m attributes $(A_1, A_2, \ldots, A_m)$ that need to be identified by decision. In the system operating process, the sensor layer first connects the received data with the specific attribute of the target according to the extracted information characteristics and the prior knowledge. It gives an output value B about the attribute. Then it computes the probability (likelihood function) of each sensor's output value when each proposition is true. According to Bayes' theorem, we compute the posterior probability when each hypothesis is true. Finally, we judge the attribute conclusion according to inference rules.

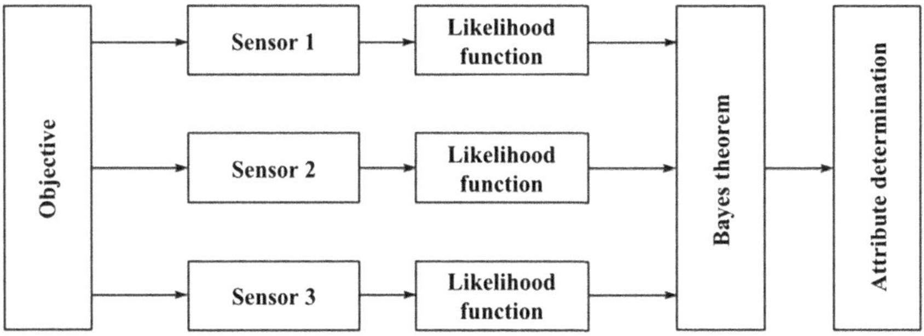

Fig. 8.3 Bayesian Inference Framework Model [1]

The conditional probability of sensor output B under attribute A is P, and B_i is mutually independent, then the joint probability function of the system is as follows:

$$P(B_1 \cap B_2 \cap \ldots \cap B_n) = \prod_{k=1}^{n} P(B_k|A_i) \tag{8.2}$$

Substituting the formula into Bayes formula:

$$P(B_1 \cap B_2 \cap \ldots \cap B_n) = \frac{P(A_i)\prod_{k=1}^{n} P(B_k|A_i)}{\sum_{j}\prod_{k=1}^{n} P(B_k|A_i)P(A_j)} \tag{8.3}$$

In the attribute judgment stage, the final decision for the posterior probability of all attributes (A_1, A_2,..., A_m) in all attribute sets can be made using maximum or threshold decision logic.

Bayes method is one of the earliest methods for uncertainty reasoning, and its major superiority is to use the probability to represent all kinds of uncertainty. It has a solid axiomatic basis and intuitive properties. It also avoids subjective bias and blind search, and computation by combining the prior and the posterior information. It is of low computational cost and high accuracy. The drawback of Bayes theorem is that it assumes all probabilities are independent of each other, which cannot reflect the reality. Furthermore, the accuracy of prior probability is crucial for the quality of Bayes method. However, there is no complete theoretical system available for determining the prior distribution, only some criteria, which poses some challenges on the applications of the Bayes method. Lastly, when the system rules change, all probabilities need to be re-computed to maintain the system relevance and consistency, which limits the system's flexibility.

Bayes fusion example [2] assumes that the identification frame set $H = \{H_1, H_2, H_3, H_4\}$, in which H_1, H_2, H_3, and H_4 represent four possible events whose prior probabilities are known.

$$P(H_i) = (0.3, 0.25, 0.28, 0.17)$$

After processing the data of the three sensors, the likelihood function obtained is

$$P(H_i) = (0.10, 0.40, 0.20, 0.30)$$
$$P(H_i) = (0.35, 0.28, 0.18, 0.19)$$
$$P(H_i) = (0.21, 0.45, 0.10, 0.24)$$

Find the probability after fusion $P(H_i| S_1, S_2, S_3)$.

According to Formula (8.3)

$$\mathrm{P(S_1, S_2, S_3)} = \frac{P(H_i)\prod_{k=1}^{3} P(S_k|H_i)}{\sum\limits_{j}\prod_{k=1}^{n} P(S_k|A_i)P(H_j)}$$

$$Assume\ \alpha_b = \frac{1}{\sum_j \prod_{k=1}^{n} P(S_k|A_i)P(H_j)},$$

$$P(S_1, S_2, S_3) = \alpha_b P(H_i)P(S_k$$

$$= \alpha_b P(H_i)P(S_1|H_i)P(S_2|H_i)P(S_3|H_i)$$

$$= \alpha_b(0.30, 0.25, 0.28, 0.17)(0.10, 0.40, 0.20, 0.30)$$

$$(0.35, 0.28, 0.18, 0.19)(0.21, 0.45, 0.10, 0.24)$$

$$= \alpha_b(0.0022, 0.0126, 0.001, 0.0023)$$

Since $P(H_i|\ S_1,S_2,S_3)(i = 1,2,3,4)$ needs to satisfy the sum of probabilities is 1, $\alpha_b = 55.1310$.

Therefore, $P(H_i|\ S_1,S_2,S_3) = (0.1216,0.6947,0.0556,0.1282)$. From the result of fusion, the probability of target H_2 is relatively high.

8.2.2 Evidence Confidence Reasoning

The confidence function theory stems from Dempster's research and was enhanced by Fisher's probabilistic reasoning method. It was mathematically formalized by Shafer under the general theory based on evidence and is now commonly known as Demster-Shafer (DS) evidence theory [3]. The confidence function theory is a theoretical framework of evidence reasoning that is appealing and widely used for handling uncertainty and imprecise data. DS evidence theory introduces the concept of assigning confidence and plausibility to potential measurement hypotheses and combining them with predefined fusion rules. Several basic concepts of DS evidence theory are introduced below [2].

(1) Basic Probability Function

Assume Ω is an identification frame, and a subset of Ω can represent all propositions in the domain. The power set 2^{Ω} contains Ω, all its subsets and the empty set. For example, if the mutually exclusive and complete object identification set $\Omega = \{a_1,a_2\}$, the power set contains $2^2 = 4$.

$2^{\Omega} = \{\varphi, a_1, a_2, a_1 \cup a_2\}$, The basic probability function is a mapping defined at $2^{\Omega} \to [0, 1]$ that satisfies:

$$M : 2^{\Omega} \to [0, 1]$$

$$\sum_{A=\Omega} M(A) = 0$$

Then, M is said to be the probability allocation function on 2^{Ω}, and $M(A)$ is called the basic probability function representing the exact trust in A, which is an element of the object identification framework (e.g., a_1).

(2) Trust function

Trust function *Bel* is defined at $2^{\Omega} \rightarrow [0, 1]$, and

$$\text{Bel}(A) = \sum_{B \subset A} M(B), A \subseteq \Omega$$

In the formula, *the Bel* function is also known as the lower limit function, indicating complete trust in *A*.

(3) Likelihood function

The likelihood function *Pl* is defined at $2^{\Omega} \rightarrow [0, 1]$, and

$$Pl(A) = 1 - \text{Bel}(\overline{A}) = \sum_{B \cap A} M(B), A \subseteq \Omega$$

where *Pl* function is also called the upper bound function and $\overline{A}$ is the complement of *A* in Ω. The relationship between the trust function and the likelihood function is as follows: $Pl(A) \geq Bel(A)$, $(Pl(A), Bel(A))$ is called the trust interval.

DS evidence fusion rules: M_1 and M_2 are Ω two probability distribution functions; the orthogonal and $M = M_1 - M_2$ are defined as follows:

$$M(\varphi) = 0$$
$$M(A) = (1 - k_c)^{-1} \sum_{B \cap C} M_1(B) M_2(C)$$

where $k_c = \sum_{B \cap C = \varphi} M_1(B) M_2(C)$ is called the conflict factor. If $k_c = 0$, it means that the two pieces of evidence are in complete conflict, then there is no orthogonality and M; If $k_c \neq 0$, then orthogonality and M are also a probability distribution function. The fusion of multiple probability allocation functions can be performed in pairs, and the final fusion result is independent of the fusion sequence. Figure 8.4 shows the fusion process of multiple sensors.

DS evidence theory example [2] (see Table 8.1) target recognition framework $\Omega = \{a_1, a_2\}$, is $2^{\Omega} = \{\varphi, a_1, a_2, a_1 \cup a_2\}$. Now there are two types of sensors S_1 and S_2, and the basic probability distribution function of target detection by the two sensors are as follows:

$$M_1 : M_1(\varphi) = 0, M_1(a_1) = 0.6, M_1(a_2) = 0.2, M_1(\Omega) = 0.2$$
$$M_2 : M_2(\varphi) = 0, M_2(a_1) = 0.4, M_2(a_2) = 0.5, M_2(\Omega) = 0.1$$

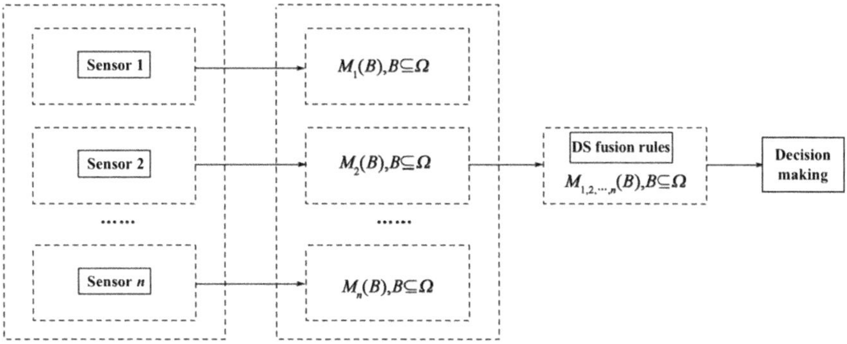

Fig. 8.4 DS evidence fusion process

Table 8.1 Computation process of DS fusion in the example

M_2	M_1			
	$M_1(\varphi)=0$	$M_1(a_1)=0.6$	$M_1(a_2)=0.2$	$M_1(\Omega)=0.2$
$M_2(\varphi)=0$	0	0	0	0
$M_2(a_1)=0.4$	0	0.24	0.08	0.08
$M_2(a_2)=0.5$	0	0.3	0.1	0.1
$M_2(\Omega)=0.1$	0	0.06	0.02	0.02

The DS evidence theory fuses M_1 and M_2, and the basic probability distribution function M_{12} is obtained after fusion.

$$\begin{aligned}
k_c &= \sum_{B\cap C=\varphi} M_1(B)M_2(C)\\
&= 0+0+0+0+0+0+0.3+0.08=0.38\\
M_{12}(a_2) &= (1-k_c)^{-1}\sum_{B\cap C=\{a_2\}} M_1(B)M_2(C)\\
&= \frac{0.1+0.1+0.02}{1-0.38}=0.3548\\
M_{12}(\Omega) &= (1-k_c)^{-1}\sum_{B\cap C=\Omega} M_1(B)M_2(C)\\
&= \frac{0.02}{1-0.38}=0.0323
\end{aligned}$$

The computation process of DS fusion in the example:

$$M_{12}: M_{12}(\varphi)=0, M_{12}(a_1)=0.6129, M_{12}(a_2)=0.3548, M_2(\Omega)=0.0323$$

8.3 Multi-source Data Fusion Driven State Assessment

The safety issue of bridges in service is becoming more and more serious in China. Firstly, a large number of bridges in service have reached the aging stage. Years of operational loading and environmental corrosion have caused various degrees of structural damages to these bridges. Secondly, due to rapid economic development, some areas have experienced greatly increased traffic volume and heavy vehicle numbers in recent years, which caused the actual operational loads on bridges to exceed the design loads and jeopardized their structural safety. Lastly, bridge structures may face extreme events (earthquakes, strong wind, fire, etc.) during services, which may result in structural damage and even collapse. This problem is especially acute for large-span cable-supported bridges and irregular-shaped bridges with complex structural static and dynamic responses. To ensure the safe operation of bridges and prevent serious accidents, bridge condition assessments are essential.

8.3.1 Data-Driven State Assessment

Large bridges have service life cycles that usually last for several decades or even centuries after their construction. However, the health condition of bridge structures will inevitably deteriorate due to the combined effects of long-term environmental erosion, unavoidable structural aging, fatigue, traffic loads, natural disasters and other factors. This will lead to significant decrease in the bridge's load-carrying capacity and shortened service life. In extreme cases, it may even cause disastrous accidents. Moreover, quality issues with large bridges due to imperfect design practices and insufficient construction techniques are also causes of bridge accidents. Many tragic accidents have demonstrated the necessity of continuous health assessment of bridge structures in order to avoid tragedies. Bridge structure health assessment has important practical implications for ensuring the safe operation of bridge structures. It is urgent to enhance real-time evaluation, inspection and maintenance for the safe performances of bridges in service.

After years of research, some achievements and progresses have been made in bridge structural health monitoring; for example, the comprehensive monitoring contents, advanced and diversified monitoring equipments, gradually improved system functions and developed bridge damage identification theories have come into practice. Some bridges have been tested from the beginning of construction to obtain as much uninterrupted structural information as possible. However, due to the multi-disciplinary nature of bridge structural health monitoring and complexity of the environmental conditions, there are still many shortcomings and defects in the current health monitoring systems for bridges, which also limit their applications in practice. These shortcomings mainly include.

① Heavy reliance on automatic monitoring and neglect of manual inspection. From a practical perspective, it is not easy to achieve effective health assessment merely relying on the monitoring and evaluation systems;

② Heavy emphasis on hardware system and limited attention on processing and evaluating the test data. At the algorithm level, the current system usually cannot extract effective information in time from the massive data. There are still gaps and flaws in damage identification and comprehensive evaluation for health monitoring;

③ Interface has no visual functions. At the software level, it is more user-friendly for maintenance and management personnel to visually obtain monitoring and analysis results; and.

④ Current structural evaluation technology cannot make timely and effective judgments on the health of bridge structures in a big data environment.

Bridge structural condition assessment can be described as a process from data collection to data management. These data include all data related to the health condition of the bridge structure. The current bridge structure health condition assessment systems used in China and other countries generally utilise various monitoring sensors to obtain data for assessing the bridge structural health condition via changes in indicators of the static and dynamic performances of the bridge in operation. The amount of data so obtained is limited and the data type is monotonous, which cannot help to accurately predict the evolution of the bridge health condition. In terms of data management, due to limitation of the data processing capabilities, potential value of the data has not been fully exploited and the data processing cycle is long. With the advent of cloud computing era, big data technology has attracted extensive attention from both the Chinese and other countries' academics under the promotion of cloud computing concept. The features of big data can be summarized as the massive amounts of data, diverse types of data and short processing cycles. The data collected during bridge structural health condition assessment fully comply with the features of big data. Therefore, how to rapidly process a massive amount of data collected by the monitoring sensors to provide a basis for bridge structural health maintenance decision making is an urgent scientific issue that needs to be addressed.

Multi-source information fusion is also called multi-sensor information fusion, where sensors refer to the generalized sensors that include not only the conventional physical sensors but also the various information resources related to the environment, such as the human perception systems. In general, multi-source information fusion is a comprehensive analysis and integration control process using logical algorithms, computers, etc., based on information resources obtained at different times and spaces to obtain more accurate and informative results. In recent years, real-time health monitoring systems and state evaluation practices driven by multi-source data fusion have attracted attention from both the academics and the practitioners. Yang Jie et al. [9] summarized and analyzed the framework model for data fusion, and discussed its establishment method for fusing various sensor information in a long-term monitoring system. Nie Gongwu et al. [10] summarized some deficiencies in maintenance management and long-term monitoring systems, and proposed several ideas concerning the complementary relationship and mutual integration.

8.3.2 State Evaluation Algorithm

(1) Conventional Evaluation Method

The conventional methods for condition assessment include expert evaluation, analytic hierarchy process (AHP) and reliability theory. The expert evaluation method involves pre-formulating the scoring criteria and the damage classification standards based on survey statistics and experimental analyses. Experienced engineers then inspect and score the existing bridges and evaluate their material qualities, damage degrees, etc. The main steps of the expert evaluation include

Dividing the bridge's state-level standard;

Determining the components involved in the state evaluation and their corresponding weights in the entire evaluation system; and

Conducting a state evaluation based on the detection results, determining the bridge's state score based on the state level of each component and its corresponding weight.

The main advantages of this method are conceptual clarity and simple calculation. However, there are also drawbacks for the approach, such as the fact that individual components with severe damages but small weight values cannot be reflected in the overall bridge evaluation. In addition, the components involved in the evaluation and their weights set are determined according to a standard, and the process lacks flexibility. The AHP uses hierarchical weighting for evaluation, and the selected items are evaluated using a percentage-based scoring system. Its basic idea is similar to that of the expert evaluation. The difference between them is that AHP uses pairwise comparison judgment matrices to determine the weights, and further introduces variable weight principles to reflect the scores of bottom-level components with poor conditions in the overall evaluation.

Reliability theory uses failure probability or reliability indicators to quantify the safety level of structures. Because it is based on probability statistics, this theory can effectively deal with uncertainties such as the load and the resistance, especially those affecting the structural reliability. Therefore, it can provide a reasonable theoretical framework for bridge evaluation. The specific safety assessments based on the reliability theory can be realized according to the following steps:

① confirming possible failure modes; failure modes can be classified as the component and the system failure modes depending on the bridge form and the load type. Available confirmation methods include failure tree analysis, model testing and structural analysis;
② selecting analysis models representing various failure modes, including linear elastic analysis, nonlinear analysis and limit analysis;
③ determining the uncertainties of the loads and the structural resistance at the component level and the system level via the load models and the resistance models, respectively;
④ using component or system reliability methods to calculate reliability indicators or failure probabilities; and

⑤ checking the safety of structures accordingly. If the reliability indicator exceeds the specification value, then the structure is safe; otherwise, it is unsafe.

(2) Data-Driven Evaluation Algorithm

The core idea of a data-driven evaluation algorithm is to evaluate bridge structures using data instead of expert's experience in the conventional evaluations. A specific index is selected as the basis of structural evaluation for this approach, and then the statistical model is utilized to predict this index's change.

(1) Time Series Models

Time series models are widely used in data-driven evaluation algorithms. Time series are ordered based on random variables of time. By modeling the time series, we can discover the underlying statistical pattern in the continuous historical data and predict its future trend according to the pattern. Common time series models include Autoregressive (AR) models, Moving Average (MA) models, Autoregressive Moving Average (ARMA) model, and Autoregressive Integrated Moving Average (ARIMA) model.

A state index of the bridge structure (such as vertical settlement of the bridge) can be chosen as the evaluation criterion. The time series model treats the data series generated by the prediction object over time as a stationary time series and then builds a regression model based on the past values and random errors of the dependent variables of the time series. In the ARMA model, we assume that the predicted value X_t at the next time is changed by the actual value at the previous p times $X_{t-1},X_{t-2},\ldots,X_{t-p}$ and the error sequence of the first q moments $e_{t-1},e_{t-2},\ldots,e_{t-q}$ is formed by linear combination, as shown in Eq. (8.4):

$$X_t = \sum_{j=1}^{p} a_j X_{t-j} + \sum_{j=1}^{q} b_j e_{t-j} \tag{8.4}$$

By estimating each parameter, q, a_j, and b_j, a time series model for predicting the future state of the bridge can be obtained, and the future state of the structure can be predicted by the model to evaluate the health of the bridge. We can further consider multiple state indicators and use the multi-variable time series model for prediction correspondingly. For estimating parameters in the time series model, readers can refer to professional books on time series.

(2) Hidden Markov Model

The hidden Markov model is a probabilistic model that assesses the bridge state by predicting the change of a certain state parameter of the structure. It uses a similar principle as the time series model to establish a prediction model to evaluate the health of the bridge structure.

We must first learn the Markov chain to understand the hidden Markov model. If a random process $\{X_t, t = 1,2,3,\ldots,n\}$, whose state X_{t+1} at time $t + 1$ is only related

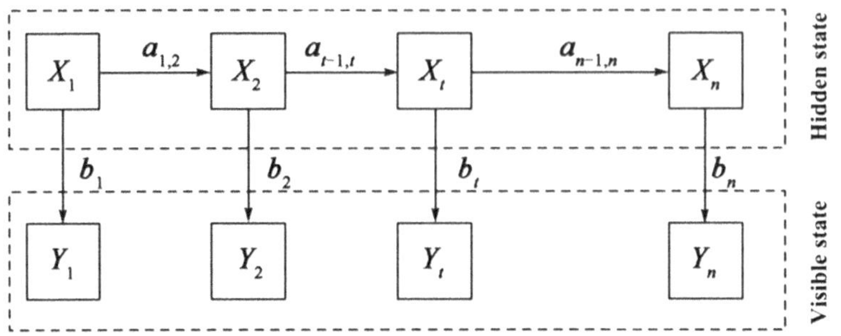

Fig. 8.5 Schematic diagram of hidden Markov model

to the state X_t at time t and not to the state before time t, then such a process is called Markov chain.

The hidden Markov model is a model consisting of two stochastic processes, namely a potentially unobservable stochastic sequence process (Markov chain) and an observable stochastic sequence process, as shown in Fig. 8.5. In Fig. 8.5, we call $a_{t\text{-}1,t}$ the transfer probability from t-1 to t, and b_t the output probability.

Using the statistical method, the transfer probability and the output probability, which are two key parameters in the hidden Markov model, can be determined based on the monitoring data (for specific calculation methods of the parameters, readers can refer to relevant books). This can help to predict the state index of the bridge and assess the bridge state at any time.

(3) Evaluation Algorithm Driven by Multi-source Data Fusion

The conventional statistical methods for bridge state evaluation often rely on specific structural features and therefore have limitations. Some algorithms that can fuse multi-source data have been developed in recent years, including the Bayesian model framework, the Kalman filtering method and the machine learning-based correlation models. The Bayesian model framework has been explained in Sect. 8.2.1. This section will focus on the principles of data fusion evaluations based on the Kalman filtering algorithm and the machine learning model. Most of the current fusion algorithms are for homogeneous or similar signals (such as image-image or vibration–vibration data). For heterogeneous signals (such as sound-image data), fusion evaluation still depends on manual methods. This is also an important future research direction in bridge evaluation.

(1) Kalman Filtering Method

The Kalman filtering method is an algorithm that uses linear system equations and input–output data to estimate the system state optimally. It is an evaluation method that uses state recursion for data estimation. The estimation process of this method is shown in Fig. 8.6.

The Kalman filter method consists of two stages: the prediction stage and the update stage. In the prediction stage, the filter uses the previous estimated state to

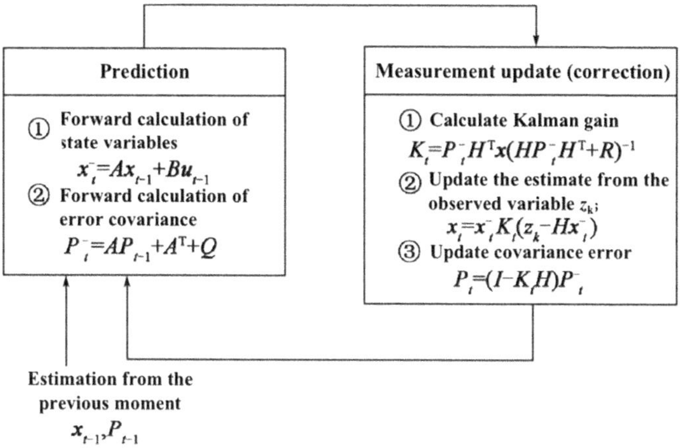

Fig. 8.6 Estimation process of Kalman Filter

predict the current state. In the update stage, the filter uses the current observed state to optimize the prediction and obtain a more accurate new estimate. After multiple updates, the system state will converge, and then we can use a Kalman filter to predict.

We can use several Kalman filters simultaneously for data fusion. As shown in Fig. 8.7, we first use a Kalman filter for each sensor data and then use a data fusion algorithm in a global filter to fuse the local optimal estimates. This can avoid the distortion of evaluation results due to a single sensor's failure.

(2) Neural Networks and Deep Learning Models

Machine learning, an important branch of artificial intelligence, has developed rapidly in recent years and has great potential in bridge health evaluation. Neural networks and deep learning models have become a research hotspot because of their strong generalization, self-learning, adaptive and robust abilities. Here we will introduce the application of neural networks and deep learning models in structural state

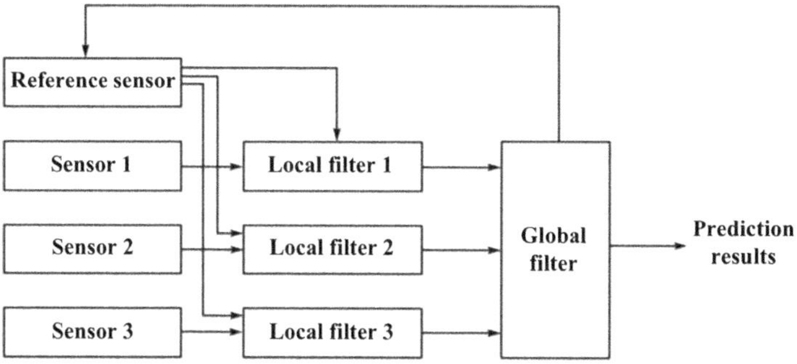

Fig. 8.7 Data fusion evaluation based on Kalman Filter

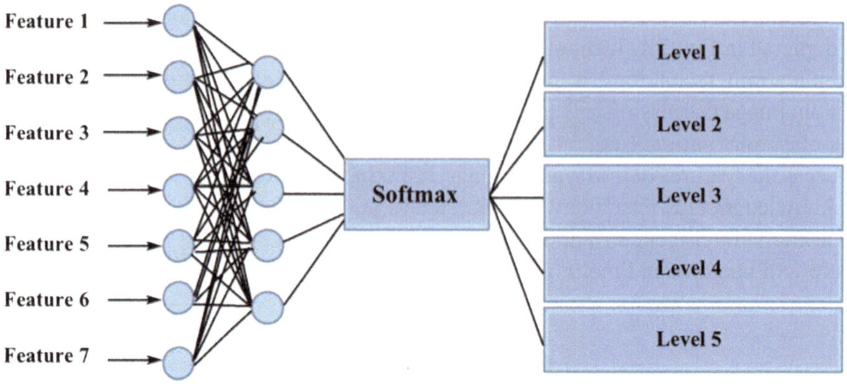

Fig. 8.8 BP neural network structure

evaluation. The specific implementation of neural networks has been introduced in Chap. 6.

One of the significant features of neural networks is that they can integrate the feature fusion and the state evaluation decision. Neural networks for structural state evaluation usually have this structure: the input layer is the several features to be fused, and the output layer is the structural state evaluation level, with the Softmax as the activation function, as shown in Fig. 8.8.

In shallow neural network models, the structural evaluation features must be manually extracted, which requires human intervention and cannot be fully automated. In recent years, researchers have tried to apply deep networks to bridge structural state evaluation. Figure 8.9 shows that the original data is extracted by a feature extractor via the feature extraction, then fused by a fusion layer, and the structural state evaluation level is finally output by a classifier. The feature extractor usually consists of several convolutional layers or fully connected layers, which adaptively extract various data features. Deep learning models can realize the fully automated process of data-evaluation results, and are evaluation methods with fine research and application prospects.

8.4 Applications of State Evaluation Based on Multi-source Data Fusion

This section introduces several specific applications of state evaluation based on multi-source data fusion in bridge engineering, so that readers can better appreciate the theories presented above.

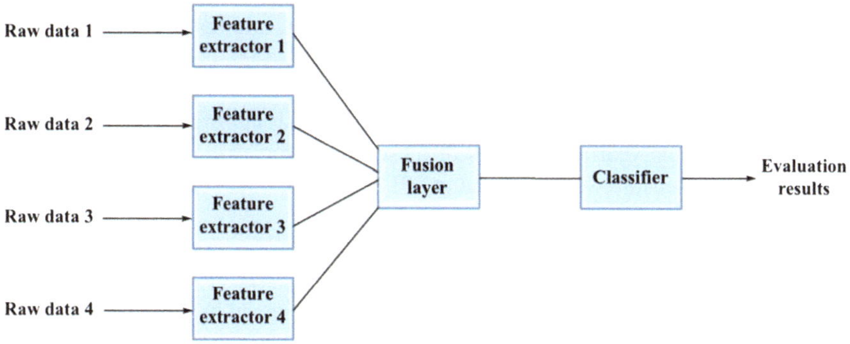

Fig. 8.9 Framework of deep learning model for feature fusion

8.4.1 Comprehensive Evaluation of a Cable-Stayed Bridge Based on Multi-indicator Evidence Fusion

This example is taken from References [12, 13]. The authors evaluated the comprehensive performance of a cable-stayed bridge utilizing multi-indicator evidence fusion.

(1) Bridge Profile

A large-span steel cable-stayed bridge was built in Jiangsu Province in 2005. The main bridge with a total length of 1288 m has a steel tower and steel box girders. The main defects detected after a detailed inspection are:

(1) For the stay cable system, 16.1% of the steel guarding barrels lost their protective paint, 16.7% of the dampers lost their coating layer, and 8.3% of the anchor heads leaked their anti-corrosion grease.
(2) For the steel box girder, 12% of the box rooms lost their coating layer, with a maximum deterioration rate of 0.13% for a single box room. 4.2% of the box rooms corroded, with a maximum corrosion rate of 0.08% for a single box room. 73.6% of the inclined bar cracked, with a maximum crack length of 60 cm.
(3) For the tower, 55.8% of the chambers lost their coating layer, with a maximum deterioration rate of 0.27% for a single chamber. 42.3% of the chambers had bolt corrosion or loosening, with a maximum breakage rate of 9.4% for a single chamber.
(4) The concrete cover beam for the pier was damaged, with a damaged area of 0.05 m^2.
(5) For the supports, 31.8% were damaged with slightly corroded bolts and partially corroded steel plates.
(6) The bridge deck pavement was marred with local oil stains and pits. Two long longitudinal cracks were detected. The expansion joint rubber devices had some aging strips.

(7) Protective paint of 2.3% of the guardrails fell off, and some guardrails were deformed.
(8) Forces of the longest cables on both sides of the south tower and the mid-span deflection of the main beam were monitored for a long time. Figures 8.10 and 8.11 shows the yearly variations of these parameters. Since all the deflection monitoring sensors were replaced in 2011, only the data before 2011 are given here.

(2) Evaluation Index

Figure 8.12 shows the indexes for evaluating the cable-stayed bridge's health condition. The target layer outputs the bridge comprehensive technical condition. The index system has three layers below the target layer. The first layer has the five

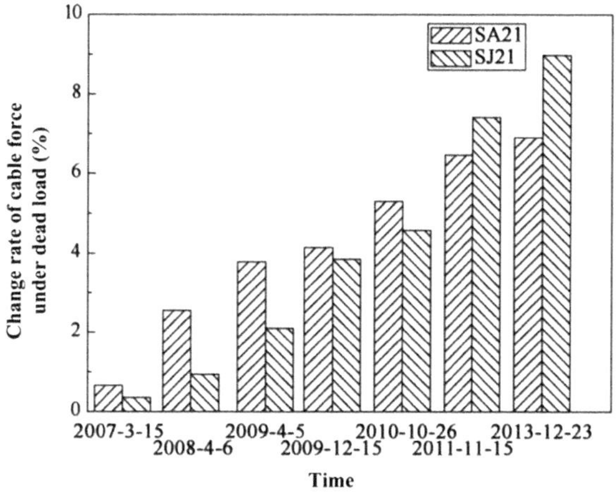

Fig. 8.10 Annual changes in cable forces of SA21 and SJ21, on both sides (revised from Ref. 12)

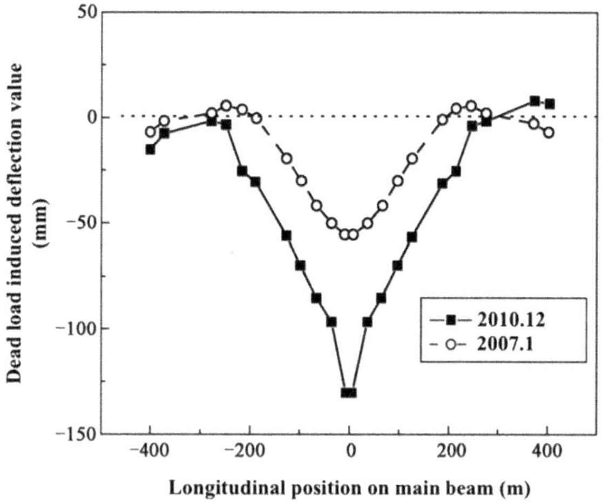

Fig. 8.11 Long-term variation of midspan deflection of main beam (revised from Ref. 12)

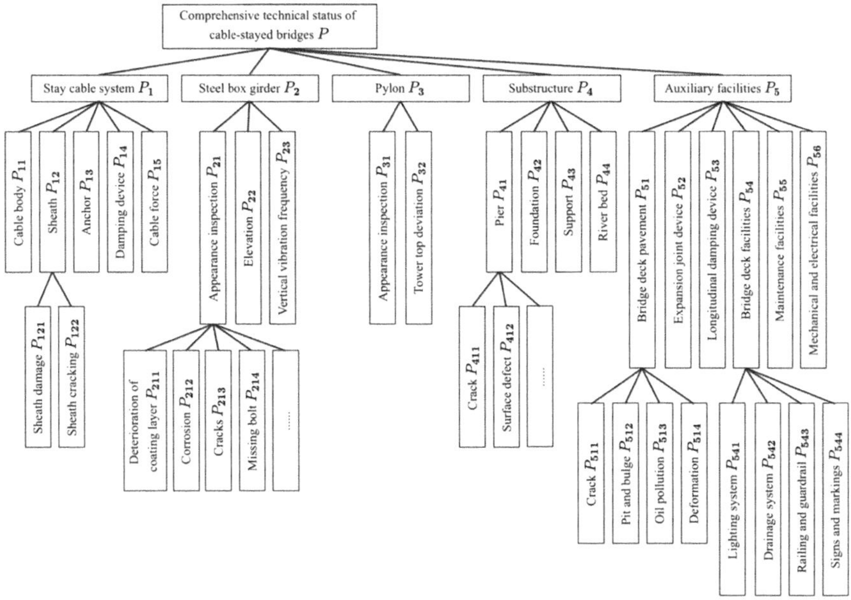

Fig. 8.12 Comprehensive evaluation index system for cable-stayed bridges (revised from Ref. 12)

major components of the cable-stayed bridge, the second layer has more detailed components, and the third layer has specific distress indexes.

(3) Fusion Evaluation Method

Figure 8.13 shows the computation process of the fusion evaluation method utilized for cable-stayed bridges. The evaluation is mainly based on the fuzzy theory and a hierarchical evaluation algorithm. The specific steps are as follows:

(1) Refer to the relevant formula for component score deduction in the evaluation standard to compute the component score index in layer two. Then, compute the membership degree. The evaluation index uses a piecewise linear ascending index evaluation standard to quantitatively define the relationship between the distress degree and the deduction value.
(2) Use the analytic hierarchy process to compute the weights of the indexes at the first and second layers in the hierarchy system. Then, combine the fuzzy membership degree and weight results to compute the basic reliability of indexes at the second layer.
(3) Use fusion rules to compute the credibility of the index of the first layer and the overall evaluation result of the top-layer cable-stayed bridge. Synthesize the basic confidence of the top and first layers from the lower layer evidence. Use Dempster's combination rule for T pieces of evidence to synthesize the basic credibility.

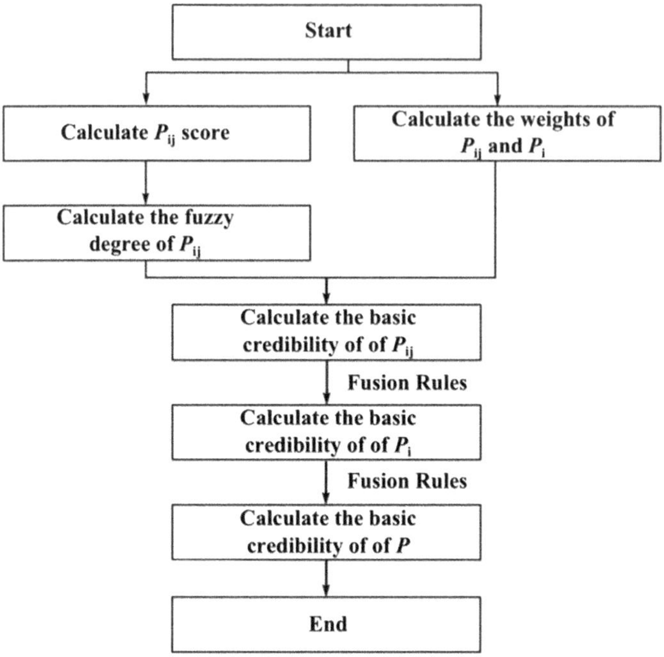

Fig. 8.13 Computation process of fusion evaluation method for cable-stayed bridges (revised from Ref. 12)

(4) Evaluation Result

Utilizing the above methods, the final basic reliabilities of the five types of components and the overall bridge were obtained, as shown in Fig. 8.14. According to Fig. 8.14, the cable-stayed bridge is generally in good condition.

8.4.2 Safety Evaluation of a Cable-Stayed Bridge Based on Neural Network Data Fusion

This case is taken from Reference [14]. The authors used a neural network to complete data fusion of the evaluation data so as to realize the safety assessment of a cable-stayed bridge.

(1) Bridge Status

A cable-stayed bridge, located 2 km downstream of Hunhe Changqing Bridge in the south of Shenyang City, was built in 2003 and is a Class I urban bridge. In April 2011, the Traffic Experiment Center of Harbin Institute of Technology conducted dynamic and static load tests and appearance inspections on the bridge. The test results are:

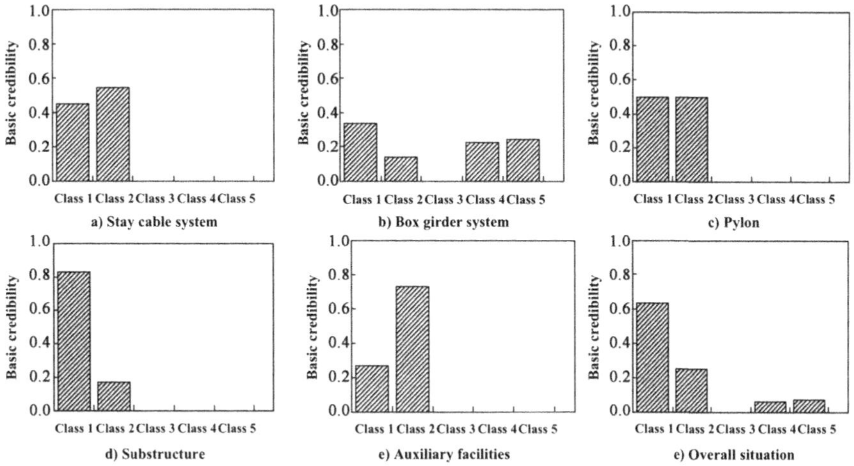

Fig. 8.14 Evaluation results of main components and overall bridge (revised from Ref. 12)

(1) The appearance survey found that the PE sheath of a middle span stay cable arranged in the main span was broken. There were cracks in the transverse diaphragms of the main beam, and most crack widths did not change much.
(2) When the bridge was completed in November 2003, the measured deck elevation in the main span was 18 mm lower than the design elevation. After seven years of operation, the theoretical middle deflection of the main span was 59 mm, but the measured deflection was 87 mm, which was 28 mm more than the theoretical deflection.
(3) Without vehicle loads on the bridge deck, data was collected on permanent strain detection points in the main beam. The comparison results showed that the maximum tensile strain of each point during monitoring was 5 μm, and no obvious tensile strain was observed.
(4) The bridge's natural frequencies were monitored, and the dynamic model of the bridge was established by using frequency domain analysis data collected in a passing vehicle scenario. The middle span of the main span of the bridge deck was a measuring point. The first eight order frequencies measured at this point were 0.51 Hz, 1.37 Hz, 2.85 Hz, 2.95 Hz, 3.09 Hz, 3.24 Hz, 3.38 Hz, and 5 Hz, respectively.

(2) Evaluation Index

The indexes shown in Fig. 8.15 were used to evaluate the overall condition of the bridge.

(3) Fusion Evaluation Method

The expert evaluation method was used to determine the main evaluation indicators, as shown in Table 8.2.

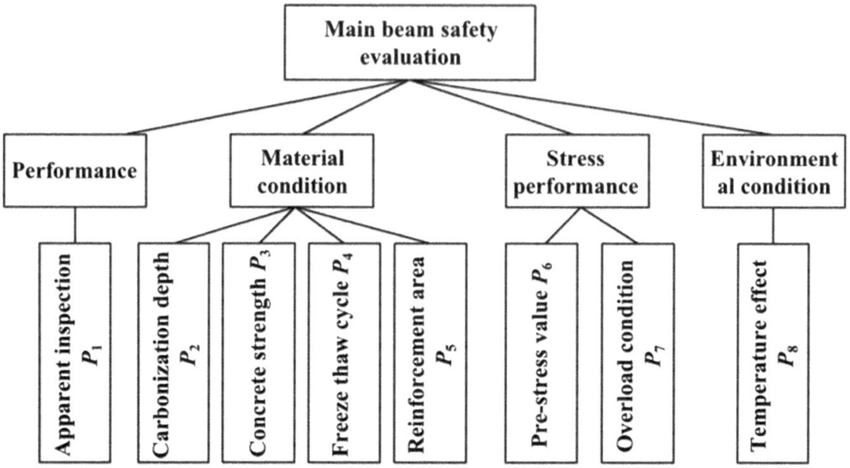

Fig. 8.15 Comprehensive evaluation index system for bridges (revised from Ref. 14)

Table 8.2 Determination of evaluation indicators (Reproduced from Ref. 14)

Indicator Name	Apparent inspection	Pre-stress value	Concrete strength (MPa)	Freeze–thaw cycle (Times)	Overload condition	Temperature effect (°C)
Value	0.8764	0.9544	63.40	512	1.00	0

The neural network shown in Fig. 8.8 was used to fuse the evaluation indicators and to evaluate the overall state of the bridge, which is divided into five levels, among which level 1 has the best security condition. A large amount of data were utilized to train the neural network in the bridge database.

(4) Fusion Evaluation Result

The bridge was identified as being at the second class safety level and having enough safety reserve for current use. This matched the inspection results and proved the model's effectiveness and accuracy.

8.4.2.1 Data Fusion Method Based on Kalman Filter Applied to Bridge Response Prediction

This case is taken from Reference [15]. The authors used the Kalman filter method to fuse the data from two sensors so as to realize the response prediction of the bridge.

(1) Bridge Status

This case is a field test, as shown in Fig. 8.16. The prestressed concrete bridge has a 11 m span and four prestressed concrete girders as the main load-carrying

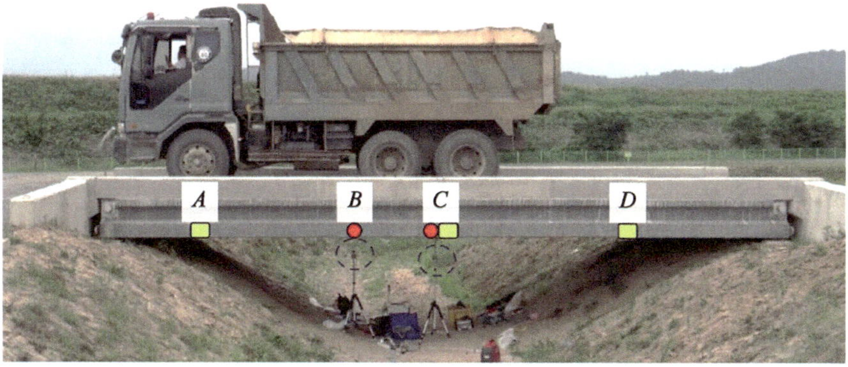

Fig. 8.16 Overview of bridge [15]

components. A truck weighing 28.93t passes the bridge at 5 km/h and 15 km/h speeds, respectively. Two laser displacement sensors and two accelerometers were installed at locations B and C on the bridge. Three strain gauges were installed at locations A, C and D.

(2) Evaluation Index

The bridge displacement was estimated by integrating the strain gauge and accelerometer data for bridge evaluation.

(3) Fusion Evaluation Method

The employed method used Kalman filter to estimate the displacement based on data from two different measurements. The basic idea was to extract the displacement information from each measurement in the best way possible. The acceleration response had more information in the high-frequency band, while the strain response contained more information in the low-frequency domain. So, by using the information selectively, this method estimated the displacement more accurately based on each measurement.

(4) Fusion Evaluation Result

As shown in Fig. 8.17, the present data fusion method can give a good estimation of the bridge displacement. It demonstrated that the bridge displacement induced by the passing vehicle was low, and the bridge safety was high.

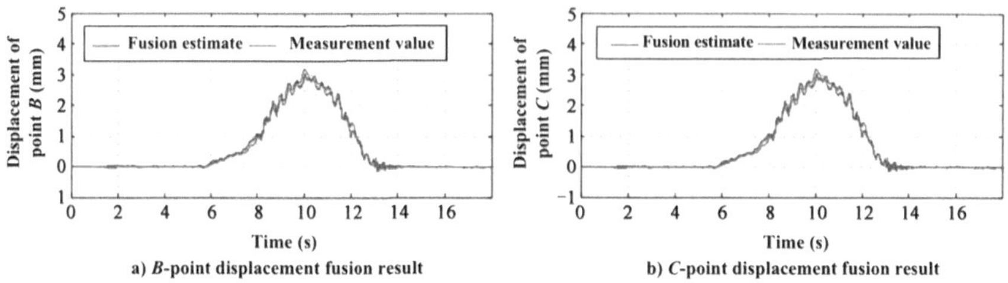

Fig. 8.17 Displacement fusion results [15]

8.5 BIM/BrIM for Bridge Management

Although in recent years, BIM technology has been widely and maturely applied in the construction industry, its promotion in bridge engineering is relatively lagging behind the housing construction industry. Currently, the application of BIM technology in bridge engineering is mainly based on a single function point. Different projects' application objects, implementation subjects and strategic focus are different. In terms of bridge management, designing and developing an information system that can realize mass data fusion and application based on the whole life cycle and accommodate multi-participant is crucial [18].

8.5.1 Bridge Information Modeling Technology Based on BIM/BrIM

(1) Concept of BIM Technology

BIM technology is now a key research direction in all fields of civil engineering. It is an important means to improve the efficiency of the industry. The National Building Information Modeling Standard released by the National BIM Standards Board (NBIMS)—United States Version defines BIM as: "BIM is the data expression of the physical and the functional characteristics of facilities or building projects, a shared knowledge resource for the decision-making of the full life cycle of the facilities or building projects from construction to demolition, and a semantic, continuous and digital building model that can substantially represent the full life cycle of the physical building [19]."

BIM technology originated in the 1970s when Charles Eastman, a scholar with both architectural and computer backgrounds, wrote "An Outline of Building Description System" in 1974. In this work, he proposed the Building Description System (BDS) to address the problems of redundancy, information updates and analysis costs in architectural drawings. BDS was the first attempt to create a data structure, hardware, graphic display and element placement for building information modeling [20]. However, due to limitations in computer power and user interfaces, BIM technology did not become popular until the 2000s, when Autodesk acquired Revit Technology Company and promoted the term BIM [21]. Since then, BIM technology has been widely used and well developed worldwide, with various applications in project management, simulation, collaboration and life cycle analysis.

In China, the development of BIM technology can be divided into four stages. The first stage (1998–2005) was the conceptual and research period based on IFC theory. It carried out application exploration for a few pilot projects. The second stage (2006–2010) took the BIM application of the Beijing Olympic Games and Shanghai World Expo as representatives. In BIM applications, the Chinese practitioners cooperated with other countries' engineers in the design stage and initially used space management and collision detection functions. The third stage (2011–2015) saw the application of BIM technology been popularized in China, mainly in the conventional and digital applications. Scholars conducted theoretical and practical research from multiple perspectives, including collaborative design, software development, collaborative construction and life cycle. The fourth stage (2016 till now) has seen national and local governments introduced a series of policies to promote the development of BIM. BIM technology has been used in more and more practical projects. Meanwhile, combined with big data, cloud computing, virtual reality and other technologies, the application of BIM technology demonstrated the characteristics of industrial and integrated applications [22].

(2) Definition and Characteristics of BrIM

Bridges are special construction projects with various structural types, such as beam bridge, arch bridge, suspension bridge, cable-stayed bridge, etc. They span long distances, have complex structures, and face challenging construction environments. Therefore, they are more complex than buildings in terms of design and construction, and require urgent maintenance management after completion. BIM technology has developed rapidly in recent years, and its theoretical research and practical application have made great progress. However, the research and application of BIM technology in the field of bridges is less than that in the construction field. In recent years, bridge practitioners have been trying to introduce BIM technologies applied in the construction field to the bridge field to form the application of BrIM technology.

BrIM (Bridge Information Modeling) is an expanded application of BIM, whose objects are bridges. BrIM technology can be interpreted from three aspects [23], as shown in Table 8.3.

To sum up, BrIM is not a kind of software nor a three-dimensional model, but a new idea with its related methods, platforms, modeling software, etc. "Br" represents bridge, reflecting the application scope of BrIM; "I" refers to information, which covers all aspects of information in the whole life cycle of the project and reflects the breadth of BrIM. "M" indicates model, modeling, and management simultaneously, reflecting the depth of BrIM [24].

BrIM technology has eight main characteristics which are introduced below:

Table 8.3 Concept of BrIM

BrIM	Description
Bridge information model	2D CAD drawings are transformed into 3D bridge information models so that the drawing information of different majors, such as structure, hydropower, decoration, etc., can be concentrated into a 3D bridge information model. It is convenient for the overall view of bridge information
Bridge Information modeling	The parametric model application is realized, and the three-dimensional bridge information model is used to realize design optimization, pipeline synthesis, virtual construction, engineering quantity calculation, and other applications. The model's value is constantly mined, and new technical problems in actual projects are solved
Bridge information management	The digital project management platform is built based on the three-dimensional information model. Design, cost, quality, and safety management are coordinated on the project management platform to realize the model-based platform, paperless office, and fine management. It improves the efficiency of project management

(1) Visualization. In the BrIM model, the whole process can be visualized. It can be used to display renderings, generate reports and enable communication and discussion, and any decision in the project design, construction and operation stages to be carried out in a visible way (Fig. 8.18).

(2) Simulability. The BrIM model can simulate the structural components and related engineering activities of bridge engineering in the real environment. In the design stage, BrIM can simulate all structural parts that need to be considered in the design. In the bidding and construction stage, BrIM can provide a 4D simulation to help formulate a reasonable construction plan and guide the construction. It can also provide 5D simulations to help achieve cost control. In the later stage of operations, BrIM can simulate emergencies, such as destructions of the bridges in extreme events such as earthquakes, ship collisions and explosions.

(3) Coordination. The BrIM model can coordinate the conflicting problems from various disciplines and generate the coordinated results in the form of data in the early bridge construction stage.

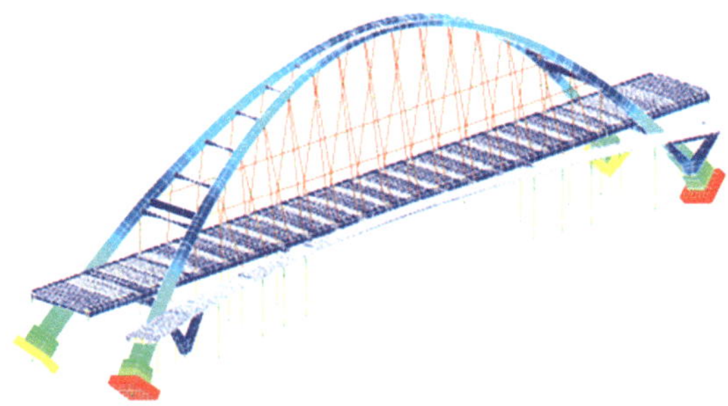

Fig. 8.18 Bridge visual model

(4) Optimisability. The BrIM model provides information on the actual existence of the bridge, including geometric information, physical information and rule information.
(5) Graphability. Through visual display, coordination, simulation and optimization of the bridge, BrIM can help the owner draws a comprehensive baseline diagram (after collision inspection and design modification to eliminate the corresponding errors), a comprehensive structural diagram (embedded casing diagram), and provide collision detection report and suggestions for improvement [25].
(6) Integration. Based on BrIM technology, integrated management can be carried out throughout the life cycle of the engineering project from design to construction and operation. The core of BrIM technology is a database established on a three-dimensional computer model, which not only contains the design information of the bridge but also can provide the whole process information from the structural design to the completion and use of the bridge, and even the end of the use cycle.
(7) Parameterization. Parametric modeling refers to establishing and analyzing the model through parameters rather than numbers. Simply changing the parameter values in the model can create a new model. The elements in BrIM take the form of components. The differences between these components are reflected by adjusting the parameters. By saving all the information of elements, a digital bridge structure is created.
(8) Information completeness. The BrIM model contains all the information about the facility, and BrIM technology can integrate the information of all parties involved in the project. Information can be shared and coordinated through the network technology based on integration. For the operation of an oversized bridge, information such as manual inspection, basic information and health monitoring system can be integrated and shared by the bridge management unit, the monitoring unit and the scientific research unit.

In nature, BrIM is the process of producing and managing bridge data throughout the project lifecycle. BrIM's mission is to use digital information to provide "simulation and analysis" of all aspects of the project for all participants. BrIM's goal is to maximize the efficiency of schedule, cost and quality, and provide the owners with professional design, construction and operation services (Fig. 8.19).

(3) Bridge Life Cycle Information Model Management Process

The Bridge Information Model Standard issued by the Federal Highway Administration of the United States defines the information flow direction of bridges in the design stage [32]. The design stage consists of the conceptual, preliminary and final design phases. The participants include the equipment management, consultant, evaluation, structural and traffic engineers.

In the conceptual design stage, the consultants and equipment managers summarize the preliminary investigation models and the conceptual design models, and

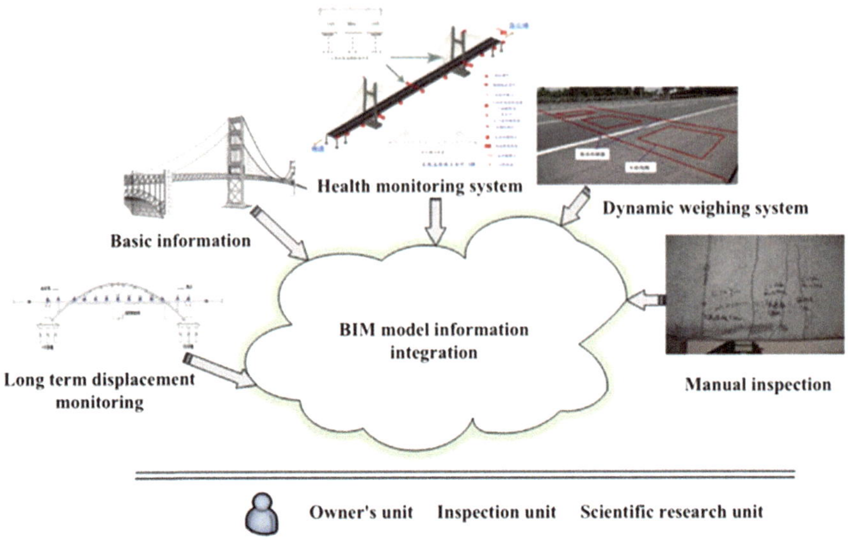

Fig. 8.19 Bridge BIM model information integration diagram [24]

exchange information with the traffic engineers to prepare for the subsequent preliminary design of the road geometry. At the same time, the project design engineers and evaluators have a full exchange of information. Then it enters the preliminary design stage, where the scheme design engineers, evaluators and structural engineers exchange information and models to produce the initial structural model and the preliminary design model. At the same time, the traffic engineers communicate with the structural engineers to complete the preliminary road geometry design based on the preliminary investigation and the conceptual design models. In the final design scheme, the preliminary design model is further developed after full communication between all parties. After each stage of the development process, the model is evaluated by the evaluators to meet the requirements, form the final design and proceed to the next stage.

Although the IFC standard applied in the bridge field started late, it developed rapidly. With the continuous progress of technology, relevant standards were constantly updated and gradually improved, which enabled the application of BIM technology in the bridge field to develop in a more standardized way.

8.5.2 BrIM Application: Dashengguan Yangtze River Bridge

This section introduces the application of BrIM technology in bridge engineering by taking the PHM (Prognostic and Health Management) system of the Dashengguan

Yangtze River Bridge, designed and developed by the China Academy of Railway Sciences, as an example.

(1) Project Overview

Nanjing Dashengguan Yangtze River Bridge on the Beijing-Shanghai high-speed Railway, which was put into operation in 2011, carries four lines of railway and two lines of light rail. When it was completed, the high-speed railway bridge had the largest span and undertook the heaviest load in steel truss arch bridges in the world. The main bridge, from north to south, has a total length of 1615 m, 128 intersections, and (84 + 84)m continuous steel girder + (108 + 192 + 336 + 336 + 192 + 108)m six-span continuous steel truss arch bridge.

To expand and upgrade the existing Nanjing Dashengguan Yangtze River Bridge monitoring system on the Beijing-Shanghai high-speed railway, the developing works of Prognostic and Health Management (PHM) system arranged for the long-span railway bridge were carried out. The integrated data mainly included the bridge's basic data, the construction monitoring data, the health monitoring data, the on-location inspection data and the maintenance data. Data is integrated from data sources at different stages, such as the completion acceptance, the test, the joint test, the trial operation and the opening operation stages. Storage management is conducted in the three-dimensional data space of "equipment, geography and time." Big data analysis and modeling technology were used to predict the trends of the bridge monitoring and detection data. Correlation analysis of data from different sources and of different types was carried out to realize the digitization of bridge maintenance and management. State diagnosis, prediction analysis and health management of the bridge were undertaken to support the bridge's maintenance and repair.

(2) System Architecture and Function Design.

The architecture of the PHM system consists of five layers: the data collection layer, data storage layer, data processing layer, function module layer and service application layer, as shown in Fig. 8.20.

Based on the value of BrIM and the advanced concept of PHM, the main functions of the long-span bridge PHM system are designed. These include the basic management, visual management, intelligent inspection, maintenance and repair, diagnosis and prediction, and system management. Figure 8.21 shows the functional architecture of a long-span bridge PHM system that uses basic BIM.

(3) Establishment of BrIM Operation and Maintenance Model

(1) Component Classification and Coding

Component classification refers to rational division of the bridge structure (BrIM model architecture). It should not only follow the bridge's structural system but also consider the special needs of the operational objects. Component division of the long-span bridge structure generally follows the hierarchical division principle from the whole to the local until the minimum component element layer is satisfied.

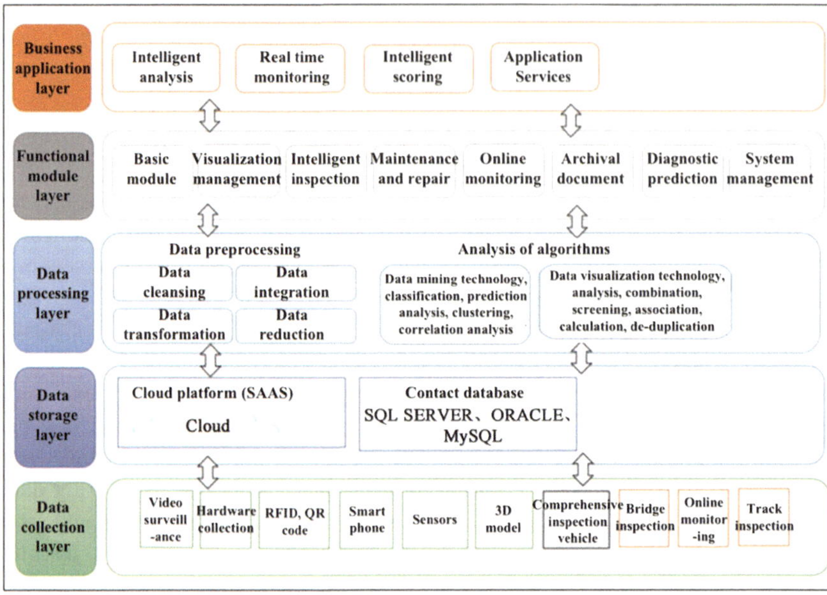

Fig. 8.20 Architecture of the PHM system

Multi-source information is associated with the minimum component element. The structure of Nanjing Dashengguan Yangtze River Bridge was first divided according to the specialty (bridge) and the structural position, and then divided according to the local structure and the detailed structure at different levels. A parallel relationship existed between components at the same classification level and a membership relationship between components at different levels. The component classification sketch is shown in Fig. 8.22.

The bridge structure was divided into levels through the classification of bridge components. In order to realize the interworking and exchange of information between the system modules and to facilitate accurate identification of the corresponding information by the computer, it is necessary to assign corresponding coding for each level and each category based on the classification. The "index" effect of coding effectively realized information interaction based on the BrIM model. Based on the classification table of railway engineering components given in the Standard for Classification and Coding of Railway Engineering Information Model (Version 1.0), a technical application framework based on IFD classification and coding was established. Table 8.4 shows the IFD classification codes of the substructure and ancillary facilities of the Dashengguan Yangtze River Bridge.

(2) Creation of BrIM Model

The PHM system used the BIM model of bridge operation and maintenance as the carrier to link data from different sources in the design, construction and operation stages. The BIM model of the Dashengguan Yangtze River Bridge was built with

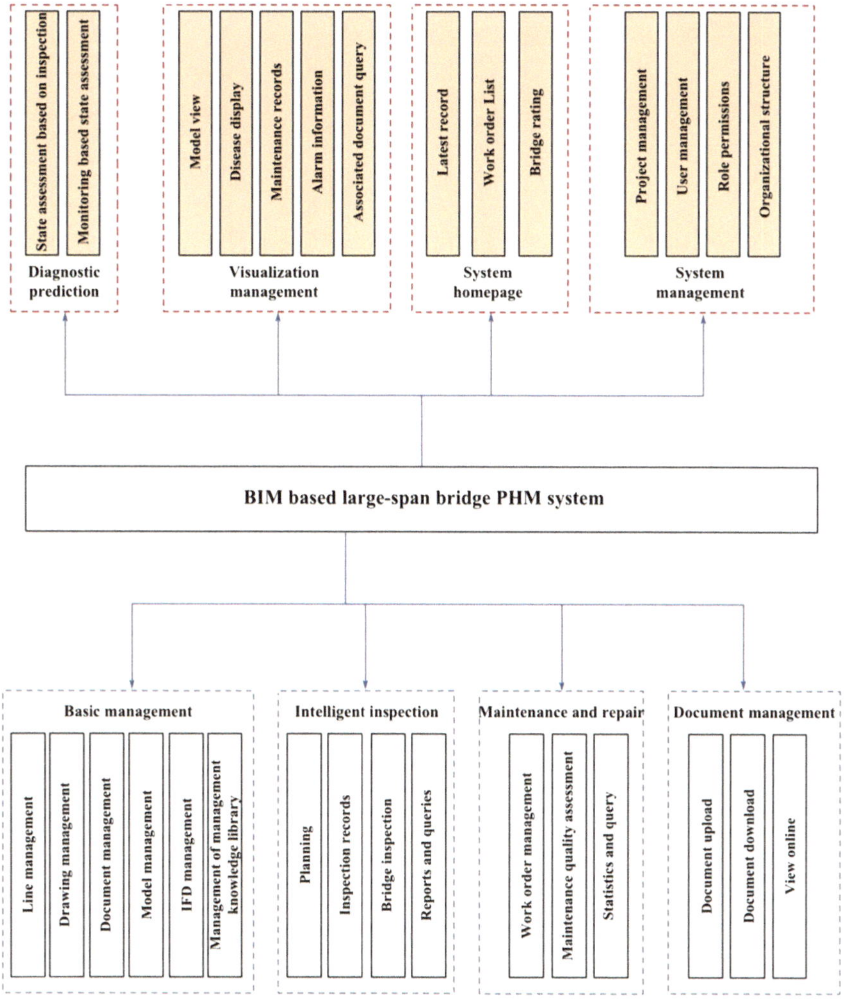

Fig. 8.21 Functional architecture of BIM based long span bridge PHM system

CATIA, a mainstream software. It used Bill of Material (BOM) and created a product structure data file that the computer can read through a secondary development plug-in. A model instantiation process established a link relationship between the component library and the data file. Finally, a bridge model for the operation and maintenance stage was created. It included 3,480 items of the superstructure and 281 items of the substructure, supports and attachments. Figure 8.23 shows the overall BrIM model of the Dashengguan Yangtze River Bridge.

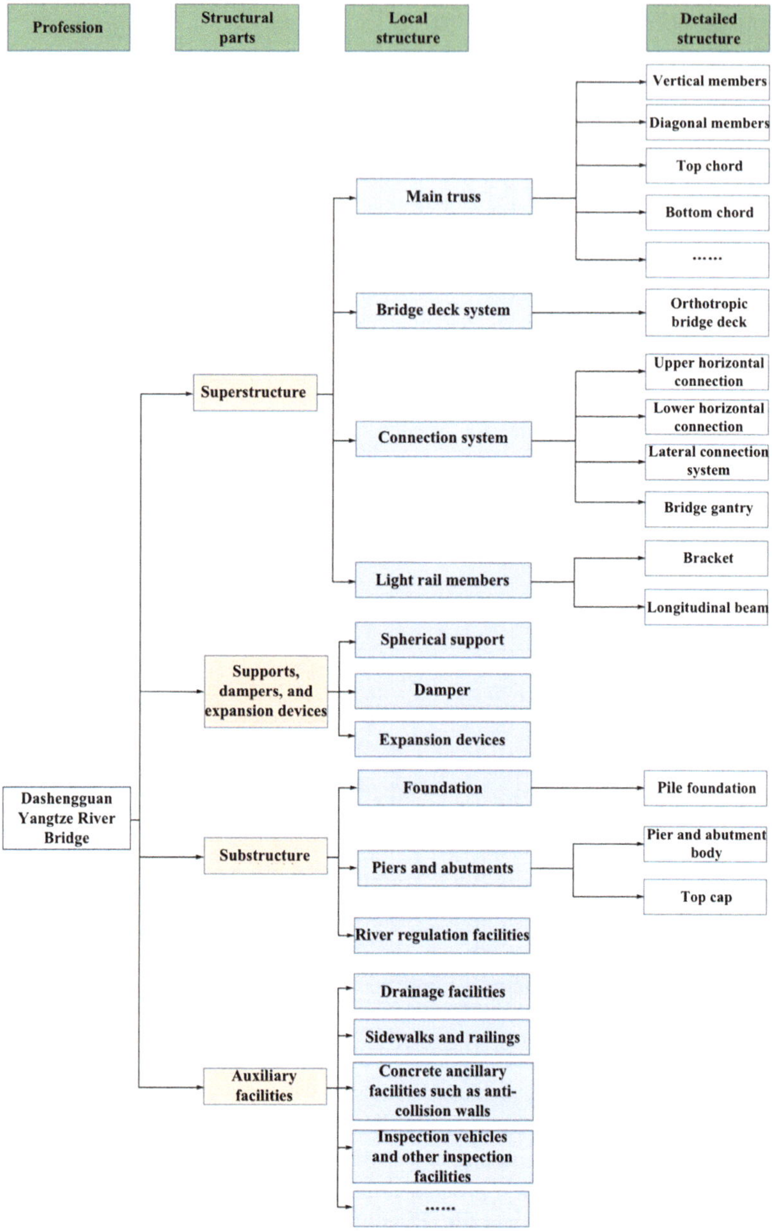

Fig. 8.22 Classification of components of dashengguan yangtze river bridge

Table 8.4 IFD classification code of Dashengguan Yangtze River Bridge

Code	First level	Second level	Third level	Fourth level
53–12 00 00	Bridge			
53–12 20 00		Substructure		
53–12 20 10			Pier	
53–12 20 10 20				Top cap
53–12 20 10 50				Pier body
53–12 20 20			Foundation	
53–12 20 20 20				Pile foundation
53–12 20 40			River regulation facilities	
53–12 40 00		Ancillary facilities		
53–12 40 15			Drainage facilities	
53–12 40 30			Sidewalk and railing	
53–12 40 40			Anti-collision walls and other concrete ancillary facilities	
53–12 40 50			Inspection vehicles and other inspection facilities	
53–12 40 85			Anti-collision facilities and other navigation auxiliary facilities	
53–12 40 95			Wind barrier facilities	
53–12 40 96			Operation channel	
53–12 40 97			Fences and other auxiliary facilities of steel components	
53–12 40 98			Lighting, signage, and others	

(4) Multi-source Information Association

Component units are the smallest units of bridge structure division and the basic carrier of information transmission, integration, fusion and circulation in the operation and maintenance BrIM model. Some of the multi-source information was directly written on the BrIM model. The rest of the multi-source information was coded

Fig. 8.23 Overall BrIM model of Dashengguan Yangtze River Bridge

according to the numbering rules to create unique components that the computer can recognize and link to the database.

The system development converted the BrIM model to HOOPS Stream File (HSF) format for storage. The SQLite database stored the design information. The basic attributes of the component BrIM model were geometric information, IFD coding and drawing number. The database batch added other design attributes, such as material and concrete strength level, to associate and display the design information and the BrIM model. The construction information code was an attribute of the BrIM model in the database.

The system linked the unstructured construction attribute information to the BrIM model. It also developed batch tools to add construction information to the database of the BrIM model attribute information. This enabled the electronic data and the BrIM model to be linked and the construction stage information to be traced and viewed easily. The database connected the operation and maintenance information with the BrIM model. Moreover, the database updated continuously to integrate and store the operation and maintenance information. Then, various professional analysis methods were used to support the bridge maintenance. Figure 8.24 shows the relationship between the information of different stages and the BrIM model.

Based on the practical application needs of operation and maintenance of the Dashengguan Yangtze River Bridge, the PHM system for long-span bridge was developed based on BrIM. It has proven that the PHM system provided a new idea for applying BrIM technology in bridge operation and maintenance management.

8.6 Digital Twin Modeling and Cyber Physical System

Digital twinning is a cutting-edge technology that emerged recently, and is one of Gartner's top 10 strategic trends for 2019. It integrates the conventional modeling and simulation technology with the Internet of Things, big data and artificial intelligence technologies. It can replicate any entity or process in the future physical world via related technologies. Digital twinning establishes a near-real-time connection between the physical and digital worlds, and enables them to interact and cooperate

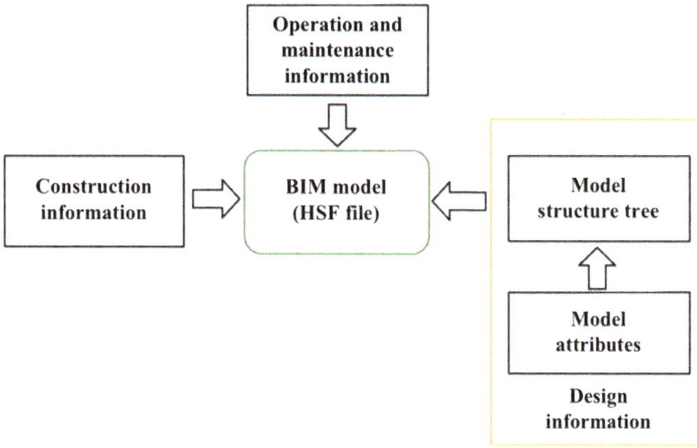

Fig. 8.24 PHM system multi-source information association

with each other. In the intelligent operation and maintenance of bridges, digital twinning can create a dynamic digital replica of the bridges throughout their life cycles. It can help bridge engineers make better operation and maintenance decisions and achieve intelligent bridge maintenance.

8.6.1 Overview of Digital Twin Modeling

(1) Definition of Digital Twinning

The concept of digital twinning first appeared in 2003. Compared with BIM technology, digital twinning emerged later, so there is no unified definition for the technology at present. Many other countries' research institutions and Chinese industry experts have offered their understanding of digital twinning. Although institutions and experts at home and abroad have different understandings and expressions of digital twinning, generally speaking, digital twinning is "digital twins." The internal structures of the two are the same, but one exists in the physical world and the other in the digital world.

In the White Paper on Digital Twin (2019) [37] issued by China Institute of Electronic Information Industry Development, digital twinning is defined as "Digital twin is the digital model of an existing or to-be-existing physical object that can sense, diagnose and predict the state of the physical object in real-time through measurement, simulation and data analysis". Digital models can regulate the behaviors of the physical objects, evolve by learning from each other, and improve the stakeholders' decision-making during the life cycle of the physical objects. In short, the physical object transmits various data (text, images, etc.) to the digital model, which in turn

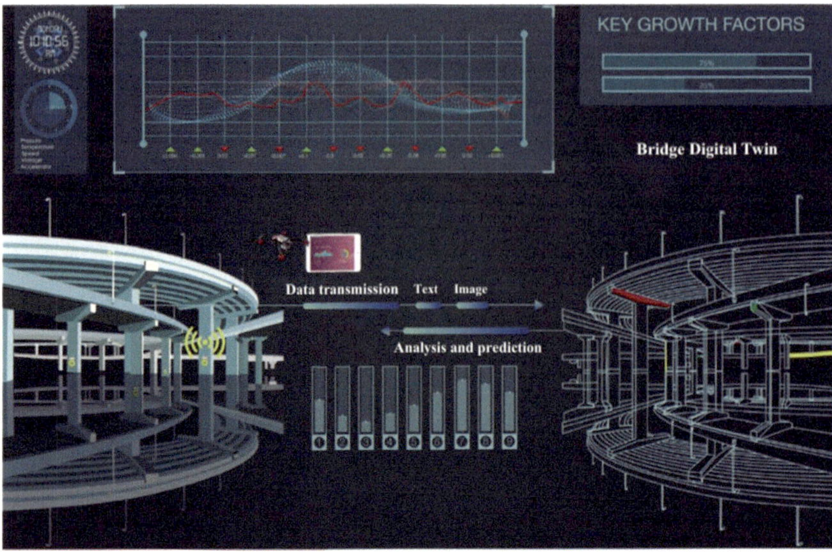

Fig. 8.25 Schematic diagram of the concept of digital twins

analyzes the data, predicts the structural behaviors, and feeds the results back to the physical object, as shown in Fig. 8.25.

(2) Characteristics of Digital Twinning

Digital twinning came from the simulation technology but differs from the conventional "simulation", and is more "realistic". It is neither a conventional computer-aided design (CAD) nor a sensor-based Internet of Things solution. Digital twinning is regarded as an important progress in simulation and optimization technologies, and a cutting-edge concept leading the new generation of simulation technology [38].

First, digital twinning integrates various data of the physical objects and is a realistic mapping of them. Digital twinning requires the virtual digital model established in the information space to be "realistic". Digital twin model should be a carrier or system that integrates multi-physical and multi-scale simulation to reflect the actual state of its corresponding physical entity. Digital twinning can link real-time data in physical space with virtual digital models, making it easier for managers to monitor the operating state of physical entities.

Second, digital twinning reflects a complete state of the environment, and co-evolves and accumulates knowledge with the environment. The digital twin is a highly dynamic system that covers the entire life cycle of the physical objects, from design and construction to operation and management. It has a unified data source and avoids the problem of data silos. All information sensed by the sensor or generated by the execution system is stored in the historical database of the virtual digital model and updated in real-time as the physical entity system changes.

More importantly, digital twinning can combine the real-time data from physical entity operation with a digital virtual model and predict the changing rules from the historical data. This allows the managers to preview and verify the impact of the control and management in the digital system corresponding to the physical system in advance, while the physical system is still running normally. They can also help deduce a good action plan based on the simulation results. This effectively improves the arrangement of activities in the physical space, avoids unnecessary losses and waste of material resources, and dynamically adjusts and timely corrects deviations [39].

As these characteristics show, digital twins are highly "realistic" compared with the physical entities. The emergence of digital twin technology will change the previous research mode where the managers could only conduct field researches on objects in the physical space. It will greatly reduce innovation risks and costs by rehearsing and verifying various innovation schemes using the virtual digital models, which provides powerful support for massive innovation.

(3) Digital Twinning and Cyber Physical Systems

Cyber physical systems (CPS) are fusion technologies that include computing, fusion and control (Fig. 8.26). In a broad sense, CPS is a controllable, reliable and extensible networked physical equipment system that deeply integrates computing, communication and control capabilities based on the environmental perception. It achieves deep integration and real-time interaction through the feedback loop of mutual influence between the computers and the physi-processes in order to add or expand new functions in a safe, reliable, efficient and real-time manner so as to monitor or control a physical entity.

CPS connects the information and physical worlds, enabling intelligent objects to communicate and interact, to create a real network world. The ultimate goal of CPS is to achieve complete integration of the information world and the physical

Fig. 8.26 CPS core

world, build a controllable, trusted, scalable, secure and efficient CPS network, and fundamentally change the way humans construct engineering systems. In a sense, CPS is a method to realize real-time interaction between information and physical worlds. Ensuring real-time performance is the basic requirement of the CPS.

On the other hand, a digital twin model (DTM) is a high-fidelity virtual model of a physical object created in a virtual space to simulate its behavior in the real world and to provide feedback. DTM reflects the bi-directional dynamic mapping process.

Both CPS and DTM enable precise management and manipulation of the physical world via interactions between the virtual and the physical worlds. However, they have different emphases. CPS emphasizes the powerful computing and communication capabilities of the information world and focuses on real-time control and interaction rather than mirror model, which leads to its poor replicability. DTM focuses more on virtual models and the mapping relationship between the physical world and the information world which is a one-to-one correspondence, so one-to-one mapping is realized in DTM, and "real-time" is not emphasized. The model is therefore highly replicable [40].

"Real-time" is one of the biggest differences between the current applications of CPS and DTM. Combination of the artificial intelligence technology and the physical information system is difficult to meet real-time requirements, resulting in great limitations on the development of CPS. However, DTM does not emphasize real-time interaction between the virtual models and physical entities so it can be used well in non-real-time situations. For example, it is used effectively to simulate the product life cycle process, to predict and evaluate product performance, etc. Moreover, when DTM requires real-time, it can be realized using the methods and principles adopted in CPS.

CPS can be understood as the method and principle of real-time interaction between the virtual and physical worlds. On the other hand, DTM can be a data fusion platform of the highest level generated by CPS (which can realize real-time interaction) [41]. According to Springer's 2020 book "Cyber Physical Systems and Digital Twins," "Cyber physical systems can be used as by-products of digital twin research," which supports this view.

8.6.2 Cyber Physical System

Cyber Physical System (CPS), integrating computing technology with communication and control technologies, realizes the combination of computing resources and physical resources. CPS is currently the leading research direction in the field of automation, which has attracted extensive attention from academia and industry, and is recognized as one of the core technologies to achieve Industry 4.0. As far as computing is concerned, the knowledge scope of CPS includes all the above-mentioned computing modes. In a broader sense, combining subsystems of CPS (including physical ontology, perceptual system, communication and control system, and computing hardware) is for data flow between computing and subsystems, therefore, the authors regard CPS as a computing model.

In recent years, application of CPS in civil engineering has also become an important research direction, which is expected to become an important driving force for the digitization and industrialization of the construction industry. This section introduces the source, concept, basic architecture and core technology of CPS, and explores its potential applications in civil engineering.

(1) Concept and characteristics of CPS

(a) Concept of CPS

As an emerging concept, Cyber Physical System (CPS) has attracted extensive attention at home and abroad, but experts and scholars in different countries have different understandings of it which are formed in its short development time. In order to unify the concepts of CPS, this book adopts the definition presented in the White Paper on Cyber Physical System issued by China Cyber Physical System Development Forum [50]: CPS integrates advanced perception, computing, communication, control and other digital technologies with automatic control technologies, constructs complex system in which human, machine, object, environment, information and other elements in physical space and information space are mapped, timely interacted and efficiently coordinated so as to realize on-demand response, rapid iteration and dynamic optimization of resource allocation and system operation.

The basic components of CPS include sensors, control execution units and computing processing units, as shown in Fig. 8.27. The sensor collects the signal from CPS; the computing processing unit computes and analyzes the collected data; the control execution unit controls the CPS according to the computed results; and the data transmission between them is accomplished by means of the communication network.

(b) Characteristics of CPS

CPS, as a comprehensive technical system, connects the physical space and the information space, drives the automatic data flow in it, and realizes the optimal allocation of resources. CPS has six typical characteristics: data driven, software definition, universal connection, virtual reality mapping, heterogeneous integration and system autonomy.

(1) Data driven: CPS can continuously transform the data from implicit form in the physical space to explicit form in the information space by building a closed-loop enabling system of automatic flow of "state perception, real-time analysis, scientific decision-making and accurate implementation" data, and constantly form a knowledge base through iterative optimization.
(2) Software definition: industrial software is codification of the laws of various industrial production links, supporting the vast majority of production and manufacturing processes. As a manufacturing oriented CPS, software has become one of the core carriers to realize its functions.

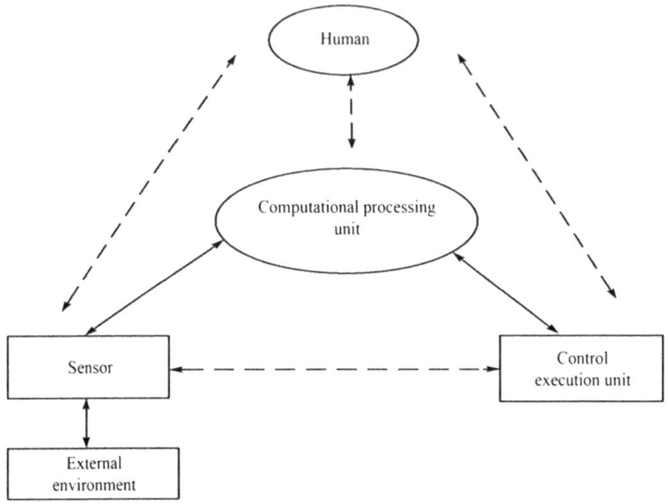

Fig. 8.27 Basic composition of CPS [51]

(3) Universal connection: network communication is the basic guarantee of CPS, which can realize interconnection between the CPS internal units with that of other CPSs.
(4) Virtual reality mapping: CPS contributes to a closed loop channel for interaction between information space and physical space data, to realize interaction between information virtual bodies and physical entities.
(5) Heterogeneous integration: there are large number of different hardware, software, data, and networks in high-level CPSs. CPS integrates these heterogeneous hardware, software, data and networks to realize automatic data flow in different links of information space and physical space.
(6) System autonomy: CPS can process and analyze in the information space according to the perceived environmental change information, and adaptively respond to external changes. At the same time, self organization between CPSs is realized through network platform interconnection (such as CPS bus and intelligent service platform) in higher level CPSs.

(2) Implementation of CPS

Cyber Physical System (CPS) is a hierarchical system. An intelligent component, an intelligent device, an intelligent production line, and an intelligent factory may all become CPSs. At the same time, a factory may contain multiple production lines, and a production line may also be composed of multiple equipments. The CPS can be divided into the following levels: unit level CPS, system level CPS, and system of system (SoS) level CPS. Their relationships are shown in Fig. 8.28.

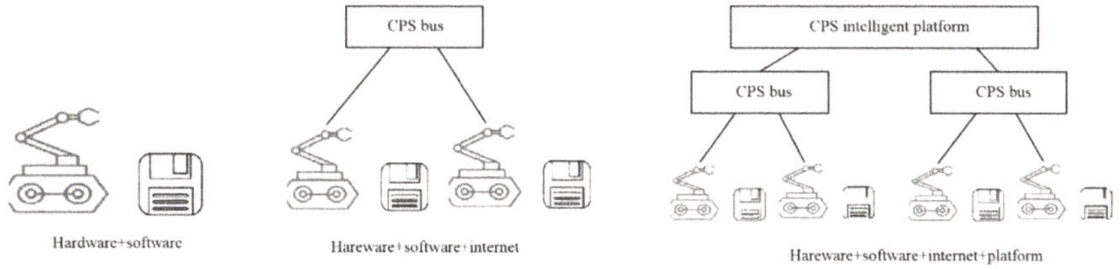

Fig. 8.28 Hierarchical evolution of cyber physical system

Unit level CPS: generally, a component (such as an intelligent bearing) and a device can constitute the smallest unit of a CPS. The unit level CPS is indivisible. In other words, it cannot be divided into smaller CPS units.

System level CPS: on the basis of unit level CPSs, and through the network, collaborative deployment of system level CPS can be realized. At this level, the integration of multiple unit level CPSs and non-CPS unit equipment constitutes a system level CPS.

SoS level CPS: on the basis of system level CPSs, the collaborative optimization between system level CPSs can be realized by building an intelligent service platform of CPS.

(1) Implementation of unit level CPS

As shown in Fig. 8.29, unit level CPS is the smallest indivisible CPS unit. It aims to conduct state perception, computation and analysis of physical entities and environments through software, and ultimately to control physical entities by building the most basic closed loop of automatic data flow, so that the physical world and the information world can integrate and interact.

(2) Implementation of System Level CPS

In actual operations, most activities are usually jointly completed by several people, machines and things, which are the results of joint activities of many intelligent

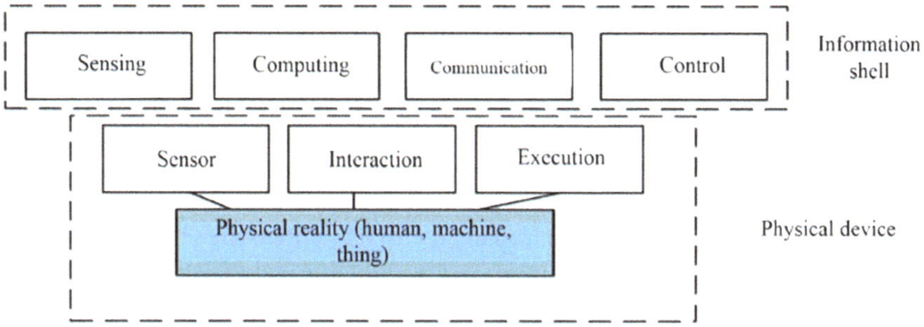

Fig. 8.29 Architecture of unit level Cyber Physical System

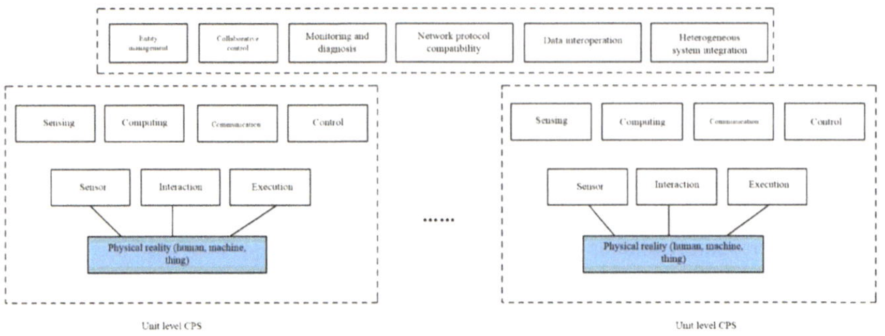

Fig. 8.30 Architecture of system level Cyber Physical System

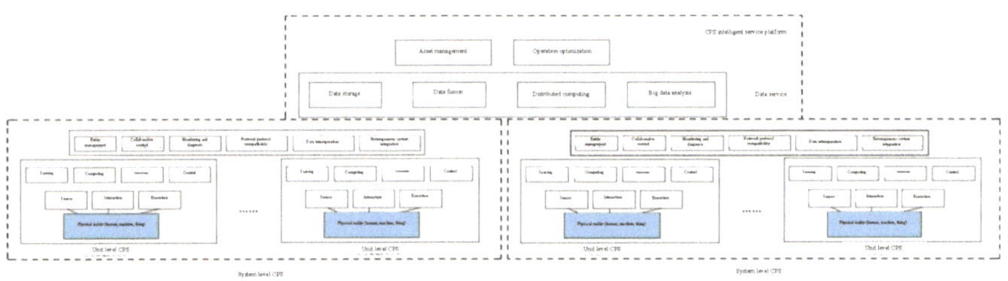

Fig. 8.31 Architecture of SoS level cyber physical system

products. These intelligent products are combined to form a system. The architecture of the system level CPS formed by the CPS bus is shown in Fig. 8.30.

(3) Implementation of SoS level CPS

As shown in Fig. 8.31, the organic combination of multiple system level CPSs constitutes the SoS level CPS. For example, multiple processes (the CPS of the system) form a workshop level CPS, or form the CPS of the entire factory.

(4) Technology of CPS

The core technologies of CPS include "one hard" (perception and automatic control), "one soft" (software), "one network" (network) and "one platform" (intelligent service cloud platform).

(1) Perception and automatic control

① The intelligent perception technology used by CPS is the sensor technology. The sensor can sense the measured information and transform the detected information into electrical signals or other required forms of information output according to certain rules.

② The virtual reality fusion control of CPS is a multi-level "perception analysis decision execution" cycle process, which is based on state perception, and the perception is often real-time, synchronous or real-time feedback to a higher level.

(2) Software: refers to those specially used to improve the R&D, manufacturing, production, service and management level of industrial enterprises, and the use value of industrial products.
(3) Network: from technical perspective, CPS network mainly involves the interconnection and plug and play of industrial heterogeneous networks.
(4) Intelligent service cloud platform: the platform processes data using edge computing technology, fog computing technology (see Sect. 5.3), big data analysis technology, etc., to provide external data services, and personalized and professional intelligent services based on the data services.

(3) Cyber Physical System and Digital Twin Technology

Digital twin is the equivalent virtual digital representation of the physical entity, which includes three parts: ① physical entity in physical space; ② virtual entity in virtual space; and ③ the relationship of data and information used to connect the physical entity with the virtual entity. This section introduces the relationship between CPS and digital twins and their combined applications. It briefly introduces the specific implementation methods of digital twins. For detailed information about digital twins, the readers can refer to this Chapter.

(1) Integration of CPS and Digital Twin

Both CPSs and digital twins promote intelligent manufacturing through closed-loop of "state perception, real-time analysis, scientific decision-making, and accurate implementation". With the help of virtual models, digital twins provide a more intuitive and effective means. Through continuous data integration, Digital Twin can provide strengthened relevant solutions. Virtual models can be used as supplements to enrich the composition and functions of CPSs. Therefore, digital twins can be regarded as the necessary basis for building and implementing CPSs.

(2) Differences between CPSs and digital twins

Although both CPS and digital twin involve the physical world and the information world with respect to their compositions, CPS and digital twin are not exactly the same. Through interaction and control of information physics, both CPS and digital twin realize accurate management and operation of the physical world. However, for the information world, CPSs and digital twins have their own emphasis. Digital Twin focuses on virtual models to achieve one-to-one mapping, while CPS emphasizes 3C functions to form a one-to-many mapping. In the function realization of CPS and digital twin, sensors and actuators support the interaction between the physical world and the information world to achieve data control and exchange. In contrast, models play an important role in digital twins, helping to interpret and predict the behavior of the physical world based on various data.

8.6.3 *Application of Digital Twin Technology in the Field of Smart City*

Digital twinning was initially applied in the military and aerospace fields, and has been used in European and American aerospace projects [42]. In recent years, digital twin technology has gradually expanded to the civil fields. In healthcare, digital twins can provide real-time monitoring, virtual surgical verification and training, doctor training, and other services to improve the accuracy of diagnosis. Dassault uses 3D Experience City, an experience platform in urban services, to create a complete "digital twin Singapore" for urban planners to solve urban energy consumption, transportation and other issues. In the field of transportation, Tesla, DNV GL Classification Society, SNC-LAVALIN and others have studied the application of digital twins in automobile, ship, train and other industries. In addition, the field of smart cities is currently a hot spot drawing the attentions of relevant enterprises and research institutions to apply digital twin technology.

For example, the digital Xiongan CIM platform has established a two-dimensional and a three-dimensional integrated urban space unified information model with the help of GIS, BIM and other technical means. It has so established a new urban pattern of accurate mapping between the "physical New District" and "digital New District," and "virtual and real integration," laying a foundation for the construction of a "digital twin city" in Xiongan. Xiongan New Area is located in Baoding City, Hebei Province, China. It covers Xiongxian County, Rongcheng County, Anxin County and some surrounding areas in Hebei Province. Xiongan New Area is another new area of national significance, following the Shenzhen Special Economic Zone and the Pudong New Area in Shanghai. It aims to relocate the non-capital functions of Beijing, explore a new model of optimal development in economically dense areas, adjust and optimize the urban layout and spatial structure of the Beijing-Tianjin-Hebei region, and foster a new engine of innovation-driven development. The design concept of Xiongan New Area is to create a hub for removing non-capital functions from Beijing and to build a new type of modern urban area led by new development concepts. The construction of digital twin cities in the Xiongan New Area is mainly reflected in the establishments of an intelligent transportation system, an intelligent emergency command center and an intelligent construction site platform.

(1) Intelligent Transportation System

According to the planning outline of Xiongan New Area, a new intelligent transportation system framework featuring real-time perception, instantaneous respons, and intelligent decision-making will be built in Xiongan New Area. A three-in-one intelligent transportation infrastructure featuring road, information and energy networks will also be built. At the same time, a holographic omnipresent interconnection perception system will also be constructed. It will focus on optimizing the layout of facilities and equipments that can sense the environment, road conditions and traffic flow information. These projects will realize the holographic perception of all elements in the road network and realize omnipresent interconnection relying on the full coverage communication network. In terms of the transportation system, intelligent shared cars will be the leading part of the intelligent transportation system in Xiongan New Area. They will greatly reduce traffic and driving accidents, improve transportation efficiency, reduce vehicle pollution emissions and increase travel convenience. Xiongan New Area will establish a data-driven intelligent collaborative control system for traffic control. It will explore the internet-connected control of intelligent driving vehicles and adopt intelligent distribution of the right of way at intersections to ensure system operation safety and improve system operation efficiency.

(2) Intelligent Emergency Command Center

In constructing the citizen service center in Xiongan New Area, the emergency command center will serve as the "brain" and command "center" of the city security system. The security of its information will be regarded as the top priority. At the same time, the stability, scalability and powerful interconnection function of the information system of the entire command center will also be highly valued by various departments. In constructing Xiongan New Area, the new area will mainly adopt a distributed intelligent management and control system, which will realize the rapid retrieval, interconnection, command and dispatch of all signals in Xiongan New Area. It will include more than 2,000 channel monitoring images and unmanned reconnaissance aircraft images. It will also include centralized intelligent management and control of the surrounding environment, which will create a "security, stability, scalability, Intelligent emergency command center integrated with connectivity functions." In this way, it will realize the real-time perception of the security situation in every corner of the city. It will comprehensively consider the development of the situation and resources to make the right decision. It will coordinate public security, traffic, fire, medical emergency and other departments for them to work together.

(3) Intelligent Site Platform

The smart site platform integrates the conventional smart site, BIM and construction information management. Its integration has reached a new level. As a data integration hub, the platform includes all the data from the virtual BIM model, the UAV aerial photography images, the monitoring images, the construction management records, the environmental monitoring, the hydropower energy consumption monitoring, and other IoT devices. It realizes automatic data collection and analysis, and early warning during the whole construction process. The platform is divided into eight functional modules: panorama monitoring module, progress management module, quality management module, safety management module, material management module, labor management module, green construction module and engineering data module. In addition, the smart site platform can also realize functions such as tower crane limit anti-collision and hook visualization, high support mold safety monitoring, and construction elevator safety monitoring.

In order to support the operation of the smart site platform, the construction site is equipped with a lot of hardware equipments. The on-site identification is carried out by the dual recognition method of "one-card + face recognition." GPS chips are installed on construction personnels' helmets to record their real-time positions and moving tracks. An environmental monitoring system that allows continuous unattended automatic monitoring and uploading of data over a 4G network has been implemented. Monitoring equipments are set up at the site, and the UAVs carry out aerial photography.

8.6.4 Application Cases of Digital Twin and CPS

(1) Application of Digital Twin in Workshops

The biggest bottleneck that hampers the realization of truly intelligent manufacturing is how to realize the interaction and integration between the physical world and the information world of manufacturing. The workshop is the basis of manufacturing activities. The intelligent level of the production process control and the global optimization ability are still insufficient for the present manufacturing practices, and further integration of workshop information and physical space is still needed. Fei et al. [49] proposed the Digital Twin Workshop (DTW) concept and studied its systematic composition and operation mechanism.

The digital twinning shop consists of the physical workshop, the virtual workshop, the Workshop Service System (WSS) and the shop twinning data, as shown in Fig. 8.32. The physical workshop is an objective entity set of workshops. It is mainly responsible for receiving production tasks issued by WSS, executing production activities, and completing production tasks following pre-defined production instructions after simulation and optimization undertaken in the virtual workshops. The virtual workshop is the digital mirror of the physical workshop. It is mainly responsible for the simulation, evaluation and optimization of the production plan and activities and the real-time monitoring, prediction and regulations of the production process. WSS is a collection of the functions of various data-driven service systems. It is mainly responsible for providing system support and services for intelligent management and control of physical workshops driven by workshop twin data, such as management and control, and optimization services for production factors, plans, activities, processes, etc. Shop twinning data is the collection of data obtained from the physical shop, the virtual shop and the WSS, as well as the data generated after fusion of the three data, which drives the operation and interaction of the physical workshop, the virtual workshop, and the WSS.

The specific production tasks in the workshop include the production factors management, production activity planning and production process control. Figure 8.33 shows the operating mechanism of the digital twinning workshop.

In Fig. 8.33, stage 1 is the iterative optimization process to achieve the production factor management in DTW. In this stage, WSS plays a leading role. When DTW receives a production task, WSS obtains real-time data concerning production factors such as personnel, equipment and materials in the physical workshop. It analyzes and

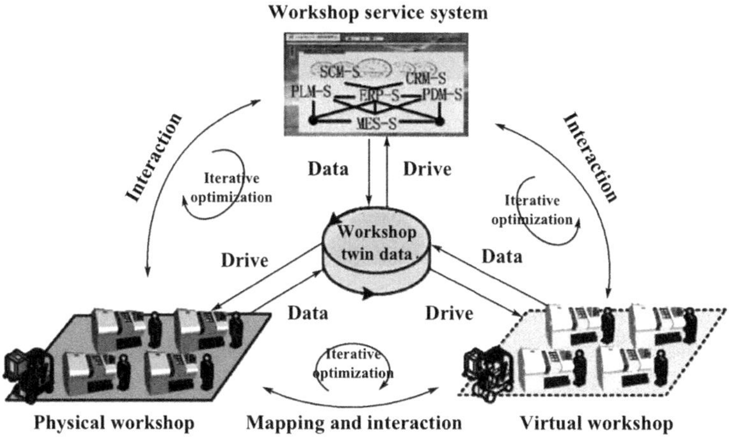

Fig. 8.32 Major system composition of digital twin workshop [49]

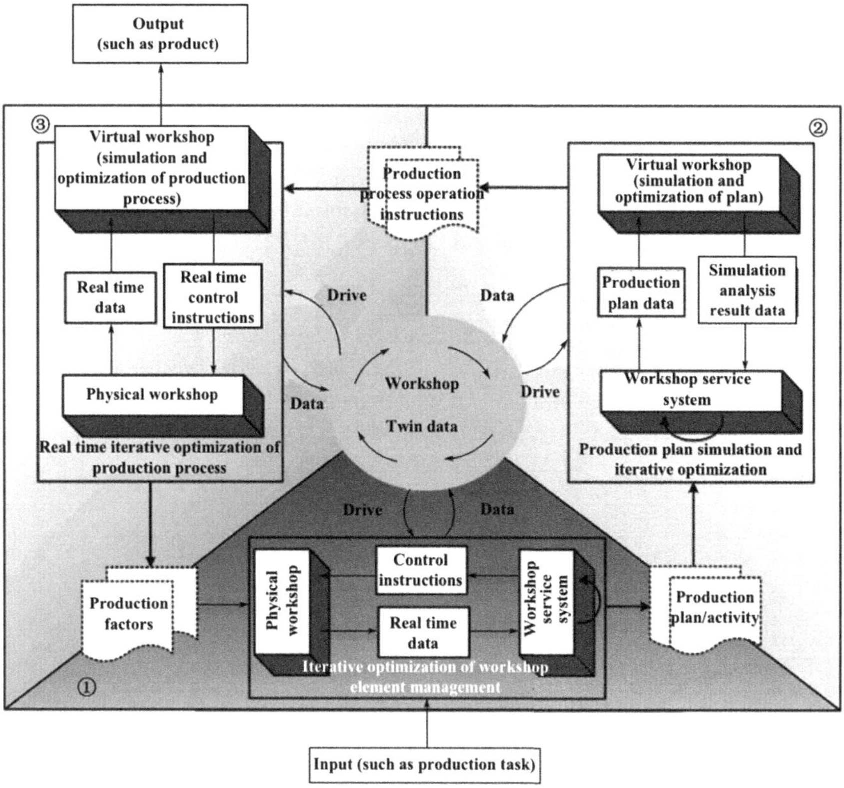

Fig. 8.33 Operation mechanism of digital twin workshop [49]

optimizes the initial resource allocation scheme and sends it to the physical workshop in the form of control instructions. The physical workshop adjusts the production factors accordingly until their optimal management is achieved.

Stage 2 is the iterative optimization process to achieve the optimal production plan in DTW. In this stage, the virtual workshop plays a leading role. The initial production plan is driven by the historical data and simulation analysis results. The virtual workshop carries out simulation and optimization of the production plan with the ensurance of the consistency among different links of the product life cycle and the predictability of internal and external disturbances. The virtual workshop feeds back the simulation results to WSS, who revises and optimizes the production plan until it is optimal.

Stage 3 is the real-time iterative optimization process to realize the optimal production process in DTW. In this stage, the physical workshop plays a leading role. The physical workshop receives the operation instruction of the production process and organizes production accordingly. In the actual production process, the physical shop transmits real-time data to the virtual shop. The virtual shop updates its own status and compares the actual running data with the pre-defined production plan data. Based on workshop twin data, the virtual workshop evaluates and optimizes the production process from a holistic perspective in real time. It controls the physical workshop using real-time control instructions to optimize the production process until it is optimized.

Through iterative optimization of stages 1, 2 and 3, workshop twinning data is constantly updated and expanded. DTW evolves and improves while undertaking its production task so as to obtain optimal products.

(2) Application of Digital Twinning in Bridge Operation and Maintenance

There are a few practical cases where the concept of digital twins was applied to bridge operation and maintenance. In 2019, South Korea proposed a prestressed concrete bridge maintenance system based on a digital twin model [50]. The system composed of three main parts (Fig. 8.34): digital twin model (DTM), three-dimensional surface model (RTM) and joint model.

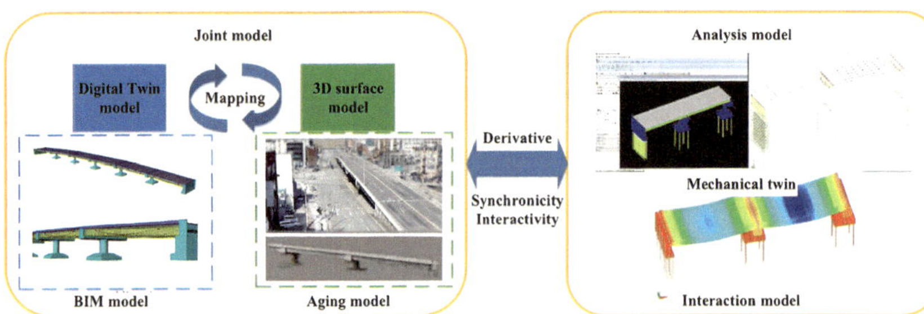

Fig. 8.34 Composition of a maintenance system for prestressed concrete bridges based on the Digital Twin model [50]

The creation of a DTM requires the establishment of data architecture that aligns with the purpose of maintenance. Data architecture includes an inventory system and object coding. First, each structural element is classified as an inventory part based on its role in the overall bridge system. It is usually classified into the superstructure (beams, plates, paving, expansion joints, etc.) and the substructure (piers, abutments, supports, etc.). It can then be subdivided into more categories depending on its characteristics. By encoding each structural element in each category with a specific code, the ID of each structural element is obtained (as shown in Fig. 8.35). After defining the inventory system and ID, a BIM-oriented information model is generated by adding the appropriate information to the specified object ID. Each structural element has two main characteristics: "attributes" and "archives." Attribute characteristics include all physical information such as geometry, material properties, load size, constraints, cost estimates, etc. Archival features contain all relevant information generated during maintenance such as inspection manuals and damage and repair records acquired continuously throughout the bridge's life cycle.

As shown in Fig. 8.36, the RTM was created using reverse modeling technology. The top (A) and side (B) of the bridge were photographed by UAV, and the surface model was obtained by image processing technology. Due to the poor flight conditions of UAV and the limited illumination conditions, the point cloud data of the bridge bottom (C) was obtained by three-dimensional laser scanning, and the surface model of the bridge bottom was obtained by relevant processing of point cloud data (such as alignment, denoising, segmentation, surface reconstruction, etc.). In addition, ground penetrating radar (GPR) data was added to the slab's top surface model to estimate the concrete's deterioration. Using the reconstructed surface model combined with 2D drawings, the whole 3D model can be parameterized to generate a 3D digital information model.

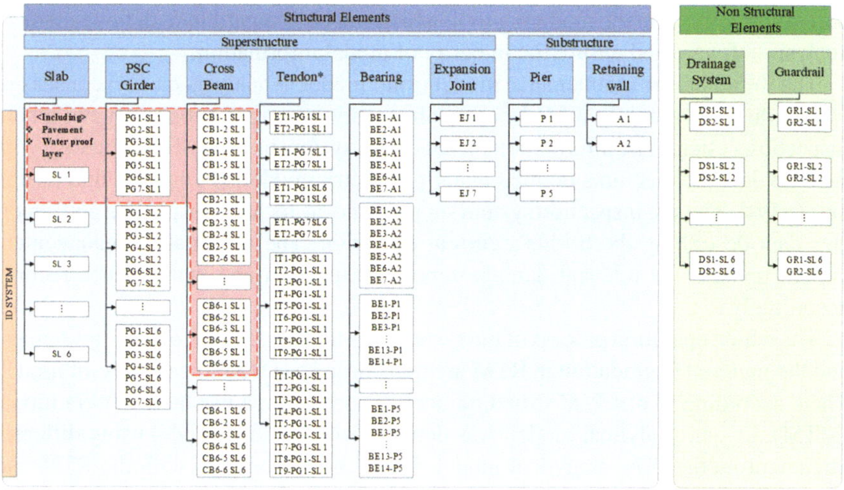

Fig. 8.35 Inventory system and object coding [50]

Fig. 8.36 Reverse modeling to generate RTW [50]

The joint model is obtained by matching the DTM and the RTM models. During the bridge maintenance stage, the initial model for the maintenance task is the physical 3D model with damage records, which is the information combination model of DTM and RTM. Each structural component in the DTM and that in the RTM can correspond individually because the inventory system and ID are defined. The RTM is separated based on the damage recording surface division, and the damage situation of each surface partition is extracted to form a damage report through image processing technology and image tracking technology. The damage information is stored in the corresponding structural component information in the DTM according to the ID system. Due to the good interactivity of BIM, the analytical model (MTM) can be derived directly from the modeling tool, which can be compatible with finite element analysis software and ensure the integrity of model information.

The initial bridge maintenance model (joint model + analysis model) consists of two modules (Fig. 8.37): a BIM-based information management system and a field inspection system. The information management system includes digital models with specific data architectures defined using parametric modeling concepts and design aids (APIs). The site inspection system supports the on-location inspection and generates data describing the bridge's current condition. The information management system updates the relevant damage reports, inspection and maintenance records accordingly.

The whole operation process of the system is as follows: Firstly, data of the damage and the material degradation in RTM are transmitted to DTM to get the joint model. Then, according to different structural analysis purposes, a mechanical twin model (MTM), i.e., an analytical model, was derived from the joint model using different alignment modes. The analytical model shared the same status with the RTM and was capable of finite element computation. An optimal maintenance solution was

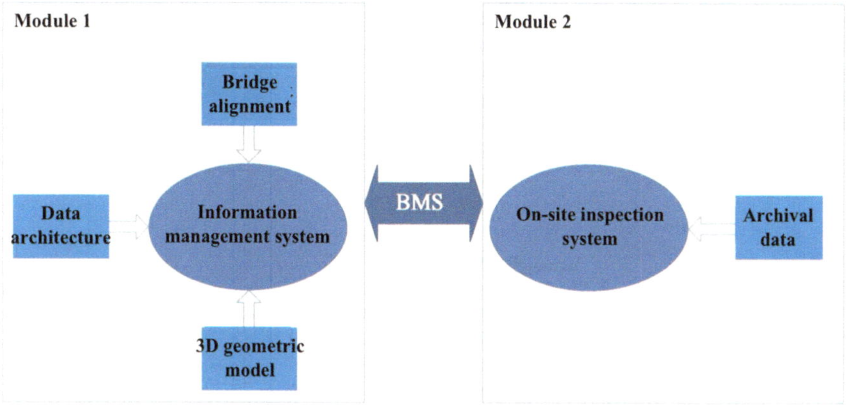

Fig. 8.37 The two modules of the bridge maintenance system [50]

obtained through finite element analysis, and on-site maintenance was performed correspondingly. Finally, after field repairs, the RTM updates the relevant data into the DTM to keep the joint model's state consistent with the bridge's actual state. The joint model is like a digital twin of a real bridge, which fully implements the digital twin concept.

(3) Application of CPS in Civil Engineering

At present, application of CPS in civil engineering is still in its infancy. Engineering application cases are rare, and more application scenarios and architectures are still needed. This section briefly introduces some of these works hoping to arouse readers' interest in the application of CPS in civil engineering.

(1) CPS for Structural Monitoring

Applications of CPSs to structural health monitoring (SHM) is a research hotspot. Bhuiyan et al. [52] thought that the conventional health monitoring system based on wireless sensor networks could be developed into a typical CPS, and proposed a CPS architecture for SHM, as shown in Fig. 8.38. The CPS model includes the physical layer and the network layer. The physical layer consists of sensors and actuators, which are respectively responsible for data collection and structural element monitoring. Various types of signals collected from the structure by the sensors using its additional sensing unit will also be processed in this layer and sent to the network layer as CPS input to the real-time decision-making system. In the network layer, after receiving the input, the decision system performs the abstract computation to check the collected signals, and then relays its decisions to the executor of the physical layer through a series of control processes.

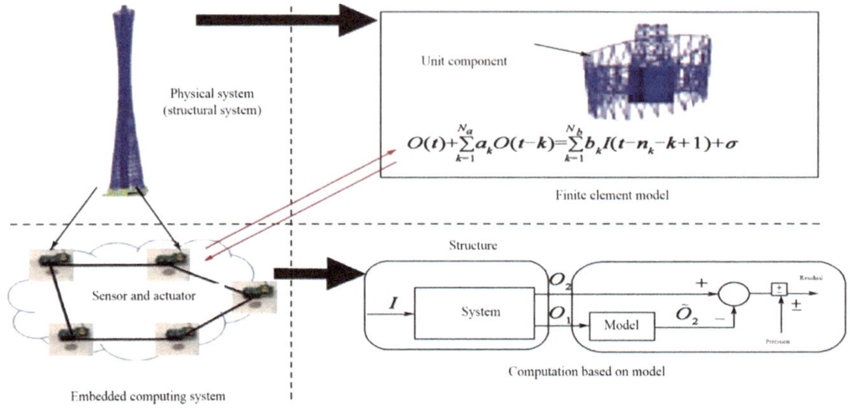

Fig. 8.38 Architecture of CPS for structural health monitoring [52]

(2) Temporary Structure Monitoring Based on CPS

Temporary structures are used for working by the construction workers and will be removed upon completion of construction. Compared with the permanent structures, safety level of the temporary structures is generally just sufficient for the construction works. Thus, daily monitoring and early warning of the temporary structures is an effective means to reduce risk. To achieve this, Yuan et al. [53] developed a temporary structure monitoring CPS. Its architecture is shown in Fig. 8.39, which includes two parts: the physical structure and the virtual model. Data sharing is realized through the database and the network communication. The physical structure included the temporary support system (such as scaffold system and formwork) and the temporary structure. Virtual model is the virtual representation of the physical elements.

The software structure and the workflow of the CPS are shown in Fig. 8.40.

In order to verify the effectiveness of the CPS, an actual environment was simulated for testing. It simulated tilting by lowering one of the rods 1/8 inch (1 inch = 2.54 cm). The four poles were inclined. Within 2 s, each pole was immediately highlighted in the virtual model. The server sent a warning signal to the mobile phone, which was displayed in the mobile app. The project director clicked the button to view the detailed situation, and immediately took corrective measures and reported to the project manager for further investigation.

8.6.5 *Key Technologies of Digital Twinning in Bridge Management*

Digital twinning is a scientific concept that combines multidisciplinary attributes, and enables virtuality and reality interaction and integration through the connection and integration of the physical and information worlds. One of its practical applications in

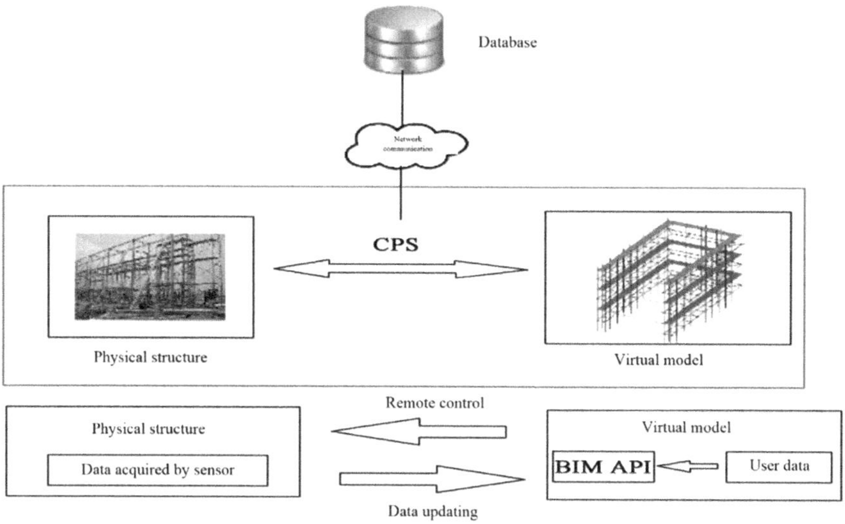

Fig. 8.39 Architecture of temporary structure monitoring CPS [53]

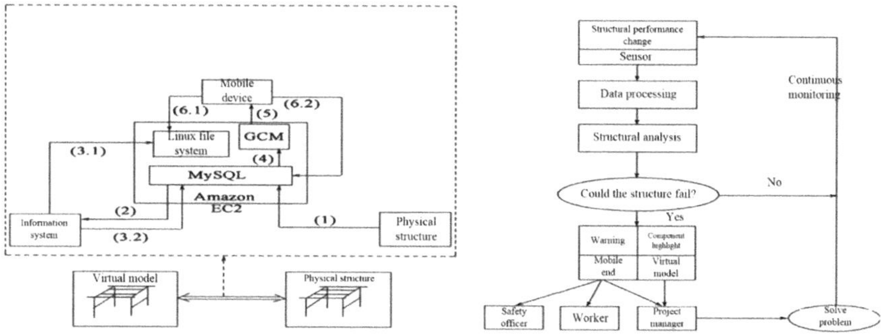

a) Software structure of cyber physical system

b) Software workflow of cyber physical system

Fig. 8.40 Software structure and workflow of the CPS [53]

bridge engineering is the use of building information model (BIM) and geographic information system (GIS) technology [51]. Maintenance of existing bridges is an important way to ensure bridge safety. Under the normal two-year inspection cycle, at least 315,000 bridge are inspected annually in the US and UK. Visual inspection is still the most common method of bridge condition monitoring. The bridge deterioration is assessed by entering the bridge status information obtained from the visual inspection into a Bridge Maintenance System (BMS) such as AASHTOWare in the USA or NATS in the UK. A similar approach is adopted in the Chinese bridge maintenance circle. However, this method does not evaluate the actual condition of specific bridge components and elements at specific locations. Because of this limitation, digital twinning will play a very important role. Based on the geometric digital twinning of the bridge, bridge material and damage information will be integrated with the geometric information on the virtual 3D representation elements of the bridge. Digital twinning focuses on the bridge's geometric expression and its components' semantic information. However, other semantic information (such as materials, defects, additional relations, etc.) will not be included [52]. The interactive correlation of bridge model elements and the reasonable data construction scheme is the bottleneck that restricts the development of digital twinning technology in bridge engineering. At present, the main solution is to propose modeling rules considering the geometry, physics and material properties that should be included in the virtual model in the three-dimensional GIS environment, analyze the infrastructure classification and characteristics of the twin data of the bridge, refine the hierarchical relationship of its components, and complete the identification coding and storage scheme for the model. Based on the above, the digital twin system is further constructed to realize the fusion of virtual model and field data, and form the application function of digital twin management of pixel, representation and decision layers.

(1) Modeling Process of Bridge Digital Twin Model

(1) BIM + GIS Modeling Scheme Analysis

A digital twin model (DTM) is a virtual model of a physical entity created digitally. In the bridge infrastructure industry, one of the basic functions of digital twinning is to draw the road horizontal and vertical design into CAD drawings and to organize the road network GIS vector element mapping into electronic maps. The rich data structure of GIS makes it possible to map the physical world in 3D. Mapping traffic structures to digital earth points using data structures such as points, lines, planes, triangulated irregular networks (TIN), grids, polyhedrons and network common data sheets (NCF) is the usual modeling practice. In this process, the key technologies are employed to process geometric modeling data, such as 3D geometric modeling technology, BIM technology, data organization and management of virtual models.

Geometric modeling data refers to the ground object measurement data and design data that can objectively reflect the physical truth, which are the basis of modeling. As the gateway connecting virtual and physical entities, digital twin geometric models are the foundation of digital twin realization. The geometric modeling methods are mainly divided into manual modeling and rule modeling. Manual modeling refers to using BIM software such as Revit to manually turn CAD drawings into 3D models and convert them into data formats supported by GIS. Finally, manually match the field measurement data to a certain coordinate on the digital earth for sharing. Rule modeling is also known as automated batch modeling. Currently, the representative software is ESRI's CityEngine, which uses Computer Generated Architecture (CGA) language to describe the geometry and the material of ground objects. The secondary development of CGA modeling code can be loaded into the abstract model of traffic ground objects to complete the three-dimensional modeling.

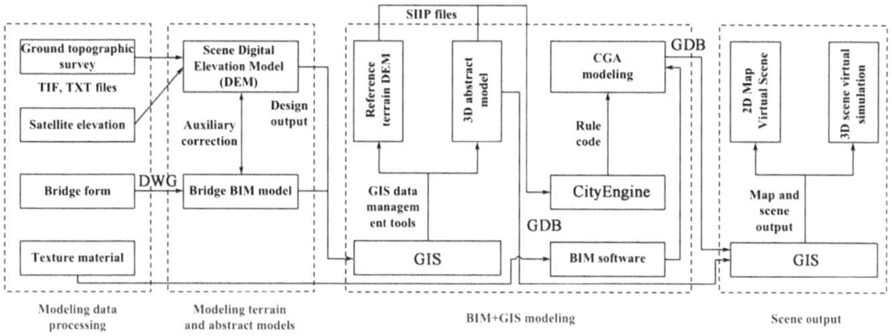

Fig. 8.41 Modeling process of BIM + GIS

(2) 3D Geometric Model Modeling Process

① Modeling Data Processing

Figure 8.41 shows the bridge digital twin model's generation, design and modeling process. Modeling data processing refers to extracting four kinds of information from the design drawings and remote sensing images:

a. Ground topographic survey data: Extract contour layers in CAD drawings.

b. Satellite elevation: Integrate with ground topographic data to create scene Digital Elevation Model (DEM).

c. Bridge deck horizontal information: derived from CAD design files; and used to establish abstract line models and to assist in correcting the DEM.

d. Texture material: Collect photos of the bridge deck reflecting the materials for the CGA modeling map in CityEngine.

② Modeling terrain and abstract model generation.

DEM data is derived from the ground topographic survey and the satellite remote sensing and scanning. Among these, topographic survey data are usually integrated into the contour layer of bridge plane design drawing (the density between contour lines intuitively represents the accuracy of terrain). In the site selection and design stages, it is generally necessary to conduct line selection and evaluation in the early stage based on ground measurement elevation. Such measurement data are obtained from an on-site survey by the engineering survey team. The measurement results generally include terrain feature drawing data and part of the accurate global positioning elevation control point data. Through digital processing of contour CAD drawings that carry elevation information and equidistant sampling along contour lines, the GIS system can output topographic point files required for TIN creation. The main factors affecting measurement accuracy are the personnel, environment and instrument. Measurement errors can be reduced or avoided by certain control methods, such as maintenance and calibration of instruments in strict accordance with specifications and strict control of the lofting accuracy. In practical engineering, although the error can be controlled to a small scale, it is difficult to ensure the continuous measurement of the terrain due to the limitations of conditions, which leads to the formation of a non-convex set of sampling points. Moreover, by using an irregular triangulation network to connect the measurement points to approximate the terrain surface, the property of a non-convex set is bound to cause a wrong assignment.

③ Digital Twin Rule Modeling

The minimum model unit proposed in Building Information Model Design Delivery Standard (GB/T 51301–2018) stipulates the degree of accuracy of virtual models in each engineering life cycle. In the design period, the model is required to reflect the object at the functional level of bridge maintenance, describe the general outline and accommodate the data of the twin planning level. In the operation period, it is required to establish the component level model such as the pile foundation model, reflecting its local geometric size and twinning more abundant data. Combined with the above analysis, the degree of refinement that can be achieved by the model depends on the modeling data. Only the modeling scheme designed according to the refinement requirement of the modeling data is feasible.

The form information, original terrain and design terrain modified by a cross-section of bridge design require to quickly transform the abstract vector model into a three-dimensional geometric model. The independent CGA modeling scheme can be compared to a "curve integration process." The geometry and texture functions of CGA language are used to describe the cross-sectional structure. The cross-sectional structure is called the "integrand function", and rapid modeling can be completed by integrating the "integrand function" along the center line. Figure 8.42 shows urban area modeling. Under terrain, roads and buildings can be generated by abstract models. At the same time, building planes that are twin to abstract models can be input into CityEngie as parameters of CGA to realize fast solid modeling.

In establishing the digital twin model for bridge custodial maintenance, realistic bridge custodial maintenance monitoring information should be decomposed into a tree structure with components as leaf nodes to form structural attribute data. Based on this structure, the realistic physical entity is transformed into an abstract

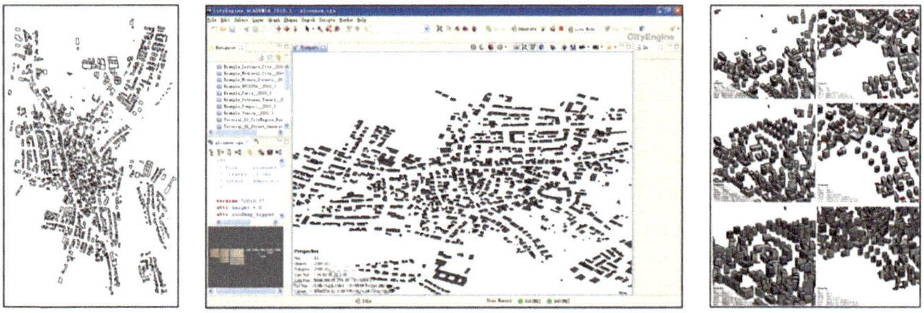

Fig. 8.42 CityEngine rule modeling

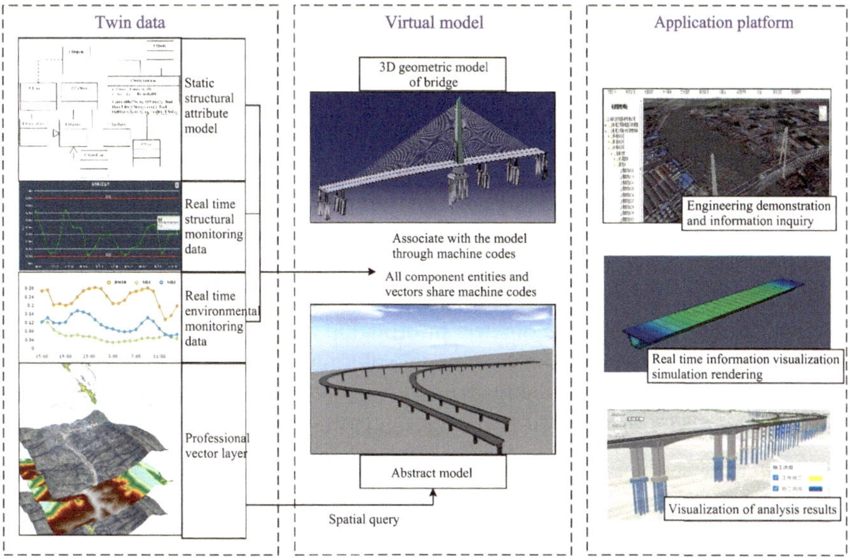

Fig. 8.43 Relationship between twin data, virtual models and application platforms

model according to the actual topological relationship of components. CGA code is written to construct the assembly, the outer frame of the assembly controls the size of components, and then BIM software models the components and generates the common 3D files. Finally, the twin data and virtual models are integrated into the application platform, as shown in Fig. 8.43.

(2) Bridge Digital Twin 3D Tiles Data Display

With the rapid development of computer and network technologies, integration of BIM and GIS in Web terminals has become the technical focus of various fields. In the development process of 3D WebGIS, 3D tile data format based on WebGL has been widely used as the standard for drawing 3D models in recent years, such as 3D Tiles used by Cesium. The 3D Tiles consists of tile set data (Tileset.json) and tile data (tile) [54]. It contains a variety of tile data formats. For 3D models, the data format is b3dm (Batched 3D Model). b3dm tiles are mainly used for 3D architectural models containing information, terrain surfaces with texture materials and mass models, which is the suitable target format for BIM data conversion. Based on the WebGL rendering mechanism at the beginning of design, this format is oriented to the fast loading of massive 3D data, so its internal structure is more in line with the fast loading of WebGL-based models. However, in terms of data format, 3D Tiles are significantly different from IFC with respect to the geometric representation of the model, semantic information, display scope, LOD hierarchy and coordinate system, which increases the difficulty in data interaction. Moreover, the current network transmission and browser rendering capabilities of the BIM and GIS integration scheme are insufficient. Therefore, the BIM model is popularly transformed from the universal IFC standard to the 3D Tiles data format, and the BIM model is quickly loaded on the Web site to support the integration of GIS and BIM on the website.

(1) IFC Conversion Approach

Since 3D Tiles data cannot be directly obtained in IFC format, IFC files must first be converted to OBJ files and then to glTF files. To ensure that the model properties after format conversion are complete and conform to the composition of components in actual bridges, the model needs to be split based on the predefined model splitting rules. The model splitting rules are based on the composition of actual bridge components. All objects in the IFC file are traversed through the specific conversion process. Objects containing geometric information are split into smaller IFC files and exported according to the above model splitting principles. At the same time, semantic attribute information of objects is preserved in JSON files. The b3dm data embed binary glTF (glb); finally, b3dm data and Tileset.json file form 3D Tiles data.

IFC to OBJ

In the IFC file, spatial structure is defined through IfcRelAggregates. Model components are placed according to local positioning. Each building component has its independent coordinate system (IfcLocalPlacement) and attribute PlacementRelto corresponds to the upper-level coordinate system, while attribute RelativePlacement corresponds to the relative position information. By repeated referencing of the coordinate systems, the coordinate system of building components is nested under multiple IfcLocalPlacements, and its position in the global coordinate system of the entire project is determined. By nesting references of multiple IfcLocalPlacement objects, the relative position of the local coordinate system where the component is located in the global coordinate system is obtained (the origin and the coordinate axis of the local coordinate system are located in the global coordinate system).

IfcOpenShell is an open-source (LGPL) software library for IFC files that provides a predefined function library for IFC file parsing and a conversion framework from IFC data to OBJ data. IfcConvert provided by IFCOpenShell library is used to convert the IFC files obtained above into OBJ files.

OBJ to glTF

The glTF is a 3D model file format designed for easy network transmission of 3D models. Its data structure is shown in Fig. 8.44. The JSON file records the entire structure, including the node hierarchy, materials, cameras and other logical structures. The BIN file records specific data, including the vertices, outer normals and texture coordinates. The GLSL file contains the shader information during model rendering.

The files obtained in the previous steps are sub-tiles after the model is sliced. Corresponding mesh and material data are written into the glTF file. In the format conversion process, it is necessary to determine the correspondence between texture coordinates and vertex coordinates, and map textures to 3D polygons by setting texture coordinates. Since the sub-tiles are not continuous, a vertex attribute batch is added to the file to distinguish them from each other. In the final file obtained, each building is

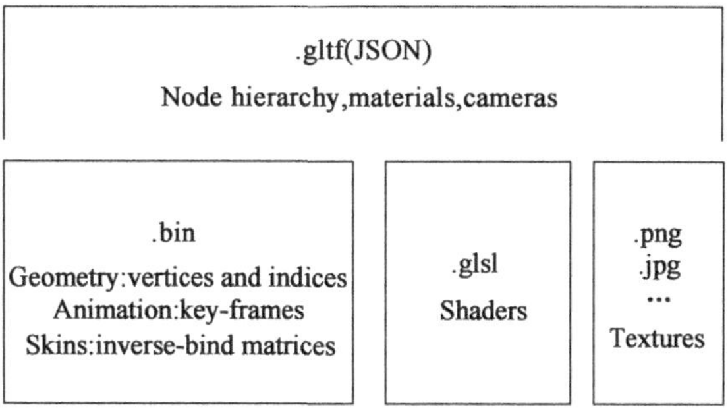

Fig. 8.44 Data structure of glTF files

composed of meshes of components, and the same components have the same identifier. The meshes are stored in the mesh array (meshes), and the mesh contains an array of primitives (primitives). The primitives contain variable attributes (attribute), which include vertex position POSITION, vertex normal NORMAL, vertex texture coordinate TEXCOORD_0, and mesh identifier _BATCHID.

glTF to b3dm

The glTF file is converted into binary glTF (glb) as part of the file body, and together with the file header, feature table and batch table, the tile data b3dm file is obtained. The batch table stores user-defined data related to the model. These data are used to interact with the users, such as filling UI or issuing REST API requests, and do not affect the display of the model.

Although there are many components in the actual buildings, walls, doors and windows at the same position on each floor are all the same building components, and a large part of them are repeated as the same components. In the file header (header), each component is assigned a batch, all IDs form an interval, and vertices with the same ID belong to the same component. The attribute information stored in the batch table in the file body is also called batchid [54], and its data structure is shown in Fig. 8.45.

The b3dm file and tileset data together constitute a 3D tile. The tileset file contains tile metadata to organize the spatial structure of 3D tiles, as shown in Fig. 8.46. The spatial bounding box (Bounding Volume) is an array that represents the spatial range of the current tile data. The tile data of all child nodes are included in the bounding box of the parent node tile data to reflect the spatial hierarchy of the tile data structure. The bounding box can be a cuboid (box), a sphere (sphere) or a region defined jointly by latitude, longitude and height to adapt to the slicing requirements

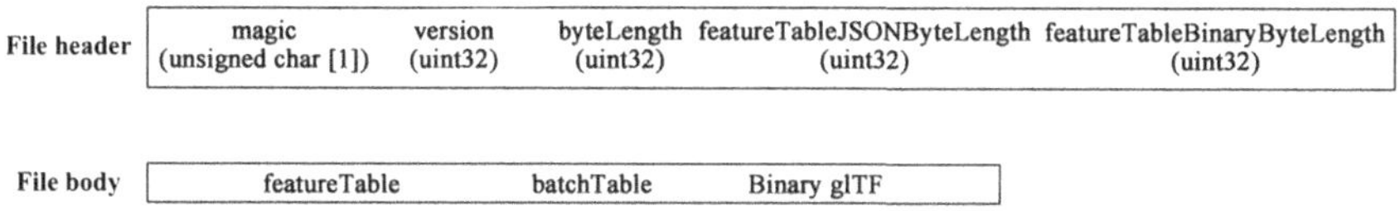

Fig. 8.45 Data structure of b3dm files

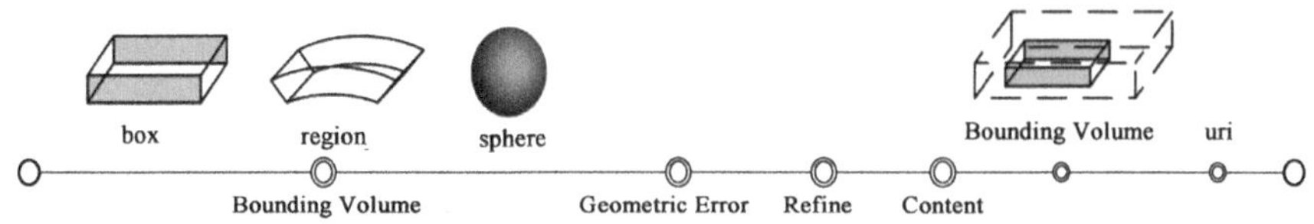

Fig. 8.46 Data structure of tile set files

of different data. Geometric error determines the level that should be loaded under the current viewpoint in LOD partitioning. Refinement indicates the refinement method for loading tile data, which supports ADD and REPLACE. Content specifies the loaded tile data.

(2) Display of IFC Data

To realize the display of IFC data, Revit software is first used to export the residential building model in IFC format and upload the exported file to the BIMServer server. The BIMServer API retrieval package is called to retrieve the entire model, which is divided into building components according to the proposed retrieval rules. The corresponding IFC file and JSON attribute file are saved for each component. Then, the BIM_to_3Dtiles format conversion tool is used to process the IFC file. In this process, the data mapping, organization and index structure are constructed to obtain the glTF file. On this basis, JSON files containing property information are added to form b3dm files. Further, by adding a tile set file (Tileset.json) to the b3dm file, the 3D Tiles data is made up that can be loaded in Cesium. By packaging the obtained 3D tile data and publishing it through a Node.js server, the BIM model of the residential building can be loaded in the Chrome browser. Figure 8.47 shows the final implementation of model loading in the Cesium framework. Figure 8.39a is the initial page of the Cesium framework, proving that the data is published successfully. Figure 8.47b is the long view of the model, and it can be seen that the transformation of the coordinate position is smooth; Fig. 8.47c is the overall display of the model, and the attribute information can be displayed smoothly.

The BIM + GIS modeling method provides a feasible way to create a digital twin model of the bridge. Terrain and bridge are finally integrated through the modeling data preprocessing, terrain modeling, abstract model generation and CGA modeling. After creating the model, BIM model is transformed from the universal IFC standard to the 3D Tiles data format, supporting the integration of GIS and BIM data at the Web end, and realizing the fast and efficient rendering interaction of data under the visualization framework.

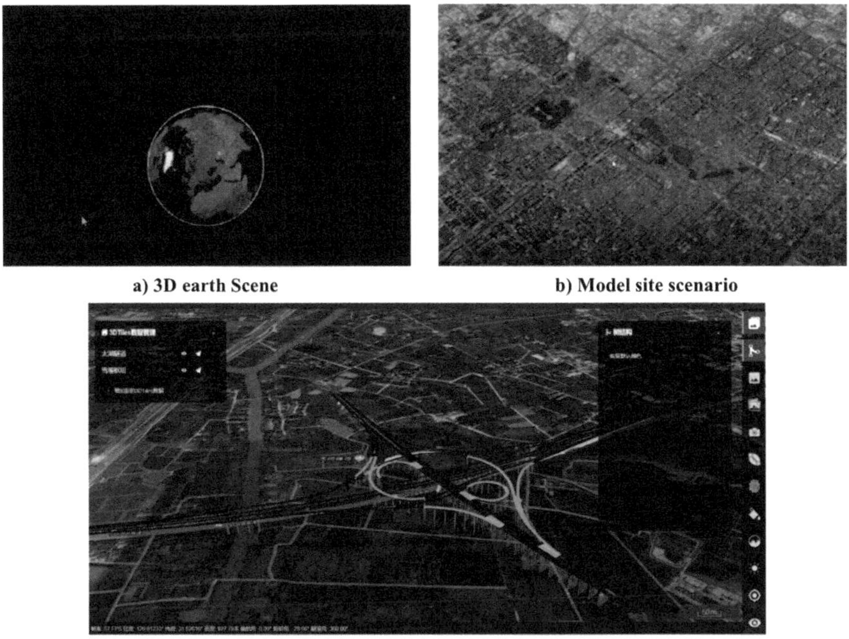

a) 3D earth Scene b) Model site scenario

c) Overall scene of bridge model

Fig. 8.47 Bridge model components and attribute scenarios

8.6.6 Challenges of Digital Twinning Technology in Bridge Management

(1) Technical Limitations and Supporting Platform Problems

Digital twin bridge is essentially the innovative and upgraded intelligent bridge, which has certain limitations and technical boundaries. Firstly, application of intelligent technology requires integrated and efficient supports from the bridge construction, operation and maintenance management. However, smart bridge projects are mostly based on domain models of specific types aimed at specific stages of business systems with their cloud computing resources. Therefore, it cannot achieve comprehensive information and data integration. Digital twin bridges can unify models in all fields into a complete information system, so it can be expanded from the original "silo-ed" project construction method to "horizontal layered" promotion. In order to avoid the emergence of separate governance and information island phenomenon, the idea of bridge information system construction is no longer divided according to the conventional stages, such as planning, design, construction, operation and maintenance, but transformed into the unified integration framework of various levels of elements of information system, such as geometric model construction, events, transactions, workflow, etc., to increase the horizontal connection on the shaft mechanism and realize the efficient utilization of bridge big data resources.

(2) Lag Between Theoretical Research and Algorithm Technology

Digital twin bridge is a new idea and an innovative way of smart bridge development driven by technology evolution and demand upgrading. "New" means frontier, but digital twin bridge still lacks systematic, theoretical research compared with the other digital technologies. Firstly, the bridge behavior depends on its functional load and natural load, both of which are uncertain. Secondly, the service cycle of bridges is measured in decades, and the performance of materials and components will degrade with time. How to describe their performance quantitatively in the system is a big challenge. Thirdly, the actual bridge has strong nonlinear behavior in unexpected events (such as explosions, earthquakes, ship collisions, etc.), so it is necessary to establish an efficient alternative model when building its digital twin. Fourthly, bridge monitoring and management are sparse in time and space.

Currently, the static modeling of digital twin bridges worldwide is only preliminary and mainly used to guide the planning and design work. The digital model must semantically express the dynamic information, such as bridge operation and maintenance events. The presentation rules of operation and maintenance data in the bridge information model, relationship between data, software and model, how to reflect the advantages of bridge management and service under the digital twin mode, and how to make decisions based on static and dynamic data simulation and optimization are all basic research directions that need in-depth studies. In addition, there is a lack of technological innovation in the hardware and software systems for intelligent construction, operation and maintenance control of bridges by "virtual control and real control."

(3) Digital Twin Bridge System will Become Mainstream of Development

Digital twin mapping bridge is an innovative technology that integrates almost all digital technologies. To construct a digital twin bridge system, it is necessary to understand and reflect on the imprecise and fuzzy self-care rules, and the operation mechanisms of bridge construction and maintenance, which will enable the integration and innovation of simulation, monitoring, diagnosis, prediction and control of physical bridges. Five aspects need to be further explored. The first layer is a simulation that involves the virtual mapping of physical objects. The technology and standard of full-scale simulation need to be further explored due to the complexity of bridge construction and operation process and the diversity of elements. The second layer is the monitoring that involves reflecting the changes in physical objects through the virtual model. The collection and transmission of physical object data cannot be separated from the Internet of Things. Its coding, positioning, standards, security and other issues have yet to be solved. The third layer is a diagnosis that requires complex multi-dimensional data processing and anomaly analysis algorithms based on artificial intelligence when the bridge structure is abnormal. The fourth layer is a prediction that involves predicting potential risks and maintaining structures and equipment reasonably and effectively. Prediction requires integrating and intensifying many technologies, which presents great challenges regarding flexibility and adaptability. The fifth layer is the control that needs the support of huge and complex software and hardware systems, especially to enable the management and service of digital bridges and physical bridges through software.

8.7 Summary

In today's era of big data analytic, integrating multi-source data is very crucial for decision-making of bridge maintenance and management. BIM and twin technology provide a visual and digital management platform for integrating multi-source data, and facilitate innovative development of intelligent bridge maintenance and management. However, limited by the distinctiveness of the bridge structure itself, many challenges need to be addressed. In the future, combined with the structural characteristics of bridges, the application mode of digital twin technology in bridge engineering should be explored so that digital twin mapping of the whole life cycle of bridges will become the mainstream of development.

References

1. Luo JH, Wang ZJ (2015) Multi source data fusion and sensor management. Tsinghua University Press, Beijing (in Chinese)
2. Li H, Bao YQ, Li SL et al (2016) Structural health monitoring data science and engineering. Science Press, Beijing (in Chinese)
3. ASG (1976) A mathematical theory of evidence. Priceton University Press
4. Sofge DA (1994) Structural health monitoring using neural network based vibrational system identification. IEEE Xplore
5. Guo JQ, Xie XB, Bie RF et al (2014) Structural health monitoring by using a space coding-based deep learning algorithm with wireless sensor networks. Pers Ubiquit Comput 18(8):1977–1987
6. Roshandeh AM, Rashed Poormizaee FSA (2014) Systematic data management for real-time bridge health monitoring using layered big data and clone computing. Int J Innovat Sci Res 29–39
7. Jeong S, Zhang YL, O'Connor S et al (2016) A NoSQL data management infrastructure for bridge monitoring. Smart Struct Syst 17(4):669–690
8. Jeong S, Zhang YL, Hou R et al (2016) A cloud based information repository for bridge monitoring applications. In: Lynch JP (2016) Sensors and smart structures technologies for civil, mechanical, and aerospace systems
9. Yang J, Li AQ, Li ZX (2008) Data fusion framework for structural health monitoring of bridges. J Disaster Prevent Mitigat Eng 28(3):292–297 (in Chinese)
10. Nie GW, Sun LM (2011) Integration of bridge maintenance patrol and health monitoring system information. J Shanghai Jiaotong Univ (Chin Ed) 45(S1):104–108 (in Chinese)
11. Lv YZ, Ren W, He SH (2008) Reliability analysis method for in-service concrete bridges based on the probability of load occurrence evaluation. Highway Transp Technol 25(6):60–64,70 (in Chinese)
12. Liu XL, Huang Q, Ren Y et al (2017) A comprehensive evaluation method for multiple indicator evidence fusion of cable-stayed bridges. J Harbin Inst Technol 49(3):74–79 (in Chinese)

13. Liu XL (2017) Application of multi-source information fusion technology in state evaluation of steel structure cable-stayed bridges. Southeast University, Nanjing (in Chinese)
14. Li YF, Yu H, Liang L (2018) Safety assessment of PC cable-stayed bridges based on multi factor data fusion. J Dalian Jiaotong Univ 39(3):95–99,114. (in Chinese)
15. Cho S, Park JW, Palanisamy RP et al (2016) Reference-free displacement estimation of bridges using Kalman filter-based multimetric data fusion. J Sens 3791856-1–3791856-9
16. Ahmed M, Moselhi O, Bhowmick A (2018) Two-tier data fusion method for bridge condition assessment. Can J Civ Eng 45(3):197–214
17. HAM (2017) Integrated NDE methods using data fusion-for bridge condition assessment. Concordia University
18. Lin ST (2018) Research on the digital architecture of transportation infrastructure for multi source data fusion. Highway Transp Technol 35(9): 122–127,145. (in Chinese)
19. Li JC (2015) The meaning and characteristics of BIM technology. In: 2015 National architecture digital technology teaching seminar (in Chinese)
20. Eastman CFD, Lafue G et al (1974) Pittsburgh: carnegie-Mellon University. Institute of Physical Planning
21. Li JC (2015) BIM applications introduction. Tongji University Press, Shanghai
22. Wang HL (2019) Research on the application of BIM and VR technology in interaction design of construction industry. Tianjin University of Technology, Tianjin (in Chinese)
23. Yao JN, Liu ZZ (2017) Application of BIM technology in construction engineering. Jiangxi Building Materials 2017(18):69,73. (in Chinese)
24. Huang ZG (2018) Construction and development of bridge maintenance management system based on BIM. Southeast University, Nanjing (in Chinese)
25. Wu L, Wang GY (2017) Fundamentals of BIM modeling and application. Beijing Institute of Technology Press, Beijing (in Chinese)
26. Zeng Q, Zhang QL, Zhang JH (2021) Research on IFC data storage based on object based hierarchical database. Civil Const Eng Inf Technol 13(01):17–23 (in Chinese)
27. Zhang Y (2009) Research on building engineering information integration and management based on BIM. Tsinghua University, Beijing (in Chinese)
28. Wang Y, Zhang JP, Hu ZZ (2011) Research on IFC data description standards for construction. Civil Const Eng Inf Technol 3(4):9–15 (in Chinese)
29. Chen LC, Lai HH, Deng XY et al (2015) Research on entity extension method for ifc standard domain layer. J Graph 36(2):282–288 (in Chinese)
30. Tan KB (2016) Research on the data structure and pricing model of engineering pricing based on BIM. Beijing Jiaotong University, Beijing (in Chinese)
31. Feng P, Song SF, Qi CL et al (2017) Research on bridge engineering data storage standard based on IFC4 standard. Rail Technol Innovat 1:36–41 (in Chinese)
32. U.S. Department of Transportation (2016) Bridge information modeling standardization, volume I: Information Exchanges
33. Yang BW (2006) Modern trendy sensor application manual. Mechanical Industry Press, Beijing (in Chinese)
34. Yang B (2018) Research on Information visualization of bridge structural health monitoring based on BIM. Chongqing Jiaotong University, Chongqing (in Chinese)
35. Wang C (2015) Research on IFC expression and integration method of monitoring information based on BIM. Harbin Institute of Technology, Harbin (in Chinese)
36. Cao ZX (2017) Research on the application of BIM technology in bridge management and maintenance. Chongqing Jiaotong University, Chongqing (in Chinese)
37. China Center for Information Industry Development (2019) Digital Twin White Paper (in Chinese)
38. Rosen R, von Wichert G, Lo G et al (2015) About the importance of autonomy and digital twins for the future of manufacturing. Ifac Papersonline 48(3):567–572
39. Zhou Y, Liu CC (2018) Logic and innovation of building Xiong'an into a digital twin city. Urban Studies 25(10):60–67 (in Chinese)

40. Tao F, Qi QL, Wang LH et al (2019) Digital twins and cyber-physical systems toward smart manufacturing and industry 4.0: correlation and comparison. Engineering 5(4):653–661
41. Bottani ECA, Murino T, Vespoli S (2017) From the cyber-physical system to the digital Twin: the process development for behaviour modelling of a cyber guided vehicle in M2M logic. In: XXII Summer School "Frances co Turco"-Industrial Systems Engineering
42. Tao F, Liu WR, Zhang M et al (2019) Digital twin five dimensional model and its applications in ten major fields. Comput Int Manuf Syst 25(1):1–18 (in Chinese)
43. Yang LY, Chen SY, Wang X et al (2019) Digital twins and parallel systems: development status, comparison, and prospects. Acta Automatica Sinica 45(11):2001–2031 (in Chinese)
44. Muilenburg D (2018) Boeing CEO talks 'Digital Twin' Era of Aviation[EB/OL]. Accessed https://www.aviationtoday.com/2018/09/14/boeing-ceo-talks-digital-twin-era-aviation/
45. BOEING (2017) Developing airplane systems faster and with higher quality through model-based engineering[EB/OL]. Accessed https://www.boeing.com/features/innovation-quarterly/may2017/feature-technical-model-based-engineering.page
46. Moneer H, Alex J, Thomas H (2018) A standards-based approach for linking as-planned to as-fabricated product data. CIRP Ann 67(1):487–490
47. Li H, Tao F, Wang HQ et al (2019) Integrated development framework and key technologies for complex product design and manufacturing based on digital twins. Comput Int Manuf Syst 25(6):1320–1336 (in Chinese)
48. Zhao D, Peng H (2017) From the lab to the street: solving the challenge of accelerating automated vehicle testing
49. Tao F, Zhang M, Cheng JF et al (2017) Digital twin workshop—a new mode of future workshop operation. Comput Int Manuf Syst 23(1):1–9 (in Chinese)
50. Shim CS, Dang NS, Lon S et al (2019) Development of a bridge maintenance system for prestressed concrete bridges using 3D digital twin model. Struct Inf Eng 15(10–12):1319–1332
51. Zhang WH, Zhou XY, Wu HP et al (2020) A digital twin system for highway traffic based on 3D GIS technology. Comput Int Manuf Syst 26(1):28–39 (in Chinese)
52. Lu RD, Brilakis I (2019) Digital twinning of existing reinforced concrete bridges from labelled point clusters. Automat Constr 105:102837.1–102837.16
53. Teng QS, Mi JZ, Sun SY (2018) A Method for 3D modeling of road networks. Sci Surveying Mapp 43(3):99–103 (in Chinese)
54. Kulawiak M, Chybicki A (2018) Application of web-GIS and Geovisual analytics to monitoring of seabed evolution in south Baltic sea coastal areas. Mar Geodesy 41(4):405–426

Chapter 9
Decision-Making Technologies for Intelligent Maintenance and Management

Abstract Upon completion of bridge inspection, monitoring and analysis, the subsequent and crucial step is to formulate decisions for bridge maintenance, which should be logic or evidence based, optimal considering economic and environmental constraints, and feasible subject to resources at hand. There exists a situation that the decision-making is straightforward if the data and information are of low complexity. The reality is often the opposite, that is, data and information from the life-cycle inspection and monitoring processes, when brought to the maintenance engineers and managers, are often non-structured, high-dimensional, mutually dependent, and still 'big' in their volume. Recognizing these challenges, this chapter focuses digital technologies that can facilitate scientific and rational decision-making for bridge maintenance and management. Among them, Artificial intelligence (AI) technologies can facilitate decision-making taking the role of assisting human-based decision-making or of driving the decision-making. Two human–machine interfacing technologies are introduced, including virtual reality (VR) and augmented reality (AR), which provide advanced visual analytics for facilitate spatial–temporal understanding, logic formulation, and rational decision-making. General applications and specific applications of these technologies in the context of bridge maintenance and management are highlighted too in this chapter.

9.1 AI Based Decision-Making

9.1.1 Overview

Currently, bridge maintenance strategies primarily rely on bridge inspection results and reference to corresponding maintenance codes, such as the "Code for Maintenance of Highway Bridges and Culverts" (JTG 5120—2021), while taking into account the professional expertise and experience of the decision-makers. For instance, when dealing with crack damage in the substructure of a bridge, the bridge maintenance decision-maker may formulate a maintenance decision based on recommendations in the code and the results of on-site crack inspections, combined with

G. Wu et al., *Intelligent Bridge Maintenance and Management*, Springer Tracts in Civil Engineering, https://doi.org/10.1007/978-981-97-3827-4_9

their own maintenance experiences. This could involve counter-measures such as applying epoxy resin sealant or pressure grouting.

It is apparent that this process is filled with subjective uncertainties. Often, different decision-makers might propose different maintenance strategies for the same situation. This implies that the maintenance strategies formulated by different decision-makers will have variations in quality, leading to differences in reliability, remaining service life and other indicators of the bridge health after maintenance. In this context, if the subjective uncertainties introduced by the decision-makers themselves in the current decision-making process are not addressed properly, even with the use of various advanced technologies to address the issues in bridge inspection, there is no guarantee that the bridge, based on the current maintenance strategy, will continue to meet the requirements of normal use, ultimate load-carrying capacity and serviceability in a satisfactory manner. Therefore, the primary goal of bridge maintenance decision-making is to achieve a structured and standardized decision-making process, reducing subjectivity and ensuring quality of bridge maintenance.

Meanwhile, due to the existence of various objective factors, such as inability of AI technology to completely replace human involvement and inherent randomness in the external environment where bridges are located, the maintenance of different bridges still relies on the decision-makers' experiences. Therefore, formulation of bridge maintenance strategies cannot completely exclude involvement of the decision-makers. However, in order to better develop a system for formulating bridge maintenance strategies, it is necessary to track the effectiveness of the final maintenance decisions for each bridge, quantitatively evaluate the decision-making level based on the assessment results of post-maintenance inspections, and compile them into a database. This will facilitate continuous improvements in the later stages.

In addition, the current formulation of bridge maintenance strategies usually focuses on individual bridges, without considering the overall management, coordinated planning and formulation of maintenance strategies for other bridges within the same road network. As a result, bridge maintenance often requires long duration and frequent traffic closures, leading to a waste of time and money.

In summary, we need to reform the current bridge maintenance decision-making process to introduce more structured, standardized and objective decision-making methods and systems, and eliminate the subjective uncertainties introduced by the decision-makers themselves. At the same time, in the decision-making process, we should shift from considering individual bridges to considering the overall road network where the bridges are located. This allows for coordinated planning, avoiding unnecessary waste, optimizing resource allocation, extending the lifespan of bridge clusters, and ensuring the safety of people's lives and properties. Additionally, it is necessary to establish a reasonable evaluation system to quantitatively assess the bridge maintenance strategies formulated for each bridge. This will facilitate further improvement of the bridge maintenance strategy formulation system in the future.

With this background, coupled with the advancement of computer science, the concept of AI based decision-making has been proposed to overcome the current deficiencies in bridge maintenance by incorporating AI technology into the bridge maintenance decision-making process.

AI based decision-making, as the name suggests, refers to the introduction of AI technologies to partially or completely replace the role of humans in decision-making, eliminating subjective uncertainties, and enhancing standardization of decision-making. AI based decision-making can be divided into two aspects: AI-assisted decision-making and AI-driven decision-making, depending on the role and level of intelligence played by AI in the decision-making process.

9.1.2 AI-Assisted Decision-Making

AI-assisted decision-making refers to the role of AI technology in the entire bridge maintenance decision-making process to provide auxiliary supports. It falls within the domain of weak artificial intelligence and can replace humans in executing relevant stages of the decision-making process. However, the ultimate decision-making for maintenance is still made by humans, similar to the conventional bridge maintenance decision-making process. The introduction of AI technology can greatly improve decision-making efficiency and accelerate the retrieval, recording, and summarization of relevant data and information during the decision-making process. Although AI-assisted decision-making cannot completely solve the aforementioned problems, it can serve as a transitional stage towards achieving AI-driven decision-making.

In AI-assisted decision-making, humans remain the core decision-makers, and AI primarily operates based on instructions issued by humans. Its subjectivity is relatively weak, but it still has prominent advantages at this stage. For example, Siri, the intelligent voice assistant in Apple smartphones, is an AI-assisted technology that utilizes Natural Language Processing (NLP). The users can interact with their phones directly through voice commands. Similarly, NLP technology can be introduced into bridge maintenance decision-making, allowing the decision-makers to no longer rely on hardware such as keyboards and mice to interact with the systems. This eliminates the need for tedious mechanical tasks such as retrieving maintenance data, inputting text, performing data calculations and storage. It greatly enhances decision-making efficiency and enables automated and standardized archiving of bridge maintenance data, facilitating subsequent decision evaluation, and tracking the generation of future automated AI-driven decision-making systems.

The Touch and Talk (TNT) human–computer interaction system released by Smartisan Technology Co., Ltd. in 2018 exemplifies this concept. Although it is not yet mature and still has various issues, its global-based gesture and voice interaction system presents a highly promising AI-assisted decision-making approach. It has revolutionized the conventional computer's graphical user interface (GUI) interaction method and significantly improved the work efficiency. Furthermore, in recent years, the popularization of VR and AR technologies can be effectively applied in AI-assisted decision-making, assisting the decision-makers in data retrieval, on-site surveys and reinforcement solution selection.

In the future, the ideal state of AI-assisted decision-making systems would resemble J.A.R.V.I.S., the intelligent butler in the movie "Iron Man", capable of seamless communication with the decision-makers and fulfilling their various needs. Currently, there are already many technologies available for bridge-assisted decision-making, including the aforementioned Natural Language Processing (NLP) technology, VR technology, AR technology, and so on.

9.1.3 AI-Driven Decision-Making

Compared to weak AI-assisted decision-making, in AI-driven decision-making, AI has greater autonomy. It can autonomously make decisions and provide maintenance plans based on the current evaluation status of the bridge and relevant maintenance data, without human involvement. This falls within the domain of strong artificial intelligence. The Go-playing robots AlphaGo and AlphaZero, which have defeated various top Go players multiple times in the past, belong to the active decision-making systems. They can actively make autonomous decisions and formulate optimized strategies based on the current state of the game.

Compared to weak artificial intelligence in AI-assisted decision-making, active AI-driven decision-making belongs to the realm of strong artificial intelligence, but currently can only address some simple system problems. In the field of bridge maintenance, achieving the ideal goal of AI-driven decision-making is relatively challenging due to the complexity of bridge systems and the presence of significant uncertainties. However, in industries such as medicine, logistics, and urban planning, there have been preliminary prototypes of AI-driven decision-making systems for addressing some simpler problems. For example, in the field of medicine, there is the Clinical Decision Support System (CDSS), which is a human–computer interactive medical information technology application system. It learns from electronic medical record systems and internet databases to accomplish clinical decision-making and provide decision support to doctors and other healthcare practitioners.

Currently, to achieve AI-driven decision-making, techniques such as expert systems and reinforcement learning can be employed. By training learning systems with relevant domain knowledge and data, the system can learn and mimic the decision-making process of bridge maintenance decision-makers, thus achieving active decision-making. The next section will introduce the technologies involved in AI-assisted decision-making and AI-driven decision-making.

9.1.4 AI Technologies for Decision-Making

Interaction between the conventional bridge lifecycle management research and the non-destructive testing and structural health monitoring research is limited. Scholars in related fields are mostly conducting research within their respective domains, with different focuses and directions. The research achievements in various fields are difficult to integrate with each other, making it challenging to truly realize comprehensive bridge lifecycle management. As shown in Fig. 9.1, based on the development of emerging disciplines in computer science, this book proposes a new concept of intelligent lifecycle management by fully integrating technologies such as digital twins, big data, Internet of Things and AI into the processes of bridge monitoring, inspection, assessment and maintenance decisions. This aims to enhance scientific, objective and timely aspects of bridge management, bridging the limitations and deficiencies in current conventional bridge lifecycle management, and truly perform bridge management from a holistic perspective. It aims to meet the requirements of safety and applicability while considering durability and cost-effectiveness.

Currently, AI based decision-making in bridge maintenance, as the final component of intelligent lifecycle management of bridges, remains a cutting-edge research field. The research outcomes in this area are still relatively limited. In this section, we will introduce the relevant technologies that can be utilized for the development of AI based decision-making, as shown in Fig. 9.2. Among them, due to the relative maturity and significant implications for the development of AI based decision-making, VR technology and AR technology will be discussed in detail in the following two sections.

(1) Natural Language Processing (NLP)

NLP is dedicated to the study of various theories and methods that enable effective communication between humans and computers using natural language to serve as an interface between humans and machines. Applying NLP techniques to AI-assisted decision-making in bridge maintenance can improve the interaction efficiency between decision-makers and computers. Additionally, it facilitates retrieval, input, storage and management of relevant bridge maintenance decision data, thereby enhancing the level of bridge maintenance decision-making.

Among all living beings, only humans possess the ability of language. Various human intelligences are closely related to language. Human logical thinking takes the form of language, and the majority of human knowledge is recorded and transmitted

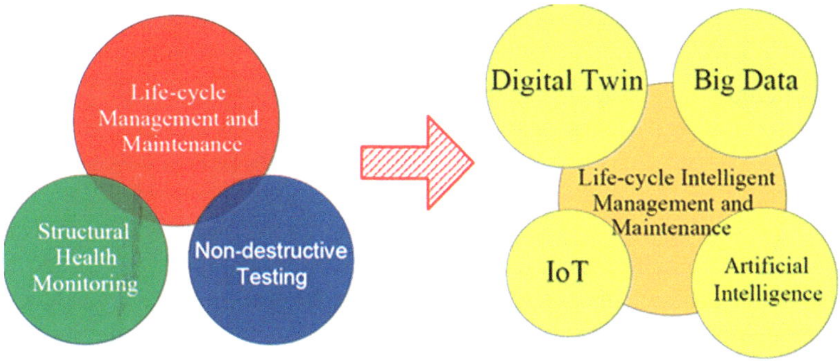

Fig. 9.1 Intelligent life-cycle maintenance and management

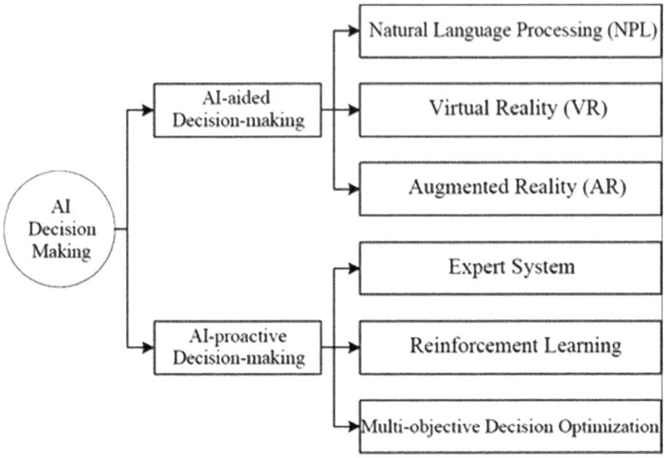

Fig. 9.2 Related technologies for AI based decision-making

in the form of language and text. Therefore, natural language processing is an important, even core, part of AI. Bill Gates has also stated that "Language understanding is the crown jewel of artificial intelligence."

The ability to communicate with computers using natural language has long been a pursuit of humanity. It holds significant practical value as it allows people to use their preferred language to interact with computers, without the need to spend a substantial amount of time and effort learning unfamiliar computer languages. Achieving natural language communication between humans and machines means enabling computers to both understand the meaning of natural language texts and express given intentions, thoughts and ideas in natural language texts. The former is known as Natural Language Understanding (NLU), while the latter is referred to as Natural Language Generation (NLG). Therefore, natural language processing involves two core tasks: natural language understanding and natural language generation.

Natural Language Understanding aims to enable machines to have the same language comprehension ability as humans. However, due to the many difficulties inherent in language understanding, machine understanding of natural language still lags far behind human-level comprehension. Firstly, language lacks strict rules and exhibits complex patterns. It allows for free combinations and can be composed into complex linguistic expressions. Secondly, language is an open set, allowing people to invent and create new ways of expression freely. Lastly, language requires the connection to practical knowledge and has a certain knowledge dependency. The use of language is based on the environment and context. Therefore, Natural Language Understanding needs to address five technical challenges: language diversity, language ambiguity, language robustness, language knowledge dependency and language context. In contrast to Natural Language Understanding, the purpose of Natural Language Generation is to bridge the communication gap between humans and machines, transforming non-linguistic data into language formats that humans can easily understand, such as articles and reports. It involves six steps: content determination, text structure, sentence aggregation, grammaticalization, reference expression generation and language implementation.

Natural language processing began in the 1950s. In 1954, researchers at Georgetown University attempted to automatically translate over 60 Russian sentences into English, claiming that the problem of machine translation could be solved within 3 to 5 years. However, the actual progress of this work fell behind expectations. It was not until the late 1980s that statistical machine translation systems were developed, marking a significant advancement in machine translation research. In the 1960s, one particularly successful natural language processing system was the SHRDLU system developed by MIT's Winograd, which was used to control the actions of robots. The system combined syntax analysis, semantic analysis and logical reasoning, greatly enhancing the system's capabilities in language analysis. Until the 1980s, most natural language processing systems were still based on a set of complex, manually crafted rules, resembling expert systems. However, starting from the late 1980s, machine learning algorithms were introduced to language processing, leading to innovation in natural language processing. Some of the earliest machine learning algorithms used, such as decision trees, were systems composed of hard rules, similar to previously crafted rules. However, research in natural language processing has increasingly focused on soft, probabilistic models for decision-making. These models are usually sufficient to handle unexpected input data, especially when there are errors in the input, and they produce reliable results when integrated into larger systems that involve multiple subtasks.

In recent years, natural language processing has entered a phase of rapid development. The increasing abundance of various resources such as lexicons, semantic grammar dictionaries and corpora, along with the rapid progress in technologies like word segmentation, part-of-speech tagging and syntactic analysis, and the emergence of new theories, methods and models, have propelled the prosperity of research in natural language processing. The trends of the internet, mobile internet and integration of the world economy and society have created an urgent demand for natural language processing technology, providing strong market impetus for the development of natural language processing research. Although achieving natural language understanding or natural language generation is currently not as simple as originally imagined, from the perspective of existing theories and technological status, a general, high-quality natural language processing system remains a long-term goal. However, practical systems with considerable natural language processing capabilities have already emerged for certain applications, some of which have been commercialized and even industrialized. Examples include multilingual database and expert system natural language interfaces, various machine translation systems, full-text information retrieval systems and automatic summarization systems. Among them, the most well-known products include Apple's widely-used personal assistant Siri, personal intelligent assistants from companies like Google, Amazon, Microsoft and Baidu, and the intelligent voice input method developed by iFlytek, which have greatly facilitated people's lives and work. From these products, it can be seen that applying natural language processing technology to bridge AI-assisted decision-making is highly feasible at the current level of technological development. Based on natural language processing technology, bridge maintenance decision-makers can directly store and retrieve bridge management data through human–machine dialogue, perform corresponding calculations and formulate final decisions.

(2) Expert Systems

Expert systems are a branch of AI that utilizes a vast amount of specialized knowledge to address problems that currently only experts can solve. Bridge decision-making based on expert systems refers to the intelligent software system that utilizes expert systems to mimic experienced experts and make maintenance decisions for existing bridges, following the relevant rules formulated in current bridge decision-making.

One of the pioneers of expert systems, Professor Edward Feigenbaum of Stanford University, defines expert systems as "intelligent computer programs that apply knowledge and reasoning to solve complex problems that only experts can solve." In other words, expert systems can be seen as a type of intelligent computer system that possesses a vast amount of specialized knowledge. They can utilize the specialized knowledge and experience provided by one or more experts in a specific field and employ AI reasoning techniques to solve and simulate various complex problems that are typically solvable only by experts. This allows experts' expertise to be accessible without being limited by time and space.

It is important to note that expert systems simulate the reasoning of experts in specific domains. For example, in the field of bridge maintenance decision-making, expert systems simulate the reasoning process of how experts derive maintenance decisions based on inspection and monitoring results, rather than simulating the problems themselves (i.e., modeling the problem domain through mathematical models). Expert systems only simulate the problem-solving capabilities of experts. Therefore, using this characteristic for bridge maintenance decision-making can effectively eliminate decision discrepancies caused by differences in decision-makers' professional knowledge and experience, thereby enhancing the standardization of bridge maintenance decisions.

In general, a basic expert system consists of six components: knowledge base, database, inference engine, explanation mechanism, knowledge acquisition and user interface, as shown in Fig. 9.3.

In an expert system, the knowledge base is its core component, consisting of factual knowledge and heuristic knowledge specific to the domain (such as the domain of bridge maintenance decision-making). The former refers to public and widely accepted facts, such as relevant indicators in bridge inspection and decision-making standards. The latter refers to the experiential and heuristic knowledge in the specific domain, such as the personal experience and decision-making expertise of bridge decision-making experts. The knowledge base of an expert system is a collection of multiple expert knowledge about a domain or specific problem. It can provide the users with expertise and knowledge of more than one expert, and can be continuously enhanced and improved over time. The performance level of an expert system depends primarily on the quantity and quality of knowledge it possesses. The

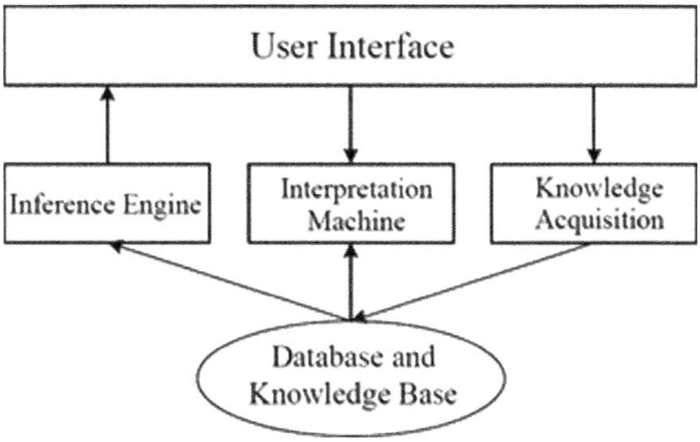

Fig. 9.3 Framework of an expert system

more higher quality knowledge an expert system has, the stronger is its problem-solving ability. The database, on the other hand, is used to store knowledge, input data, initial states and inference processes within the expert system, serving as the working memory area of the expert system.

Knowledge acquisition is an important learning function of expert systems. It is used to modify and add knowledge to the knowledge base, incorporating new knowledge to continuously enhance the system's expertise. It is similar to continuously "inputting internal force" into the system, allowing its "skill" to become more profound.

A reasoning engine is a set of programs used to control and coordinate the entire expert system. It utilizes the knowledge from the knowledge base, based on user input data (such as bridge defect data obtained through inspections and monitoring in bridge maintenance), and follows certain reasoning strategies (such as forward reasoning, backward reasoning and mixed reasoning) to solve current problems, interpret user requests and ultimately provide conclusions. When designing a reasoning engine, it should conform to the reasoning process of experts. Generally, the reasoning engine of an expert system is separate from the knowledge base, which not only facilitates knowledge management but also enables system generality and scalability.

The primary role of the explanation mechanism is to explain how an expert system deduces inferential conclusions based on input information and to answer user queries, enabling users to understand the entire reasoning process, and the knowledge and data applied during the process. The main reason for incorporating an explanation mechanism in expert systems lies in the fact that expert knowledge, in general, is mostly experiential knowledge accumulated by human experts in practice, and therefore, usually only the experts themselves have an understanding of it. Moreover, experiential knowledge is often derived from experience and lacks theoretical guarantees. Typically, this knowledge is not documented in textbooks or other professional books. Consequently, this knowledge remains unfamiliar to others and essentially constitutes proprietary knowledge of the experts. If an expert system only provides final conclusions without any explanations, it will inevitably affect the level of trust users have in those conclusions, especially when the system's conclusions contradict the users' own perspectives. Therefore, an expert system should possess explanatory functionality, allowing it to answer user questions and inform users about how it solves problems, which knowledge it utilizes, the content of that knowledge, its sources, its rationale, and so on, making the expert system "transparent" to the users. Good transparency also facilitates knowledge validation and modification, enabling the users to evaluate and discern the conclusions drawn by the system.

The final user interface provides a user-friendly interactive environment for the users, facilitating the exchange of information between the expert system and the users. The aforementioned six components constitute an integrated whole of the expert system. Through computer programs and initial knowledge learning and training, the expert system can simulate a highly skilled bridge maintenance and management expert, and achieve AI driven decision-making for bridges based on input inspection and assessment data. However, the current bridge decision-making based on expert systems can only address individual bridge decision problems and cannot yet achieve comprehensive decision-making for a group of bridges, failing to reach the level of holistic planning for the entire road network.

In an expert system, expert knowledge, as the core of the system function, has the following problems:

- Instability of expert knowledge. Expert knowledge is often heuristic in nature. Therefore, compared to logical knowledge, it is unstable. When faced with new situations or problems, human experts can readily revise existing knowledge or induce new knowledge to deal with these new issues. However, expert systems find it difficult to exhibit the same flexibility, which ultimately leads to ineffective decision outcomes.
- Difficulty in acquiring expert knowledge. The knowledge acquisition of expert systems generally involves a series of interactions between the domain experts (such as in the field of bridge maintenance decision-making), software engineers and computers. In this process, if the software engineer lacks expertise in the field of bridge maintenance, it becomes difficult to grasp the concepts, relationships and problem-solving methods used by the experts. As a result, the knowledge of the experts cannot be accurately input into the system, making it challenging for the expert system to replicate the logic and thinking process of the experts during decision-making.

More importantly, in most cases, the experts rely primarily on experience and intuition when making bridge maintenance decisions, making it difficult to accurately describe them using mathematical theories or other models. Furthermore, in order to solve problems in a specific field, the experts need to know much more than just principles and facts, including a significant amount of common knowledge from daily life. Therefore, it is impossible to expect the domain experts to compile all the knowledge they have in a short period of time.

To solve the aforementioned two types of problems, a relatively simple and effective method is to input the knowledge compiled by the domain experts into the system, continuously update and learn from it, and make the development of the expert system an iterative process. This also means that developing a successful expert system often requires extensive testing with practical engineering cases to identify and eliminate erroneous knowledge, as well as to continuously modify and expand the system to achieve high performance.

HP captured 1000 rules from 12 product series through two knowledge engineers, one programmer and three customer service experts, which further led to the development of the company's customer equipment diagnostic system. This system assists the customer service representatives in handling millions of phone inquiries each year. XCON, an expert system developed by Digital Equipment Corporation (DEC) and Carnegie Mellon University in the 1970s, is a commercial system used to devise hardware configurations for computer systems, resulting in significant economic benefits. Regarding the application of expert systems in civil engineering,

Kostem [1] developed an expert system for highway bridge evaluation in 1986. However, this system was essentially an interface for a bridge classification program developed using the conventional software approaches. Later et al. [2] developed a fuzzy expert system for bridge damage assessment, which combines genetic algorithms with neural networks and can be used for durability evaluation of reinforced concrete bridges. Melhem et al. [3] also developed an expert system for bridge condition assessment, which utilizes multi-dimensional decision models and fuzzy set theory to evaluate the condition of bridges. Mikami et al. [4] developed a learning-based expert system for repairing steel bridges in 1994, using a novel inference engine that can learn from user feedback. Subsequently, Wilson et al. [5] developed a decision support system for the entire life cycle of bridges, which evaluates various aspects of bridge design, construction, operation and maintenance.

The system proposes evaluation methods, calculation models, multi-parameter databases and knowledge systems with distributed interactive capabilities, considering factors such as bridge functionality, maintainability, operational cost, safety and environmental impact within its expected service life. However, due to various existing issues in the aforementioned expert systems and the complexity of bridge maintenance decision-making itself, there is currently no expert system that has been truly applied to bridge maintenance decision-making.

(3) Reinforcement Learning

Reinforcement Learning (RL) is a type of machine learning that enables computers to learn and solve problems autonomously. In the process of computer learning, humans only need to set a "small goal" for the computer, and the specific optimal strategy is obtained by the computer through continuous trial and error without human intervention. More precisely, reinforcement learning refers to the learning process of an intelligent agent that interacts with the environment and guides its behavior based on rewards obtained, aiming to maximize the overall reward gained by the agent.

A common model for reinforcement learning is the standard Markov Decision Process (MDP), where the next state of an agent depends not only on the current state but also on the action taken. The goal of reinforcement learning is to adjust the strategy to find the policy that maximizes the rewards and accumulates the highest total reward. Its basic elements include states, actions, state transition probabilities, rewards and policies (see Fig. 9.4). The problem-solving process of reinforcement learning can be roughly divided into two steps: (1) prediction, where a given policy

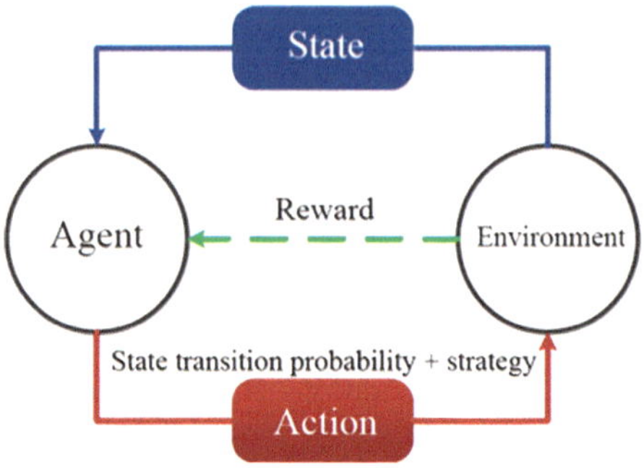

Fig. 9.4 Framework of reinforcement learning

is evaluated to estimate the corresponding state value function and the accumulated value obtained from rewards, and (2) action, where the optimal action for the current state is determined based on the value function. Depending on the given conditions, reinforcement learning can be categorized into model-based RL and model-free RL.

(1) Model-Based Reinforcement Learning

Model-based reinforcement learning is a method that uses dynamic programming to solve Markov Decision Processes. Depending on the policy, it can be divided into two methods: policy iteration and value iteration, both of which belong to dynamic programming. These methods require a comprehensive understanding of the rewards and state transition probabilities in the entire problem, and it is possible to store all the information of the Markov decision processes. However, in games like Go, the total number of states is 3 to the power of 19, which is extremely large and cannot be stored, making model-based reinforcement learning methods unsuitable.

The main idea behind policy iteration is as follows: first, randomly initialize a policy and calculate the value of each state under this policy. Then, based on these state values, derive a new policy. Calculate the value of each state under the new policy and repeat this process until convergence. In this process, calculating the value of each state under a policy is known as policy evaluation, and obtaining a new policy based on state values is called policy improvement. Compared to the two-stage policy iteration of policy evaluation and policy improvement, value iteration combines these two steps more closely. Firstly, through one round of policy evaluation, calculate the expected value of future states for each possible action in the current state. Select the action that leads to the state with the highest expected value and use this maximum expected value as the value of the current state for the next round of policy evaluation. Repeat this process until the value function converges. Value iteration is significantly faster than policy iteration.

(2) Model-Independent Reinforcement Learning

Compared to model-based reinforcement learning, model-free reinforcement learning refers to the approach where we have incomplete knowledge about the reward function and transition probabilities, and need to explore them ourselves. There are three main algorithms in model-free reinforcement learning: Monte Carlo

(MC) method, SARSA (State-Action-Reward-State-Action) method and Q-learning method.

The MC method generates policy samples randomly and calculates the corresponding state-action values based on these samples. It keeps track of the values and visit counts for each state-action pair. With an increasing number of random samples, the values and visit counts for each state-action pair are continuously updated. In contrast to the MC method, which requires complete policy samples, SARSA is a form of temporal difference (TD) learning that utilizes the Markov property and updates the value of the current state based only on the next state information. SARSA guides the system to explore according to a given policy and updates state values at each exploration step. The framework of Q-learning is similar to SARSA. Q-learning also guides the system to explore according to a policy and updates the state values at each exploration step.

The aforementioned methods mainly focus on solving reinforcement learning problems with discrete and small-scale state spaces. When facing more complex state spaces, including continuous ones, the conventional methods become computationally expensive. In such cases, deep reinforcement learning methods should be employed, with Deep Q-learning (DQN) being one of the classical algorithms. The basic idea of DQN algorithm originated from Q-learning, but it approximates the value function. Instead of directly calculating the Q-value based on the state-value (s) and action (a), it employs a Q-network composed of neural networks to compute the Q-values. Consequently, DQN has the capability to handle large-scale reinforcement learning problems. However, the DQN algorithm encounters an issue: it does not guarantee convergence of the Q-network. In other words, we cannot necessarily obtain converged Q-network parameters, leading to poor performance of the trained model. To address this problem, various variants of the DQN algorithm have been developed, such as Nature DQN, Double DQN and Dueling DQN. These methods still fall within the category of value-based learning algorithms. Although they have been successfully applied in many fields, they still exhibit limitations in handling continuous actions, addressing problems with restricted states, and resolving stochastic policy problems. To overcome these challenges, performance-based deep reinforcement learning algorithms can be employed, such as the Monte Carlo Policy Gradient (reinforce) algorithm, as well as methods that combine policy and value, such as Actor-Critic algorithms, A3C algorithm, and Deep Deterministic Policy Gradient (DDPG) method, to address the aforementioned issues.

In the field of reinforcement learning applications, the most well-known example is the Go AI program called AlphaGo, which has defeated numerous human Go masters. AlphaGo utilizes the deep reinforcement learning DQN algorithm to maximize reward accumulation through self-play, thereby determining the best moves in various game states. This algorithm aligns more closely with human decision-making in the real world and finds widespread applications in intelligent robot control, board game play, game completion, autonomous driving, and other decision-making and control problems. Similarly, in the domain of bridge maintenance, reinforcement learning methods possess remarkable potential with their human-like autonomous decision-making capabilities, enabling proactive maintenance decision advantages.

In the maintenance process, reinforcement learning algorithms (agents) act as bridge maintenance decision-makers, continuously interacting with the targeted bridge structures (environment), providing different maintenance strategies, and calculating corresponding bridge performance reliability indicators as environmental feedback and state evolution. Eventually, optimal bridge maintenance strategies are obtained to minimize the life-cycle cost while maximizing the bridge's service life. Currently, scholars have already attempted to optimize maintenance decisions for simply supported beam bridges and large-span cable-stayed bridges using the deep reinforcement learning DQN algorithm. The results have shown that the DQN algorithm can efficiently and autonomously select the optimal maintenance strategies for bridge structures with varying degrees of complexity. These cases illustrate the significant position and value of deep reinforcement learning technology in future bridge AI based decision-making.

It is worth mentioning that visualizing the decision results allows for a more intuitive observation of the changes occurring in the bridge before and after maintenance. Generative models are a novel and reliable approach to solving such simulation generation tasks. Generative models include auto-encoder network models and generative adversarial network models. The output of these models is no longer a classification or prediction result but rather a sample individual that conforms to the distribution space of the input samples. For example, it can generate 3D models or images of bridges after maintenance.

Generative Adversarial Networks (GANs) are primarily inspired by the idea of zero-sum games in game theory. They were first proposed by Ian Goodfellow in 2014 and have attracted widespread attention in the academic community. The principle of generating images using GANs is illustrated in Fig. 9.5.

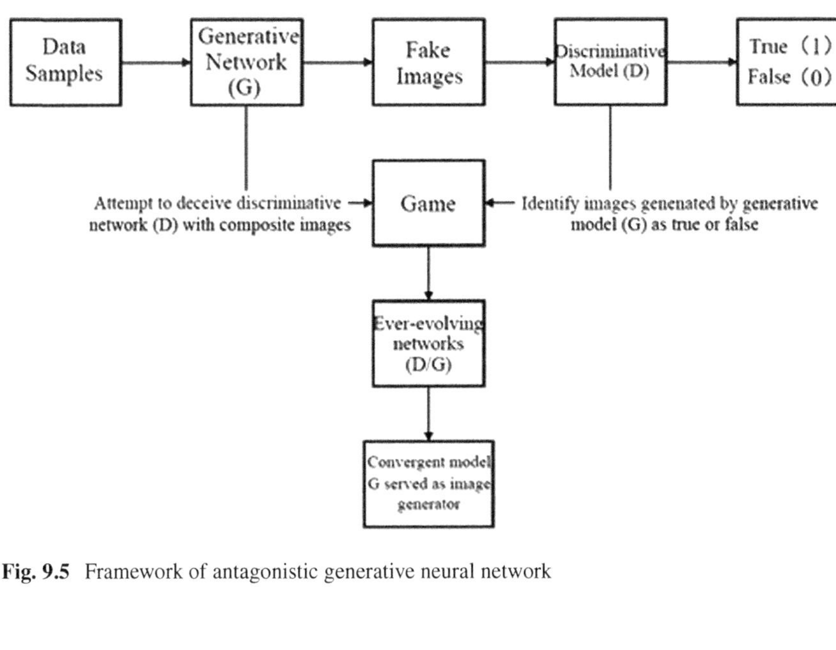

Fig. 9.5 Framework of antagonistic generative neural network

A GAN consists of two components: the generator model (G) and the discriminator model (D). The generator model learns the distribution of real data, while the discriminator model is a binary classifier used to distinguish between real and generated data. The generator model (G) generates data based on the sample data, and then both real and generated data are input to the discriminator model (D), which outputs the predicted class. The discriminator model is typically a binary classifier, assigning a label of 1 to the real images and a label of 0 to the generated images.

During the training process, the objective of the generator model G is to synthesize images to deceive the discriminator model D, making D classify them as real. On the other hand, the objective of D is to accurately distinguish between G's generated fake images and real images. Thus, G and D form a dynamic "game" where they continuously evolve. The generated images by G become increasingly similar to the real images, while the discriminator model D improves its ability to judge the authenticity of images. After a sufficient number of iterations, the generator model G reaches a converged state, leaving G as the sole generator of images. The generated images are statistically indistinguishable from the real images.

(4) Multi-objectives Optimization

The aforementioned technical methods mainly focus on AI-based bridge maintenance decision-making for individual bridges. However, in more practical scenarios, when formulating bridge maintenance decisions, it is necessary to consider the entire network of bridges. During the construction of the road network, the construction time of bridges along the route is relatively concentrated, resulting in similar service periods for these bridges. In this context, within a certain time period after completion of bridge construction, bridges of the same type in the road network may experience similar types of damages, leading to "shock-like" traffic paralysis and extensive maintenance demands. This highlights the significant importance of maintenance decision-making for a group of bridges.

Unlike the maintenance of individual bridges, bridge networks involve multiple maintenance targets, large maintenance volumes, diverse structural forms and complex environments. Simply applying the maintenance decision-making models for individual bridges to the maintenance decision-making for bridge networks would result in higher maintenance costs and waste of resources. To address this issue, it is necessary to optimize the maintenance decision-making for bridge networks. This optimization involves considering the different structural forms and environmental effects of multiple bridges within the road network and optimizing the maintenance decision-making accordingly.

Optimization of maintenance decision-making for bridge networks refers to selecting the optimal maintenance timing and measures at regular analysis intervals or within a certain time period. After maintenance, the degradation process of the structural performance of bridges is delayed or temporarily halted, resulting in a postponement for reaching the threshold of serviceability. The degradation process of structural performance after maintenance is generally considered to be a repetition of the degradation prior to maintenance. When optimizing maintenance decision-making for bridges, the extension of service life has an impact on maintenance benefits and costs: the greater the extension of service life, the better the maintenance benefits, but the required maintenance measures often come with higher costs. The purpose of optimizing maintenance decision-making is to comprehensively consider various factors that affect the optimization objective function based on the concept of intelligent bridge lifecycle management, such as the location of the bridge within the road network, degradation of the bridge's performance, assessment of the bridge's performance state, maintenance costs (including direct and indirect costs), etc. This allows for formulation of the optimal maintenance strategy under the current optimization objective. Depending on the number of objective functions in the maintenance decision-making optimization, it can be classified as single-objective optimization or multi-objectives optimization.

In mathematics, a single-objective optimization problem refers to computing the optimal solution for a specific objective under certain constraints. In the optimization of maintenance decision-making for bridge networks, the objective function is typically set as the minimum maintenance cost over the entire lifecycle of the bridge network. Common constraints would include the threshold value of structural failure probability, critical state level of the bridges, carrying capacity level or remaining service life. Alternatively, the maximization of maintenance benefits can be used as the objective function, with the constraint being the budget allocation for bridge network maintenance. This is because the fundamental and crucial question faced by bridge management authorities in formulating maintenance decisions for bridge networks is how to effectively utilize limited funds for maintenance. Renowned scholar Frangopol from Lihai University also believed that the goal of bridge network management is to achieve a balance between the total life-cycle cost and the overall life-cycle reliability by efficiently utilizing limited funds. In some existing bridge management systems like BRIDGIT and PONTIS, the minimum equivalent expected average cost (the ratio of total maintenance cost to expected service life) is adopted as the objective function for decision optimization. From the perspective of the number of objective functions, the optimization of bridge maintenance decisions using the cost–benefit ratio can be considered a single-objective optimization problem, which inherently takes into account both maintenance effectiveness and costs.

In contrast to single-objective optimization problems, multi-objectives optimization problems involve more than one objective function that needs to be simultaneously achieved. For example, in bridge maintenance decision-making, there are two optimization objectives: minimizing maintenance costs and maximizing maintenance effectiveness. Compared to single-objective optimization, multi-objectives optimization allows for considering more practical bridge management requirements and aligns better with the actual needs of bridge managers. Therefore, it represents the future research and development trend of bridge maintenance strategy optimization.

Compared to single-objective optimization, in practical problems, the goal of multi-objectives optimization is to obtain optimal values for different objective functions within the same value range. These objectives are generally interdependent, and even conflicting. To improve the adaptability of one objective function, it is necessary to compromise the optimality of other conflicting objective functions. In other words, it is impossible to simultaneously optimize all objective functions when solving multi-objectives optimization problems. The so-called optimal solution to multi-objectives optimization problems is a set of non-dominated solutions obtained through coordination among these objective functions. Therefore, the number of optimal solutions in multi-objectives optimization problems is often infinite, and the set of all optimal solutions is called the non-dominated Pareto solution set. Compared to the optimal solution in single-objective optimization, which appears as a point in the objective space, the Pareto solution set generally represents a continuous or dispersed surface in the objective space. In this context, the key challenges in solving multi-objectives optimization problems lie in ensuring that the obtained Pareto optimal surface is closest to the true optimal surface, that the distribution of optimal solutions is uniform and covers a comprehensive range. Additionally, since there are infinitely many Pareto optimal solutions, they cannot be directly applied. This requires decision-makers to select the most satisfactory solution from these non-dominated solutions based on practical considerations.

There are three main approaches for selecting the final solution: the first approach involves obtaining the final solution based on the requirements of the decision-maker using the non-dominated solution set already obtained; the second approach is the interactive method, where the final solution is gradually determined through dialogue between the analyst and the decision-maker; and the final approach involves transforming the multi-objectives problem into a single-objective problem through weighted combination, based on the decision-maker's prior knowledge and then solving it, which is essentially the same as the approach used in single-objective optimization mentioned earlier.

Currently, with the continuous development of mathematics, operations research, systems theory and other related disciplines, the methods used to solve multi-objectives optimization problems have become increasingly diverse and mature. These include goal programming, dynamic programming, linear and nonlinear programming, artificial neural networks, and various evolutionary algorithms such as genetic algorithms and particle swarm optimization.

Liu et al. [6] conducted multi-objectives optimization research on minimizing both bridge deck inspection and maintenance costs, and deterioration costs using genetic algorithms, which were applied to the optimization of bridge deck maintenance in a transportation network. Miyamoto et al. [7] proposed a multi-objectives maintenance optimization approach that minimizes inspection and maintenance costs, and maximizes structural durability based on an expert rating system using genetic algorithms for optimization. Furuta et al. [8] considered the lowest life cycle cost, maximum service life and maximum target safety performance in the optimization of civil engineering infrastructure inspection and maintenance. Peng Jianxin et al. developed maintenance strategies that minimize maintenance costs within the structural life cycle and maximize structural condition indicators, while satisfying structural performance requirements and budget constraints. Kim et al. [9] considered uncertainties in structural inspection and maintenance and optimized the inspection and maintenance time, plus measured by maximizing the structural service life and minimizing the inspection and maintenance costs. Frangopol and Liu [10] proposed a road network bridge management decision optimization framework based on multi-objectives optimization problems. They used a two-stage stochastic dynamic programming approach that considered the remaining service life of each individual bridge and its importance in the road network. In the first stage of the framework, optimal maintenance decisions for each bridge were obtained based on the objective of minimizing the life cycle maintenance cost. In the second stage, the available maintenance budget was allocated according to the importance of each bridge, while satisfying the optimal maintenance decisions obtained in the previous stage as much as possible.

9.1.5 AI-Based Decision-Making in Bridge Engineering

As shown in Fig. 9.6, Yuanfu et al. [11] applied the Bridge Maintenance Intelligent Assistant Decision Support System (BMIADSS) to the Xihoumen Bridge, the world's first segmented steel box girder suspension bridge in the Zhoushan Mainland-Continental Bridge Project. BMIADSS includes professional databases such as user professional database, information management system professional database and professional database for intelligent decision support system. Based on the inspection data and non-destructive testing results of the bridge, a complete set of bridge intelligent decision support system was established using advanced information processing technologies such as neural networks, fuzzy evaluation, genetic algorithms and immune genetic algorithms.

Based on the Jiangyin Bridge Structural Health Monitoring System, Zhenan [13] transferred the collected massive raw data to the database in the monitoring center for safety assessment. However, during the data acquisition process, there may be noise

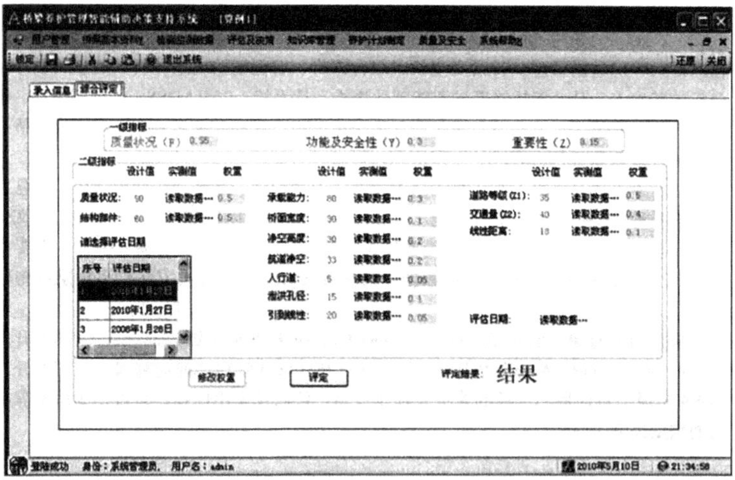

Fig. 9.6 Integrated rating user interface of BMIADSS [12]

interference, sensor failures or network transmission failures, resulting in a large amount of noise and faulty data in the raw data. To efficiently identify data anomalies and to distinguish structural anomalies from system anomalies, a deep learning-based method for data anomaly analysis in structural health monitoring systems was proposed. This method utilized the LSTM model to learn the feature relationships between time series, using historical monitoring data from the normal operation phase of the structure to train the model. Then, based on this model, the predicted response was obtained using the structural historical response, and combined with Dynamic Time Warping (DTW) distance calculation and Isolation Forest anomaly detection to successfully identify abnormal data in the bridge acceleration data. This method can differentiate structural anomalies from system anomalies through logical correlation analysis and multidimensional anomaly detection. It improved upon the previous methods that required manual threshold setting for data filtering, enabling continuous 24-h data monitoring and anomaly analysis to obtain higher-quality data content. It also enhanced the monitoring system's ability to provide early warnings and significantly improved the automation level of the structural health monitoring system.

Huang [14] applied statistical analysis to reveal the limitations of the Markov chain method, identified important factors affecting bridge performance deterioration, and developed an application for predicting the future state of bridges. Based on the historical inspection records, maintenance records and database data of bridge deck components in Wisconsin, bridge deck deterioration was selected as a significant factor. A bridge deck damage prediction model based on artificial neural networks was established using pattern classification methods, considering the potential nonlinear relationships between observation conditions and identification factors. The model consisted of an 11-5-5-5-5-5 neural network structure, involving 11 bridge attributes such as maintenance history, service time, design load and length. The structure has 5 hidden layers, with each layer containing 5 hidden neurons, and one layer served as the output layer with 5 output labels. In the Three Forks Cross-Validation, the neural network prediction model achieved classification rates of 84.66% for the training set and 75.39% for the test set, indicating that the backpropagation multilayer perceptron (BP-MLP) performed well in handling pattern classification problems and accurately predicted the bridge deck conditions, providing useful information for maintenance planning and decision-making. The research also highlighted the importance of maintenance and upkeeping records for effective bridge maintenance decision system, emphasizing the need for institutions to strengthen the collection and archiving of such data.

As shown in Fig. 9.7, Liqun et al. [15] designed a bridge structural health monitoring model (KPCA-SVM) based on data mining techniques. The model utilized wireless sensors to collect bridge structural health data. Kernel Principal Component Analysis (KPCA) was applied to preprocess the bridge structural health data, removing redundant features and reducing the dimensionality of the detection features. Subsequently, Support Vector Machines (SVM) were employed to learn from the bridge structural health data. SVM is a machine learning algorithm that minimizes structural risk while incorporating modern statistical theory. By fitting

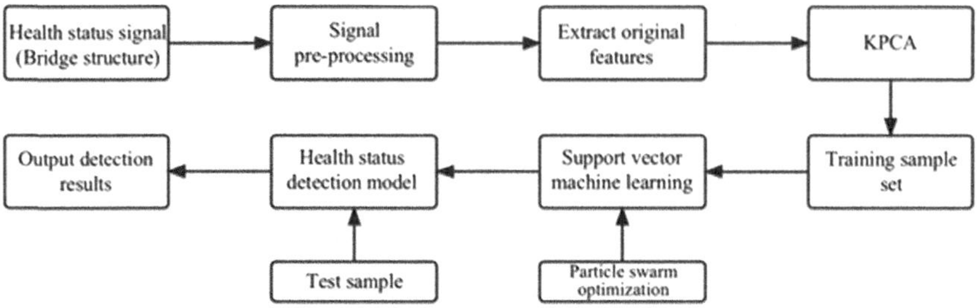

Fig. 9.7 Workflow of a data mining-based bridge structure health condition detection model

the relationship between the bridge structural state and feature values using SVM learning and training samples, the parameters of the bridge structural health monitoring model were determined using Particle Swarm Optimization (PSO) algorithm. This approach established an optimal bridge structural health monitoring model.

Guixuan et al. [16] developed a bridge intelligent decision support system (expert system) based on a bridge in Yamaguchi Prefecture, Japan. The system followed the maintenance management steps of inspection, evaluation and counter-measures. It consisted of three interconnected subsystems: bridge database management system, bridge aging evaluation system and bridge maintenance management plan optimization system. The bridge database management system was primarily responsible for including basic information about the bridge, inspection records, and maintenance and strengthening records. The bridge aging evaluation system transferred expert knowledge and experience into the computer, and data such as the bridge's basic elements, environmental conditions, traffic volume and inspection results. By setting up a hierarchical evaluation structure centered on the durability of the main beam, a five-layer feedforward neural network was used for inference. Deep learning was then performed using backpropagation algorithm to evaluate and predict the aging degree of the bridge based on durability and load-bearing capacity indicators. The bridge maintenance management plan optimization system employed genetic algorithms (GA) to simulate higher biological inheritance and evolution, as well as immune algorithms (IA) to emulate the immune system of living organisms so as to perform combined optimizations. It provided the optimal maintenance management plan within a certain time frame. The specific strategies included recommending appropriate maintenance and strengthening methods, determining suitable timing based on system judgment, and estimating the required budget. This system not only assisted bridge managers in handling large amounts of inspection data and evaluating the aging characteristics of existing bridges but also provided the best bridge maintenance management plan based on budget and construction conditions.

As shown in Fig. 9.8, Sidi [17] leveraged AI algorithms to address the problem of preventive maintenance decision-making for transportation facilities. The pavement condition index (PCI), which comprehensively indicated the functional state of the pavement structure, was selected as the core indicator for segment division. Seven types of pavement distress were considered as influencing factors for segment division, and a segment division model was established. A comparison was made between the original PCI values and the PCI values calculated by the probabilistic neural network. The correlation between the two reached 96.96%, and the correlation continued to improve with an increase in the number of samples. This clearly demonstrated that the accuracy of the preventive maintenance segment division model constructed based on the probabilistic neural network could fully met practical engineering needs. The probabilistic neural network classification method blurred the classification boundary values of the segments, automatically grouping similar distress types together, which facilitated the implementation of maintenance work. This method could quickly handle segment division cases with a large amount of data and allowed for real-time updating of data and results. It addressed the limitations of the conventional segment division methods, such as a small applicable scope, complex operations and low accuracy. It was suitable for large-scale network-level pavement segment division.

Furthermore, as shown in Fig. 9.9, a preventive maintenance decision-making model was constructed based on the multi-objectives backbone particle swarm optimization algorithm. Economic, social and environmental benefits were evaluated as the objective functions of the decision model, and an optimal maintenance strategy selection method based on fuzzy functions was proposed. Chen Sidi, based on the LTTP database, constructed a network-level preventive maintenance decision-making case with 20 sets of segment data. Through analysis of the case, the constructed intelligent pavement maintenance management model has achieved optimal choices for maintenance timing, maintenance segments and maintenance measures as it formulated the most scientific and efficient preventive maintenance plan within a specified period.

Ju [18] established a network-level funding optimization decision model based on the characteristics of rural road maintenance funds in Kunming City. The model aimed to maximize the effectiveness of fund allocation under the constraint of limited

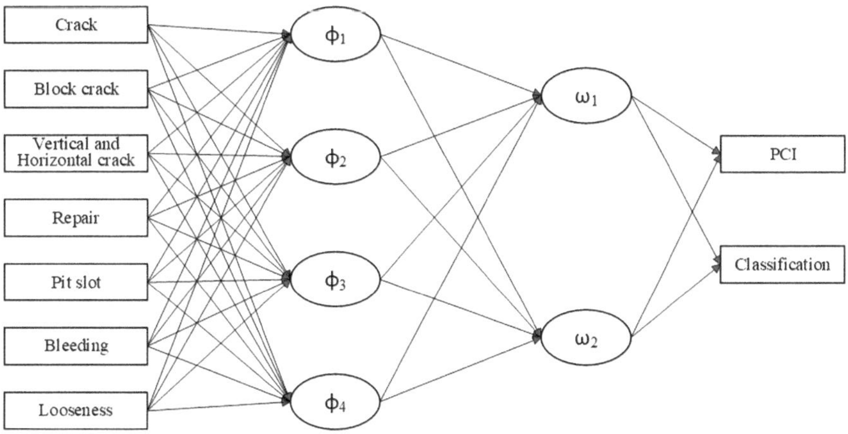

Fig. 9.8 Road segmentation model

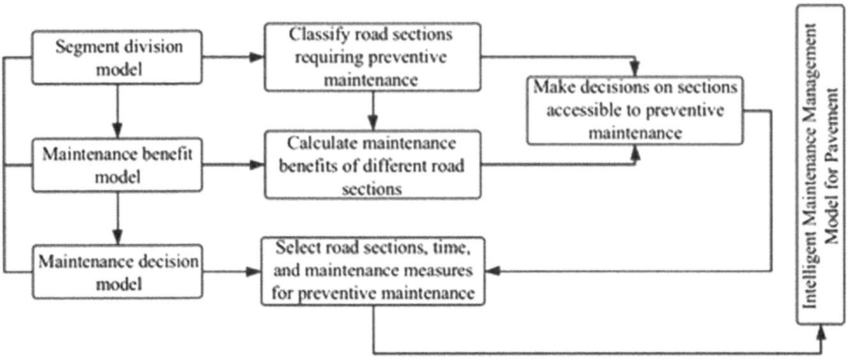

Fig. 9.9 Intelligent road surface maintenance management model

total maintenance funds by considering the pavement performance and the importance of roads in the road network as parameters. The Analytic Hierarchy Process was used to evaluate the maintenance benefits, including economic, social and environmental benefits. Considering the large total mileage and dispersed distribution of rural roads, as well as the road maintenance goals of "every road must be maintained" and "maintenance prioritization," various optimization decision-making methods were analyzed and compared. Ultimately, a multi-objectives optimization decision-making approach was adopted to establish the funding optimization decision model for rural road maintenance. For the case study of Kunming City's rural roads, two forms of funding optimization decision models were constructed: single-year model and multi-year model. The particle swarm optimization algorithm was used to solve the models and to provide the most reasonable maintenance plans, corresponding investment amounts, Pavement Quality Index (PQI) and social benefit scores, as shown in Fig. 9.10. The results indicated that the single-year funding optimization decision model yielded 14 Pareto optimal solutions, from which the maintenance department could directly select appropriate solutions for implementation based on the actual situation. The multi-year maintenance funding optimization model produced 20 Pareto optimal solutions, but most of the implementation plans were concentrated in the first year, which might lead to uneven allocation of management resources. However, considering the parameters set in this model, it represented the optimal maintenance decision-making plan. Finally, the results demonstrated that this model format could be further developed, and is worth applying in highway maintenance.

Deep Reinforcement Learning (DRL) is considered an important component of AI systems and has been applied to decision-making and control tasks in various engineering applications. From the perspective of AI, the problem of formulating maintenance decisions can be analyzed as a case of reinforcement learning. Efficient solutions can be achieved through a series of sampling algorithms to minimize maintenance costs and obtain optimal maintenance decisions. As shown in Fig. 9.11, Wei et al. [19] proposed a DRL-based model for automated decision-making in bridge maintenance. The input dataset of the model can be historical data from real

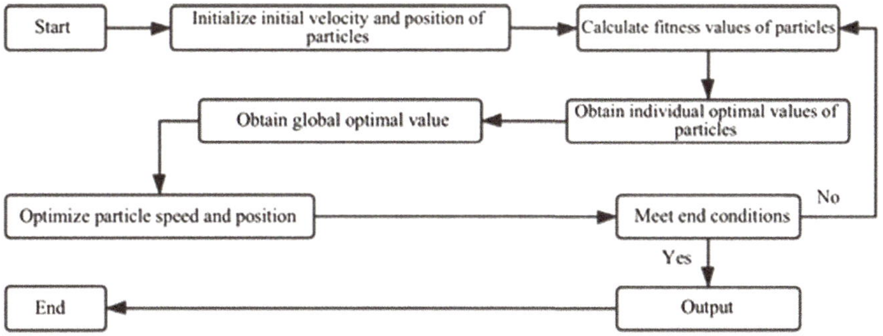

Fig. 9.10 Flow chart of particle swarm optimization algorithm

Fig. 9.11 Schematic diagram of the mapping of enhanced learning and bridge maintenance tasks: reward—economic cost of maintenance decisions and risk in some cases; action—some kind of bridge maintenance behavior

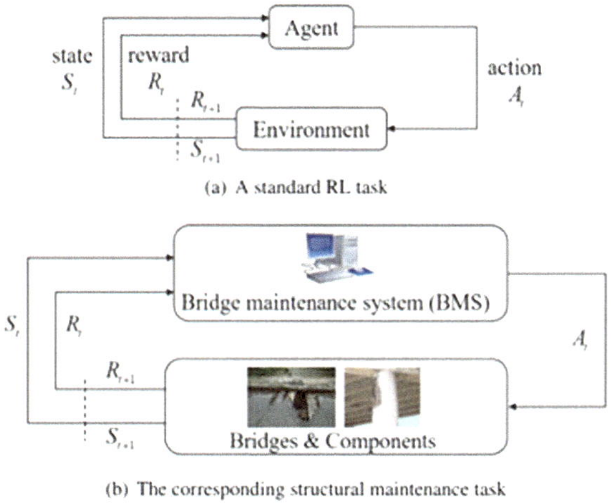

bridges or bridge data from virtual models. These two types of data models can also be merged into the same framework. In the model architecture, deep learning is responsible for evaluating the value function, while reinforcement learning is responsible for improving the mechanism. Through iterative learning in the evaluation and improvement stages, the model learns the optimal strategy. The structured agent of the deep neural network allows the model to be applicable to various bridge cases with minimal modifications.

9.2 Virtual Reality Technology

The conventional communication method between humans and computers involves viewing the results of information processing through the conventional output interfaces such as printing or screen display. The information obtained by humans from computers is often one-dimensional, resulting in a situation where humans have to adapt to computers and revolve around them. However, with the rapid development of social productivity and scientific technology, more challenges have been posed to adapt to the future information society. One of these challenges is to establish a multi-dimensional and visualized information integration environment that allows humans to participate in the process of information processing and decision-making through sight, hearing, touch and bodily forms. VR serves as the key technology that supports this multi-dimensional and visualized information space.

Virtual reality (VR), as an emerging technology that enables the creation and experience of virtual worlds, is a convergence of various technologies such as computer technology, sensor technology, multimedia technology, simulation technology, network technology and human–computer interaction technology. It is a challenging interdisciplinary field and a frontier research area, often hailed as the "new technology of the twenty-first century." VR showcases its broad application prospects and has been widely applied in various industries, steadily maturing over time. The application of VR in bridge engineering is also an important part of smart maintenance. Therefore, this section will provide a detailed introduction to the relevant concepts, characteristics, and applications of VR, as well as explore its application in bridge engineering.

9.2.1 Overview of VR

(1) Definitions and Features

In 1989, Jaron Lanier first proposed the concept of "virtual reality" and conducted defining and research on its content. The term "virtual" indicates that the environment or scene is not real but exists within the computer as an artificially constructed virtual world. "Virtual reality" immediately attracted the attention of scholars giving rise to a series of synonymous terms, such as "virtual environment" (VE), "artificial reality," "cyberspace," and so on. In 1992, experts and scholars at the US National Science Foundation Workshop suggested using "virtual environment" instead of "virtual reality." However, to this day, the academic community still commonly uses the term "virtual reality."

Brooks [20] defines "virtual reality experience" as any experience that effectively immerses the users in an interactive virtual world. This means that the users can "enter" this virtual environment, use a variety of new interactive devices and natural physical skills (such as turning their heads, waving their arms, etc.) to issue various commands to the computer, autonomously and dynamically control their viewpoint, and receive real-time feedback from the environment in terms of visual, auditory, tactile and other sensations, thus creating a sense of immersion in the corresponding real environment. Its essence lies in being an advanced computer user interface that provides the users with intuitive and natural real-time perceptual interaction methods, maximally facilitating user operations, thereby reducing user burden and improving the overall efficiency of the system. The fundamental goal of VR is to achieve realistic experiences and natural skill-based human–computer interaction. As a new comprehensive information technology, VR, with its real-time three-dimensional spatial representation capabilities, interactive operating environment and immersive experiences it brings to the users, has changed the tedious, rigid and passive nature of interaction between humans and computers.

Burdea and Coiffet [21] proposed the concept of "Virtual Reality Technology Pyramid," which succinctly describes the fundamental characteristics of VR systems—the three "I"s: immersion, interaction, and imagination, which ultimately distinguish VR against 3D animation:

Immersion refers to the ability to immerse and engage users in the computer-generated virtual environment. Users can see, hear, touch and even smell everything in the virtual scene just as they would in the real world, creating a sense of "presence."

Interaction refers to the ability for users to interact with objects in the virtual space, including manipulating various objects in the virtual scene and receiving real-time feedback from the virtual environment [22]. It encompasses degree of interactivity with objects, naturalness of feedback received from the environment and reasonable movement of objects in the virtual scene, among others.

- Imagination highlights the vast imaginative space offered by VR technology, expanding human cognitive capabilities. By immersing themselves in a "real" virtual environment, users can interact with and gain sensory and rational understanding from the virtual environment, deepening concepts and inspiring new ideas.

The 3I characteristics emphasize the pivotal role of humans in VR systems, where users are immersed in a virtual world created by computer software and hardware. Through the interactive means provided by the system's software and hardware, the users can interact with the system, fulfilling their real-world ideas while also triggering virtual imaginations. In addition to the 3I features mentioned above, VR also possesses multi-modality. It emphasizes that besides visual perception, which general computers possess, VR also includes auditory perception, tactile perception, motion perception and even includes gustatory and olfactory perception.

(2) Evolution of VR

The evolution of VR is shown in Fig. 9.12.

First stage (1950–70s): the exploration stage. The origin of the VR concept can be traced back to the simulator developed by Morton Heilig in 1956, which integrated a 3D display, stereo speakers, odor generator and vibrating seat. However, its bulky size prevented it from being commercially viable. In 1965, Ivan Sutherland, the founder of computer graphics, presented a report titled "The Ultimate Display" at the IFIP (International Federation for Information Processing) conference, proposing the use of a display screen as a window to observe the world, providing the observer with a sense of immersion. This was a milestone in the development history of VR technology. In 1968, Ivan Sutherland developed a helmet-mounted stereoscopic display using two cathode-ray tubes that could be worn on the eyes. This display

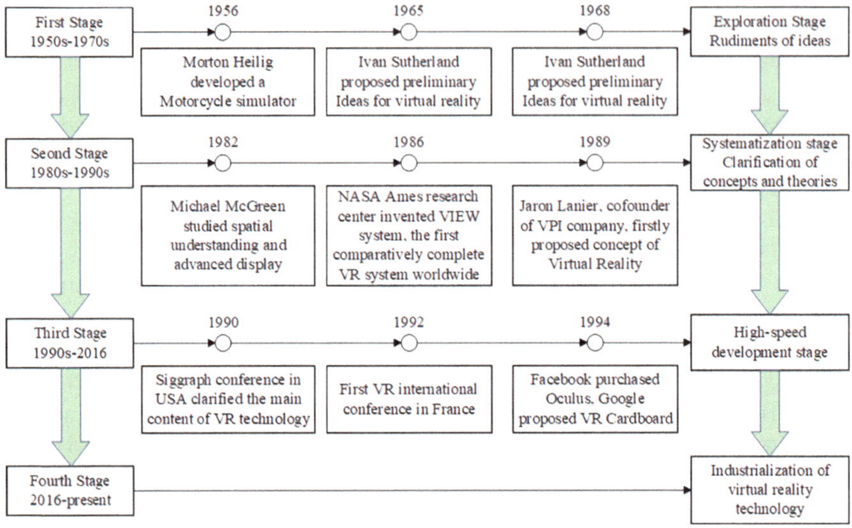

Fig. 9.12 Evolution of VR

became a foundational achievement in stereoscopic 3D display technology, providing a prototype and reference for the development of VR devices in the future. Overall, this stage of VR technology development progressed slowly and was in the stage of conceptual brewing.

Second stage (1980–90 s): the systematic stage. During this stage, the basic concepts, features, related theories and research on VR were gradually clarified. In 1981, Michael McGreevey began his research on the “spatial understanding and advanced display”. In 1984, the first commercially available VR device, the RB2, was introduced. The development of NASA Ames Experiment Center’s VIEW system in 1986 marked the world’s first complete multi-purpose, multi-perception VR system. Jaron Lanier, one of the founders of VPL, introduced the term “virtual reality” in 1989.

Third stage (1990–2016): the rapid development stage. After the 1990s, with breakthroughs and rapid developments in scientific fields such as computer technology, high-performance computing, human–computer interaction devices, computer networks and communication, as well as the significant demand in important application areas like military training, aerospace, etc., VR technology entered a stage of rapid development. In 1990, the American Siggraph conference discussed VR technology and clearly defined its main components: real-time 3D image generation technology, multi-sensor interaction technology and high-resolution display technology, which set the direction for the development of VR technology. In 1992, the first international conference on VR technology was held in France, playing an important role in establishing the independent research status of VR.

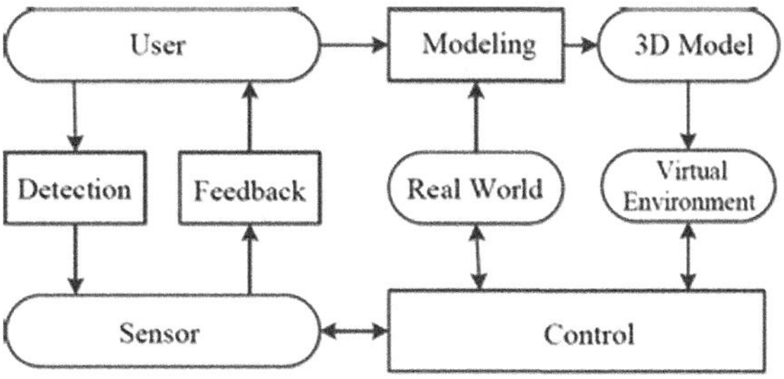

Fig. 9.13 Functional modules of VR systems

Fourth stage (2016 to the present): the stage of industrialization and development. 2016 was a milestone year in the development of VR technology. Its products gradually gained popularity and started to infiltrate various vertical industries. At CES 2016, Oculus officially released the Oculus Rift head-mounted VR device, and other devices like HTC Vive and Samsung's Gear VR also emerged. Intel and Qualcomm started providing support for VR at the chip level. Game engines such as Unity, Blender, CryEngine, etc., announced comprehensive support for VR technology. In the gaming and entertainment industry, major game companies such as EA, Ubisoft, NetEase, Tencent, etc., released their representative works. From this year onwards, a large amount of capital poured into the VR market, leading to the emergence of more diverse VR device products. The VR content industry and technological support became more mature, and the user base continued to expand.

(3) Technology Components, Classification and Trends

Figure 9.13 illustrates the six major functional modules of VR systems.

Sensors are an important part of achieving human–computer interaction and serve two main tasks: (1) receive commands from the user and apply them to the virtual environment; and 2) provide the results of the operations to the user in various forms of feedback. The detection module is primarily used to detect the user commands and then apply them to the virtual environment through the sensor module. The feedback module is responsible for receiving information from the sensor module and providing real-time feedback to the user. The control module is mainly used to control the sensors, enabling interaction between the user, the virtual environment and the real world. The modeling module acquires three-dimensional data from the real world and establishes their three-dimensional models. 3D models are three-dimensional representations of the real world and constitute corresponding virtual environments.

A typical VR system consists of computer, input/output devices, application software and databases. The computer plays a crucial role in the VR system, responsible for real-time rendering calculations of the entire virtual world and real-time interaction calculations between the users and the virtual world. To achieve effective human–computer interaction in VR systems, special input devices are used to recognize various forms of user input, and real-time feedback generated is delivered to the users through the output devices. The implementation of a VR system also relies on various auxiliary software. For example, during early data collection and image organization, 2D software such as AutoCAD and Photoshop, as well as architectural drawing software, are used. For audio and video preparation, software like Audition and Premiere are utilized. In a VR system, the role of the database is primarily to store various data required by the system, including terrain data, scene models and various created models. Every object appearing in the VR system needs to have a corresponding model in the database. The data in a VR system can be classified into four categories based on their sources and roles, as shown in Table 9.1.

VR systems can be classified into four main types based on their implementation method: immersive, desktop, augmented and distributed VR systems [23]

- Immersive VR systems primarily utilize interactive devices such as head-mounted displays and data gloves to encapsulate the user's visual, auditory and other sensory experiences, enabling the user to truly become a participant within the VR system. Through these interactive devices, the users can operate and navigate the virtual environment, generating a sense of presence. Common immersive VR

Table 9.1 Classification of data in VR systems

Data	Definition	Generation
Platform data	Public platform data and administrative data to support system operation	Public support platform (computer system, network system, etc.)
Model data	The image of things in the real world in the digital space, the subject data in the VR system	Scanners, algorithms, axiomatic systems
Sense data	It is also called rendering data, which is applied to various output devices to make users see, hear, touch, smell, etc.	Rendering algorithm
Control data	User control and influence over the virtual environment through the data generated by the input device	Human–computer interaction devices (data gloves)

Fig. 9.14 Helmet-based VR display system

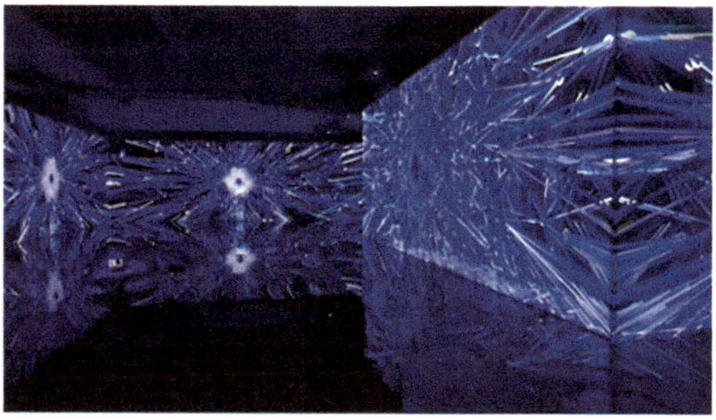

Fig. 9.15 Projected VR system

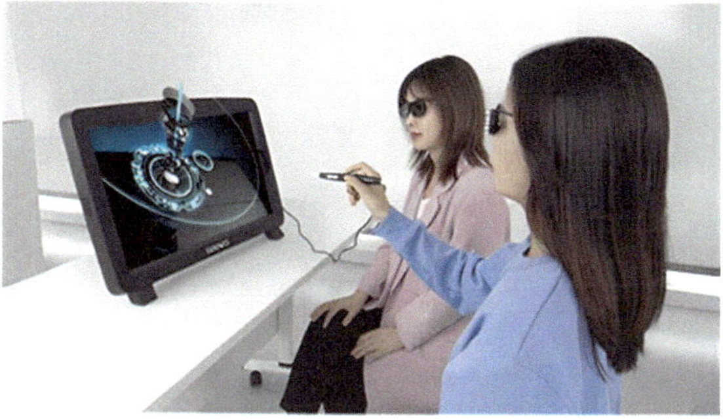

Fig. 9.16 Desktop VR system

systems include head-mounted display-based systems and projection-based VR systems (Figs. 9.14 and 9.15).

- Desktop VR systems (Fig. 9.16), also known as window VR systems, utilize devices such as personal computers or entry-level graphics workstations, using the computer screen as a window through which the users observe the virtual world. Although lacking the immersive effects of wearable displays, these systems are more widely used due to their relatively lower cost.
- Augmented VR systems (Fig. 9.17) do not require isolation from the real world. They allow the users to see the real environment while simultaneously viewing the virtual objects overlaid on the real world. It is a system that combines the real and virtual environments.
- Distributed VR systems (Fig. 9.18) are a combination of VR technology and network technology. The goal is to connect geographically distributed users or multiple virtual worlds through a network, building upon immersive VR systems. This allows multiple users to participate simultaneously in the same virtual space and interact with other users through networked computers, aiming to achieve collaborative work.

Recent research in VR technologies has followed the principle of "low cost, high performance" and has made rapid progress, exhibiting new characteristics and development trends, mainly manifested in the following aspects:

Fig. 9.17 Augmented VR system

Fig. 9.18 Distributed VR system

Real-time 3D graphics generation and display technology: The technology for generating 3D graphics has become relatively mature. However, the key focus of research lies in how to generate graphics in real-time and improve the refresh rate without compromising the quality of graphics. Additionally, VR technology relies on the development of stereoscopic display and sensor technology. The existing virtual devices are unable to meet the system's requirements, thus necessitating the development of new techniques for 3D graphics generation and display.

Development of new interaction devices: The foundation for achieving immersive experience in VR technology, which enables free interaction between humans and virtual world objects, lies in the interaction devices. Common input/output devices currently include helmet displays, data gloves, data clothing, 3D position sensors and 3D sound generators. However, most VR devices are high-end products and come with a hefty price tag. Whether it is software or hardware, VR technology will only become mainstream when it becomes affordable for the ordinary users. Although recent versions of devices have entered the market at cheaper prices, they are still not sufficiently inexpensive as consumer-grade products. Therefore, the development of new, affordable, and robust data gloves and data clothing is an important direction for future research.

- 5G + VR: There are several reasons why VR has not yet become mainstream. One reason is that VR devices have always been expensive, and another reason is the lack of connectivity. In order to create a more realistic virtual experience, devices need to operate quickly and require high levels of storage and low latency. Sometimes, even a delay of a few milliseconds can significantly impact the effects presented in VR. With the arrival of the 5G era, network latency can be reduced to at least one-tenth of its original level, improving network efficiency and increasing traffic capacity by at least 100 times. This allows for more refined and smooth VR visuals and better integration with real-world products and services. Therefore, the introduction of 5G will provide developers with tools to greatly expand the scope of VR projects and create truly immersive and remotely accessible products.

(4) Application of VR

The emergence of VR technology has triggered significant changes in various techniques and methods, transforming outdated conventional technologies and improving the means of product design and development. It has significantly increased work efficiency, greatly reduced the difficulty of tasks, and effectively mitigated operational risks. VR technology was initially applied in the gaming industry. Currently, its application has expanded to various fields, including the military domain, industrial circle, agricultural field, medical field, education and architecture.

Military requirements have been the driving force behind the development of VR technology. Since the early 1990s, the United States has been at the forefront of using VR technology in the military domain. Subsequently, military organizations in many countries worldwide have adopted VR technology for military applications. In the military field, military training and exercises are considered essential measures to enhance combat capabilities. VR technology can effectively improve the quality of military training, reduce casualties in actual combat, save training costs and resources, minimize environmental pollution and destruction, and enhance command decision-making capabilities. VR technology is also applied in the development process of weapons to shorten the weapon equipment development cycle.

In industrial simulation, VR technology is used to perform various dynamic performance analyses on models and improve design solutions. It replaces conventional physical prototyping with digital representation, reducing product development costs and expenses while enhancing product quality and performance. The renowned Boeing 777 aircraft, for example, was successfully designed using VR technology.

VR technology finds broad applications in the agricultural sector, including scientific research, teaching, production, planning, agricultural resource allocation, commodity circulation, and agricultural machinery design and manufacturing. It enables the development of agricultural machinery, and the simulation of real environments and growth processes of living organisms. By utilizing sensors, it can collect biological information, reconstruct life processes, reproduce the occurrence and control of pests and diseases in crop production, and calculate pollution levels to eliminate pollution sources in crop production. This holds significant implications for food safety and security.

The application of VR technology in the medical field involves various aspects such as surgical simulation, skill training, surgical guidance and psychotherapy. For example, through surgical simulation, doctors can plan and train for surgeries, gain a deeper understanding of specific surgical procedures, and learn and explore new techniques. Medical students, for instance, can experience the actual process of operations through skill training, enabling them to quickly grasp clinical knowledge such as human anatomy.

In the field of education, VR technology can transform conventional teaching methods from passive learning to active learning, where learners acquire knowledge and skills through their interaction with the informational environment. Virtual laboratories allow students to experience activities such as space travel and the display of compound molecular structures, making the learning process more engaging and persuasive than the conventional teaching methods.

VR systems also play a significant role in the field of architectural engineering. By using VR technology, architects can visualize and "touch" design outcomes, facilitating convenient modifications at any time. Additionally, VR systems can be quickly and easily adjusted to accommodate changes in project plans, assisting architects in making decisions, thereby accelerating the speed and improving the quality of design solutions while saving substantial funds. Furthermore, VR finds extensive applications in the assessment of major engineering projects. Large-scale public buildings or important structures, such as stations, airports, bridges, ports and dams, often have a significant impact on the landscape and environment of a particular area once constructed. Due to the high construction costs and significant societal impact of these projects, evaluating their safety, economic viability and functional feasibility becomes even more crucial.

9.2.2 Applications in Bridge Maintenance and Management

(1) Bridge Safety Accident Restoration

With the rapid development of the economy, China's bridge industry has also flourished in recent years. However, as traffic volume increases and various natural and human factors come into play, the safety of bridges is receiving increasing attention from the general public [24]. While the construction of bridges in China has been rapidly advancing, there have been recurring incidents of bridge collapses, resulting in significant social and economic losses as well as casualties, and often causing severe negative social impacts.

On October 1, 2019, at 9:30 a.m., a bridge accident occurred at Nanfang'ao Bridge in Yilan County, Taiwan. The accident resulted in the collapse of three fishing vessels, causing six serious injuries, four minor injuries and five missing persons. Nanfang'ao Bridge, completed in 1999, was the only single steel arch bridge in the Taiwan region, with a length of 140 m and a width of 15 m. It was built at a cost of 250 million New Taiwan Dollars. The bridge featured a tourist viewing platform, an important local landmark. Preliminary investigations indicated that the accident was caused by the collapse of a bridge pier on the seaward side, resulting in the bridge structure dropping and fracturing. However, the exact cause of the incident has yet to be determined. Therefore, it was essential to conduct identification and analysis of bridge safety accidents in order to identify the main causes of such accidents and actively prevent the recurrence of similar bridge safety incidents.

The identification and analysis of bridge safety accidents require the use of computer simulation technology to accurately reproduce the process of bridge safety accidents in a scientific manner. Generally, researchers simulate the accident process using mathematical approximation methods, such as finite element analysis [25], whose accuracy has been proven by numerous successful cases. However, mathematical approximation simulations have certain limitations. Firstly, the relevant information about the bridge's environmental factors (such as terrain and landforms) is usually not considered in the mathematical approximation process. In many cases, after an accident occurs, it is necessary for professional engineers and experts to conduct on-site investigations to determine the cause of the bridge accident and take appropriate rescue measures. Additionally, the results presented through mathematical analysis simulations are simplified and abstract, making it difficult for researchers to have a comprehensive and intuitive understanding. Therefore, in the process of identifying and analyzing bridge safety accidents, in addition to necessary scientific and mathematical analysis, it is also crucial to provide a realistic presentation of the process and strive to accurately depict the entire accident sequence.

In terms of disaster simulation and reproduction, VR technology is playing an incredible role, such as in the simulation and analysis of mining accidents, fire reconstruction, aircraft crash simulation, traffic accident reconstruction and crime scene reconstruction. The "reproduction" and analysis generated by these VR technologies are of great significance in mitigating and preventing disasters. Introducing VR technology into the process of identifying and analyzing bridge safety accidents can effectively compensate for the limitations of the aforementioned mathematical approximation simulation methods. It can comprehensively integrate numerical analysis simulation, terrain, landforms and the surrounding landscape of the bridge, thus more comprehensively and accurately reproducing the process of the accident. Additionally, it can provide a sense of realism through various human–computer interactions and special effects production. Furthermore, the functionality of virtual scene simulation and scene roaming can maximize the satisfaction of observation and contemplation needs during the analysis process of bridge safety accidents, assisting in the identification of safety accidents and conducting more in-depth analysis. Researchers can focus on a specific process based on their own ideas, and depending on their individual expertise or experience, with each person's focus may varying. This is more conducive to the development of thinking and finding the main causes of bridge safety accidents (Fig. 9.19).

(2) Assistance in bridge maintenance and management

(1) VR assistance for engineers to quickly familiarize themselves with the basic information

Fig. 9.19 Flowchart of bridge safety accident simulation process

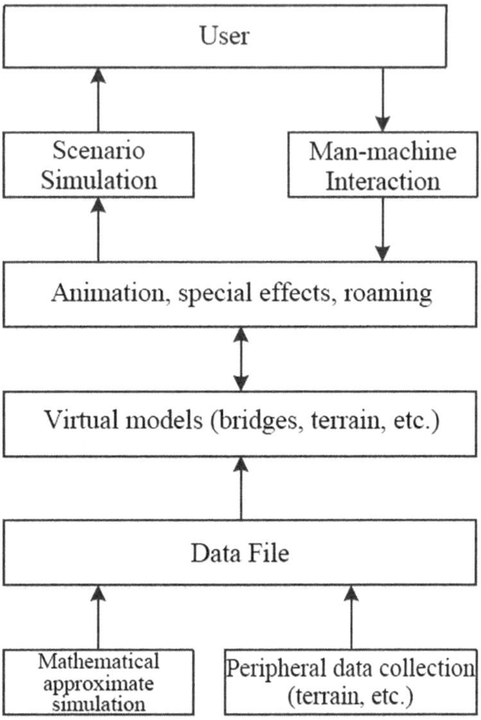

① Application Background

After a bridge sustains damage, it often requires professional engineers and experts to conduct on-site investigations and monitoring to determine the cause of the bridge damage, and to make relevant repair and maintenance decisions. However, making accurate bridge repair and maintenance decisions based solely on on-site monitoring information is difficult. It is necessary to have prior knowledge of the basic information of the damaged bridge, including the bridge type, span and structural arrangement, and then combine it with the data obtained on-site to make effective decisions. For a long time in the design and construction process of bridges, plan drawings have been predominantly used. Plan drawings can help bridge maintenance engineers understand the structure and basic information of the bridge. However, plan drawings have two shortcomings. First, plan drawings lack intuitiveness, and engineers often need to combine multiple drawings to have a basic understanding of the bridge's structure, which makes the process of obtaining information complex and time-consuming. Especially for bridges that were constructed a long time ago, their design drawings and data may not be readily available, delaying the time and efficiency of on-site repair and maintenance decision-making. Second, plan drawings have poor visualization and cannot convey the detailed aspects of bridge design to the engineers. The damaged parts of the bridge are often in the details, such as bearings, cables and anchors, which are not well represented in plan drawings and lack visual characteristics.

② Application Route

In an era where intelligent design is constantly emerging, the emergence of VR technology can compensate for the shortcomings of plan drawings. As shown in Fig. 9.20, when designing bridges, engineers can use BIM technology and VR technology as the foundation to combine the immersive nature of VR with the digitization of BIM. Based on the BIM model environment, engineers can create a virtual "real" scene of the designed bridge using VR immersive perception. Subsequently, relevant data of the designed bridge's VR real scene can be stored in the smart bridge maintenance system to establish a bridge's big data VR real scene, truly realizing digitization, mobility and visualization of bridge design information. This provides convenience when making maintenance decisions for bridges. For a bridge that requires maintenance, engineers and experts can access the VR real scene of the bridge in the smart bridge maintenance system on their computers or mobile devices. By using VR glasses and other equipment, they can quickly familiarize themselves with the relevant information of the bridge and then combine it with the on-site surveys. This significantly improves the efficiency of bridge repair and maintenance.

(2) VR applied to skills training of bridge maintenance and management engineers

① Application Background

Before bridge maintenance engineers enter the actual bridge maintenance site, they should undertake job training. The current widely adopted training mode is "theory learning + mentorship." "Theory learning" refers to the engineers first engaging in theoretical learning through written materials or videos, while "mentorship" means

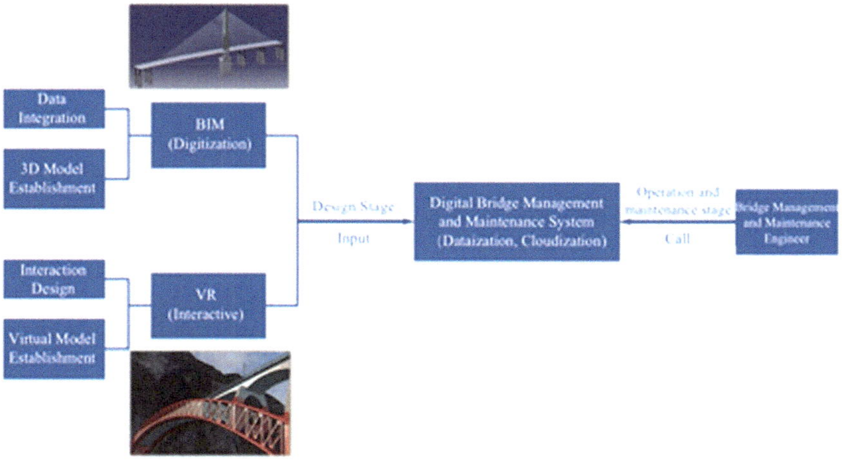

Fig. 9.20 VR assists engineers in familiarizing themselves with bridge information

that after completing the theoretical learning, experienced engineers will guide the new engineers in practical on-site operations. This mode has three major drawbacks: First, it can lead to a "disconnect between theory teaching and practice," where theoretical teaching and practical application are not synchronized, resulting in lower efficiency in skills training. Second, it can lead to a "reduction in on-site maintenance efficiency." Skills often require 10 to 20 practical exercises to become truly proficient. For more complex skills, the number of exercises may even reach 50 to 100 times. Therefore, newly employed bridge maintenance engineers may inevitably make mistakes, leading to a decrease in on-site maintenance efficiency. Third, the safety of this training mode is difficult to guarantee.

② Practical Training of Skills Using VR Technology

Nowadays, with the advancement of digital technology, digital technology based VR technology is highly favored and widely applied in various fields, especially in the areas of education and training. VR skill training refers to the use of various simulation environments and devices, combined with specialized VR courseware, to provide the trainees with highly realistic, systematic and professional training, enabling them to fully grasp the training content and practical operation essentials in a short period of time. Compared to the conventional skill training, VR skill training can significantly reduce the need for teaching resources, lower the requirements for training environments, and greatly avoid safety issues in certain hazardous skill training scenarios. Taking the conventional bridge maintenance operation training as an example, conventional training requires large space, multiple equipment investments, and high maintenance costs. Moreover, there are certain hidden risks during the training process. However, VR skill training only requires a single simulated device to provide training for multiple skills.

Compared to the conventional skill training, the advantages of VR training are self-evident. VR training can provide the learners with vivid and realistic learning environments, allowing them to interact with the virtual environment and to facilitate learning and consolidation. By incorporating VR technology into skill teaching, the cost of practical education can be effectively reduced, while significantly lowering the risks associated with practical training. By bringing immersive experiences into classrooms and other training environments, VR can enhance knowledge retention.

Today, major global companies have already started utilizing VR technology to train their employees, as they believe this technology can make training safer and more efficient. Moreover, this technology has tremendous potential in training employees' soft skills, such as empathy and teamwork. Retail giant Walmart has deployed nearly 200 VR training centers to prepare for Black Friday each year. Through VR devices, trainees can experience the busiest day of the year and learn how to address potential issues, such as inventory shortages, crowd control and customer conflicts. VR technology makes training more engaging and enjoyable, which is difficult to achieve in the real world.

Contrary to intuition, VR can even be used for soft skills training. For businesses, assessing and training soft skills, such as empathy, perceptiveness and teamwork, can be challenging, as these factors often determine individual success. VR technology can place the users in the perspective of another person and help them practice empathy. Doctors use perspective-based VR technology to understand patients' healthcare experiences, which can help change their approach to patient care.

③ Application Route

Based on VR skill training and the VR training system proposed by existing scholars [26], and in response to the practical needs of bridge maintenance, the authors present a bridge maintenance training platform based on VR technology, as shown in Fig. 9.21.

The VR training platform for bridge management comprises two major modules: platform structure module and platform function module.

1. Platform Structure Module

(a) VR Support Platform

The training platform is a software system that is created using a specialized VR development support platform, such as VRML. The VR support platform is the

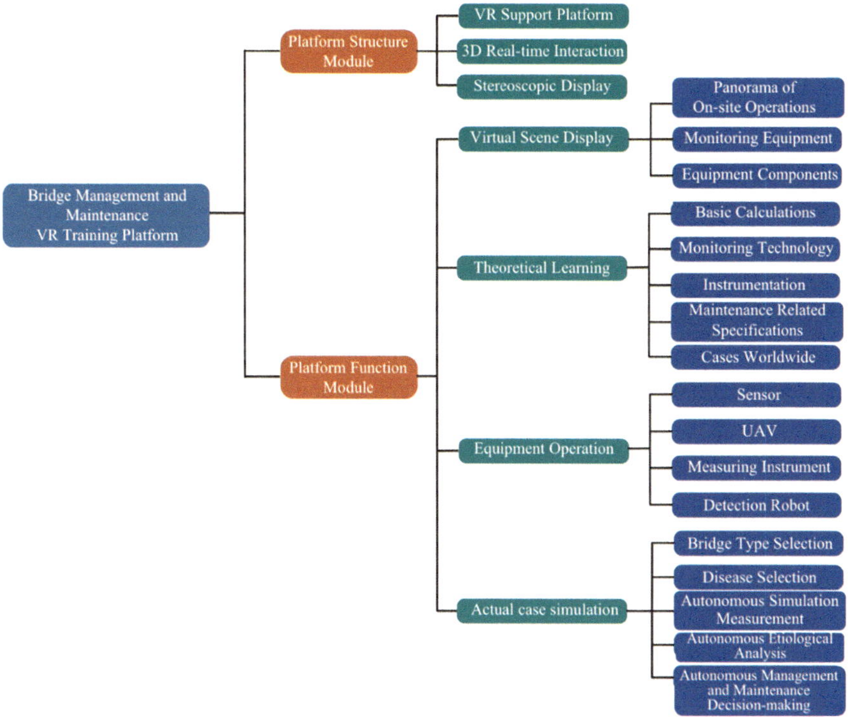

Fig. 9.21 VR-based training platform for bridge Maintenance and Management

essential component of the training platform, responsible for managing the data operations of the entire system.

(b) 3D Real-Time Interaction Part

The interactive system is a combination of conventional display devices and real-time motion capture systems, with data handling, position trackers, force feedback devices and other functions.

(c) Stereo Display Section

The stereoscopic display section is the display end of VR, providing the users with highly realistic, clear and immersive virtual environments, allowing the users to directly immerse themselves in a three-dimensional information space and freely use various information, thereby controlling the computer. The stereoscopic display section adopts a digital helmet, which can display and observe the 3D VR graphic signals in VR applications, and can also receive 3D VR graphic signals from the host by connecting to it. The usage mode is head-mounted, and the users can freely move within a three-dimensional space and observe the VR output effect through a three-degree-of-freedom spatial tracker.

2. Platform Function Module

The training platform encompasses four major functions: the virtual scene display, the theoretical learning, the equipment operation, and the actual case simulation.

(a) Virtual Scene Display Function

The user has the freedom to adjust their walking speed and viewpoint, allowing them to freely navigate and inspect the virtual scene. This scene encompasses a panoramic view of the bridge site operations, monitoring equipment utilized and even the individual components of the equipment.

(b) Theoretical Learning Function

In this feature, the users can systematically learn the theoretical knowledge related to bridge maintenance, including basic calculations of bridges, various commonly used bridge monitoring techniques, commonly used bridge inspection instruments, relevant specifications for bridge maintenance, and some typical domestic and international bridge maintenance cases. This feature eliminates the previous dull and inefficient theoretical learning mode, namely the "written + video" learning mode, and strives to present theoretical knowledge in a three-dimensional manner to the users, making it interactive, interesting and efficient.

(c) Equipment Operation Function

In this module, the users can learn the specific operation methods of bridge inspection instruments, including sensors, drones, measuring instruments, inspection robots and more. Learning the equipment operation methods in the VR training platform not only reduces the number of devices required, but also lowers maintenance costs. It also reduces the requirements for training time and physical space, achieving the true concept of "learning anytime, anywhere."

(d) Actual Case Simulation Function

The users can utilize this function to accumulate experience in on-site operational decision-making. The users can select different bridge types and different defects according to their needs, and the system will simulate realistic virtual scenarios for them to learn from. The users can independently conduct measurements, analyze bridge conditions and make maintenance decisions within this scenario, aiming to gain an immersive experience throughout the entire process.

9.3 Augmented Reality Technology

VR technology brought us the new human–computer interaction that provides the users with a strong sense of realism and presence in completely virtual environments. On the other hand, augmented reality (AR) technology overlays virtual objects onto the real world, offering a lightweight and highly immersive augmentation technology. Due to its realistic integration capabilities, AR technology serves as a powerful complement and enhancement to the real world, with vast application prospects and value. In this section, starting from the consideration of AR technology's application in future bridge engineering, we mainly introduce the relevant concepts, characteristics and applications of AR technology, and discuss and illustrate its application scenarios in bridge operation, maintenance and decision-making.

9.3.1 Overview of AR

(1) Definition and Features

The main scientific issues of VR include modeling methods, rendering techniques, human–computer interaction and devices; but there are currently common problems such as high workload of modeling, high simulation costs, insufficient match with the real world and low credibility [27].

In response to these situations, various VR augmentation technologies have emerged to combine the virtual and real environments for enhancement [28]. Among them, the technique of overlaying real objects onto a virtual environment is called Augmented Virtual (AV), while the technique of superimposing three-dimensional virtual objects onto the real world is called augmented reality (AR). These two types of technologies can be vividly distinguished as "virtual with reality" and "reality with virtual." It is widely recognized in the industry that there are two types of VR augmentation technologies, namely augmented reality and augmented virtual environment, that bridge the gap from the real world to the virtual environment. Internationally,

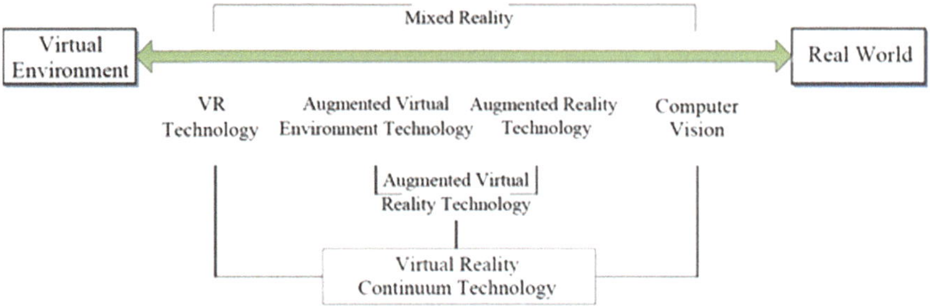

Fig. 9.22 VR Continuum

the real world (computer vision), AR, augmented virtual environment, and VR are commonly referred to as the VR continuum [29], with AR being considered as a part of mixed reality environments, as shown in Fig. 9.22. In this context, AR differs significantly from VR, as listed in Table 9.2.

Azuma [30] defined AR as a system with three essential characteristics:

- Integration of real and virtual worlds in a three-dimensional space. Unlike VR technology, AR technology does not separate users from the real world. Instead, it overlays computer-generated virtual objects and information onto the scenes of the real world, allowing the real world and virtual objects to coexist. This enables a more intuitive and in-depth understanding and interpretation of the real scene, facilitating comprehension of knowledge related to the real world within limited time and limited scenes. The augmented information can include non-geometric information related to real objects, such as videos and text, as well as geometric information, such as virtual 3D objects and scenes.
- Real-time human–computer interaction function. Using interactive interface devices in AR systems, users can interact with AR environments in a natural manner. This interaction needs to be real-time and achieve synchronous updates

Table 9.2 VR and AR comparison

	VR	AR
Technology	The scene is completely virtual	Overlaying digital images on the basis of real environment
	Emphasizing immersion and requiring complete isolation between one's feelings and the environment in which they are located	Allow users to perceive the real world while seeing the virtual environment
Device	Usually quite 'bulky'	Generally, it is mainly exquisite and compact
	More sensors, position positioning systems, motion capture systems and other equipment are needed	Cameras are essential, and products with cameras such as mobile phones, tablets, etc. can use AR technology

between the virtual world and the real world. In this way, AR technology can provide the users with a genuine experience of simulated objects in a virtual space within the real world, enhancing the enjoyment and interactivity of the experience.

- Three-dimensional registration. Three-dimensional registration refers to the process of locating objects in a scene, establishing a correspondence between computer-generated virtual objects and the real environment. It involves overlaying virtual objects onto the real scene based on accurate spatial relationships. Furthermore, the alignment between virtual and real objects is maintained as the user moves within the real environment. Detailed explanations about three-dimensional registration technology will be provided later.

(2) Evolution of AR

The research on AR technology can be traced back to 1968 when Ivan Sutherland became the first person to create an AR system using an optical head-mounted display [31]. This device, known as the first optical head-mounted display in the world, allowed real-time overlay and fusion of computer-generated images with the real scene. In the mid-1990s, Tom Caudell and his colleagues at Boeing introduced the term "augmented reality" while designing an AR system to assist workers in aircraft wiring systems. In the same year, L. B. Rosenberg developed one of the first operational AR systems called "virtual fixtures." It was not until 1994 when Paul Milgram and Fumio Kishino defined VR continuum, as shown in Fig. 9.22, as a continuous spectrum from the real environment to the virtual environment. AR and augmented virtual environment lie between the two, with AR being closer to the real environment and augmented virtual environment being closer to the virtual environment. In 1997, R. T. Azuma provided the widely accepted definition of AR in the industry.

At that time, due to various limitations, AR was not yet a widely popular technology. In 1999, the AR Toolkit, a marker-based AR system development kit, was successfully developed through collaboration between the University of Washington in the United States and Hiroshima City University in Japan. This development marked the emergence of AR from the laboratory and its gradual transition towards mass applications. The AR Toolkit has been maintained to this day and has greatly popularized AR technology, leading to the emergence of more related software applications and development systems.

In 2000, the first AR game, AR-Quake, was born. It was an extension of the popular computer game Quake and brought AR into real outdoor scenes. In 2012, Google released Google Glass, which was the world's first hardware device to truly implement AR technology, and it garnered widespread attention to mobile wearable AR devices. Following that, Microsoft released the HoloLens holographic glasses in 2015.

With the development of technology, especially the continuous updates of mobile devices such as smartphones, AR technology has gained more possibilities for application. In 2017, Apple introduced a new AR component called ARKit in iOS 11. This application is compatible with iPhone and iPad platforms, making iPhone the world's

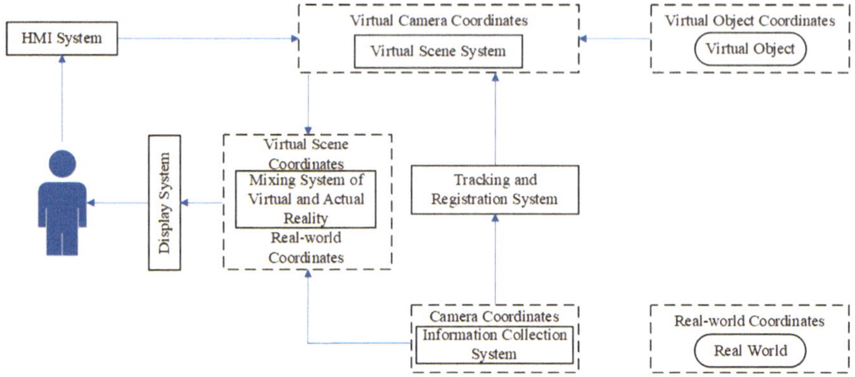

Fig. 9.23 Architecture of AR system

largest AR platform at that time. In 2023, Apple debuted their first-generation AR device – Apple Vision Pro.

(3) Technology Components

The typical structure of an AR system is illustrated in Fig. 9.23, which comprises six subsystems: information acquisition system, tracking and registration system, virtual scene generation system, virtual-real synthesis system, display system and human–computer interaction system [32]. Among them, the information acquisition system is responsible for collecting and acquiring information from the real world (such as the external environment); the tracking and registration system is used to track the user's head orientation and line of sight; the virtual scene generation system is responsible for generating the virtual objects to be added; and the AR system is intended for aligning and merging the virtual scene with the real scene.

As shown in Fig. 9.23, in an AR system, the input image is processed and used to establish the real-world space. The computer-generated virtual objects are embedded into the real-world space based on geometric consistency, forming an AR environment where the virtual and real are merged. This environment is then presented to the user through a display system, and the user interacts with the scene using interactive devices. Among them, the tracking and registration step that accurately combines the virtual and real is crucial. It serves as the foundation and key technology for AR.

Three-dimensional tracking and registration technology refers to the process of tracking and locating images or objects in the display scene, and then establishing the corresponding relationship between the virtual world and the real world coordinate system. As shown in Fig. 9.24, the mainstream three-dimensional tracking and registration technology can be divided into three types: hardware sensor-based tracking and registration technology, computer vision-based tracking and registration technology, and hybrid tracking and registration technology.

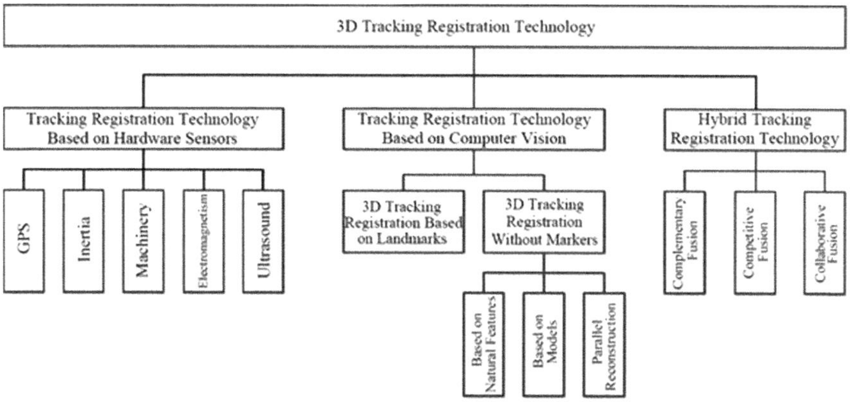

Fig. 9.24 3D registration tracing technology

(1) Hardware sensor-based tracking and registration technology relies on the coordinate information collected by various sensors to measure the position coordinates and orientations of the camera, thereby achieving camera calibration. It can be roughly divided into registration based on the Global Positioning System (GPS), inertial registration based on inertial sensors, mechanical registration based on mechanical sensors, electromagnetic registration based on electromagnetic sensors, and ultrasound registration based on ultrasound sensors. Hardware-based registration methods rely on external sensor support, having higher costs, larger size, poorer mobility performance and generally lower accuracy. Therefore, their development has been relatively slow.

(2) Computer vision-based tracking and registration technology is mainly realized through the use of computer vision algorithms [33]. The general process involves capturing real-world scenes using a camera and extracting feature points from the live scene using relevant algorithms. Then, based on the coordinate transformation, the position of virtual information in the real scene is determined, thus achieving the goal of virtual-to-real registration. The advantages of this approach include strong applicability, low cost, high accuracy and real-time performance. Depending on the presence or the absence of markers, computer vision-based registration methods can be divided into two types: marker-based 3D tracking and registration technology and markerless 3D tracking and registration technology.

① Marker-based 3D tracking and registration technology is a widely used and mature technology in current AR systems. It has low requirements for hardware processors and exhibits high robustness [31]. For this registration technique to take place, predefined markers are placed in the real scene. The camera identifies these markers and obtains their vertex information. Then, based on the principle of affine invariance, the pose change matrix between the predefined marker coordinates and the current scene marker coordinates is reconstructed to achieve the tracking and registration of virtual information. ② Markerless 3D tracking and registration technology can be roughly divided into three types: natural feature-based tracking and registration technology, model-based tracking and registration technology, and simultaneous localization and mapping (SLAM) based tracking and registration technology. Natural feature-based tracking and registration technology compensates for the limitations of marker-based 3D tracking and registration technology. It does not require predefined markers in the real environment. Instead, it computes the camera's pose information by utilizing natural features in the scene to achieve tracking and registration. Model-based tracking and registration technology uses virtual model information corresponding to the tracking and registration target as prior knowledge. It addresses the tracking and registration issues in environments with insufficient or no texture that natural feature-based tracking and registration technology may encounter. Simultaneous localization and mapping (SLAM) based tracking and registration technology does not rely on prior knowledge of the scene. It simultaneously tracks and reconstructs the 3D structure of the unknown scene during the registration and localization process.

(3) Mixed tracking and registration technology is a method that combines different types of tracking and registration techniques to obtain object poses. It effectively solves the technical challenges of achieving both high-precision tracking and robustness in AR systems but increases the complexity of the system. From the perspective of multi-sensor fusion classification, mixed tracking and registration technology can be divided into complementary, collaborative and competitive sensor fusions.

(4) Applications

AR can present virtual information in conjunction with the real world, forming a realistic integration that serves as a powerful supplement to the real world. It has broad application prospects and high commercial value. With the maturation of AR technology and the increasing number of applications, AR systems have become a new medium, gradually penetrating various fields, and are changing the way we shop, entertain and work.

(1) Digital Marketing: AR technology has opened up a new mode for digital marketing, allowing consumers to discover, understand and experience various products from a fresh perspective. Amazon is one of the brands that introduced AR technology early on, using AR to enable consumers to "try on" clothes online, providing unprecedented convenience for online shoppers. IKEA has developed an AR mobile application that allows customers to "place" IKEA furniture in their homes through the app before making a purchase, enabling them to see if the colors and sizes are suitable, thereby reducing the hassle of returns and exchanges.

(2) Architecture: In the field of architecture, AR technology allows architects, construction workers, developers and clients to visualize three-dimensional buildings and interior designs at any stage of construction. Additionally, AR technology can help to identify errors and issues during construction, pointing them out before they become difficult to resolve. AR technology can also assist in maintenance of buildings and facilities, providing remote assistance to clients during repair or maintenance processes.
(3) Tourism: With the help of AR technology, tourism brands can provide immersive experiences for their potential travelers. For example, through AR solutions, agents and destinations can offer visitors more information and directional signage. AR applications can help vacationers explore resorts and learn about destinations, and so on. Gansu Provincial Museum has introduced AR interactive technology into their exhibitions. When visitors use their mobile phone cameras to identify cultural relics, the relics can be presented in a more dynamic way, such as fish patterns "swimming" on painted pottery from the Yangshao culture, providing a better exhibition experience.
(4) Education: AR technology can assist educators in allowing students to use dynamic 3D models in the classroom, helping students understand relevant knowledge and stimulating their learning interests in an engaging way. Students benefit from the visual capabilities of AR, bringing concepts to life through digital rendering and accessing information anytime, anywhere, without the need for any special equipment.
(5) Healthcare: AR can provide real-time 3D digital images and key information to surgeons, allowing them to access important information during surgical procedures without leaving the operating area. The video optical perspective AR system developed by the Department of Information Engineering at the University of Pisa can overlay X-ray data onto a patient's body.
(6) Navigation Systems: By combining GPS technology with AR, navigation applications can improve safety and provide more comprehensive navigation information. Navion offers true AR navigation, being the first holographic AR navigation system designed for automobiles. As the surroundings of the vehicle change, the system continuously updates information to provide real-time navigation guidance. Hyundai Motor Group from Korea collaborated with a Swiss high-tech startup to introduce the world's first holographic AR navigation system at the 2019 International Consumer Electronics Show (CES).

9.3.2 AR Applications in Bridge Engineering

As one of the core components of bridge structural health monitoring systems, the data acquisition system is responsible for collecting relevant data for designers, managers and decision-makers. However, the data acquisition system is costly, and its operation is quite complex for untrained inspectors. Therefore, one of the major challenges in practical engineering is how to adopt new technologies for infrastructure maintenance and management, such as low-cost wireless smart sensors, AI approaches, machine learning algorithms, unmanned aerial systems, etc. However, these technologies cannot establish a direct connection between the inspectors and the structures. AR technology can effectively address this issue.

(1) Bridge Maintenance

Currently, there is a wide variety of bridges of different sizes across the country, and maintenance is still primarily conducted through paper-based records. Paper-based records suffer from issues such as inconsistent legends and inaccurate drawing information, which pose significant challenges to maintenance work, especially in the case of complex projects, presenting greater difficulties for the inspectors. Presently, inspection work typically involves manual on-site inspections, fault identifications and plan formulations. However, for bridge projects with incomplete drawings and insufficient existing data, correlation between the information and the actual site conditions is poor, making it difficult to conduct statistical analysis and identify hidden risks, thus reducing operational efficiency and increasing maintenance costs. By employing AR technology, data can be automatically collected, recorded, transmitted, processed and analyzed. Coupled with remote control capabilities, AR can help address on-site challenges and achieve intelligent bridge inspection and maintenance management. Furthermore, AR technology can be used to improve drawing materials, enhance the digitization of drawing information and enhance the guidance of remote inspection operations.

K. E. Ammar and A. Hammad proposed a BIM-based markerless mixed reality system framework (BIM3R), which integrates maintenance management systems, building information models and video-based feature tracking technology. This framework enables on-site information retrieval, visualization of maintenance work, and collaboration between on-site inspectors and office engineers. This AR application involves dividing the bridge into different small components and adding sequential identification numbers to these components during construction. During on-site inspection and maintenance work, the AR software confirms the identification numbers using the mobile device's camera, loads the 3D model of the component on the device screen, and allows gesture-based interaction to display relevant non-graphical data, including the component's dimensions, materials, construction information, model and more. Additionally, inspection and maintenance information can be recorded in text form within the software.

For the inspection of concrete bridge defects, Karaaslan et al. [34] proposed an AI-integrated AR system and developed a human-centered intelligent approach. They created an intelligent AR framework that can be integrated into wearable holographic headset devices, as shown in Fig. 9.25. The main objective was to assist the inspectors by expediting their routine tasks, such as measuring all cracks in defect areas, calculating spalling areas, conducting condition assessments and processing data for management systems. This potentially reduces the time and cost of infrastructure inspections. The system combines the inspector's professional judgment as

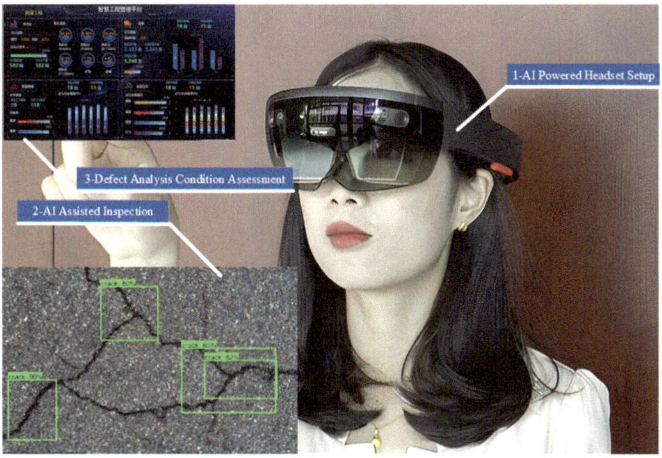

Fig. 9.25 Visual representation of an AI-driven MR system

being human-centric. Artificial intelligence helps the inspectors gather more quantitative and objective data, and the human-centered AI interacts with the inspectors. The collaboration between inspectors and AI aims to improve visual inspections rather than completely replacing human intervention in the inspection process. This research has made significant contributions to infrastructure inspection, maintenance, management practices and safety for bridge owners.

Based on the aforementioned related studies, a general bridge maintenance process based on AR technology can be summarized as shown in Fig. 9.26. Firstly, bridge inspectors use AR headsets during routine inspections of the infrastructure and walk around the bridge according to system guidance for regular or scheduled inspections. While the inspector performs routine inspection tasks, the integrated AI system in the headset continuously guides the inspector and displays potential defect locations. Once a defect is found, if the inspector confirms its location, the system proceeds with defect segmentation, characterization, determining the specific type of defect, and initiates analysis. If the defect boundaries require any correction or further subdivision, human inspectors can intervene and calibrate the analysis, displaying their annotations in the AR scene. Secondly, the inspectors record and mark defect elements or inspection information. If the inspector chooses to mark defect elements, the defect location information is automatically inherited from the BIM model and added as attributes. These attributes include the type of defect, severity of the defect, and any other annotations or observation results. All these data are saved in a database and immediately shared with the office for planning and consultation using the AV mode.

(2) Visualization of Structural Damage

In the conventional bridge maintenance methods, the detection of concrete cracking and spalling often relies on manual close-range visual inspection. This approach is time-consuming, labor-intensive and subject to significant subjectivity, making it

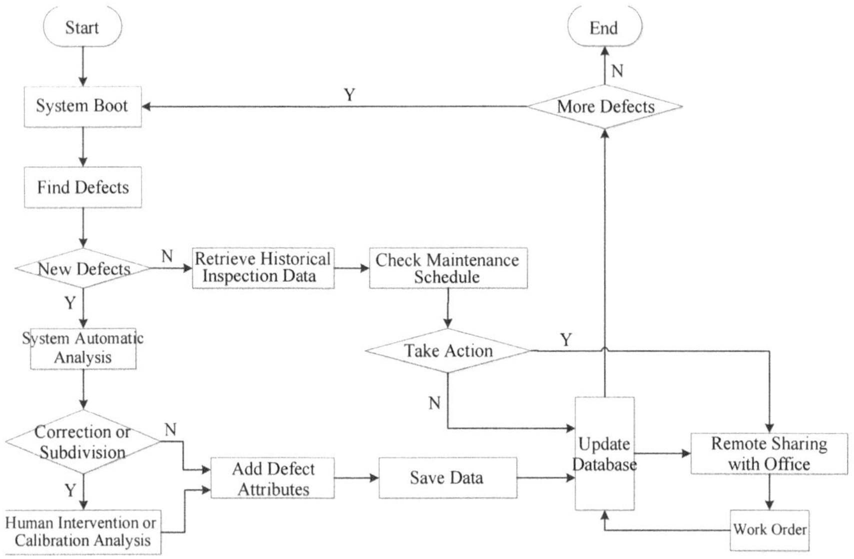

Fig. 9.26 Bridge maintenance process based on AR technology

difficult to identify small cracks and defects. In the future, with the development and maturation of AR technology, it can be combined with AI techniques and applied in bridge maintenance to achieve visualization of structural damages. This can assist the bridge inspectors in making faster and better decisions, as shown in Fig. 9.27.

Bridge inspectors need to conduct regular inspections of bridges, and in this case, they only need to wear a head-mounted AR display integrated with AI technology to go to the site, without the need for other complex equipment and tools. Upon arrival at the site, the inspector puts on the AR display, and the integrated AI system will guide the inspector to identify potential damage locations and annotate them using boundary wireframes in the AR device. The inspector then reviews the damage locations to check for any missed areas. If there are any, the system can re-annotate the damaged areas by manually adjusting the threshold. Once confirmed, the AI system performs damage segmentation within the boundary box and visualizes it in the AR system. For example, by highlighting the damage with colors, the inspector can visually identify the precise locations of the damage. Finally, through automatic analysis and computations by the AI system, the damage is quantified, and the results are added to the real environment through AR. Simultaneously, the on-site data is transmitted back to a remote terminal for analysis, to determine the development trends of concrete cracks and spalling. This information is then transmitted to the head-mounted display, allowing the inspector to understand the next steps in the development process of the cracking and spalling, thereby assisting them in making informed decisions based on scientific insights.

Compared to the conventional inspection methods, this method has three advantages: firstly, through AI searching and AR boundary line annotations, even small

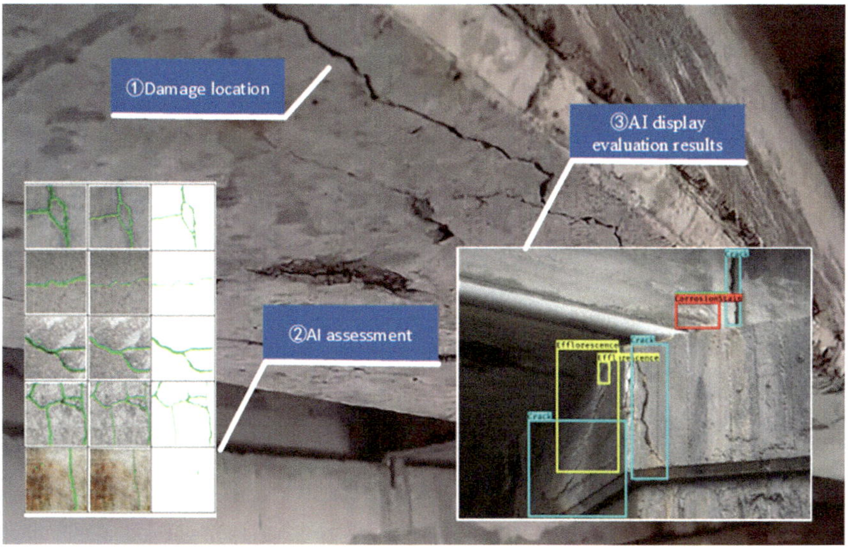

Fig. 9.27 AR technology for structural damage visualization

cracks and defects can be detected, avoiding the possibility of overlooking them due to negligence. Secondly, the AR zoom-in feature allows the inspection of damages from a distance, which is particularly significant for areas that are difficult to reach (such as bridge piers). Inspectors do not need to rely on other tools (such as bridge inspection vehicles or ladders); they only need to magnify and examine the relevant areas in the head-mounted AR display. Thirdly, by displaying the development trends of structural damages analyzed by a remote terminal through AR, the inspectors can quickly assess whether action needs to be taken.

The visualization of bridge structural damages and their development process using AR technology is just one of the many application scenarios in bridge maintenance. With technological advancements, bridge maintenance will undoubtedly progress towards greater intelligence.

9.4 Summary

AI, VR and AR, as representative technologies of the Fourth Industrial Revolution, are gradually changing our lives. The future decision-making and management of bridge maintenance will undoubtedly be digitized, visualized and intelligent. However, currently, these three intelligent technologies have not been widely applied due to maturity of the technology and cost, plus many issues still need to be addressed. In the future, it is hoped that more new technologies and concepts can be combined with the conventional techniques and applied to intelligent Maintenance and Management of bridges.

References

1. Kostem CN (1986) Design of an expert system for the rating of highway bridges. Expert Systems in Civil Engineering. ASCE
2. Furuta H, He J, Watanabe E (1996) A fuzzy expert system for damage assessment using genetic algorithms and neural networks. Comput Aided Civ Infrast Eng 11(1):37–45
3. Melhem HG, Aturaliya S (1996) Bridge condition rating using an eigenvector of priority settings. Comput-Aided Civ Infrastruct Eng 11(6):421–432
4. Mikami I, Tanaka S, Kurachi A (1994) Expert system with learning ability for retrofitting steel bridges. J Comput Civ Eng 8(1):88–102
5. Wilson JL, Wagaman SJ, Veshosky DA et al (1997) Life-cycle engineering of bridges. Comput-Aided Civ Infrastruct Eng 12(6):445–452
6. Liu C, Hammad A, Itoh Y (1997) Multiobjective optimization of bridge deck rehabilitation using a genetic algorithm. Comput-Aided Civ Infrastruct Eng 12(6):431–443
7. Miyamoto A, Kawamura K, Nakamura H (2000) Bridge management system and maintenance optimization for Existing Bridges. Comput-Aided Civ Infrastruct Eng 15(1):45–55
8. Furuta H, Kameda T, Fukuda Y et al (2003) Life-cycle cost analysis for infrastructure systems: life-cycle cost vs. safety level vs. service life. Third IABMAS workshop on life-cycle cost analysis and design of civil infrastructures systems
9. Kim S, Dan MF, Soliman M (2013) Generalized probabilistic framework for optimum inspection and maintenance planning. J Struct Eng 139(3):435–447
10. Dan MF, Ming L (2007) Bridge network maintenance optimization using stochastic dynamic programming. J Struct Eng 133(12):1772–1782
11. Li YF, Fan M, Hou JL et al (2011) Research on intelligent decision support system for maintenance management of large and complex bridge engineering. Southwest Jiaotong University Press, Chengdu (in Chinese)
12. Bao JM (2019) Image synthesis based on generative adversarial network. University of Science and Technology of China, Hefei (in Chinese)
13. Zhang ZN (2018) Complex structure flexibility identification and performance evaluation based on subspace technology. Southeast University, Nanjing (in Chinese)
14. Huang Y-H (2010) Artificial neural network model of bridge deterioration. J Performance of Const Facil 24(6):597–602
15. Wu LQ, Zhang LL (2020) Bridge structure health status detection based on data mining technology. J Jilin Univ (Eng Edn) 50(2):565–571 (in Chinese)
16. Wang GX, Gongben WS, Yan BF et al (2004) Development of intelligent decision support system for bridges. Highway Transp Technol 21(9):70–73 (in Chinese)
17. Chen SD (2019) Research on decision model of pavement preventive maintenance based on multi-objective Particle swarm optimization algorithm. Beijing Jiaotong University, Beijing (in Chinese)
18. Wang NJ (2017) Particle swarm optimization algorithm based network level decision model for optimization of rural road maintenance funds in Kunming. Kunming University of Science and Technology, Kunming (in Chinese)
19. Wei SY, Bao YQ, Li H (2020) Optimal policy for structure maintenance: A deep reinforcement learning framework. Struct Safety 83(6):101906.1–101906.13
20. Brooks FP (1999) What's real about virtual reality? IEEE Comput Graphics Appl 19(6):16–27
21. Burdea G, Coiffet PVR, Presence: teleoperators and virtual environments, 12(6): 663–664
22. Foley JD (1987) Interfaces for advanced computing. Sci Am 257(4):126–135
23. Zhao QP (2009) Overview of virtual reality. Science China (Part F: Information Science), 39(1):2–46. (in Chinese)
24. Liu F (2014) In-depth investigation and analysis of recent bridge safety accidents. Central South University, Changsha (in Chinese)
25. Kumar P, Bhandari NM (2005) Non-linear finite element analysis of masonry arches for prediction of collapse load. Struct Eng Int 15(3):166–174

26. Liu YJ, Zhao J, Lin Y et al (2019) VR based 110 kV skill training system for substation operation and maintenance. Electron Measurement Technol 42(21):131–136 (in Chinese)
27. Zhao QP (2009) A survey on virtual reality. Sci China: Ser F (English Edition). 52(3):348–400
28. Azuma R, Baillot Y, Behringer R et al (2001) Recent advances in augmented reality. IEEE Comput Graphics Appl 21(6):40–47
29. Zhou Z, Zhou Y, Xiao JJ (2015) Overview of virtual reality enhancement technology. Sci China (Inf Sci) 45(2):157–180 (in Chinese)
30. Azuma RT (1997) A survey of augmented reality. Presence Teleop Virt 6(4):355–385
31. Arth C, Grasset R, Gruber L et al (2015) The history of mobile augmented reality. Comput Sci
32. Zhong HJ, Liu XL, Wu XL (2008) Research on augmented reality system and its key technologies. Comput Simul 25(1):252–255 (in Chinese)
33. Hoff WA, Nguyen K, Lyon T (1996) Computer-vision-based registration techniques for augmented reality. Intell Rob Comput Vis XV 2904:538–548
34. Karaaslan E, Bagci U, Catbas FN (2019) Artificial intelligence assisted infrastructure assessment using mixed reality systems. Transp Res Rec 2673(12):413–424

Chapter 10
Closing Remarks and Our Vision

Abstract In this final chapter, we provide a quick review of digital technologies covered in this book. A different view is given from the systems engineering perspective about the disciplinary nature of intelligent maintenance and management of bridges. We provide a positive outlook for the next five years on how intelligent maintenance and management for bridges will be implemented with broader sensing, deeper learning, and higher automating capabilities. Last, although it is a bold move, the authors provide a vision for a long-term future, wherein intelligent maintenance and management might have become a small fraction of our connected societies enabled by general artificial intelligence and other relevant technologies.

10.1 Remarks

10.1.1 Summary of the Book

The book proposes a new paradigm for bridge engineering—Intelligent Bridge Maintenance and Management, developed to meet the imperative need of managing bridge structures to achieve lifecycle sustainability and resilience. Although its disciplinary classification is within Civil Engineering, the general approach to treating this sub-disciplinary branch falls under the domain of Systems Engineering.

According to the International Council on Systems Engineering (INCOSE) [1], Systems Engineering studies how to provide a transdisciplinary and integrative approach to enabling the successful realization, use, and retirement of engineered systems, using systems principles and concepts and implementing scientific and technological methods.

Regarding the implementation of scientific and technological methods, this book shifts its contents away from methods directly based on the first principles of physics, which include novel materials and innovative structural systems. Although they continue to emerge, engineering structures, including bridges, still follow the laws of classical mechanics when adopting these material and structural-system technologies. Instead, this book embraces modern computer, electronics, communication, and

G. Wu et al., *Intelligent Bridge Maintenance and Management*, Springer Tracts in Civil Engineering, https://doi.org/10.1007/978-981-97-3827-4_10

data science technologies, collectively, digital technologies. Throughout the book, we follow such an approach and focus on the implantation of digital technologies, especially emerging ones, to enable sustainable and resilient services that are supposed to be provided by bridge systems.

As such, we introduce our readers to nearly all aspects of modern digital technologies. In Systems Engineering, although many scholars and practitioners have heralded digital transformation across all industry sections when proposing systems solutions [2, 3], bridge systems differ significantly from other systems, manufactured or fabricated. Simply put, bridge systems are, as well as other infrastructure systems, *constructed*. When compared with other mechanical or electrical products, constructed bridge systems have three unique attributes: (1) being critical links in a massive transportation network that supports modern social and economic activities; (2) serving a long time with a typical duration of 70 or more years; and (3) higher uncertainties from multiple dimensions.

We target the first two attributes directly as the technologies presented in this book serve to realize sustainability and resilience in lifecycle bridge operation. High uncertainties are partially targeted mostly at the physical levels, as bridges are subjected to different levels of operational loads, various sources of environmental degrading, and extreme events at unknown times with variable intensities or magnitudes. That said, we have little emphasis on uncertainties from the social and economic dimensions. However, they do exist and can significantly interfere with the practice of bridge management.

In the social dimension, uncertainties arise from differences in organizational preparation before adopting technologies and human judgment and intervention during employing technologies. In the economic dimension, civil infrastructure systems, including bridges, are mostly initiated and then constructed by means of governmental or public funds. Implementing lifecycle maintenance and management would inevitably aggravate the portion of the infrastructure budget beyond their construction cost. With policy-making uncertainties, locally and nationally, financial variances are well expected for lifecycle bridge maintenance and management. In this book, we hold a positive view toward these social and economic uncertainties, and readers may argue that we idealize excessively about the real world. To this end, we acknowledge this likely predisposition and insist on our positive outlook toward next-generation intelligent bridge maintenance and management.

To approach this goal, we categorize digital technologies into two groups, which overlap to some degree: ones crucial for realizing intelligent maintenance and those relevant for intelligent management. For the former, they include IoT-enabled sensing, inspection automation via robotic equipment, big data-enabled computing, machine learning, and AI for data understanding and predictive modeling. For the latter, we emphasize that intelligent bridge management is a process that connects end-users—bridge engineers, managers, and decision-makers. As such, digital modeling technologies, including data and information fusion, visualization-enabled digital modeling (e.g., building information modeling, BIM; digital twin modeling, DTM), and, finally, cyber-physical modeling, are synthesized in this book. To enable advanced scientific decision-making, AI-assisted decision-making and immersive front-end technologies, including virtual reality (VR) and augmented reality (AR), are further surveyed.

10.1.2 Short-Term Outlook

It is expected that intelligent maintenance and management of bridges, as an emerging systems-engineering solution paradigm, would confront questioning at the onset. Simple and typical ones include: Is it feasible? And is it better that it can replace existing technologies?

Various emerging digital technologies presented in this book have experienced the typical hype cycle, as shown in Fig. 10.1 [4], including the so-called 'Peak of Inflated Expectations' and the 'Trough of Disillusionment.' We argue that although intelligent maintenance and management for bridges is an emerging methodology and its visibility starts with an academic concept and evolves very slowly, its development and adoption would fully embrace the phases of the 'Plateau of Productivity' of other emerging technologies (as shown in the engineering adoption curve in Fig. 10.1). Hence, it avoids the technology trough of maturity. This advantage is rooted in a common nature of the Civil Engineering sector in that it tends to adopt relatively mature technologies if compared with other industries, e.g., aerospace, automobiles, and even consumer electronics.

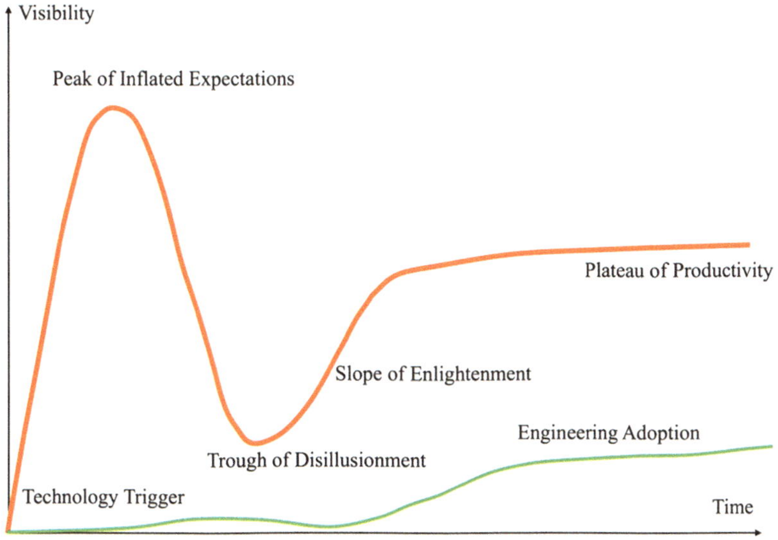

Fig. 10.1 Typical technology hype cycle (Recreated based on concepts developed by Gartner Inc. [4]) and engineering adoption curve likely occurred in the Civil Engineering sector

Regarding whether intelligent bridge maintenance and management can replace conventional engineering practice, it remains not fully answered, as its globally wide adoption is still in its infancy or growing stage. As described in Chap. 2, light has been shed that a systematically implemented intelligent bridge maintenance and management platform has been brought into practice, which adopts nearly every technology suite presented in this book. Figure 10.2 illustrates the technology configuration and the decision-making dashboard for data analysis and information fusion.

We are positive about the outlook for the next five years or potentially the immediately impactful duration of implementing intelligent maintenance and management for bridges proposed in this book. We assert that the future platform, compared with the one in Fig. 10.2, will be equipped with broader sensing modalities, deeper learning-enabled predictive modeling, and a higher level of automated interaction between physical and decision-making layers. The platform, as demonstrated in Fig. 10.2, will be architected with augmented design for different layers. The major ones may include:

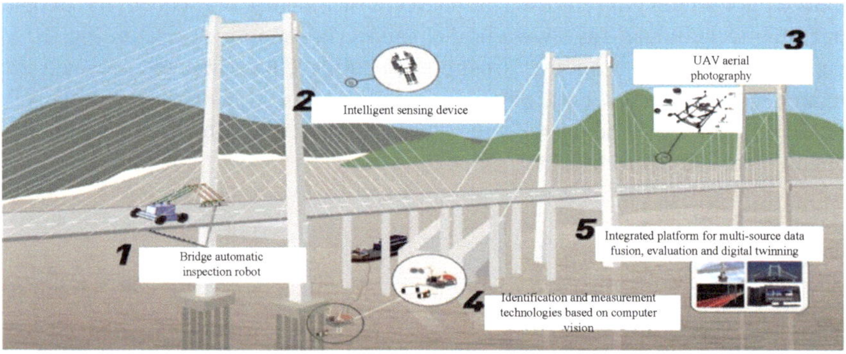

a) Single bridge

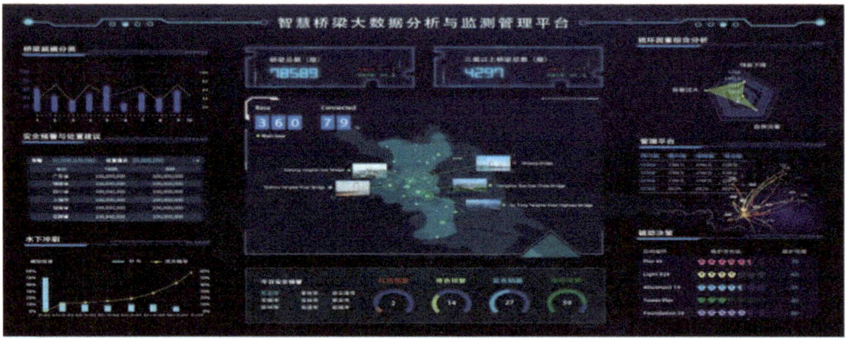

b) Bridge group

Fig. 10.2 Demonstration of bridge maintenance technologies and a dashboard for management and decision-making

Layer for deeper and more comprehensive replication of bridges and other physical assets. Two suites of supportive technologies hold promise in the next five years.

5G and 6G supported Internet of Things (IoT) sensing technology. As 5G is defusing the markets and societies globally, the next-generation communication technologies or 6G are on the horizon. Such a prospect will greatly enable the maturity of the Internet of Things, including their more ubiquitous density, much lower latency, and much wider bandwidth.

Robotic automation. Bridge inspection automation will be achieved beyond the use of 'automatic' machines or equipment toward an automated operation, and autonomous operation may be in trial for a limited number of bridge inspection cases. For example, aerial inspection using UAVs can be operated in a way that they can search for structural damage and defects without human intervention [5]; research has been conducted about deploying structural health monitoring sensors through aerial robotics [6].

Layer for AI-enabled data understanding, modeling, and prediction. As generative AI and large language and vision models become more robust, accurate, and adaptive to many industrial situations [7], the current training–testing-based AI practice has been transformed into a prompt-based engineering practice. For example, time-series data, visual data, and other multi-media data from IoT sensors and imagers collected from a bridge can be analyzed by providing a simple user-orientated query with natural human input (e.g., texts, voice commands, or hand gestures). For this matter, given the accelerating advance in the AI domain today, we hold our positive view that it can be realized in the next three to five years.

Layer for human infrastructure and machine interfacing. It is believed that digital modeling of bridges and other civil structures will move closer to unleashing the full expectation of digital twin modeling (DTM) as a real-time digital replica of physical facilities. To conduct decision-making, immersive VR/AR headsets will become a commodity that connects these DTMs and bridge engineers and decision-makers who readily pick up VR/AR devices anywhere and anytime.

Last, we note that the three layers above are not isolated and statically connected; rather, they must be dynamically interacted by design.

10.2 Vision

Envisioning the long-term future of intelligent maintenance and management for bridges and other constructed facilities is better characterized as visionary insight rather than mere imprudence. However, this is an audacious task. In the following, we attempt to share our vision for the future; to be less imprudent, our vision is presented by raising research questions for the community of researchers and forward-thinking practitioners, followed by our opinions. In addition, our scope of vision goes further beyond constructed facilities, enveloping the complete built environment.

What are the future sensing modalities for any built environment?

We have SHM, NDE, and IoT-based sensing today. Toward the future sensing modalities, our vision is that multi-scale and multi-dimension sensing with cognitive edge and cloud computing will be fused and ready for much enhanced understanding. A number of sensing modalities are described herein:

Embedded sensing. Fiber optics and other emerging embedded sensing technologies (e.g., graphene sensors) can become much more cost-effective or scientifically proven. With such embedded sensing technologies, self-aware and even self-diagnosis engineering structures are likely to be constructed.

Robotic sensing. We mentioned previously that robotic inspection or monitoring is on the horizon, including automated inspection through unmanned aerial or ground vehicles and automatic deployment of structural health monitoring sensors through aerial robotics. We argue if robotic vehicles with embedded sensors or deployable sensors can plan, navigate, search, detect, and assess structural integrity, it is necessarily a great leap, thanks to the advances in next-generation communication, AI-enabled control, scene understanding, and actuation. When publishing this book, the world is witnessing the debut of several humanoid robots from leading industry players and innovators.

Geospatial sensing. It has been an active research field in the past 20 years that remote sensing, via different platforms (satellites, aircraft, and UAVs), has been adopted for many tasks that demand geospatial coverage, including monitoring and mapping of natural hazards and disasters. With RS technologies penetrating the consumer markets (e.g., StarLinks for global communication, BeiDou GPS for global two-way positioning), we foresee a possibility that optical and Radar remote sensing data can be 'ordered' for use anytime and anywhere for individuals and for specific constructed facilities in a built environment.

Social crowdsourcing. To be elaborated more later, social crowdsourcing has been researched extensively as well in the past decades. However, its impact has not been fully revealed in assessing and managing constructed facilities. As driverless and electric vehicles penetrate more extensively, we envisage that a transformative change is impending.

Towards robotic sensing for autonomous inspection and monitoring, should it be a completed machine-dominated autonomation or, better yet, human-in-the-loop autonomation?

We have moderate confidence that a fully cognitive mode for the inspection and monitoring of constructed facilities may be partially realized in limited situations. Instead, we tend to believe that in the relatively long term (5 to 10 years), robotic platforms (including aerial and ground vehicles and humanoids) would work with humans with a high-level collaboration, while industry leaders are proposing the near future where humanoids can work with human side-by-side. In our recent work, we attempted to advocate such a mode, termed human-in-the-loop autonomous inspection [8]. The conservative motivation in this work is that in a foreseeable feature (5–10 years), engineering data (i.e., pertinent to structural damage, defect, and general integrity information against complex service situations and natural scenes) remain scarce or small. As such, human-in-the-loop automation is an optimal approach to balance risky automation and high-fidelity task completion.

What is the endgame of AI for the management of physical facilities?

A different way to ask is whether General Artificial Intelligence (or GAI) or GAI-enabled technologies can provide the most robust prediction and optimal solution with ‘good intentions.’ Not attempting to hold a humble opinion, we cannot answer such philosophical questions. As generative AI technologies are at their absolute hype and advance exponentially today (‘inflated expectations’), some worry seriously about the ‘doom’ possibility of AI, and many hold optimistic views and treat it as produced by humans, hence reined by humans.

On the other hand, it is worth mentioning that besides the success of the current generative AI featured by the recent large language and vision models, many other types of AI may have important roles in facilitating cognitive automation in advancing intelligent maintenance and management of constructed facilities. A few potential GAI methodologies are listed below, and the selection criterion is based on whether it can mitigate a strong premise in the current practice that the state-of-the-art models, at least those adopted in Civil Engineering, require sufficient data for learning, testing, and validation.

Reinforcement Learning (RL). RL has been proven successful in many control and game-theoretic applications, including robotics and board games. Related to the human-in-the-loop AI mentioned above for managing engineering structures, we believe there is a vast unknown space to explore the application of RL technologies.

Capsule networks and other DL models accommodate small datasets. Capsule Networks can address the limitations of Convolutional Neural Networks (CNNs), particularly their sensitivity to spatial hierarchies and viewpoint variations, which preserve spatial hierarchies and are more robust to variations in the input. With such properties, its significance may be shed in situations of understanding complex 3D scenes in a built environment (e.g., through robotic vehicles).

One-shot or even zero-shot learning. To advance to the next level, namely, to predict or conduct inference with little or no data (e.g., one-shot or zero-shot learning), the condition is that a sophistically complex model encoding sufficient engineering knowledge should be trained in the first place. One-shot or zero-shot learning for simple applications (e.g., inpainting) is active research today, and it is expected that it can be extended to engineering fields.

Symbolic AI and physics-informed AI. Both symbolic AI and physics-informed AI are being researched actively today as they can discover new physical laws from data or predict dynamic system behavior based on classic governing equations (e.g., partial differential equations). Specifically, neuro-symbolic AI fuses two historically divergent AI methods, neural networks and symbolic AI, which provide high-level representations of data-based empirical information and first-principle-based logic. We envision that such an AI approach may bring meaningful enhancement to future digital twin modeling.

With these potential GAI technologies, however, we keep one game-changer variable behind—quantum computing. If quantum computing hardware is eventually fabricated with a significant number of qubits ('Quantum supremacy'), it may totally change the landscape. To this end, we observe that researchers in Civil Engineering have started to adopt quantum computing circuits for solving specific problems, e.g., damage detection in ground photos [9] and building detection in remote sensing images [10].

Could intelligent maintenance and management of built environments form a symbiotic relationship with our society?

This question is asked with a grander vision that the future urban centers and rural habitats are truly smart—implying that the goals of IoT, or specifically infrastructure-to-everything and society-to-everything connections, are realized. If yes, it means that the proposed intelligent maintenance and management of bridges and constructed facilities become a fraction of such connected societies and built environments. Our research question is how to facilitate the flow of information in such connected huge systems. We speculate that an incentive mechanism needs to be fulfilled such that sensing, computing, and decision-making from the public and stakeholders need to be appraised or valued. The answer to this question, using today's technology, is blockchain [11], which is not included in this work.

The complexity arises when considering sensing and platform automation, machine-dominated AI, human-in-the-loop AI, and human-dominated crowd-sourcing and decision-making. How do we appraise their different contributions? Physical sensors alone are 'expendable'; however, advanced cognitive sensing may require high-cost maintenance, which ultimately affects human expectations. To human contributors, AI-assisted or other digital technologies-assisted decisions (e.g., through digital twins and augmented reality), should they deserve more incentives? We leave these questions to our readers.

10.3 Summary

In this chapter, we end the book by providing a summary and, more importantly, a short-term outlook and a long-term vision for intelligent maintenance and management of bridge structures. With this treatment, we state that either in the short or long term, this intelligent process will incorporate all constructed facilities or our complete built environment. With many variables and uncertainties in mind, we audaciously provide a vision of the future with several transformative technologies described. However, it emphasized that the vision is presented by questioning, and the answers are largely left to the readers subject to the future direction of our societies.

References

1. Walden DD, Shortell TM, Roedler GJ, Delicado DA, Mornas O, Yip YS, Endler D (2023) INCOSE systems engineering handbook (Version 5). Willey.
2. Legner C, Eymann T, Hess T, Matt C, Böhmann T, Drews P, Mädche A, Urbach N, Ahlemann F (2017) Digitalization: opportunity and challenge for the business and information systems engineering community. Bus Inf Syst Eng 59:301–308
3. Matt C, Hess T, Benlian A (2015) Digital transformation strategies. Bus Inf Syst Eng 57:339–343
4. Linden A, Fenn J (2003) Understanding Gartner's hype cycles. Strategic Analysis Report N° R-20–1971. Gartner, Inc 88:1423
5. Peng X, Su G, Chen Z, Sengupta R (2021) Structural damage detection, localization, and quantification via UAV-based 3D imaging. Struct Health Monit 2021 (2021). https://doi.org/10.12783/shm2021/36235
6. Chen J, Dai Z, Chen ZhiQiang (2019) Development of radio-frequency sensor wake-up with unmanned aerial vehicles as an aerial gateway. Sensors 19(5):1047
7. Edelman DC, Abraham M (2023) Generative ai will change your business. Here's How to Adapt. Harvard Business Review. https://hbr.org/2023/04/generative-ai-will-change-your-business-heres-how-to-adapt
8. Chen Z, Zeng Q (2022) Human-in-the-loop robotic inspection-framework and point cloud assessment. AI・データサイエンス論文集, 3(2):35–46
9. Bhatta S, Dang J (2023) Multiclass seismic damage detection of buildings using quantum convolutional neural network. Comput-Aided Civ Infrastruct Eng (2023)
10. Delilbasic A, Cavallaro G, Willsch M, Melgani F, Riedel M, Michielsen K (2021) Quantum support vector machine algorithms for remote sensing data classification. In: 2021 IEEE international geoscience and remote sensing symposium IGARSS. IEEE, pp 2608–2611
11. Marsal-Llacuna M-L (2018) Future living framework: is blockchain the next enabling network? Technol Forecast Soc Chang 128:226–234

GPSR Compliance

The European Union's (EU) General Product Safety Regulation (GPSR) is a set of rules that requires consumer products to be safe and our obligations to ensure this.

If you have any concerns about our products, you can contact us on ProductSafety@springernature.com

In case Publisher is established outside the EU, the EU authorized representative is:

Springer Nature Customer Service Center GmbH
Europaplatz 3
69115 Heidelberg, Germany

Batch number: 10703867

Printed by Printforce, the Netherlands